LINEAR
CIRCUITS

Analysis and Synthesis

A. RAMAKALYAN

Department of Instrumentation and Control Engineering
National Institute of Technology
Tiruchirappalli

OXFORD
UNIVERSITY PRESS

OXFORD
UNIVERSITY PRESS

Oxford University Press is a department of the University of Oxford.
It furthers the University's objective of excellence in research, scholarship,
and education by publishing worldwide in

Oxford New York
Auckland Cape Town Dar es Salaam Hong Kong Karachi
Kuala Lumpur Madrid Melbourne Mexico City Nairobi
New Delhi Shanghai Taipei Toronto

With offices in
Argentina Austria Brazil Chile Czech Republic France Greece
Guatemala Hungary Italy Japan South Korea Poland Portugal
Singapore Switzerland Thailand Turkey Ukraine Vietnam

Oxford is a registered trade mark of Oxford University Press
in the UK and in certain other countries.

Published in India
by Oxford University Press

ISBN 0-19-567001-9

Typeset in Times Roman
by Archetype, New Delhi, 110063
Printed in India by Radha Press, Delhi 110031
and published by Manzar Khan, Oxford University Press
YMCA Library Building, Jai Singh Road, New Delhi 110001

Dedicated with all my love to
my daughter
SAVERI SUMADYUTI
who inspires me with her sketches of networks
and my son
VIRINCHI
an independent source in my network

A. RAMAKALYAN

Preface

> 'The instruction we find in books is like fire. We fetch it from our neighbours, kindle it at home, communicate it to others, and it becomes the property of all.'
>
> — VOLTAIRE
> Philosopher and writer, 1694–1778

Linear Circuits: Analysis and Synthesis is designed to serve as a textbook for a core course in electrical circuit theory/network theory. This course is usually taught during the second year (third semester) in the electrical and electronics, electronics and communications, computer science, electronics and instrumentation, instrumentation and control branches in an engineering programme. The emphasis of this book is on the development of the basic ideas of electrical engineering, rather than on giving bland definitions. An additional objective of this book is to encourage the reader to develop hierarchical thinking wherein he/she can see more complex systems as a generalization of simple circuits and techniques. There are no specific prerequisites for studying this book; an understanding of elementary calculus and simple matrix algebra is necessary and sufficient.

It is true that there are some good books on this subject. However, I felt that these books fail in certain aspects. For one, a phasor is defined in every book; the students, in their third semester of BE/BTech programme, would appreciate the idea if it were developed from the basic complex numbers. Further, a mathematical coherence is required in the presentation, for the subject teaches *'problem solving'* rather than electrical engineering per se. An oft repeated simple statement—the basic operations such as addition, multiplication, and division can be generalized over the field of complex numbers—makes a profound impact on the student's aptitude for problem solving.

Organization of the book

What is it that draws me, and many more, to this subject? I discovered that it is the model building and problem-solving strategies that we learn during this course. I have experienced that even when one is correct in stating 'what' happens, people tend to accept it only if he/she can also explain 'how' it happens. Perhaps a small, one-year-old child would know that a bulb glows if a switch is put on. But, 'Why should it glow?', 'How does it glow?', 'When doesn't it glow?', etc. are the questions that make a difference in anyone's outlook and attitude. My inspiration for this book comes from the following quotation:

> 'India ought to educate her citizens in moving from a knowledge-based system to an inquiry-based system while ensuring education for all.'
>
> — SABEER BHATIA
> Hotmail founder

The subject has been organized into chapters that can be studied continuously without any jumps. It is a cohesive presentation that is suited to the present fast-paced, one-semester course on circuit theory. Although the contents of this course have, more or less, become standard, the rationale for the organization of chapters in this book is as follows.

Chapter 1 provides a general introduction to the subject. The development of electricity, magnetism, and electromagnetism is traced through some important historical data. The connections between electromagnetic phenomena and circuit theory are also presented. These connections, of course under certain assumptions, allow us to derive Kirchhoff's rules, which are otherwise known as laws. A very brief introduction to graph theory and mathematical modelling is also provided.

Chapter 2 introduces the basic elements of circuit theory. The voltage–current relationship is emphasized for each of the elements. The sign convention for active elements and passive elements is introduced, and the law of conservation of power is illustrated. The ideas of graphs and mathematical modelling have been further extended from Chapter 1.

Chapter 3 begins the course formally with standard definitions of networks, circuits, series and parallel connections, etc. Next, a detailed summary of the notation used in this text is presented. The industry standard computer program, SPICE, is briefly introduced, and a data structure to describe circuits is presented. Details on SPICE have been given as an appendix, since it is important to concentrate on the subject before trying computer-generated solutions to problems. Towards the end of the chapter, two typical circuits, the ladder and the bridge, have also been presented. The chapter concludes with the basic ideas of circuit analysis, the core of this text.

Chapter 4 presents several techniques, such as nodal analysis, mesh analysis, the principle of superposition, and Thevenin's and Norton's theorems, for the analysis of circuits. Emphasis is laid on obtaining a system of linear equations and packaging them in the matrix-vector form. To focus on the technique, simple circuits containing dc sources and resistances alone are considered. Consequently, the linear system of equations involves real algebra. Various popular methods of obtaining solutions to linear systems of equations and their quick review are presented in Appendix C. Circuits with controlled sources are also presented, and the subtle differences in the resulting linear system of equations are illustrated. The reader is advised to thoroughly learn all the techniques presented in this chapter before moving onto later chapters.

Chapter 5 introduces the transient response observed in the circuits containing inductors and capacitors, when subjected to sudden change in the sources. An important property, called time-invariance, of the elements is discussed. Elementary signals such as the unit impulse and the unit step are defined. Waveform synthesis, pulse train in particular, is also discussed. This chapter is of fundamental importance. Both first-order and second-order circuits are presented in this chapter, and the nuances are highlighted. The reader will encounter ordinary differential equations (with constant coefficients) for the first time in this chapter. Several intuitive examples have been provided to make the presentation interesting. The reason for presenting both first-order and second-order circuits in the same chapter is to maintain a continuous flow of ideas.

Chapters 6 and 7 take the reader to circuits operating with ac sources. The idea of a phasor is developed from the basic complex numbers and the reader is encouraged to appreciate the inherent beauty of this technique vis-à-vis the time-domain approach to solve differential equations. For the sake of simplicity, the transient response is ignored in this chapter. The techniques of Chapter 4, such as nodal analysis, etc., are invoked again in this chapter, with a generalization to complex algebra. Some basic ideas of circuits as filters are also presented in terms of transfer functions and Bode plots. Chapter 7 may be completely omitted without loss of continuity, if the time allotted to the first course on circuit theory is not adequate.

Chapter 8 introduces the computations pertaining to power. Real power, reactive power, and complex power are defined and an elegant mathematical framework is illustrated. Power systems are also introduced, albeit briefly; the focus is on balanced three-phase power systems. Chapter 9 unifies the transient response of Chapter 5 and the steady-state response of Chapter 6 using Laplace transformation. The presentation is circuit analysis oriented. Rational network functions having zeros and poles have been illustrated. The reader can see a generalization of several ideas of circuit analysis in this chapter. Chapter 10 presents the circuits from an input–output point of view, in terms of the two-port parameters. Six sets of parameters and their interrelationships have been dealt with. In particular, the relationship between the driving point and transfer functions is discussed in detail. Several examples illustrating the applications, such as amplifiers and magnetically coupled coils, as well as the interconnections of two-port networks have also been presented.

In many universities, *synthesis of networks* is not covered owing to its mathematical rigour. However, it is strongly felt that an engineer has to be motivated towards designing networks using the techniques he/she has learnt for analysis. Accordingly, Chapter 11 is devoted to the design of networks with a good balance of intuition (developed over the past 10 chapters) and mathematical rigour. Of course, the background of a second/third semester student is kept in mind. The rational network functions have to satisfy certain important properties, and these are enumerated as obvious observations made on standard networks. Further, the numerator and the denominator polynomials of a given network function both have to be Hurwitz polynomials. Without loss of continuity, a definition of a Hurwitz polynomial is presented, but the criterion, called the Routh–Hurwitz criterion, for testing whether a given polynomial is Hurwitz has been deferred to Appendix D; usually this criterion is studied in detail in a later course on control systems. For the sake of simplicity and brevity, the synthesis of RL, RC, and LC networks given the driving-point functions is broadly dealt with. A brief discussion on synthesis given transfer functions is also presented.

Chapters 12 to 14 contain answers to certain questions of fundamental importance such as: Why is it that we study about sinusoids so much and not about triangular functions? The treatment is meant to make the reader look ahead and pave the way for more advanced courses. Several features, particularly the linearity, make the operational amplifier (Op-Amp) a very interesting device. Chapter 12 is devoted to studying circuits that contain one or more ideal Op-Amps. The study of Op-Amps required a course by itself and little justice can be done in a single chapter. Chapter 13 is devoted to showcasing Fourier analysis, another important subject by itself. However, the material in these chapters will give some limited exposure to these subjects and would come in handy to the reader while studying signals and systems, control systems, communications, and DSP.

Examples, exercises, and end-of-chapter problems

There are about 350 solved examples, gradually varying in their presentation from completely solved to moderately solved, throughout the text. In addition, there are about 350 'do-it-now' exercises, which demand the reader's careful attention. I strongly recommend the reader (the teacher as well as the taught) to solve the exercise problems as and when they appear concurrently along with the text. Answers have been provided to almost all these exercises. Of course, there are also about 350 end-of-chapter problems; these may be considered to belong to the 'do-it-anytime' category. In addition to the problems to be worked out using paper and pencil, a set of ten experiments, which can be easily conducted with a sense of accomplishment in the laboratory, is provided in Appendix B.

The CD-ROM accompanying this book

A CD-ROM is provided as an ancillary material along with this book. It contains an easy-to-use version of SPICE, called the Orcad's PSPICE 9.2 lite edition, which can be downloaded free of cost from www.orcad.com. An interested reader may quickly copy this program to his/her PC and start working with the programs given in Appendix A. The SPICE, *Simulation Program with Integrated Circuit Emphasis*, is introduced in Chapter 3 as a data structure. A typical SPICE program has three parts—circuit description, analysis request, and output request. The circuit description can be either tabular or graphical. In Chapter 3, the tabular description in pursued; in an introductory course, the tabular description is better of the two. On the CD-ROM is also available a graphical interface called **Capture** that draws circuits using an intuitive menu. However, I would like to caution the reader that there could be some 'bugs' in this program. For easy reference, extensive on-line manuals are also available on the CD-ROM in the printable document format (PDF).

In spite of presenting this CD-ROM, I would strongly recommend the following. There is no substitute for understanding circuit behaviour, which the reader gets only through hand analysis. Further, in an introductory course, circuit complexity is purposefully kept at its lowest level and hence computer power is not mandatory. It makes sense if the reader develops an inquiring attitude and questions the outcome of the program. Of course, the reader finds the program quite useful while doing courses on microelectronics circuits and operational amplifiers in a later semester.

Answers to the end-of-chapter problems are also provided on the CD-ROM in a separate folder.

Finally, on the lighter side, I quote the French novelist Gustave Flaubert:

'It is splendid to be a great writer, to put men into the frying pan of your imagination and make them pop like chestnuts.'

Acknowledgements

I would express my sincere thanks to my educators—my students—for the past eight years or so. They include the present final year students Chandra Sekhar, Jaideep Jaikar, Srinath, Karikalan, and many more. I would sincerely appreciate my research assistant Ms Vanitha for her enthusiasm and patience in preparing the laboratory experiments and the solutions manual.

My family has supported me in all my moods. I owe a lot to my parents, my wife, and my children; they think I have already lost the battle!

I would like to pay my humble respects to Prof. M. Vidyasagar, whose lectures and books give a fillip to my research and teaching and provoke me to consider writing. Once he said, 'I am more interested in educating the reader than impressing the reader; I think I am a good expositor; good exposition makes the research seem better than it is.'

I express my heartfelt gratitude to the directors of NIT Tiruchirappalli for their passion to put the institute at greater heights, and to my colleagues in the ICE department for constantly encouraging me to think what education means. I thank my HoDs, past and present, for the facilities and the academic environment they envisioned.

Dr N. Kalyanasundaram, Professor of ECE Department, NIT Tiruchirappalli, has been an ideal current source to my neural network! I have no words for his affection.

It is time for me to remember Prof. G.V. Padma Raju for teaching me this subject for the first time way back in 1987. He introduced me to Guillemin's book, thereby fixing this beautiful and fascinating subject permanently in my mind.

I appreciate the effort and commitment of the editorial team at OUP and offer my sincere thanks for goading me constantly to the completion of the project.

And, forgetting my dear friend Dr Abraham Mathew of NIT Calicut is forgetting myself! He has fuelled my enthusiasm to write this book with his constructive criticism on several pedagogical issues.

Above all, I prostrate before the Almighty for giving me the courage and energy to take up projects like this.

— A. RAMAKALYAN

Contents

List of Symbols

A   ampere, unit of current

A   circuit matrix

B   susceptance, measured in siemens

$\overline{B}$ magnetic field

C   capacitance, measured in farads

$\mathcal{C}$ set of complex numbers

$d(t)$ any desired signal; instantaneous voltage or current

$\mathcal{D}$ phasor for $d(t)$

$\delta(t)$ unit impulse

$\underset{=}{\triangle}$ by definition

$\overline{E}$ electric field

E   output matrix in the state-space model; SPICE symbol for VCVS

f frequency, measured in hertz

F   farads

F   SPICE symbol for CCCS

G   conductance, measured in siemens; SPICE symbol for VCCS

ζ damping factor

$h(t)$ impulse response

H   henry, unit of inductance

H   SPICE symbol for CCVS

$H(s)$ [or $H(j\omega)$] transfer function

$\mathcal{H}$ transfer matrix

θ phase angle, measured in degrees

$i(t)$ instantaneous current, measured in amperes

I   constant current

$\mathcal{I}$ phasor for $i(t)$

j $\sqrt{-1}$

J   joule, unit of energy

K   kilo, standard abbreviation for 10^3

k residues in partial fraction expansion

L   inductance, measured in henrys

m   milli, standard abbreviation for 10^{-3}

M   mega, standard abbreviation for 10^6

M   mutual inductance, measured in henrys

$\mathcal{M}$ Mesh incidence matrix

μ micro, standard abbreviation for 10^{-6}

n   nano, standard abbreviation for 10^{-9}

$\mathcal{N}$ node incidence matrix

$p(t)$ instantaneous power

P   average power, measured in watts

π pi

q charge, measured in coulombs

Q   reactive power, measured in volt-ampere-reactive (VAR); selectivity factor

r dynamic resistance, measured in Ω

R   resistance, measured in Ω

$\mathfrak{R}$ set of real numbers

s Laplacian variable, measured in nepers/s

S   apparent power, measured in volt-ampere (VA)

S   complex power, measured in VA

σ real part of the complex variable s

t time

T   period of a periodic signal; transmission matrix in two-port parameters

$\mathcal{T}$ time set

τ time-constant, measured in seconds

$u(t)$ any instantaneous signal; voltage or current

U   any constant signal

$\mathcal{U}$ phasor for $u(t)$

ϕ phase angle, measured in degrees

$v(t)$ instantaneous voltage, measured in volts

V   constant voltage, measured in volts

V   volt, unit of voltage

$\mathcal{V}$ phasor for $v(t)$

ψ phase angle, measured in degrees

W   watt, unit of power

ω frequency, measured in rad/s

Ω ohm, unit of resistance

$x(t)$ instantaneous state variable

$\overline{x}$ state vector

X   reactance, measured in Ω

$y(t)$ any arbitrary instantaneous signal, such as $u(t)$

$\mathcal{y}$ admittance, measured in siemens (Ω^{-1})

$\mathcal{Z}$ impedance, measured in ohms

CHAPTER 1

Introduction

This text on circuit theory is written to serve two basic purposes—(i) to introduce the *principles of electrical engineering* in a systematic way and (ii) to *develop model-building and problem-solving strategies* in the minds of young engineering students in their formative semesters (second and third). Once we mention 'electrical engineering' there is a potential danger that it might restrict the audience. However, we do not propose to look into the subject as it was done some time ago; instead, we shall give a fresh approach with an emphasis on mathematical model building and problem solving. The presentation in this fashion would pave a conceptual highway to advanced subjects such as control systems, communication systems, and digital signal processing.

In general, engineering students would have acquired familiarity with electricity and magnetism, as two sections in physics, a little earlier. The treatment would have been strongly biased from the point of view of certain fundamental laws, such as Coulomb's law and Ohm's law, that the electrons and/or the components have to necessarily obey. Put it briefly, making use of electricity and magnetism to design interesting things is far away from such an orientation. The reason is, perhaps, that the students at that time possess a rather limited view of the world around them; that it is virtually the same thing being taught in the name of electricity, magnetism, and mechanics is certainly beyond the comprehension of a high-school student. It is time that the student expands his view to garner several beautiful things. And, we believe that this textbook would help him in this *fait accompli*.

1.1 A Historical Perspective

Historically, dynamics (yet another section of physics) is the most important intellectual discipline. It is a scientific discipline concerned with 'the change of state', i.e., movement or evolution of some sort, of a system; recall Newton's first law—every body 'continues' to be in its state of rest or of uniform motion ... . This subject has its foundations in questions such as 'What causes the rivers to flow?', 'What causes the Sun[1] to rise, move, and set?', 'What causes the Moon to *hang* up in the sky?', etc. Thus, dynamics deals with the spatial motions of interacting bodies under the influence of one or more forces, i.e., movement with respect to position coordinates, e.g., the inverse square law of gravitational force. Edwin Hubble (1889–1953) observed in 1929 that the galaxies are all rushing away from one another. Ever since, we have been believing that the universe is an evolving, i.e., a dynamic, entity rather than a static one, and that time, matter, energy, and space began about fifteen billion years ago with the *big bang*. Thus, a galaxy is a *dynamic model* of the universe that has been built based on a set of logical observations. We also have a model for our solar system—the earth

---

[1] Does it really?

and other planets revolve around the sun. When we study the *motion* or 'rates of change[2] of state' of these dynamic systems, we get a set of differential equations as the primary mathematical description; recollect the lessons in the first course on calculus. We call the set equations of motion as the *mathematical model* of the system. It is widely accepted that dynamics has become mathematical (and hence more methodical) with Galileo's kinematics and its culmination with the laws of Newton and Euler, and the works of Leibniz, Bernoulli, d'Alembert, Lagrange, and many more.

Nature obeys unchanging laws that mathematics can describe!

The fundamental hypothesis of Newtonian revolution, as it might be called, is that the principles of nature can be expressed in terms of mathematics, and physical events can be *predicted and designed* with mathematical certainty. What we mean by prediction is the following. A mathematical model represents a physical phenomenon in terms of a set of differential equations. Let us take up a simple example.

Example 1.1 Let the differential equation be

$$\frac{d}{dx} f(x) = f(x)$$

The solution would be, quite naturally, a function of the spatial coordinate x. In this example, it is easy to see that

$$f(x) = ce^x$$

where c is some constant. We might then substitute certain values of x where we have made our earlier observations to build the model and check for the consistency of the model. For instance, we might set $x = 0$ and see if we get the 'initial condition $f(0) = c$'. Once we find that the model is consistent, we may substitute any arbitrary value of x, say x_1, and we readily have an answer as to how the phenomenon would appear at position x_1. This value x_1 need not be among the set of coordinates we have utilized in building our model. It is in this sense that we talk about the prediction of events.

We can turn this around and design a function $f(x)$ that satisfies our observations at several positions x_i, $i = 1, 2, \ldots$. The reader may look at the weather bulletins regarding the forecast of sunrise and sunset, rainfall, cyclones, eclipses, etc. for a more concrete example. Yet another example is the well-known prediction about Halley's comet that appears every 76 years. The reader might also recall Mendeleev's periodic table in chemistry and the prediction of properties of elements yet to be discovered. It is exciting to note that certain actions of different elements in biological systems are apparently determined by their positions in the periodic table—gradation in some form, either vertical or diagonal, seems to be significant and calls for a wider study so as to include other elements about which very little or nothing is known.

---

[2] with respect to the spatial coordinates

After dynamics, several natural sciences, including electricity, magnetism, and thermodynamics, followed suit. Philosophical thinking and the motion of (and conjectures about) heavenly bodies are not the only source and destination of dynamics. Indeed, in addition to the spatial motion of bodies, dynamics encompasses time-varying phenomena such as electricity (we are more interested in the 'time rate of change' of the charge of an electron, rather than its physical displacement), thermal and chemical activities, bridge design, supply-chain management, automotive suspension, town planning, transportation, optimization of natural resources, and so on. In other words, dynamics includes the study of 'time behaviour' of the state of a system, e.g., it takes 24 h for the earth to complete one rotation on its axis. Thus, we might model several engineering systems mathematically as differential equations. We just need to learn which variable represents what physical object; for example, in the familiar Ohm's law in electricity, if we replace the voltage with force, and replace the current with displacement, we get Hooke's law of mechanical systems: the spring constant is *analogous* to resistance. We would remind the reader that Newton's second law, which speaks about the 'time rate of change' of momentum of a body, gives us differential equations and enables us to *predict* the motion of the body with respect to time.

We end this section with the following comments[3] on mathematical modelling. In our daily lives our actions are performed invariably with certain expectations in view. For example, when we heat water, we expect it to boil and not to freeze. Thus, belief is implicit, in all such actions, that events in the world do not happen at random, but that they take place in an orderly manner. In other words, order objectively exists in nature, and mathematical models arrived at on the basis of the accepted canons of scientific enquiry do help us to predict natural happenings.

1.2 Dynamics and Electrical Engineering

The initial observations of electricity and magnetism occurred in antiquity (around 600 BC), and were associated with such simple phenomena as (i) amber, when rubbed with wool, attracting light objects; (ii) electric eels stunning their prey; (iii) atmospheric lightning; and (iv) lodestone attracting iron objects. The word 'electron' is derived from the Greek word for amber, and the word 'magnet' is derived from Magnesia, a place in Greece, where lodestone was first believed to be found. Nevertheless, electricity and magnetism as sciences were substantially created in the seventeenth and eighteenth centuries. We should acknowledge two prominent events that laid the foundation for these sciences. One is the paper *De-Magnete*, published by the British scientist William Gilbert (1544–1603) and the other is the discovery by the Italian scientist Alessandro Volta (1745–1827) that a continuous electric current can be produced using a *voltaic pile*. It is interesting to

---

[3] Interested readers are urged to refer to the insightful article: J.R. Lakshmana Rao, 'Scientific Laws, Hypotheses, and Theories: Meanings and Distinctions', *Resonance*, vol. 3, no. 11, pp. 69–74, 1998. Resonance is a journal of science education and can be accessed at **http://www.iisc.ernet.in/academy/resonance**.

note that these two sciences, electricity and magnetism, developed independently of each other until the early nineteenth century.

In 1820, the Danish scientist Hans Christian Oersted (1777–1851) demonstrated that electric currents could produce magnetic fields. In 1831, the English enthusiast Michael Faraday (1791–1867) demonstrated that time-varying magnetic fields could produce electric currents. In other words, there is a very strong 'coupling' between the electric and magnetic fields—an electric field produces a magnetic field, which in turn produces an electric field, which in turn produces a magnetic field, so on. These two demonstrations led the Scottish physicist James Clerk Maxwell (1831–1879) to *predict* the existence of 'electromagnetic waves'. Based on the experiments of Charles-Augustin de Coulomb (1736–1806) and John Michell (1724–1793) Maxwell discovered that light is an electromagnetic wave. Subsequently, he gathered the works of Frenchmen André-Marie Ampere (1775–1836), Jean-Baptiste Biot (1774–1862), and Felix Savant (1791–1841), the German scientist and mathematician Karl Friedrich Gauss (1777–1855), and several others and compiled in 1864 a summary of *electromagnetic theory* as a set of four 'differential equations', known as Maxwell's equations, and laid the foundations for radio, television, and several modern devices and systems; modern electrical engineering in short. Within the next 25 years, the German scientist Heinrich Hertz (1857–1894) experimentally demonstrated radio waves.

In the following subsection we shall look at Maxwell's equations from the circuit theory perspective. We assume that the reader is already familiar with the basic physics since we are not going to teach anything here. We shall just show an alternative derivation which couples dynamics and circuit theory.

1.2.1 Maxwell's Equations and Circuit Theory

Let us recollect that our earlier knowledge of electricity and magnetism was centered around

- electric-field intensity $\vec{E}$ measured in newton per coulomb,
- magnetic-flux density $\vec{B}$ measured in tesla,
- current density $\vec{J}$ measured in amperes per meter square, and
- charge density ρ measured in coulomb per meter square.

In terms of these quantities, the differential forms of Maxwell's equations are given as follows.

Gauss's law for electricity:

$$\nabla \cdot \epsilon_0 \vec{E} = \rho \tag{1.1}$$

Gauss's law for magnetism:

$$\nabla \cdot \vec{B} = 0 \tag{1.2}$$

Ampere's law (with Maxwell's correction):

$$\nabla \times \vec{B} = \mu_0 \epsilon_0 \frac{\partial \vec{E}}{\partial t} + \mu_0 J \tag{1.3}$$

Faraday's law of induction:

$$\nabla \times \vec{E} = -\frac{\partial \vec{B}}{\partial t} \tag{1.4}$$

where ∇, called *nabla* or *del*, is the differential operator in three (spatial) dimensions given by

$$\nabla = \left(\frac{\partial}{\partial x} \hat{x} + \frac{\partial}{\partial y} \hat{y} + \frac{\partial}{\partial z} \hat{z} \right) \tag{1.5}$$

The scalar (or dot) product of nabla and a vector gives the *divergence*, i.e., a measure of how much the vector spreads out from the point in question. Likewise, the vector (or cross) product of nabla and a vector gives the *curl*, i.e., a measure of how much the vector *curls around* the point in question. It is easy to see that each of the four equations is a partial differential equation. We recommend the reader to verify the dimensions on either side of these equations. These equations clearly suggest the strong interaction between the electric and magnetic fields.

Exercise 1.1 Maxwell's equations are given in the 'differential form' above. Rewrite them in the 'integral form'.

The four Maxwell's equations represent what may be called as *electrodynamics*, owing to the 'time-varying' behaviour (i.e., the wave behaviour) of the electric field and the magnetic flux. Observe that there is a *coupling* between the electric and magnetic fields—to solve for $\vec{E}$ we need to solve for $\vec{B}$ and vice versa. We recollect that ϵ_0 is called the permittivity and μ_0 is called the permeability, which are related to the velocity of light as

$$c^2 = \frac{1}{\epsilon_0 \mu_0} \quad \text{where } c = 3 \times 10^8 \text{ ms}^{-1} \tag{1.6}$$

The 'static' forms of Maxwell's equations are derived from the complete forms Eqns (1.1)–(1.4) by imposing the conditions that there are no time-varying fields $\vec{E}$ and $\vec{B}$. These conditions may be mathematically represented as

$$\frac{\partial \vec{E}}{\partial t} = 0 \quad \text{and} \quad \frac{\partial \vec{B}}{\partial t} = 0 \tag{1.7}$$

Looking at Maxwell's equations with these conditions, we observe that each of the equations is a function of either the electric field $\vec{E}$ or the magnetic field $\vec{B}$, but not both. We may say that there is no 'coupling' between the electric and magnetic fields. Perhaps this is the reason why electricity and magnetism grew as independent sciences until early nineteenth century. Gauss's law for electricity and Faraday's law form what may be called *electrostatics* and Gauss's law for magnetism and Ampere's law form *magnetostatics*.

   In between the dynamics and statics there is an important case of *quasistatics*. We shall explore this in some detail now.

1.2.2 Electroquasistatics

Let us consider, arbitrarily, a pair of conducting surfaces separated by an open space, wherein the influence of electric field is much higher than that of magnetic field. Equivalently, we may say that the current density $\vec{J}$ (primarily due to the transportation of charge) is negligible compared to the current *induced* by the electric field. For the sake of convenience, we consider the fields in only one dimension, say along the x axis. Accordingly, the fourth equation of Maxwell (Faraday's law of induction) may be written as

$$\frac{d\vec{E}}{dx} = -\frac{\partial \vec{B}}{\partial t} \tag{1.8}$$

Looking at this equation dimensionally, we get

$$\frac{\vec{E}}{L} \propto \frac{\vec{B}}{T} \tag{1.9}$$

At this point, imposing the condition that the influence of electric field is much higher than that of magnetic field, we get

$$\frac{\vec{E}}{L} \gg \frac{\vec{B}}{T} \tag{1.10}$$

Looking at Ampere's law in the same way, we get

$$\frac{\vec{B}}{L} \propto \epsilon_0 \mu_0 \frac{\vec{E}}{T} \tag{1.11}$$

and imposing our requirement from Eqn (1.10) and eliminating $\vec{B}$, Ampere's law may be written as

$$\frac{\vec{E}}{L} \gg \frac{\vec{E}L}{c^2 T^2}$$

which simplifies to

$$\frac{L}{cT} \ll 1 \tag{1.12}$$

Equation (1.12) tells us lots of things. For instance, if we consider T as the period of electromagnetic wave propagation, then with the velocity of light c, we get $cT = \lambda$, the wavelength. Accordingly, Eqn (1.12) suggests that our assumption, Eqn (1.10), holds at wavelengths much higher than the characteristic length L of the electromagnetic process. Or equivalently, we may rewrite Eqn (1.12) in terms of frequency f in hertz as

$$\frac{Lf}{c} \ll 1$$

For instance, if we consider the separation of the two conductors, such as the separation between the two parallel plates of a capacitor, as 1 cm, then the frequency of the voltage (the potential difference between the two conductors) must be much less than $30 \times 10^9 = 30$ GHz. In other words, if we are looking at an electromagnetic phenomenon where the operating frequencies are much smaller than a few gigahertz, the phenomenon may be considered as quasistatic.[4] Thus, Maxwell's equations for electroquasistatics are as follows:

$$\nabla \cdot \epsilon_0 \vec{E} = \rho$$

$$\nabla \cdot \vec{B} = 0$$

$$\nabla \times \vec{B} = \epsilon_0\mu_0 \frac{\partial \vec{E}}{\partial t} + \mu_0 \vec{J}$$

$$\nabla \times \vec{E} = \vec{0} \tag{1.13}$$

Clearly, we take into account the magnetic field induced by the electric field (as in Oersted's observation), but we ignore the current produced by the magnetic field (as in Faraday's observation).

1.2.3 Magnetoquasistatics

Let us now look at a situation where the influence of magnetic field is much higher than that of electric field, i.e., we shall take into account Faraday's observation, ignoring Oersted's observation. In this case, Ampere's law may be written as

$$\frac{\vec{B}}{L} \gg \epsilon_0\mu_0 \frac{\vec{E}}{T} \tag{1.14}$$

Using Eqn (1.14) in Faraday's law [Eqn (1.9)] and eliminating $\vec{E}$ we get, once again,

$$\frac{L}{cT} \ll 1 \tag{1.15}$$

The result that Eqns (1.12) and (1.15) are identical suggests that when the operating frequencies of the fields are much smaller than those of the electromagnetic waves, the electric and magnetic fields are decoupled. We may choose one of the fields to be much stronger than the other. Maxwell's equations for magnetoquasistatics may be summarized as follows:

$$\nabla \cdot \epsilon_0 \vec{E} = \rho$$

$$\nabla \cdot \vec{B} = 0$$

$$\nabla \times \vec{B} = \mu_0 \vec{J}$$

$$\nabla \times \vec{E} = -\frac{\partial \vec{B}}{\partial t} \tag{1.16}$$

Comparing Eqns (1.13) and (1.16) we notice that the decoupling is done only along the spatial coordinates; the time variations, $\partial \vec{E}/\partial t$ in Eqn (1.13) and $\partial \vec{B}/\partial t$ in Eqn (1.16), are maintained.

We shall now concentrate on two important consequences of the above quasistatics. What we derive in the following would lay the foundations of the modern circuit theory. First, we look at the way energy in electroquasistatics and magnetoquasistatics is stored. This would throw more light on what is happening.

---

[4] The word 'quasi' means 'almost'.

Let us consider Fig. 1.1. The energy balance for a volume V bounded by the surface S is sketched in this figure.

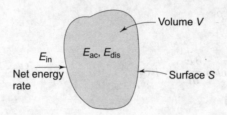

Fig. 1.1

The law of conservation of energy tells us that the rate of energy flow E_{in} into the volume V through the surface S equals the sum of the rate of energy accumulation E_{ac} in the volume V and the rate of energy dissipation E_{dis} in the volume V. This may be expressed mathematically as

$$-\int_S \vec{P} \cdot d\vec{S} = \frac{\partial}{\partial t} \int_V w \, dV + \int_V p_d \, dV \tag{1.17}$$

where $\vec{P}$ is the rate of flow of energy per unit area through the surface S into the volume V, w is the energy density within V, and p_d is the rate of dissipation of energy density within V. Owing to the directions of the flow of energy considered in Fig. 1.1, we put a negative sign to the surface integral on the left-hand side of Eqn (1.17).

Let us now consider electroquasistatics:

$$\nabla \times \vec{B} = \epsilon_0 \mu_0 \frac{\partial \vec{E}}{\partial t} + \mu_0 \vec{J} \tag{1.18}$$

$$\nabla \times \vec{E} = \vec{0} \tag{1.19}$$

Here we need some manipulations. First, we shall *dot* multiply Eqn (1.18) by $\vec{E}/\mu_0$ and Eqn (1.19) by $\vec{B}/\mu_0$, and then subtract to get

$$\frac{\vec{E}}{\mu_0} \cdot (\nabla \times \vec{B}) - \frac{\vec{B}}{\mu_0} \cdot (\nabla \times \vec{E}) = \epsilon_0 \vec{E} \cdot \frac{\partial \vec{E}}{\partial t} + \vec{E} \cdot \vec{J} \tag{1.20}$$

We use the general vector identity

$$\vec{C} \cdot (\nabla \times \vec{A}) - \vec{A} \cdot (\nabla \times \vec{C}) = \nabla \cdot (\vec{A} \times \vec{C})$$

and the divergence theorem to rewrite Eqn (1.20) as

$$-\int_S \left(\frac{1}{\mu_0} \vec{E} \times \vec{B} \right) \cdot d\vec{S} = \frac{\partial}{\partial t} \int_V \frac{1}{2} \epsilon_0 \vec{E} \cdot \vec{E} \, dV + \int_V \vec{E} \cdot \vec{J} \, dV \tag{1.21}$$

so that we may identify the *Poynting vector* $\vec{P}$, w, and p_d from Eqn (1.17).

Exercise 1.2 Derive Eqn (1.21) from Eqn (1.20) using the divergence theorem of vector calculus.

We are more interested in

$$w = \frac{1}{2}\epsilon_0 \vec{E} \cdot \vec{E} \tag{1.22}$$

which gives us the electromagnetic energy density accumulated in a volume V; we notice at once that this energy accumulation is accomplished via the electric field $\vec{E}$ alone. Equation (1.22) tells us that *in an element/system characterized by the electroquasistatic approximation, energy is stored in the electric field*. In the following chapter we will introduce the element *ideal capacitor* that exemplifies this characteristic. We urge the reader to verify the units and dimensions on either side of Eqns (1.17), (1.20), and (1.22).

We leave it as an exercise to the reader to show that *in an element, such as an ideal inductor, characterized by magnetoquasistatic approximation, energy is stored in the magnetic field* as

$$w = \frac{1}{2\mu_0} \vec{B} \cdot \vec{B} \tag{1.23}$$

Exercise 1.3 Derive Eqn (1.23).

The electric and magnetic fields are seen to be *localized* in the capacitive and inductive elements, respectively, and are negligible elsewhere. The second consequence is more subtle as we see now. We may define a *node* as a point at which no fields exist. Accordingly, for an arbitrary surface S enclosing a node, we may consider Ampere's law, in its integral form, and observe that

$$\int_S \vec{J} \cdot \vec{S} = 0 \tag{1.24}$$

By definition, $\vec{J}$ is the current density, i.e.,

$$i_T = \int_S \vec{J} \cdot d\vec{S} \tag{1.25}$$

where i_T represents the net or total current passing through the surface S. Clearly, from Eqns (1.24) and (1.25), we have

$$i_T = 0 \tag{1.26}$$

or, more conventionally, if there are n electrical leads passing through the surface S and joining at the same node N, we have

$$\sum_{k=1}^{n} i_k = 0 \tag{1.27}$$

Put in words, Eqn (1.27) tells us that

the algebraic sum of the currents entering any node must be zero.

We emphasize the word *algebraic* to account for the arbitrary directions in which the current would be flowing in the leads connected to the node. This is popularly called *Kirchhoff's current law* (KCL) after the German physicist Gustav Kirchhoff (1824–1887). However, this is *not* an independent law of physics; it is rather a convenient approximate consequence of Ampere's law for physical systems in which the electric and magnetic field effects substantially uncouple. Accordingly, in this text, we would prefer to call this as *Kirchhoff's current rule* (KCR). This rule is illustrated in Fig. 1.2.

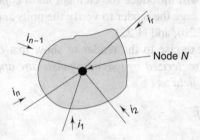

Fig. 1.2

We may also consider Faraday's law in an analogous fashion. This time we shall select a hypothetical path joining several nodes, thereby neglecting the effect of the magnetic field $\vec{B}$. We illustrate this in Fig. 1.3. It is not too difficult to see that Faraday's law, in its integral form, reduces to

$$\int_C \vec{E} \cdot d\vec{l} = 0 \tag{1.28}$$

where the integration is performed over the chosen contour, such as the closed *path* starting at node N_1 and terminating on the same node N_1 as shown in Fig. 1.3. Such a closed path or contour is called a *loop*.

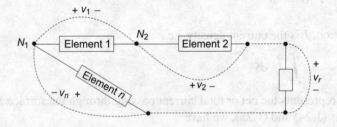

Fig. 1.3

By definition, the potential difference between any two points a and b is given by the *line integral*

$$v_{ab} = \int_a^b \vec{E} \cdot d\vec{l} \tag{1.29}$$

From Eqns (1.28) and (1.29), we simply have

$$\sum_{k=1}^n v_k = 0 \tag{1.30}$$

where v_k is the potential difference across the kth element of a total of n elements in the closed contour of Fig. 1.3. Put in words, Eqn (1.30) tells us that

> *the algebraic sum of the voltage drops around any loop must be zero.*

Here we place the emphasis on the words 'algebraic' and 'loop'. Owing to the same reason as before, we call this statement *Kirchhoff's voltage rule* (KVR).

We wish to bring the following to the reader's attention: in addition to the quasistatic assumption, we have the assumption of a negligible time rate of change of the magnetic field $\vec{B}$ in Faraday's law. Consequently, we should not call the statements in Eqns (1.27) and (1.30) as 'laws' in the fundamental sense; they are simply theorems or corollaries or rules. We shall refer to them as Kirchhoff's rules throughout this text.

To sum up, electricity and magnetism are natural phenomena that were initially observed to be independent. Subsequently, experimental evidence has been established that there is a strong coupling between the electric field and the magnetic field. This led to electromagnetism and was elegantly packaged into four differential equations by Maxwell. Dynamics has a tremendous influence in this development. With Maxwell's equations as the starting point, we have made some simple derivations, called the quasistatics. Further, we have examined two important consequences of these quasistatics which lay the foundations of the modern circuit theory; we have, in some sense, foreseen two elements—the capacitor and the inductor—where the electric field and the magnetic field, respectively, are localized. It is time to build several interesting things, of great practical value, on the foundations of circuit theory. We shall do it in the following chapters.

1.3 Overview of the Text

In Chapter 2, we introduce the electrical circuit elements, the fundamental laws that govern the behaviour of these elements, and much more. In Chapter 3, we shall look at the basic definitions, the interconnections among two or more elements, and develop circuit rules from quasistatics. We shall also look at some preliminary ideas of circuit-analysis techniques. In Chapter 4, we shall develop an almost exhaustive set of techniques for problem-solving. These techniques, as they appear in Chapter 4, employ real algebra, i.e., we work with real numbers. The circuits may be called static since the voltages and currents do not vary with time.

Chapter 5 introduces the energy-storage elements—capacitor and inductor—into the circuits together with switches. Transient response is the most important idea one has to master. Accordingly, we spend a considerable amount of time here, not just defining, but *developing* several things. In Chapter 6 we introduce the time-varying sources, namely, the sinusoidal sources. We will be employing complex algebra as a generalization of real algebra; a real number is a complex number with zero imaginary part. This allows us to generalize the techniques of Chapter 4 to solve *any* circuit. Chapter 7 is an extension of Chapter 6, wherein we introduce the ideas of frequency response, semi-logarithmic plots, filters, etc. Although the circuit exhibits transient response, we ignore it in this chapter. As an extension and a popular application of Chapters 6 and 7, Chapter 8 is devoted to three-phase circuits and power systems. Since this is neither a text on power engineering nor a text for core electrical engineers alone, the treatment is not very rigorous.

Chapter 9 is a further generalization of the frequency response, wherein we shall address both the transient and steady-state responses in a unified framework. This can be done with the help of Laplace transformation. Once again, we keep the discussion on Laplace transformation to a minimum and the reader is expected to have learnt it in an earlier or concurrent course in mathematics. Chapter 10 generalizes the notion of circuits to that of systems. The input–output paradigm is introduced in this chapter along with certain basic properties of general systems. Chapter 11 is devoted to the synthesis, i.e., the design, of networks given certain specifications, such as driving point impedance. This is a very interesting subject in its own right and is only discussed superficially in this chapter. Chapter 12 introduces the workhorse of modern electronic era—the operational amplifier (Op-Amp in short). We shall not present the theory of Op-Amps here, rather we introduce this as yet another ideal circuit element, much like the resistor, or inductor, or capacitor, and exemplify certain important circuits that *need to* employ Op-Amps. In Chapter 13, we shall discuss Fourier's contribution to signals. We conclude the text with a discussion on general linear systems in Chapter 14.

As we go through, we emphasize on the mathematical descriptions, model building, and problem-solving. In so doing, we believe that the reader would acquire a *broad view of the world*, without which he/she would forever remain incomplete. In other words, the reader should, at least, ask questions such as

- how to describe the antibiotic immunity;
- how to describe the flickering of a flame of a candle;
- how to describe the breaking of glass or melting of ice;
- how to describe the sound of a piano;

and many more. The answers would range from simple science to intensive research areas. We recommend an interesting column 'Ask the experts' in the magazine *Scientific American*.

To elaborate our theme, let us look at the following everyday problems.

Example 1.2 Suppose we wish to travel from station A to station B. We look at the road map to first locate the stations, and then decide upon the route. This simple idea may be *abstracted* as shown in Fig. 1.4.

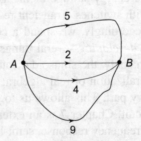

Fig. 1.4

What we have shown in Fig. 1.4 is technically called a 'graph.' The two dots representing the stations are called *nodes* or *vertices*[5] of the graph, and the curves showing several routes from *A* to *B* are called *branches* or *edges* of the graph. Observe that we have shown arrows to indicate our proposed travel from *A* to *B*. Further, adjacent to each of the arrows, we put a number corresponding to the distance between *A* and *B* via that particular route; equivalently, we may interpret the number as the cost involved in travelling from *A* to *B* via that route. We need not say explicitly that we choose the route that is shortest in distance, or equivalently the cheapest in cost.

This is one of the most trivial examples of *modelling*. By modelling[6] we mean to say that some abstract idea of a problem is put in black and white on a piece of paper so that we 'see' the essential features of the problem. Solving the problem follows this.

It is interesting to note that the theory of graphs has been independently discovered by several eminent people such as Leonhard Euler (1707–1783), Arthur Cayley (1821–1895), and Gustav Kirchhoff. Let us next move on to the more popular Euler's graph.

Example 1.3 The seven-bridge problem. There are two islands linked to each other and to the banks of a river by seven bridges as shown in Fig. 1.5(a). The problem is to begin at any of the four land areas *A, B, C, and D*, walk across each bridge 1, 2, ..., 7 exactly once, and return to the starting point.

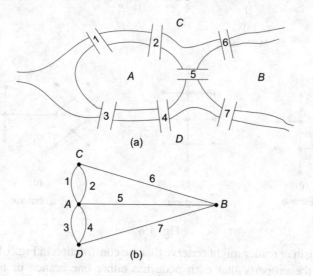

Fig. 1.5

---

[5] *junction* is another name.
[6] Let us keep overselves away from wax models of famous personalities, and the commercial ad-world!

With our experience in Example 1.2, it is not too difficult to translate the present problem into a graph shown in Fig. 1.5(b), with each of the four land areas represented as a node and each of the seven bridges represented as a branch. We might be prompted to solve the problem by tracing the branches. And, it so happens that we would never succeed. In other words, there is no solution to this problem. For a problem like this to have a solution, each node should have an even number of branches connected to it. Clearly, in the graph shown in Fig. 1.5(b) none of the nodes satisfies this property. In fact, Euler *generalised* the seven-bridge problem and developed the rule that for any graph to be traversable, each node should have an even number of branches connected to it.

Exercise 1.4 Assume that there is one more branch between A and C and one more between D and B in the graph [Fig. 1.5(b)]. Show that the problem is solvable by tracing out at least one possible route.

Exercise 1.5 Suggest another alternative to Exercise 1.4 so that the problem may be solved.

Example 1.4 Isomers of C_nH_{2n+2}.

Cayley showed the application of an important class of graphs, called trees, in the enumeration of isomers of the saturated hydrocarbons C_nH_{2n+2}. A few isomers are shown in Fig. 1.6.

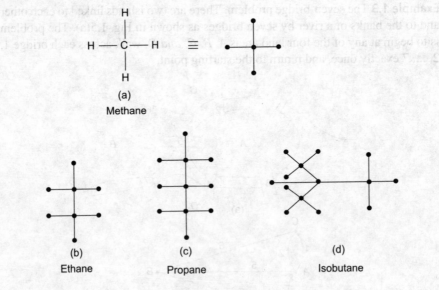

Fig. 1.6

An inquisitive reader might observe that the construction in Fig. 1.6 enumerates trees with the property that each node has either one branch or four branches connected to it.

Example1.5 Kirchhoff's rules and circuit equations.

Here is a fascinating piece of history. Physically speaking, Kirchhoff derived his rules from quasistatics. These rules help us develop standard procedures (in later chapters) to analyse circuits via linear systems of equations (LSE). In an attempt to solve these linear equations, Kirchhoff developed the theory of trees. Mathematically speaking, he abstracted electrical circuits and replaced them by their underlying graphs. One example is shown in Fig. 1.7 with a network in Fig. 1.7(a), and its underlying graph in Fig. 1.7(b). He proved that the 'independent cycles' of a graph determined by any of its 'spanning trees' will suffice to solve the LSE. One spanning tree of the graph shown in Fig. 1.7(b) is shown in Fig. 1.7(c).

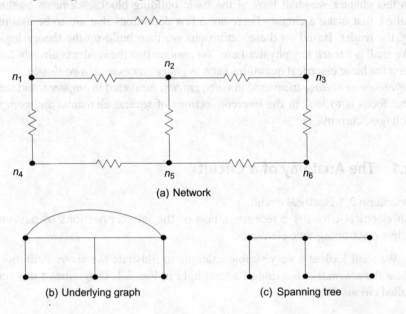

(a) Network

(b) Underlying graph   (c) Spanning tree

Fig. 1.7

We will not define these technical terms—trees, spanning trees, independent cycles, etc.—here. Instead, we shall first see what they mean in Chapter 3 in terms of networks and then provide appropriate definitions. Our approach is as follows. Graph theory is a powerful tool that finds applications in physics, chemistry, engineering, civil engineering, economics, and many more fields. In other words, if we choose to become mathematicians and learn graph theory, we are ready to learn many more useful things. However, since we chose to become electrical engineers, let us learn electrical network theory and henceforth learn to 'abstract', i.e., model, many more practical things, not necessarily belonging to electrical engineering, as electrical networks. As far as the subject of graph theory goes, there is no uniformity in the terminology. One alternative to the word graph is 'network'. For instance, the road map of Fig. 1.4 may be depicted as a bunch of resistors connected in parallel; since the current chooses the path of least resistance, so do we. How nice it is to identify ourselves with electric current. The graph for the seven-bridge problem resembles a popular circuit called the $R–2R$ ladder as we see in Chapter 3. As we go ahead, we shall look at several interesting problems that can be represented by circuits. We shall also develop elegant problem-solving strategies.

CHAPTER 2

The Elements

In this chapter we shall look at the basic building blocks, *elements* as they are called, that make a circuit. There are a few definitions that are to be assimilated by the reader. Based on these definitions we then build up the theory logically. We shall not teach any physics here. We assume that the reader is already familiar with the basic electrical quantities such as *charge* measured in *coulombs*, *potential difference or voltage* measured in *volts*, *current* measured in *amperes*, and the like. Our focus is to look at the interconnections of several elements, and computing voltages, currents, etc.

2.1 The Anatomy of a Circuit

Definition 2.1 Electrical circuit
An electrical circuit is a representation of the interconnections of passive and active electromagnetic elements.

We shall look at a very simple example to illustrate the above definition. We show the *schematic diagram* of a torchlight in Fig. 2.1. Diagrams of this sort are called *circuit diagrams*.

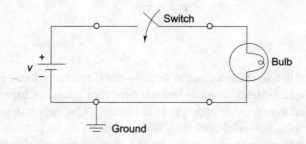

Fig. 2.1

In what follows, we advise the reader to pay particular attention to the words in italics. There are three *circuit elements* depicted in Fig. 2.1. First, we have the battery that provides the required energy. Since we understand that it is the potential 'difference' that matters to us, we show two *terminals*—one marked '+' (positive) and the other marked '–' (negative). It is customary to connect one of the terminals (in this example the negative terminal) of the battery to the *ground* (also shown in Fig. 2.1) whose potential is *defined* to be 0 V. This allows us to say that the battery yields the required voltage.

Secondly, this battery is *connected* to the bulb with the help of a pair of wires. These wires, more appropriately called the connecting wires or leads, are shown in the diagram as simple lines (or curves), with unfilled circles at their ends, and are expected to satisfy the following conditions:

1. These are made up of a good electrical conductor, e.g., copper.
2. There is no potential difference between any two arbitrary points on a given wire, i.e., the wires do not oppose the flow of current.
3. All current entering one end of a given wire exits from the other end, i.e., the wires do not accumulate charge.

Next, we show the *switch* as a small break along one of the leads to show that the battery delivers energy to the bulb only when needed and is disconnected otherwise. Lastly, we show the bulb as yet another *two-terminal* device, the ends of which are connected to the '+' (via the switch) and '–' terminals of the battery.

Let us recollect that energy can neither be created nor destroyed, but can be *converted* from one form to another. It is easy to 'see' that, when the switch is closed, the battery causes the current (generated by conversion from chemical energy) to flow out of its positive terminal, through the leads and the bulb, and back into its negative terminal, from where the current returns to its positive terminal, thereby completing a closed path or a *loop*. While the current flows through the bulb, the electric energy gets converted into light and heat.

2.1.1 Sign Convention

It is conventional to say that the battery *releases* energy and the bulb *absorbs* energy. More generally, circuit elements that release energy are called *active elements* and those that absorb energy are called *passive elements*. In general, an active element is called the *source* and a passive element is called the *load* or *sink*. Further, we adopt the following sign convention to distinguish between the active and passive elements:

If in an element the current flows from a point of lower potential to that of a higher potential, i.e., the current leaves the '+' terminal and enters the '–' terminal, then it is an active element, otherwise it is a passive element.

Here is a word of caution in applying the above sign convention. We are quite familiar with *rechargeable* batteries for cellular telephones, walkmans, and other portable devices. Let us consider one such battery. It is clear from the sign convention that it is an active element as long as it supplies energy to the device it is connected to. However, while we are recharging the battery, it becomes a passive element since it now absorbs energy from another source (active, of course). Hence, we need to exercise care, by carefully examining the direction of flow of energy, while calling an element an active device in a given circuit.

2.2 Circuit Elements

We may further generalize the circuit diagram of Fig. 2.1 to contain many more elements, both active and passive. A *source* may deliver either a voltage, such as the battery in Fig. 2.1, or a current. We shall now learn the standard symbols we use in drawing circuit diagrams. We are not much bothered about the manufacture of sources in this book.

2.2.1 Independent Sources

In Fig. 2.2(a) we show the standard symbol of an *independent voltage source*, and we show an *independent current source* in Fig. 2.2(b). Let us explain these terms. An independent source is one that delivers energy on its own for any length of time, irrespective of the elements connected to it. For instance, a (fully charged) dry cell delivers a voltage of 1.5 V whether it is connected to the bulb of a torchlight, the dc motor of a walkman, or to some other device. By definition, an independent source is an *idealistic* assumption. After a period of time, we observe that the torchlight gets dimmer since the terminal voltage across the dry cell decreases. However, as a first step, we shall not pay attention to these aspects at this moment; we shall address this in a few paragraphs from now. Observe that the polarities of voltage and the direction of current are shown in the figure according to the sign convention we have adopted.

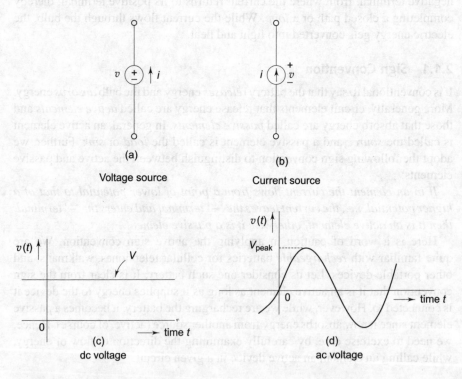

Fig. 2.2

The independent sources are generally of two types—direct current (dc) and alternating current (ac). Theoretically, dc sources give us voltage/current of constant magnitude indefinitely, for example, 1.5 V for a dry cell. We denote the magnitude with an uppercase italic letter V or I. The time function of a dc voltage[1] source is shown in Fig. 2.2(c) as a horizontal straight line parallel to the time axis. The time function of a dc current source is identical to that of a dc voltage source except that the vertical axis represents current. An ac source is one whose magnitude varies with time, in particular, varies between a positive peak and a negative peak. Our household voltage is sinusoidal with a magnitude of 220 V and a frequency of 50 Hz. A typical sinusoidal voltage source has the time function shown in Fig. 2.2(d). An ac voltage is denoted by a lowercase letter such as $v(t)$ along with the argument t. For example, the household supply is written as

$$v(t) = 220 \sin(100\pi t) \text{ V}$$

Other ac signals such as rectangular and triangular also exist. We shall elaborate these concepts in later chapters.

2.2.2 Controlled Sources

In contrast to the independent sources we shall have *controlled sources*. A source is said to be 'controlled' if its energy is controlled by some voltage or current located elsewhere in the circuit. For instance, if we look at a public address system we observe that the (electrical) energy delivered to the loudspeaker is controlled (indirectly, of course) by the (mechanical) energy supplied to the microphone. To keep things simple, we assume that the energy supplied by a controlled source is *directly proportional* to that of the controlling variable. Accordingly, we see that there are four possible controlled sources as shown in Figs 2.3(a)–2.3(d).

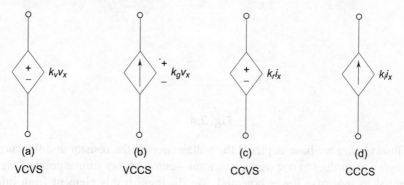

| (a) | (b) | (c) | (d) |
| VCVS | VCCS | CCVS | CCCS |

Fig. 2.3

The first one is a *voltage-controlled voltage source* (VCVS) wherein the voltage supplied by the source depends on a voltage v_x appearing elsewhere in the circuit. In this case, the proportionality constant k_v is dimensionless. The second one is a *voltage-controlled current source* (VCCS) wherein a voltage v_x elsewhere in the circuit controls the current delivered by the source; the proportionality

---

[1] It does not sound nice to read this as 'direct-current-voltage' or 'direct-current-current'. However, for a long time, it has become a standard practice to call a constant (or steady) voltage/current by the short form dc, and to call a time-varying voltage/current by the short form ac.

constant k_g has the dimensions of ampere/volt. Likewise, we may enumerate the other two possibilities as *current-controlled voltage source* (CCVS) and *current-controlled current source* (CCCS) with appropriate dimensions for the proportionality constants k_r and k_i, respectively. Observe that we have adhered to the sign convention.

Next, we shall examine the passive elements. In fact, there are several elements that serve as passive elements. Of these we are interested in just three—resistance, inductance, and capacitance.

2.2.3 Resistors

Earlier we mentioned about leads as those which do not oppose the flow of current. As such, they may be called *conductors* of electricity. However, there are certain elements called *resistors* specifically designed to oppose the current flow. These elements are made up of conductive material such as carbon composition. The circuit symbol of a resistor is shown in Fig. 2.4. The parameter R denotes the ability to oppose current flow, and hence it is called *resistance* and is measured in *ohms*, denoted by Ω, named after the German physicist Georg S. Ohm (1789–1854).

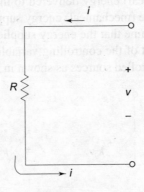

Fig. 2.4

Observe that we have depicted the voltage *across* the resistor and the current *through* it according to our sign convention—current flows from a point of higher potential to that of a lower potential. We distinguish this element from others with the help of the voltage–current relationship it possesses. When subjected to a voltage across its two terminals, the resistor allows a current i that is directly proportional to the applied voltage of v volts; the proportionality constant is the inverse of the resistance R. Put as an equation, we have

$$i = \frac{1}{R} v \tag{2.1}$$

Strictly speaking, any resistor should be manufactured to satisfy this relationship, referred to as *Ohm's law*. Notice that we call this as a *law* unlike Kirchhoff's rules in the preceding chapter. Since this v–i relationship is meant for the resistor alone, Ohm's law is an *elemental law*. Equation (2.1) may be pictorially represented as a straight line passing through the origin of the v–i plane as shown in Fig. 2.5.

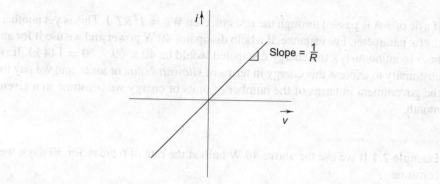

Fig. 2.5

The slope of the line indicates the reciprocal of resistance, called *conductance*, measured in siemens, denoted by Ω^{-1}, after German-born English inventor Karl W. Siemens (1823–1883). If the resistance offered by the element is zero ($R = 0$) the slope is infinite and the voltage $v = 0$ irrespective of the current i. This is the characteristic of a *short circuit*. On the other hand, if the element offers an infinite resistance ($R = \infty$) the slope is zero and the current $i = 0$ irrespective of the voltage v. This is the characteristic of an *open circuit*. It is easy to see that the switch in our circuit of Fig. 2.1 behaves like a short circuit when closed and like an open circuit when opened. A fuse in the household wiring behaves like a short circuit under normal conditions, i.e., when rated current flows through it. However, if the current exceeds the rating, the excess current results in an increase in the heat dissipated and the fuse *blows*, i.e., it becomes an open circuit.

Another property of resistors is their ability to dissipate power, i.e., absorb electrical energy and convert it into heat. At any instant of time t, the power dissipated by a resistor is given by

$$p(t) = v(t)i(t) \text{ W} \tag{2.2}$$

where $v(t)$ is the voltage applied across the resistor at a given instant of time t and $i(t)$ is the current passing through it at the same instant. Using Ohm's law, Eqn (2.2) may be rewritten as

$$p(t) = \frac{v^2(t)}{R} = i^2(t)R \text{ W} \tag{2.3}$$

Commercially available resistors have a *wattage rating* representing the maximum power they can safely dissipate. This means that, according to Eqn (2.3), given a resistance R we should not connect it across a voltage source that exceeds $\sqrt{p(t)R}$ at any given instant of time t; equivalently, the current passing through the resistance should not exceed $\sqrt{p(t)/R}$ at any given instant of time t.

Owing to these properties, we may *model* several devices, for instance the bulb in Fig. 2.1, as resistors. The total amount of power dissipated by a resistor in an interval $[0, T]$ of time gives the energy dissipated,

$$W_R = \int_0^T p(t)\, dt = R \int_0^T i^2(t)\, dt \text{ J} \tag{2.4}$$

If a dc of I A is passed through the resistor, then $W_R = I^2RT$ J. This is yet another useful parameter. For instance, if a bulb dissipates 40 W power and we use it for an hour (continuously), the energy dissipated would be $40 \times 60 \times 60 = 144$ kJ. It is customary to express this energy in terms of *kilowatt-hours* or *units*, and we pay to the government in terms of the number of units of energy we consume in a given month.

Example 2.1 If we use the above 40 W bulb at the rate of 6 h/day for 30 days, we consume

$$40 \times 6 \times 60 \times 60 \times 30 = 25.92 \times 10^6 \text{ J}$$

of energy. We may convert this into kilowatt-hours by using the relationship

$$1 \text{ kW h} = 1000 \text{ J/s} \times 60 \times 60 \text{ s} = 3.6 \times 10^6 \text{ J}$$

That is the energy consumed is

$$25.92 \times 10^6 \text{ J power} = 7.2 \text{ kW h energy}$$

We may calculate the bill accordingly.

Exercise 2.1 Obtain the above result directly by multiplying the power in kilowatts with time in hours.

Many typical appliances such as toasters, electric heaters, and irons may be modelled as resistors with appropriate wattage ratings. We shall discuss power and energy in detail in a later chapter.

In passing, we note that the resistor is a *linear* element owing to the straight-line-like relationship between the voltage across and the current through it. We might say that the resistor we have been using so far is also an ideal one. This is due to the fact that we ignore the dependence of the resistance parameter on temperature or time. We simply assume that an element of a given resistance R *never* changes its value.

Example 2.2 Suppose that we have a 10 kΩ resistor with a 0.25 wattage rating. Then, the maximum possible voltage that can be applied across the element is

$$v_{max} = \sqrt{pR} = 50 \text{ V}$$

and the maximum current that can be allowed to pass through the element is

$$i_{max} = \sqrt{p/R} = 0.05 \text{ A} = 50 \text{ mA}$$

Any excess voltage across or current through it would simply burn the element.

Exercise 2.2 If a resistor dissipates a power of 20 mW while carrying a current of 5 mA through it, what is the resistance? What is the voltage applied?

(*Ans.* 800 Ω, 4 V)

At this point, we add the following to our sign convention. The product of voltage v and current i gives *power*. And, we understand that only the sources *release* power while resistors (and other passive elements we are going to study next) *absorb* power. Somehow we need to distinguish between these two. We do it simply by assigning a positive sign to the product $v \times i$ for the sources and by assigning a negative sign to the passive elements.

In other words, if the current leaves an element via its positive terminal, then

$$p = +v \times i$$

and the element is said to release energy to the surrounding circuit. Otherwise, the element absorbs energy from the surrounding circuit.

2.2.4 Inductor

The ability of a circuit element to produce magnetic-flux linkage in response to a current is called *inductance*; such an element is called an *inductor*. An inductor consists of a coil of insulated wire around a core. To some extent, all conductors exhibit induction effects. Ferromagnetic material, such as iron, cobalt, or nickel, is employed for the core. The circuit symbol for an inductor is shown in Fig. 2.6 with the appropriate sign convention. We use the letter L to denote the inductance of an element. This is measured in *henrys* after American physicist Joseph Henry (1797–1878).

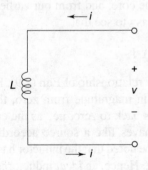

Fig. 2.6

The voltage–current relationship may be derived, unlike Ohm's law, as follows. First, in tune with Oersted's experiment, a current i would create a magnetic field in the core, and hence a magnetic flux ϕ. If the coil has N turns, then

$$\lambda = N\phi \quad \text{Wb turns} \tag{2.5}$$

is the flux linkage. Inductance L is defined as the rate at which λ varies with the applied current, i.e.,

$$Nd\phi = d\lambda = Ldi \tag{2.6}$$

Given a core with N number of turns, the inductance parameter L is related to the permeability μ of the core material, the cross-sectional area S of the core, and the mean length l of the flux path through the core as

$$L = \mu \frac{N^2 S}{l}$$

Since the differential flux and the differential current are *linearly* related, an inductor is also a linear element. According to Faraday's law, the time rate of change in flux linkage induces a voltage given by

$$v = N \frac{d\phi}{dt} \tag{2.7}$$

Putting Eqns (2.6) and (2.7) together, we get

$$v = L \frac{di}{dt} \tag{2.8}$$

Equation (2.8) tells us that an inductor develops a non-zero voltage across its terminals if and only if the current passing through it changes with time. This is strictly in accordance with the sign convention described earlier, i.e., if the current flows from top to bottom, as shown in Fig. 2.6, then the top terminal has a higher potential than the lower one. It is easy to see that a direct current, which does not vary with time, produces a zero voltage across the inductor, and the element behaves like a perfect conductor, i.e., a short circuit. On the other hand, if we pass a sinusoidal current through the element, a sinusoidal voltage develops across its terminals. Thus, an inductor stores energy in the form of potential energy in the magnetic field inside the core, and from our earlier derivation [Eqn (1.23)] in magnetoquasistatics, it is easy to see that

$$w_{\text{in}}(t) = \frac{1}{2} Li^2(t) \text{ J} \tag{2.9}$$

In addition to the energy relationship of Eqn (2.9), Eqn (2.8) tells us something else. If the current grows in magnitude from zero, the inductor absorbs energy and if the current decreases back to zero, i.e., as the current flows in the opposite direction, the inductor behaves like a source according to our sign convention earlier. However, it should be noted that the inductor has to first absorb some energy and store it before releasing. Hence, an *ideal* inductor is still a passive element, but we say that it is non-dissipative. Notice that, for the energy dissipation of a resistor, in Eqn (2.4) we have used W_R, with a capital letter W and without the argument t, as the symbol.

We shall now turn around Eqn (2.8) to derive a more useful expression. Given a voltage $v(t)$, we may compute the current carried by the inductance as

$$i(t) = \frac{1}{L} \int_0^t v(x)\,dx + i(0) \tag{2.10}$$

Equation (2.10) tells us the following. Suppose we take an inductance L henrys and apply a (time-varying) voltage $v(t)$ volts from a reference time $t = 0$ onwards. Then, the current carried by the inductor is the definite integral *plus* a residual current $i(0)$. We need to explain the term $i(0)$. It is possible that the inductor might have been used in some other circuit, at an earlier time, wherein it absorbed some energy and then it was disconnected. Owing to its non-dissipative nature, the absorbed energy gets *trapped* in the core. If, now, the inductor is brought again and connected across a voltage source at the current reference time $t = 0$, then the inductor finds a *way out* to release its earlier trapped energy. It is this trapped, or *stored*, energy that is represented by the term $i(0)$. It is customary to specify an inductor with inductance L as well as *initial condition* $i(0)$. Thus, an inductor is an *energy-storage element*.

An important application of inductors is in voltage converters and switching power supplies where the voltages and currents are of significant magnitude, for instance the domestic power supplies with 220 V and 5 A. Other applications include automobile ignition and radio receivers. However, when we talk about electronic circuits, inductors are not preferred owing to their bulky size. There are some clever circuits which *simulate* inductors and we prefer these in electronic circuitry; we shall look at such circuits in a chapter on operational amplifiers towards the end of this book.

2.2.5 Capacitors

There is yet another energy-storage element, the capacitor. This is an element where electroquasistatics prevail, in contrast to an inductor where magnetoquasistatics dominate.

The ability of an element to store charge in response to an applied voltage is called *capacitance*; such an element is called a capacitor. Figure 2.7 shows the circuit diagram of a capacitor with the appropriate sign convention. We use the letter C to denote the capacitance of the element. This is measured in *farads* after Michael Faraday.

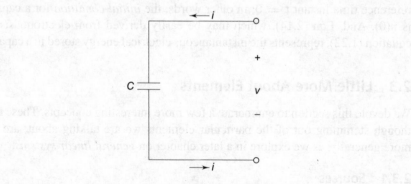

Fig. 2.7

By construction, a capacitor (also called 'condenser') consists of two conductive plates separated by a thin insulator. Applying a voltage across the plates causes positive charge q to accumulate on the plate at a higher potential and an equal amount of negative charge on the plate at a lower potential. Capacitance is the *rate* at which the charge accumulated varies with the applied voltage, i.e.,

$$C = \frac{dq}{dv} \tag{2.11}$$

The capacitance of an element is also given by

$$C = \epsilon \frac{S}{d}$$

where ϵ is the permittivity, S is the cross-sectional area of the plates, and d is the distance between the plates. Thus, plates of a large cross-sectional area and a small separation have a large capacity to store charge. In practice, wherever a pair of conductors is separated in free space, such as the electrical/telephone transmission lines, we see capacitance manifested.

Equation (2.11) tells us that the capacitor is another linear element. From Eqn (2.11) we may derive the following useful expressions:

$$i(t) = C \frac{dv}{dt} \tag{2.12}$$

$$v(t) = \frac{1}{C} \int_0^t i(x)\,dx \; + \; v(0) \tag{2.13}$$

$$w_C(t) = \frac{1}{2} C v^2(t) \tag{2.14}$$

Equation (2.12) shows the differential relationship between the applied voltage $v(t)$ and the current carried by the element. Here, to sustain a non-zero current through the capacitor, the applied voltage should be time-varying. In other words, a constant (dc) voltage causes no current to pass through the element, i.e., the capacitor behaves like an *open circuit*. Equation (2.13) suggests that it is customary to represent the initially stored charge, if there is any, in terms of a voltage $v(0)$ at the reference time instant $t = 0$; in other words, the *initial condition* for a capacitor is $v(0)$. And, Eqn (2.14), which may be easily derived from electroquasistatics equation (1.22), represents the instantaneous electrical energy stored in a capacitor.

2.3   Little More About Elements

We devote this section to enumerate a few more interesting concepts. These ideas, though stemming out of the particular elements we are talking about, are valid more generally, as we explore in a later chapter on *general linear systems*.

2.3.1   Sources

A voltage source is said to be independent if it maintains a prescribed voltage across its terminals, no matter how much current flows out of it. In other words, a source of 5 V (dc), for example, connected across a resistor of 1 Ω would deliver a current of 5 A (dc); when it is connected across a resistor of 5 Ω, it would deliver a current of 1 A, and so on. Accordingly, the v–i characteristic of an independent voltage source is shown in Fig. 2.8.

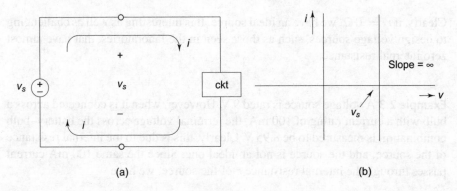

(a)   (b)

Fig. 2.8

We notice that the slope of the straight line is infinite and hence the *internal resistance* of an independent voltage source is zero. In this sense, we say the voltage source is *ideal*. In practice, however, the terminal voltage changes with the current drawn by the surrounding circuit. Let us recollect the observation we all would have made while travelling in a vehicle. We notice that when we start the engine with the lights on, the lights flicker. A battery is the source of energy in a vehicle for ignition as well as headlights and, ideally, it should supply the required energy; the flicker is due to a small voltage drop at the terminals of the battery when, in addition to the starter motor, the lights are also connected across it. In other words, the internal resistance of the voltage source is not exactly zero. A practical voltage source is shown in Fig. 2.9 with the internal resistance r *in series* with the source. We use a lowercase letter r since this resistance is not a fixed quantity, but varies from circuit to circuit.

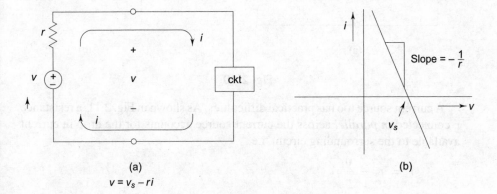

(a)   (b)

$$v = v_s - ri$$

Fig. 2.9

If a current i passes through the source, and hence through r, the net voltage available across the terminals is

$$v = v_s - ri$$

Clearly, if $r = 0\,\Omega$, we have an ideal source. It is interesting as well as challenging to design voltage sources, such as those seen in the laboratories, that have almost zero internal resistance.

Example 2.3 A voltage source is rated 9 V. However, when it is connected across a bulb with a current rating of 100 mA, the terminal voltage across the battery–bulb combination is measured to be 8.95 V. Clearly, this is due to the internal resistance of the source, and the source is not an ideal one. Since the same 100 mA current passes through the internal resistance r of the source, we have

$$9.0\,\text{V} - r\,\Omega \times 100\,\text{mA} = 8.95\,\text{V}$$

or

$$r = 0.5\,\Omega$$

In an analogous way, we may also look at the v–i characteristic of an independent current source, as shown in Fig. 2.10, as a straight line with zero slope, i.e., an independent current source would deliver the same current irrespective of the voltage across its terminals. Hence, the *internal resistance* of an *ideal* current source is infinite.

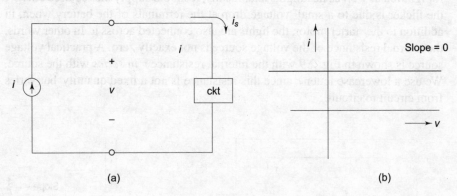

Fig. 2.10

A current source too has practical difficulties. As shown in Fig. 2.11, a resistance r connected *in parallel* across the current source accounts for the drop in current available to the surrounding circuit, i.e.,

$$i = i_s - \frac{v}{r}$$

We may quickly observe that, unlike a voltage source, a practical current source becomes an ideal one if $r = \infty$.

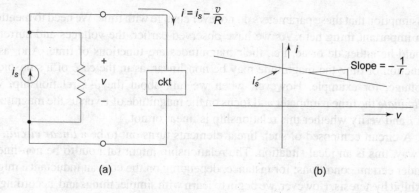

Fig. 2.11

Example 2.4 Suppose we have a current source of $i_s = 1$ mA. Let us see what happens if the surrounding circuit forces a voltage of -1.5 V across the terminals. Referring to Fig. 2.11, owing to the polarity and sign convention, the current source actually *absorbs* energy from the surrounding circuit. The surrounding circuit actually draws a current

$$i = 1\,\text{mA} - \frac{-1.5}{r}\,\text{A}$$

That is, the surrounding circuit draws more current than what is available from the source. This result is not surprising since, owing to the polarity of the voltage forced by the surrounding circuit, the current source absorbs energy rather than releasing it.

While voltage sources are readily available as batteries and regulated power supplies, current sources are not as readily available. In the first place, they are a mathematical abstraction. Further, we may design electronic circuits, with metal-oxide-semiconductor field-effect transistors (MOSFETs) or operational amplifiers, that emulate the behaviour of a current source.

Exercise 2.3 A practical current source is connected across two 1 kΩ resistances in parallel and the terminal voltage is measured to be 2.475 V. If one of the two 1 kΩ resistances is removed from the circuit, the terminal voltage goes up to 4.902 V. Sketch the circuits corresponding to the two situations and estimate the source current i_s and its internal resistance r.

(*Ans.* 5 mA, 50 kΩ)

2.3.2   Linear, Lumped, and Bilateral Nature

As we have already seen, there exists a *linear* relationship between voltage and current for a resistance, voltage and charge for a capacitance, and so on. In other words, the parameters R, L, and C are constant with respect to the variation of the magnitude of voltage, current, or charge. In addition, we also have an implicit assumption that these parameters do not ever change with time. We need to mention an important thing here. As we have observed earlier, the voltages and currents could be either dc or ac, i.e., their magnitudes are functions of time. And, as a function of time the magnitude may be non-linear, as in the case of a sinusoidal voltage, for example. However, when we talk about the $v-i$ relationship, we *eliminate* the time parameter and focus on the magnitude of v versus the magnitude of i and verify whether this relationship is linear or not.

A circuit composed of such linear elements turns out to be a *linear circuit*. In a way, this is an ideal situation. The relationship might turn out to be non-linear under certain conditions; for instance, depending on the core an inductance might exhibit hysteresis. However, we begin to learn with simpler things and, accordingly, we consider only the linear behaviour. Further, most of the time we might be interested in only a very restricted range of operation, such as few millivolts or a few microamperes, in which case we might make a *linear approximation*. We shall elaborate on this later.

Next, we prefer to work with passive elements that are relatively smaller in size. For example, an electrical/telephone transmission line may extend to several hundred kilometers. As it is, it may be modelled as a conductor with small resistance. When connected to a source, the energy is transported via the transmission line with a finite velocity (at most, with the velocity of light) and there could be a delay. Accordingly, we say that the resistance is *distributed* in space, i.e., the resistance is to be accounted as a function of distance as $R(l)$. On the other hand, if the elements are not so physically distributed in space that the circuit parameter is *concentrated* at just one point in space, the element is said to be *lumped*. This is exactly what we have assumed, in terms of the operating frequencies/wavelengths and the length of the electromagnetic phenomena, to derive quasistatics in Chapter 1. A circuit composed of such lumped elements may be called a lumped circuit. In this book we consider only lumped circuits.

We shall consider another aspect of the passive elements. For instance, let us take a (two-terminal) resistor and send current through it in either direction. There is no reason to say that the resistor would oppose the flow of current more in one direction than in the other. Same is true for an inductor and a capacitor. In other words, the resistance, inductance, or capacitance of an element is independent of the polarity of the voltage across it and the direction of current through it. Accordingly, we call these elements as *bilateral*. In contrast, we also have *unilateral* elements such as diodes.

Put it briefly, we confine ourselves to *linear*, *lumped*, and *bilateral* passive elements in this textbook.

2.3.3 Duality

While doing the quasistatics in Chapter 1, and also while learning about inductance and capacitance, we have come across several analogous situations. The statements of Kirchhoff's rules, for example, were almost word for word with voltage and current interchanged. An inductance differentiates current and a capacitance differentiates voltage; once again interchanging L and C along with current and voltage illustrates the similarity. This repeated similarity is only a part of a larger similar behaviour pattern in circuit theory. We call this similarity, with all its implications, the *principle of duality*. A careful look at what we have done so far would reveal the following pairs of dual quantities:

1. voltage and current
2. across and through
3. *R* and *G*
4. *L* and *C*
5. open circuit and short circuit

We would be looking at few more pairs as we go ahead. We shall take advantage of this principle of duality in several situations throughout this text.

2.3.4 Commercial Availability

Practically, the resistors obey Ohm's law (the linear v–i characteristic) only as long as i and v are confined within certain limits. Usually, these limits are available to us as a power rating. With an appropriate choice of the resistivity ρ, surface area S, and length l, resistors can be manufactured in a wide variety of values and power ratings. Generally, resistors are available as discrete components for laboratory purposes. They have a cylindrical shape and are available in a wide range, from fractions of ohms to several megaohms. Resistors are also available as *integrated circuits* (ICs) and are fabricated by depositing a thin layer of conducting material on a common substrate. We have another variety of resistors called *potentiometers*, or *pots* for short, wherein we have a provision for varying the resistance by varying the length parameter l. This device is a bar of resistive material with a sliding contact, called a wiper. Varying the wiper varies the distance from one extreme where the resistance is, say, high to another extreme where the resistance is low. The circuit symbol of a potentiometer is shown in Fig. 2.12.

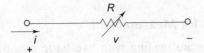

Fig. 2.12

Inductors are available ranging from microhenrys to henrys. While an inductor of a few henrys could be extremely big, commonly used inductors are no more than few millihenrys. Variable inductances are also available. These practical inductors satisfy the linear relationship only approximately. The windings around the core offer some resistance and hence a practical inductor may be thought of as an ideal inductance in series with an internal resistance, as shown in Fig. 2.13(a).

On one hand, all circuit elements, including a plain wire assumed to be a perfect conductor, exhibit a small amount of inductance. This is called *stray inductance*. On the other hand, it is the least popular element owing to its bulkiness. Of course, we may not find power supplies and high-frequency signal-processing circuits without inductors. In modern electronics, we design clever circuits that simulate the inductance behaviour.

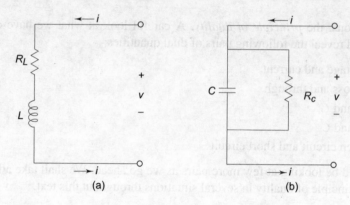

Fig. 2.13

Capacitances are commercially available in a wide range from picofarads to millifarads. Once again, larger the capacitance, larger is its size. It is impossible to imagine a capacitor of a few farads as a circuit element. Capacitance arises naturally whenever two conducting surfaces, such as wires, come in close proximity. This is called *stray or parasitic capacitance*. If a practical capacitor is charged via a voltage source and is then disconnected, we cannot expect it to retain its charge indefinitely. Instead, it gets discharged gradually, though practically very slowly. Accordingly, it is customary to think of it as having a large resistance across its terminals as shown in Fig. 2.13(b).

While we are happy doing lots of things with ideal elements, the reader has to keep in mind that in the laboratory these simple elements have a tendency to behave erratically. We need to make sure that the power rating, ambient temperature, humidity, etc. are within desirable limits. It is interesting to note that these little elements too have a *life* in the sense that repeated usage of the element wears it out.

2.3.5 Instruments for Measurement

It is apparent that we are dealing with five basic quantities—voltage, current, resistance, inductance, and capacitance. Of these, we measure voltage, current, and resistance using *voltmeter*, *ammeter*, and *ohmmeter*, respectively. For our convenience, all three meters are available as a single unit, called a *multimeter*. Throughout our discussion, the reader would have noticed that we have emphasized upon *voltage across* and *current through*. The other choice, such as current across, is grossly incorrect owing to the fundamental definitions of potential difference and transportation of charged particles. Accordingly, we connect a voltmeter across the element under question to measure the voltage across it, and we connect an ammeter in series with the element, i.e., next to it, to measure the current through it.

An important issue in making connections is the choice of correct polarity. Since an ammeter is connected next to the element under observation, it is the same current being measured that passes through the meter. The meter, which may be considered another passive element, is usually provided with a colour code such as red for the positive terminal and black for the negative terminal. A *positive current reading* means that the current enters the instrument via its red terminal and leaves it via the black terminal. In a similar fashion, a voltmeter too is provided with red and black terminals so that a *positive voltage reading* indicates that the potential of the red terminal is higher than that of the black one. From the circuit diagram, in general, it is possible to identify the direction of current and hence the voltage polarities, and we may connect the meters accordingly.

Another important issue is the reliability of the meters. For instance, if an ammeter is connected in series with an element, ideally there should not be any voltage developed across it; otherwise the circuit is altered. Hence we say that an ammeter should have an internal resistance that is small enough (as close to zero as possible). Likewise, a voltmeter is expected to have a large (ideally infinite) internal resistance so that when it is connected across an element in a circuit, no current flows through it.

A multimeter is an easy-to-carry all-in-one instrument, and is a companion of every electronics hobbyist. We have older analog multimeters wherein a coil is deflected in proportion to the quantity being measured. We also have modern digital multimeters that use electronic circuits and liquid-crystal displays. Since voltages and currents could be either dc or ac, a suitable meter is to be employed. Apparently, it is straightforward to measure dc quantities as they do not vary with time. However, it is little trickier to measure a sinusoidal voltage or current since the voltage would be zero at one instance, maximum at another instance, and so on. For such time-varying quantities we define, in a later chapter, *average value* and *rms value*. A dc meter would give the average value of a time-varying signal and an ac meter would give the rms value.

To measure voltages, ac voltages in particular, we have an important meter called the *oscilloscope*. This instrument allows us to watch how the voltage varies as a function of time, and compute the maximum, average, and rms values, periodicity, and phase angle. The reader would have expected that this instrument should also be connected across the element under observation.

We also have *watt meters* to measure the power absorbed/released by a certain element.

2.4   Models of Physical Systems

As we have claimed at the beginning of this text, our focus is not just learning some principles of electrical engineering. Our scope is rather broad and encompasses some good practice of model building and problem solving. While we have seen some examples as abstract graphs at the end of Chapter 1, we shall see some more examples in this section with appropriate circuit elements.

To begin with, we emphasize that, owing to the lumped parameter models of R, L, and C being studied throughout this subject, it is customary to interchangeably use the words resistance and resistor, capacitance and capacitor, and inductance and inductor since circuit theory is more concerned with the *mathematical behaviour* of the interconnection of several circuit elements. For instance, we may use the phrase 'a circuit has an inductance' in lieu of 'a circuit has an inductor' and vice versa. In other words, we need not necessarily restrict ourselves to electrical circuits alone; rather we may think about, i.e., *model*, several physical systems as circuits.

Example 2.5 Let us consider our household wiring. We notice that, in general, all the appliances, such as fans, lights, and audio systems, have a voltage rating of 220 V, 50 Hz (in India). And, it is possible to operate almost all of these devices simultaneously. This may be modelled as a series connection of all these devices along with a source given to us by the electricity board. This is shown in Fig. 2.14(a).

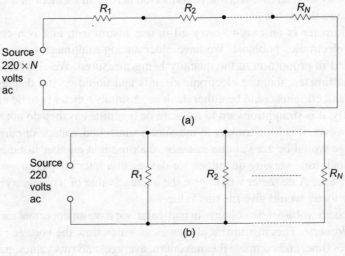

(a)

(b)

Fig. 2.14

Observe that we represent each device with a resistance R_i. However, we may quickly notice the absurdity in this model. First, when we start using a new device at home we do not *break* the circuit and introduce it. And, if there are N such devices the total requirement would be $N \times 220$ V, which works out to a big, rather dangerous, source to be given to us by the electricity board. Adding several houses to be catered by the board, it is just impossible to meet the demand. Then, what is the right model? As one would have guessed, we connect all these devices *across* a fixed supply of 220 V as shown in Fig. 2.14(b) and the government has just to give this to us. Observe that there is no need to break or intrude into any circuit. Of course, as we keep on adding more and more devices, the internal resistance of the source may also increase and hence there could be a physical limit on the number of devices. This problem is called *loading* in popular jargon. Clearly, the second model is more logical.

Example 2.6 Consider the circuit in Fig. 2.15. Historically, Dr. Woodbury of the University of Utah School of Medicine made use of this circuit to study convulsions. In this circuit, the capacitance C_1 represents the volume of drug-containing fluid, for instance the intravenous fluid. The switch, the resistance R_1, and the inductance L_1 represent the passage of the drug from the source to the body via tubes and needles. C_2 represents the volume of the blood stream of the body, and R_2

(conductance, rather than resistance, is a more appropriate term here) represents the body's excretion mechanism. The concentration of the drug dose is represented by the initial voltage V_0 across C_1. The time-varying voltage $v_b(t)$ represents the concentration of the drug in the blood stream. This circuit has the advantage that the elements may be easily changed and the effects may be readily studied using the expression for $v_b(t)$ in terms of C_1, C_2, L_1, R_1, and R_2 and without much physical experimentation. We suggest the reader to appreciate the idea of using the terms 'resistance', 'inductance', and 'capacitance' so intuitively as to mean an opposition to the flow, injection into the body, and the volume of fluid.

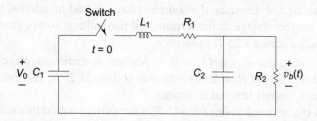

Fig. 2.15

We shall see many more models as we go through the book. Of course, we shall come back to this example again after we learn some problem-solving techniques. Meanwhile, we encourage the readers to look around for similar phenomena and build their own circuits; we suggest traffic systems, corporation water supply for a city, and the like.

Summary

In this chapter we have first defined what an electric circuit is and discussed the constituent elements of a circuit at length. These include sources (independent and dependent), resistors (fixed and variable), inductors, and capacitors. We have emphasized that the elements, and hence the circuits are, at a primary level, *linear*, *lumped*, and *bilateral*. We have thrown some light on the idea of duality. We discussed the commercial availability of the elements and instruments for measuring the fundamental quantities—voltage, current, and resistance. Further, we concluded the chapter with two examples showing how electrical circuits serve as *models* for several physical phenomena. In no way is the discussion of various topics complete in this chapter. We revisit these ideas again and again in later chapters. In other words, this chapter (and, to a large extent, Chapter 3) serves to just provide a snapshot of the entire subject.

Problems

2.1 Find the energy change in an electron as it flows from the positive terminal of a torchlight battery through the battery and out of its negative terminal. Is this energy increasing or decreasing? Comment on whether the battery releases or absorbs energy?

2.2 What are the maximum voltage and current that can be safely applied to a 1 kΩ resistance with a power rating of 1/8 W? What if the resistance is just 10 Ω?

2.3 What is the minimum 1 W power-rated resistance that can be safely connected across the household powerline of 220 V ac?

2.4 If a 2 kΩ resistance can conduct a current of 10 mA safely, what is the maximum power that this element can safely dissipate, and at what voltage?

2.5 With an element absorbing 1 W, the voltage at the terminals of a voltage source is found to be 10.0 V. If the element is removed, the terminal voltage rises to 10.2 V. Estimate the source voltage v_s and its internal resistance r. Estimate the voltage at the terminals if this voltage source (including r) is connected across a 25 Ω resistance.

2.6 A current source supplies $i = 0.45$ A when its terminal voltage is 10 V, and supplies $i = 0.4$ A when its terminal voltage is 25 V. Estimate the source current i_s and its internal resistance.
    (a) If the terminal voltage is +15 V, what current would this source supply?
    (b) If the terminal voltage is −15 V, what current would this source supply?

2.7 A 9 V battery is being charged by forcing a current of 0.4 A into its positive terminal for a duration of 90 min. Find the energy as well as the charge delivered to the battery.

2.8 Assuming the household electrical energy costs Rs. 3.00 per unit, obtain the monthly (assuming 30 days) bill for running a 100 W TV set for 6 h per day, and switching on a 40 W bulb for 8 h per day.

2.9 An element has a current of $i = 5e^{-t}$ A leaving its positive terminal and the terminal voltage is $v = 5 \times i$ V. In a time span of 0–50 μs, determine the energy absorbed (or released) by the element.

2.10 If, in the above problem, the current leaves the negative terminal of the element, what is the energy absorbed (or released) by the element?

2.11 If, as in Problem 2.9 above, the element has a voltage of $v = 5e^{-t}$ V across its terminals and the current leaving its positive terminal is $i = 3 \times v$ A, find the energy absorbed (or released) by the element in a time span of 0.1–0.6 s.

2.12 Consider the circuit shown in Fig. 2.16. If the 2 V source on the extreme right is found to deliver a power of 6 W, determine the current flowing in the series circuit. Does the 6 V source release or absorb energy? How much is the energy absorbed by element X in the black box?

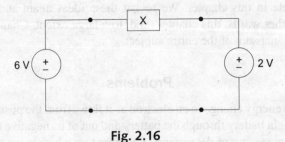

Fig. 2.16

2.13 If, in Problem 2.12, the 2 V source is found to absorb a power of 6 W, what happens?

2.14 Consider the circuit shown in Fig. 2.17. If the 5 A source is found to release a power of 10 W, determine the power absorbed/released by the other elements. What is the algebraic sum (i.e., adding up including the sign) of the three powers in the circuit?

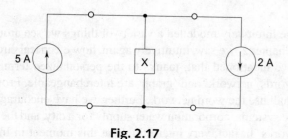

Fig. 2.17

2.15 What happens if, in the above problem, the 2 A source is found to absorb a power of 10 W?

2.16 Consider the circuit shown in Fig. 2.18. Determine which source is releasing power and which source is absorbing?

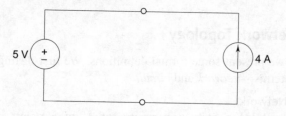

Fig. 2.18

2.17 Repeat the above problem with the polarity of the voltage source reversed.

2.18 The current at the terminals of a current source is measured with the help of an ammeter having an internal resistance of 10 Ω and is found to be 11.965 mA. Adding a 1 kΩ resistance between the source terminals (in parallel) causes the ammeter reading to drop to 11.695 mA. Determine the source current i_s and its internal resistance r.

2.19 The voltage at the terminals of a voltage source is measured with a voltmeter having an internal resistance of 10^7 Ω and is found to be 12.5 mV. Adding a 5×10^6 Ω resistance across the voltmeter causes the reading to drop to 9.5 mV. Find the source voltage v_s and its internal resistance.

2.20 Obtain the model of a household two-way switch. Use two switches, a voltage source, and a resistance (for the bulb). Enumerate the number of ways, the bulb can be turned on and off.

CHAPTER 3

Graphs, Networks, and Circuits

In Chapter 1 we intuitively modelled a variety of things we see around as graphs. At the end of Chapter 2, we saw, intuitively again, how electrical circuits resemble graphs. We have mentioned that, thanks to the personalized terminology of the theorists, the words 'network' and 'graph' are interchangeable. From this chapter onwards, we shall use the word network. Further, we have encouraged the readers to look at traffic systems, corporation water supply for a city, and the like and build their own networks. In fact, very interestingly at this moment in India, there is a rage about a network of rivers to meet the water scarcity. In this chapter we shall put all such intuitive ideas on a firm mathematical and physical foundation. Although we present the ideas in terms of electrical elements, we emphasize that there exist analogies in other fields as well. These analogies are left to the imagination and enthusiasm of the reader; of course, we would be providing appropriate examples as and when found necessary.

3.1 The Network Topology

In this section we present some formal definitions. We first distinguish between two prominent terms—*network* and *circuit*.

Definition 3.1 Network
A network consists of a finite non-empty set $\mathcal{N}$ of n points, called *nodes*, together with a prescribed set $\mathcal{B}$ of q unordered pairs of distinct points of $\mathcal{N}$. A *branch* is a line or an arc whose end points belong to $\mathcal{B}$.

We give some examples of networks in Fig. 3.1.

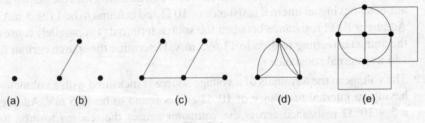

(a)   (b)   (c)   (d)   (e)

Fig. 3.1

Let us make the following observations on networks in some more detail. First, it is clear that any branch has two nodes. Second, it is not necessary that every node of a network be connected to another node via a branch. This is shown in Figs 3.1(b) and 3.1(c). Next, it may be readily observed that none of the networks in Fig. 3.1 has a branch whose end points coincide. In the definition above, we have clearly indicated that set $\mathcal{B}$ consists of pairs of distinct nodes. Since we have also mentioned the unordered pairs, we allow multiple branches between the same pair of nodes as in Figs 3.1(d) and 3.1(e). Next, although we allow two branches to intersect, as in Fig. 3.1(e), the point of intersection is not a node. And, as a mathematical generalization, we do have graphs with a non-empty set of nodes together with an empty set of branches as shown in Fig. 3.1(a).

Definition 3.2 Electrical network

It is now quite trivial to see that each of the electrical elements we have introduced in the preceding chapter, active as well as passive, qualify as networks—each element has a pair of (distinct) nodes with a branch connecting them.[1] We call each of these elements as a *simple network*. We may recursively use these simple networks to build a more complex network. An *interconnection* of two or more simple networks is called an *electrical network*.

Based on the definition above, we may imagine a variety of electrical networks. Accordingly, we shortlist the following networks which we are concerned with in this textbook. To this end, we need to address the *topological properties*[2] of networks. First, we show three electrical networks in Fig. 3.2. These appear to be

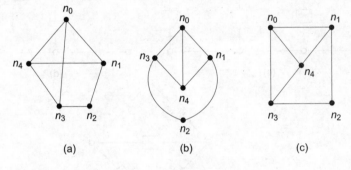

(a)   (b)   (c)

Fig. 3.2

different from each other. But, once we enumerate the nodes as n_0, n_1, n_2, n_3 and n_4, it is easy to see that all the networks are identical. In other words, a network may be stretched, twisted, or distorted in its size and shape. From this point of view the three networks in Fig. 3.2 are said to be *topologically isomorphic*. In this context it is useful to make an important observation: from a mathematical perspective a network contains several dots and lines connecting the dots; however, an electrical network contains several dots (called nodes) and electrical elements, including the connecting wires, between the nodes. We emphasize the following with reference to Figs 3.2(a) and 3.2(c): while drawing a network, sometimes it is convenient to draw two or more branches crossing each other as shown in Fig. 3.2(a); however, unless there is a thick dot shown at the intersection as in Fig. 3.2(c), the crossing branches are independent and have no common node.

---

[1] Recall that in Chapter 1, we have defined a node as a point at which there exists no field.

[2] Topology is a branch of geometry that might be credited to Euler, who tried to solve the seven-bridge problem we have shown in Chapter 1. Kirchhoff used it to study electrical networks.

An ardent reader should not be worried hereafter that the schematic wiring diagrams provided in the manuals for TVs, radios, automobile electrical circuits, etc. are different from the actual wiring in the appliance; an untrained eye tries to fool us many a time. Common sense tells us that unless the two are isomorphic topologically the appliance need not carry a manual along with it.

Secondly, there are certain standard electrical networks we frequently come across. Some of these are shown in Fig. 3.3.

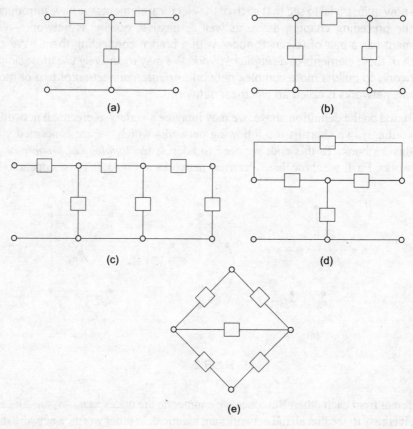

(a)

(b)

(c)

(d)

(e)

Fig. 3.3

The networks in Figs 3.3(a)–3.3(e) are, respectively, called the *T network*, the Π *network*, the *ladder network*, the *bridged T network*, and the *bridge network*.

Thirdly, all the networks we have shown so far, except Fig. 3.1(e), may be drawn in the plane of a paper without crossing the lines. Such networks are called *planar networks*. The one shown in Fig. 3.1(e) is a *non-planar* network. We focus on planar networks in this text.

Next, we will be interested in node pairs. Specifically, a node pair is called a *port* across which an electrical element may be fixed, such as the household electrical socket we are readily familiar with. We make use of the word 'port' quite frequently in this text. Next, suppose we begin a tour of a given network at a node, say n_i, and move through the element to the next node n_{i+1}, and then continue our tour passing through several nodes and elements. If we ensure that no node is touched more than once, then the set of all nodes and elements we have passed through is called a *path*. Further, if the last node of the path happens to be n_i, the node at which we had started, then it is called a *simple closed path* or a *mesh*. A *loop* is one that contains several meshes within it; equivalently, a mesh is a loop that contains no meshes. In other words, a mesh is a loop but a loop is not necessarily a mesh. These concepts are illustrated in Fig. 3.4.

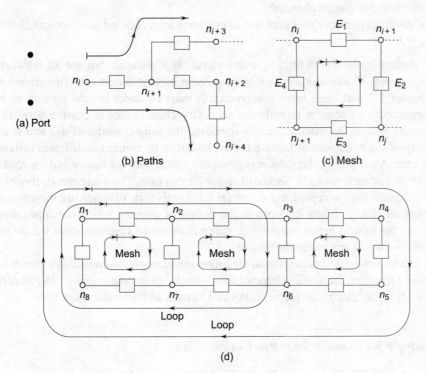

(a) Port

(b) Paths

(c) Mesh

(d)

Fig. 3.4

Figure 3.4(a) shows a port. Two possible paths are shown in Fig. 3.4(b). A simple closed path starting at node n_1 traversing through elements E_1 through E_4 and terminating at node n_1 is illustrated in Fig. 3.4(c). The eight-node network of Fig. 3.4(d) depicts three meshes—$\{n_1\text{-}n_2\text{-}n_7\text{-}n_8\text{-}n_1\}$, $\{n_2\text{-}n_3\text{-}n_6\text{-}n_7\text{-}n_2\}$, and $\{n_3\text{-}n_4\text{-}n_5\text{-}n_6\text{-}n_3\}$, and three more loops—$\{n_1\text{-}n_2\text{-}n_3\text{-}n_6\text{-}n_7\text{-}n_8\text{-}n_1\}$, $\{n_2\text{-}n_3\text{-}n_4\text{-}n_5\text{-}n_6\text{-}n_7\text{-}n_2\}$, and $\{n_1\text{-}n_2\text{-}n_3\text{-}n_4\text{-}n_5\text{-}n_6\text{-}n_7\text{-}n_8\text{-}n_1\}$. To avoid much clutter, we have shown only two loops in the figure. The reader can easily trace out the third loop. Altogether, there are six loops, out of which three are meshes.

In practice, we need not enumerate all the loops in a planar network; it is sufficient to correctly count the number of meshes. Each of the meshes m_j is depicted in the circuit with a small rectangular box enclosing the symbol 'm_j'. What is even more important is that we, rather conventionally, traverse a mesh in the *clockwise direction*. Of course, there is absolutely no harm assuming a counterclockwise traversal. But, in this text, let us speak a uniform language throughout.

Definition 3.3 Active and passive electrical networks

An electrical network that contains at least one active element, such as an independent voltage source, is called an *active network*; otherwise it is called a *passive network*.

We are now ready to define an electrical circuit in a more formal (and hence a more general) way as follows.

Definition 3.4 Electrical circuit

An electrical network that contains at least one loop is called an *electrical circuit*.

According to this definition, every circuit is a network, but not all networks are circuits. Conventionally, we build passive electrical networks (the theme of Chapter 11) and leave open (unconnected) pairs of nodes for the sources to be connected; i.e., sources provide the necessary closed paths in passive networks, thereby making it a circuit. Strictly speaking, the subject matter of this text is the study of the behaviour of passive networks driven by sources at different instants of time. Accordingly, the subject is generally called *network theory* and the reader may find several books in his/her library with this title. The name *circuit theory* is also equally well accepted and we prefer this in this text. We often use these words interchangeably. When it comes to the design of circuits, it is more appropriate to call the subject *network synthesis*, rather than *circuit synthesis* since we are not bothered about any particular source.

Let us now turn our attention to the following examples to consolidate the above ideas. Hereafter, the words 'element' and 'branch' are interchangeable, and we drop the adjective 'electrical' for networks and circuits as it is understood.

Example 3.1 Consider the network in Fig. 3.5(a).

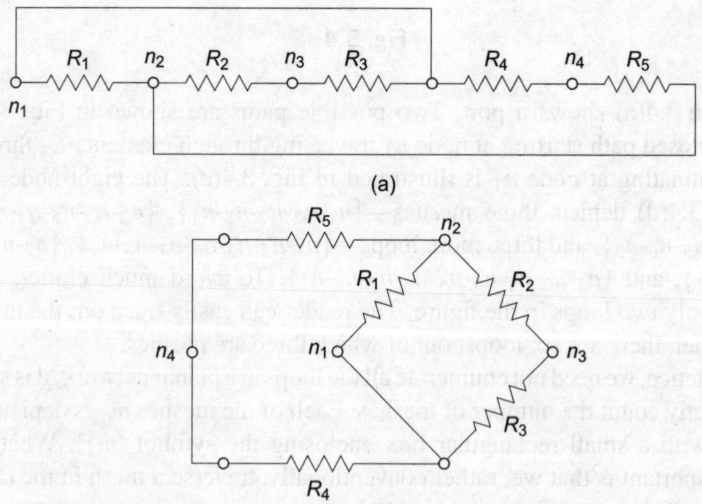

(a)

(b)

Fig. 3.5

Clearly this is a circuit since there are two meshes. As it is, it appears to be quite messy and the reader would have been already made familiar with this kind of circuits in an earlier course. There are five elements connected end to end and hence there appear to be six nodes. However, in addition to the resistors, there are connecting wires (which may also be thought of as $0\,\Omega$ resistors), one between the leftmost end of R_1 and the node joining R_3 and R_4, and another between the rightmost end of R_5 and the node joining R_1 and R_2, thereby forming the two meshes. By definition (in Chapter 2), the connecting wires have the same potential throughout and hence the apparently different nodes coalesce. Thus, effectively there are only four nodes as shown in the isomorphic circuit of Fig. 3.5(b). Counting the number of nodes and meshes is extremely important and we devote some more time on these examples and exercises.

Example 3.2 Consider the network in Fig. 3.6.

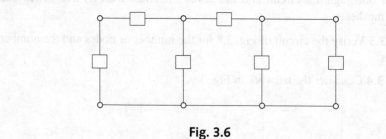

Fig. 3.6

Once again, this is a circuit since there is at least one closed path. This circuit has six branches, four nodes, and three meshes (and six loops including the three meshes).

Exercise 3.1 Verify the number of branches, number of nodes, and the number of loops and meshes in Fig. 3.2.

Exercise 3.2 Count the number of nodes and the number of meshes in the circuit shown in Fig. 3.7.

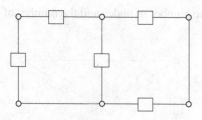

Fig. 3.7

Example 3.3 Consider the network in Fig. 3.8.

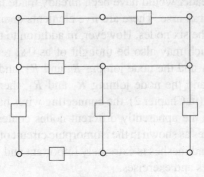

Fig. 3.8

This is once again a circuit and has seven branches with (i) five nodes and (ii) three meshes.

Exercise 3.3 Verify the circuit of Fig. 3.8 for the number of nodes and the number of meshes.

Example 3.4 Consider the network in Fig. 3.9.

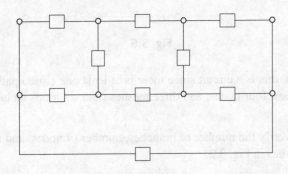

Fig. 3.9

This circuit has nine elements with (i) six nodes and (ii) four meshes.

Exercise 3.4 Obtain the number of nodes and the number of meshes in the network shown in Fig. 3.10.

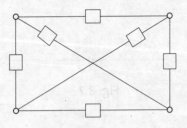

Fig. 3.10

Look at the crossing branches. Unless there appears a dot at the intersection, the two branches are independent.

Now we shall move on to define the voltages and currents in a network. Before that, we emphasize that it is extremely useful and hence important to

- label all nodes in a given circuit, marking all redundant nodes connected by uninterrupted wires;
- focus on neatness and clarity in drawing circuit diagrams with pen/pencil on paper; and
- be aware of topologically isomorphic circuits which need not look the same on paper (as long as the elements and their interconnections are the same, the circuits are identical).

3.1.1 Voltages and Currents in a Network

The definition of a *node* in an electrical network has come quite naturally for us. In the preceding chapter, we have defined it to be a point at which no field exists. From a network point of view, it would be handy to observe that all the branches/elements (including connecting wires) converging to a particular node n_i are at the same potential, called the *node potential*. It is denoted by v_i. The potential difference, or equivalently the voltage, across two distinct nodes n_i and n_j, i.e., across the branch connecting the two nodes, may be defined as

$$v_{ij} \triangleq v_i - v_j$$

or

$$v_{ji} \triangleq v_j - v_i \tag{3.1}$$

Notice the order in which the subscripts appear on the left-hand side of Eqn (3.1). Obviously, v_{ij} is positive if $v_i > v_j$; else v_{ji} is positive. Clearly, the node potential is different from the branch voltage defined above.

Since, it is only the potential *differences* that have a practical meaning, it is convenient to select a *reference node* in the given network with respect to which the potentials of all other nodes may be defined. Conventionally, we call this reference node the *ground node* and set its potential equal to zero volts. We have used this idea in Fig. 2.1 of the previous chapter. The symbol used to denote this node is shown in Fig. 3.11. In general, we choose this ground node at the bottom of a given network and call it node n_0. Of course, there is no such restriction since, isomorphically, we may draw an identical circuit that shows the ground node, say, at the top.

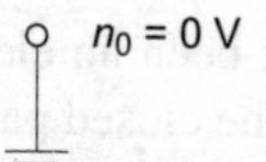

$n_0 = 0 \text{ V}$

Fig. 3.11

To avoid clutter (and hence promote neatness and clarity) in the circuit diagrams, hereafter we avoid designating the ith node as n_i; instead we use the symbol v_i which indicates both the node number as well as its potential with respect to the ground.

To delve a little deeper into the voltages and currents in a network, consider Fig. 3.12.

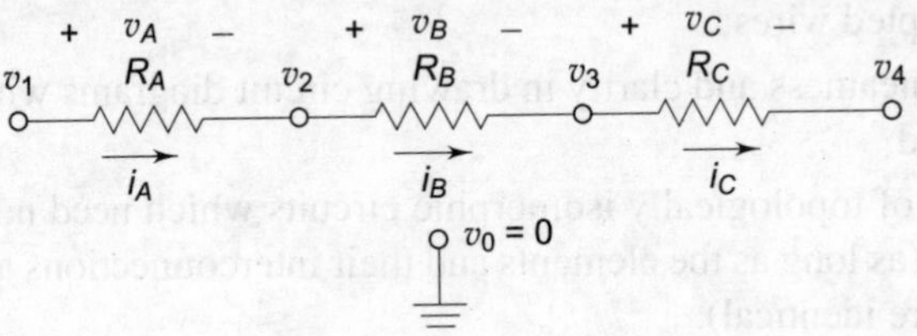

Fig. 3.12

The network (not a circuit!) in Fig. 3.12 has five nodes, out of which the isolated bottom node is designated as the ground v_0. The *node potentials* of the other four nodes are, respectively, v_1, v_2, v_3, and v_4. Without loss of generality, we may assume that an external voltage (or current) source causes the node potentials. In terms of these node potentials, the three elements have the voltages given by

$$v_A = v_1 - v_2 \text{ V}$$

$$v_B = v_2 - v_3 \text{ V}$$

$$v_A = v_3 - v_4 \text{ V}$$

Since there is a path from v_1 to v_4, the net potential difference we encounter traversing along this path may be obtained as

$$v_{14} = v_1 - v_4 \tag{3.2}$$

$$= v_1 - v_2$$

$$+ v_2 - v_3$$

$$+ v_3 - v_4$$

$$= v_A + v_B + v_C \tag{3.3}$$

Observe that we may rewrite Eqn (3.3) as

$$v_A + v_B + v_C + v_{41} = 0 \tag{3.4}$$

which suggests us that, had there been an element across v_1 and v_4 to complete the circuit, the net voltage along the closed path is zero, conforming to our earlier discussion on Faraday's law and the subsequent derivation of Kirchhoff's voltage rule (KVR) in Eqn (1.30).

Further, since there is a voltage across each of the resistances, Ohm's law tells us that there should be some current passing through it. These *branch currents*, designated as i_A, i_B, and i_C, respectively, may be computed as

$$i_A = \frac{v_A}{R_A}$$

$$i_B = \frac{v_B}{R_B}$$

$$i_C = \frac{v_C}{R_C} \tag{3.5}$$

According to the sign convention, i_A leaves node v_1, passes through R_A, and enters node v_2; i_B leaves node v_2, passes through R_B, and enters node v_3; and i_C leaves node v_3, passes through R_C, and enters node v_4. Our discussion on Ampere's law in Chapter 1 and the subsequent derivation of Kirchhoff's current rule (KCR) in Eqn (1.27) tells us that

$$i_A = i_B = i_C \tag{3.6}$$

i.e., it is the same current, say i_R, that flows through all the resistances. It is not difficult to see the rationale for this. For instance, at node v_2 the current that enters the node is i_A and the one that leaves the node is i_B and there is no other current in the picture. According to the sign convention, once again, a current $+ i$ that enters a node is identical to the current $- i$ that leaves the same node. Hence $i_A - i_B = 0$ according to KCR. Similar reasoning holds good at node v_3, which tells us that $i_B = i_C$. By transitivity, we get the result in Eqn (3.6).

Thus, we see that the electromagnetic phenomena, of course under certain frequency restrictions with reference to quasistatics, may be modelled in terms of electrical networks and Maxwell's equations may be directly applied to the networks in terms of Kirchhoff's rules. Based on these observations, we next define two standard forms of networks wherein homogeneous elements may be effectively combined to reduce a complex network into a simpler one.

3.1.2 Series and Parallel Networks

Definition 3.5 Series network
An electrical network with a chain of two or more elements connected end to end such that the same current passes through all the elements is called a *series network*.

An example of a series network is already given in Fig. 3.12. It is clear that a node is shared by *not more than* two elements. Using the principle of duality we may define another standard network wherein the voltage across each of the elements is same.

Definition 3.6 Parallel network
An electrical network with two or more elements connected across the same pair

of nodes v_i and v_j such that the same voltage v_{ij} appears across all the elements is called a *parallel network*.

We illustrate the idea in the following example.

Example 3.5 Consider the circuit in Fig. 3.13.

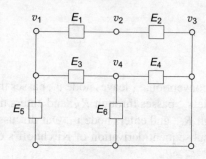

Fig. 3.13

This is a circuit having six elements with four nodes and three meshes.

- We immediately observe that the elements E_1 and E_2 are in series since the current leaving node v_1 passes through both the elements and leaves node v_3.
- We may also observe that the elements E_3 and E_4 are *not* in series. This is because node v_4 is shared by the element E_6 also and hence the current passing through E_4 is not the same as the one passing through E_3.
- Next, the elements E_4 and E_6 are in parallel across the same node pair v_3-v_4.

We may redraw the above circuit as shown in Fig. 3.14.

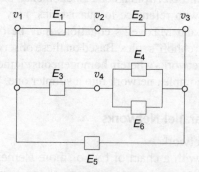

Fig. 3.14

We immediately observe that there are three parallel branches across the same node pair v_1-v_3: the top branch has a *sub-network* of two elements connected in series, the middle branch has a sub-network of two elements E_4 and E_6 connected in parallel, which in turn is connected in series with E_3, and the bottom branch has E_5 connected across v_1-v_3.

3.1.3   Equivalent Passive Elements

It would be helpful to identify if, at least, certain parts (aptly called sub-networks) of a given network have *homogeneous* elements connected either in series or in parallel. This would reduce a complex network into a simpler one with fewer nodes and/or branches. To see how it works, let us consider a series network of resistances. Consider the network in Fig. 3.12 once again. Using Eqn (3.6), we may rewrite the expression for v_{14} in Eqn (3.3) as

$$v_{14} = v_A + v_B + v_C$$
$$= R_A \times i_A + R_B \times i_B + R_C \times i_C$$
$$= (R_A + R_B + R_C) \times i_R \tag{3.7}$$

In other words, the voltage across the extreme nodes v_1 and v_4 is proportional to the current i_R passing through all the resistances and the proportionality constant is the sum of resistances. Accordingly, we may reduce the given network to one that has three nodes—v_1, v_4, and v_0 — and a single element $R = R_A + R_B + R_C$. We may generalize this expression and say that a series (sub-)network of N resistances $R_1, \ldots , R_N$ may be reduced to a single resistance with

$$R = \sum_{j=1}^{N} R_j \tag{3.8}$$

Similarly, a series (sub-)network of inductances, as shown in Fig. 3.15, may be reduced as follows.

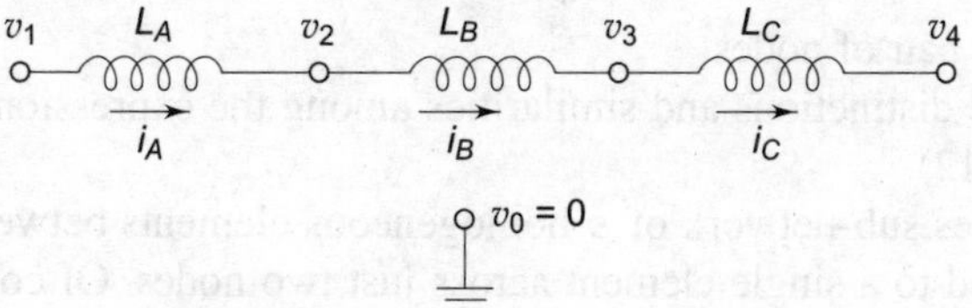

Fig. 3.15

Analogous to the case of the network of resistances, we have, using Eqn (2.8),

$$v_{14} = v_A + v_B + v_C$$
$$= L_A \times \frac{d}{dt} i_A + L_B \times \frac{d}{dt} i_B + L_C \times \frac{d}{dt} i_C$$
$$= (L_A + L_B + L_C) \times \frac{d}{dt} i_L \tag{3.9}$$

since it is the same current, say, i_L that passes through all the inductances. Clearly, extending this argument to N inductances in series, we have a single equivalent inductance

$$L = \sum_{j=1}^{N} L_j \tag{3.10}$$

across a single pair of nodes.

Let us now look at the series network of capacitances in Fig. 3.16.

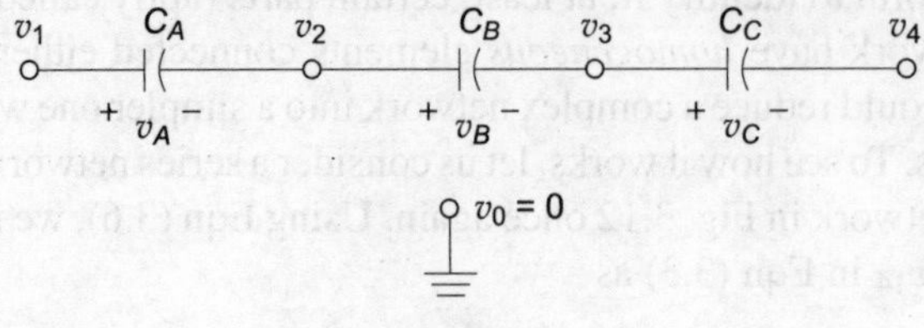

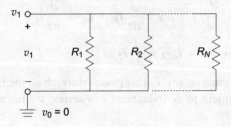

Fig. 3.16

In this case, we have, using Eqn (2.12),

$$v_{14} = v_A + v_B + v_C$$

$$= \frac{1}{C_A} \times \int i_A \, dt + \frac{1}{C_B} \times \int i_B \, dt + \frac{1}{C_C} \times \int i_C \, dt$$

$$= \left(\frac{1}{C_A} + \frac{1}{C_B} + \frac{1}{C_C} \right) \times \int i_C \, dt \qquad (3.11)$$

since it is the same current, say, i_C that passes through all the capacitances. Clearly, extending this argument to N capacitances in series, we have a single equivalent capacitance

$$\frac{1}{C} = \sum_{j=1}^{N} \frac{1}{C_j} \qquad (3.12)$$

across a single pair of nodes.

Observe the distinctions and similarities among the expressions in Eqns (3.8), (3.10), and (3.12).

Thus, a series sub-network of p homogeneous elements between $p + 1$ nodes may be reduced to a single element across just two nodes. Of course, we cannot do anything if the network has a series connection of a resistance, an inductance, and a capacitance. Such heterogeneous combinations are dealt with from Chapter 5 onwards.

Now let us look at the parallel connection of homogeneous elements. First, let us take up the case of resistances as shown in Fig. 3.17.

Fig. 3.17

All the N resistances have the same voltage v_1 across them, by definition. Since the resistances are assumed to be distinct, the currents i_1, i_2, $\ldots$, i_N, through them should be distinct by Ohm's law. The net current leaving node v_1 is then

$$
\begin{aligned}
i &= i_1 + i_2 + \cdots + i_N \\
&= \frac{v_1}{R_1} + \frac{v_1}{R_2} + \cdots + \frac{v_1}{R_N} \\
&= \left(\frac{1}{R_1} + \frac{1}{R_2} + \cdots + \frac{1}{R_N} \right) \times v_1
\end{aligned}
\tag{3.13}
$$

According to Kirchhoff's current rule, this is the current that enters node v_1 and should be related to the voltage v_1 as v_1/R by Ohm's law, where R is the equivalent resistance 'as seen by' the node pair v_1-v_0. Thus, we have

$$
i = \frac{v_1}{R} = \left(\frac{1}{R_1} + \frac{1}{R_2} + \cdots + \frac{1}{R_N} \right) \times v_1
$$

or the equivalent resistance R of a parallel combination of N resistances as shown in Fig. 3.17 is

$$
\begin{aligned}
\frac{1}{R} &= \left(\frac{1}{R_1} + \frac{1}{R_2} + \cdots + \frac{1}{R_N} \right) \\
&= \sum_{j=1}^{N} \frac{1}{R_j}
\end{aligned}
\tag{3.14}
$$

The parallel combination of N resistances is conventionally denoted as

$$
R_1 \parallel R_2 \parallel \cdots \parallel R_N
$$

where the symbol $\parallel$ stands for 'parallel to'. Compare this equation with Eqn (3.12) for the series combination of N capacitances.

For the case of just two resistances R_1 and R_2 in parallel, the following formula may be memorized easily.

$$
\begin{aligned}
R_1 \parallel R_2 &= \frac{1}{\dfrac{1}{R_1} + \dfrac{1}{R_2}} \\
&= \frac{R_1 R_2}{R_1 + R_2}
\end{aligned}
$$

In terms of the conductance $G = 1/R \ \Omega^{-1}$, we may rewrite Eqn (3.14) as

$$
G = \sum_{j=1}^{N} G_j
\tag{3.15}
$$

Next, let us look at the parallel combination of N inductances as shown in Fig. 3.18.

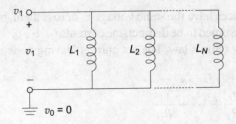

Fig. 3.18

Analogous to the case of resistances, we may derive that

$$\frac{1}{L} = \left(\frac{1}{L_1} + \frac{1}{L_2} + \cdots + \frac{1}{L_N} \right)$$

i.e., $L = L_1 \parallel L_2 \parallel \cdots \parallel L_N = \dfrac{1}{\sum_{j=1}^{N} \frac{1}{L_j}}$ (3.16)

Exercise 3.5 Derive Eqn (3.16).

Lastly, let us look at the parallel combination of N capacitances as shown in Fig. 3.19.

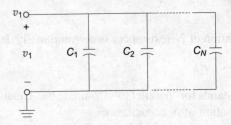

Fig. 3.19

Adding all the branch currents that leave node v_1, we get

$$i = i_1 + i_2 + \cdots + i_N$$

$$= C_1 \frac{d}{dt} v_1 + C_2 \frac{d}{dt} v_1 + \cdots + C_N \frac{d}{dt} v_1$$

$$= (C_1 + C_2 + \cdots + C_N) \times \frac{d}{dt} v_1 \qquad (3.17)$$

Since, according to KCR, the current entering node v_1 is also i, we have the following expression for the equivalent capacitance of a parallel combination of N capacitances:

$$C = C_1 \parallel C_2 \parallel \cdots \parallel C_N = \sum_{j=1}^{N} C_j \qquad (3.18)$$

Let us summarize all the above relationships in the following table.

Element	Series	Parallel
Resistance	$R = \sum_{j=1}^{N} R_j$	$\dfrac{1}{R} = \sum_{j=1}^{N} \dfrac{1}{R_j}$
Inductance	$L = \sum_{j=1}^{N} L_j$	$\dfrac{1}{L} = \sum_{j=1}^{N} \dfrac{1}{L_j}$
Capacitance	$\dfrac{1}{C} = \sum_{j=1}^{N} \dfrac{1}{C_j}$	$C = \sum_{j=1}^{N} C_j$

3.2 Element and Circuit Rules

We are now ready to analyse electrical circuits in a very sophisticated manner. We emphasize that there are just three basic tools that are required to accomplish our job successfully. They are listed below.

- **Ohm's law**

$$v = R \times i$$

  where v is the voltage across a branch of resistance R (could be just one element, or a series/parallel combination of several resistances) and i is the current through it. This law will be further generalized in later chapters to take the differential relationship between v and i of inductances ($v = L\, di/dt$) and capacitances ($i = C\, dv/dt$) also into account. It is customary to call this law as an *element rule*.

- **Kirchhoff's rules**

  KCR: *The algebraic sum of all the currents entering any node must be zero.*
  KVR: *The algebraic sum of all the voltages along any closed path must be zero.*

  The terms 'node' and 'closed path' (or 'mesh') are very clearly defined in the beginning of this chapter. Since we come across several nodes/meshes in circuits, it is customary to call these rules as *circuit rules*.

Equipped with these tools we plough the wonder land of circuits right away and reap several hidden treasures. Of course, we shall make a modest beginning in this chapter to warm up our little grey cells.

Example 3.6 Let us consider the circuit of Exercise 3.2 in Fig. 3.7 again. It is redrawn in Fig. 3.20 for convenience.

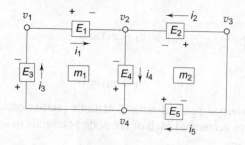

Fig. 3.20

It is easy to see that there are four nodes and two meshes. The voltages across the five elements are arbitrarily shown in the figure, and the directions of the currents are assigned according to the sign convention; for instance, the elements E_1, E_2, and E_3 are 'assumed' to be passive elements so that the currents flow from a higher node potential to a lower node potential, while E_4 and E_5 are 'assumed' to be active elements. This is only an initial assumption; once the actual values are computed, the sign associated with the magnitude determines whether the element satisfies or contradicts our assumption.

We shall now look at the circuit rules applied at each node and in each mesh as follows. We shall designate the currents entering a given node as *positive* and the currents leaving a given node as *negative*. Of course, the reader may choose the opposite designation.

KCR at node v_1:

$$+ i_3 - i_1 = 0$$

In fact, since the elements E_1 and E_3 are in series, $i_1 = i_3$ in accordance with KCR.

KCR at node v_2:

$$i_1 + i_2 - i_4 = 0$$

KCR at node v_3:

$$- i_2 - i_5 = 0$$

Notice that this is also expected since the elements E_2 and E_5 are in series. The negative sign appears owing to the directions defined.

KCR at node v_4:

$$+ i_4 + i_5 - i_3 = 0$$

From the above equations, given a sufficient number of known currents, the remaining unknown ones may be easily computed. For instance, given two out of the three currents—i_1, i_2, and i_4—we can solve for the third.

To account for the potential difference across an element while traversing a path, we take the sign (positive or negative) of the node potential we encounter first. This was already done in Eqns (3.6) and (3.7) with reference to the circuit in Fig. (3.12). And, we traverse a mesh in the clockwise direction, by default. The reader is advised to go through the following equations with rapt attention.

KVR for the mesh m_1:

$$+ v_1 - v_2 - v_2 + v_4 + v_4 - v_1 = 0$$

rewritten as

$$(+ v_1 - v_2) - (v_2 - v_4) + (v_4 - v_1) = 0$$

or, equivalently,

$$+ v_{12} - v_{24} + v_{41} = 0$$

where we have grouped pairs of node potentials (across each of the elements). Notice that we have accounted each of the node potentials twice (why?).

KVR for the mesh m_2:

$$- v_{23} - v_{34} + v_{42} = 0$$

Example 3.7 Consider the circuit in Fig. 3.8, redrawn here as Fig. 3.21 for convenience.

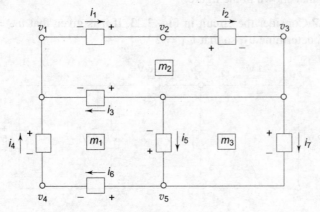

Fig. 3.21

KCR: At each of the nodes v_1, v_2, v_3, v_4, and v_5 it may be easily verified that the current rule gives the following equations, respectively.

$$-i_1 + i_3 + i_4 = 0$$

$$+i_1 - i_2 = 0$$

$$+i_2 - i_3 - i_5 - i_7 = 0$$

$$-i_4 + i_6 = 0$$

$$+i_5 - i_6 + i_7 = 0$$

In this circuit, according to the sign convention, the elements E_1, E_4, and E_5 are 'assumed' to be sources and the remaining elements are 'assumed' to be passive elements. KVR in each of the meshes gives us the following equations, respectively:

$$-v_{13} - v_{35} + v_{54} - v_{41} = 0$$

$$-v_{12} + v_{23} + v_{31} = 0$$

$$+v_{35} + v_{53} = 0$$

Exercise 3.6 Consider the circuit shown in Fig. 3.22. Suppose that $i_5 = 3$ mA and $i_6 = 5$ mA. Determine the current i_3 so that $i_1 = 1$ mA. Is the direction of i_3 shown in the figure correct? Comment.

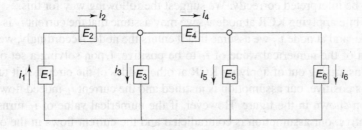

Fig. 3.22

(**Ans.** $i_3 = -7$mA, meaning that the actual current in the circuit is in the direction opposite to that shown in the figure.)

Exercise 3.7 Consider the circuit in Fig. 3.23. If it is given that $v_{12} = -1$V and $v_{34} = 5$ V, determine v_{23} so that $v_{41} = 2$ V.

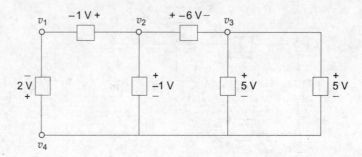

Fig. 3.23

(**Ans.** $v_{23} = -6$ V. While we apply the KVR for mesh m_1, we get that the node potential at v_2 is lower than that at v_4, unlike what is shown in the figure. Accordingly, to satisfy the KVR for mesh m_2, the node potential at v_2 must be lower than that at v_3. Hence the polarity of the voltage across the element E_2 is opposite to that shown in the figure.)

The two examples, Eqns (3.6) and (3.7), and the two exercises, 3.6 and 3.7, above suggest us the following important issue in terms of the sign convention. To illustrate this, we shall take up a typical element in Fig. 3.24.

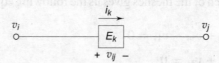

Fig. 3.24

We have fixed the polarities for the node potentials and the current direction entirely arbitrarily, *assuming* that it is a passive element. While such an element sits inside a circuit, the results of solving the equations obtained from KCR or KVR have to be interpreted correctly. We suggest the following way for this.

1. While applying KCR at node v_i, we may assume that the current i_k is leaving the node, and at node v_j we assume that it enters the node. Accordingly, we expect the sign of the numerical value of i_k to be positive. Upon solving a set of linear equations arising out of applying KCR at other nodes of the circuit, if it turns out that i_k is positive, our assumption is justified and the current i_k indeed flows in the direction shown in the figure. However, if the numerical value of i_k turns out to be negative, our assumption is contradicted and the current flows in the opposite direction.

2. While applying KVR in the mesh containing the element E_k, we may assume that the node v_i is at a higher potential than the node v_j and traverse the mesh in the clockwise direction. Accordingly, we expect the numerical value of the voltage v_{ij} to be positive. Later, when we solve a system of linear equations, if it turns out that the numerical value of v_{ij} is positive, our assumption is correct. Otherwise, our assumption is contradicted and, in fact, the node v_j is at a higher potential than the node v_i.

3. If both i_k and v_{ij} satisfy or contradict our basic assumptions, i.e., either both of them conform to the direction and polarity shown in the figure or are just opposite, then we may conclude that E_k is a passive element; it absorbs power. On the other hand, if one of them contradicts our assumption indicating that the current actually flows from the node at a lower potential to the node at a higher potential, then we may conclude that E_k is an active element; it delivers power.

A subtle observation that should be made is that there are only two options—an initial assumption is either justified or contradicted; a branch voltage is either positive or negative, and the direction of current is either one way or opposite. In other words, there exists no element that is partially passive and partially active. Accordingly, we use this reasoning to devise the following, quite interesting and amusing, problem-solving strategy.

- Assume one out of the two possibilities.
- Work out the details *in toto*.
- If everything fits the bill, then our assumption is correct and the problem is solved; else, the other possibility leads to the solution.

This is the basic principle behind what may be called *Boolean logic* or *digital logic*, which would be introduced to the reader as a separate subject. At this moment, we profess that this strategy works for many problems we come across around us. We develop several interesting and important algorithms from next chapter onwards based on this logic.

Of course, if we *a priori* know the elements in a given circuit, then we may 'fix' the directions and polarities straightaway rather than making any assumptions. However, at times, as was illustrated in some examples in Chapter 2, it may so happen that an active element such as a current source may actually absorb power rather than deliver it.

We end this section with an important note. As has been mentioned in the preceding chapter, sources may be varying with time, such as a sinusoidal source we use for our domestic usage. Accordingly, all the voltages and currents would also be varying with time in a given circuit. However, rules are rules and they have to be obeyed for all times. In other words, we need to slightly amend the rules above and say that

at any given instant of time, the element and circuit rules must be obeyed.

We shall illustrate this in the following example.

Example 3.8 Consider the circuit in Fig. 3.25. There is a lot more to discuss about this circuit, but we defer that to the following section.

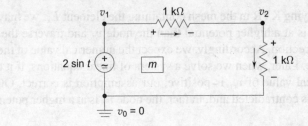

Fig. 3.25

In the one-mesh circuit shown in Fig. 3.25, we assume that the ground v_0 is 'always' at zero potential so that v_1 varies with time. Applying KVR, we get

$$v_{12} + v_{20} = 2 \sin t$$

For instance, at time $t = 0$, we have

$$v_{12} + v_{20} = 0$$

and at time $t = \pi/4$, we have

$$v_{12} + v_{20} = \sqrt{2}$$

and at time $t = \pi/2$, we have

$$v_{12} + v_{20} = 2$$

and so on.

Since the two resistances are in series, the same current i_R passes through them and also through the source conforming to the KCR applied at all the nodes. Thus, we get

$$(1 + 1) \times 10^3 \times i_R = 2 \sin t$$

or, equivalently,

$$i_R = \sin t \text{ mA}$$

Hence, each of the voltages v_{12} and v_{20} is

$$v_{12}(t) = \sin t \text{ V} = v_{20}(t)$$

It is customary to indicate the time as an argument t, as shown in the above equation, along with the variable.

3.2.1 Circuit Rules and Sources

Kirchhoff's rules have something very important to say regarding the connection of sources. Let us look at this carefully. Suppose we connect two voltage sources in series and then close the path through a resistance R_L as shown in Fig. 3.26. The polarity of v_{12} is fixed arbitrarily and may be changed relative to the relative magnitudes of v_1 and v_2.

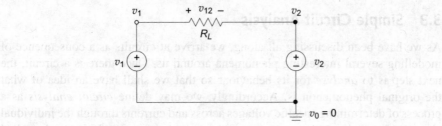

Fig. 3.26

Around the mesh, applying KVR gives

$$- v_1 + v_{12} + v_2 = 0$$

i.e., at any given instant of time, the voltage across the resistance R_L is the algebraic sum of the voltages of the sources. This is perfectly valid. Now, consider a parallel circuit in Fig. 3.27 with the same elements as above.

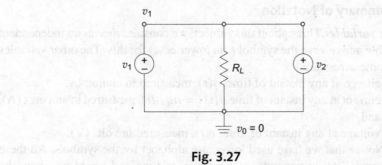

Fig. 3.27

It is easy to see that, unless $v_1 = v_2$, the circuit does not obey Kirchhoff's voltage rule. Violating Kirchhoff's rules implies violation of the *natural laws* of electromagnetic phenomena. Therefore, we should never connect two different ideal voltage sources in parallel. Of course, if the voltage sources have internal resistances in series, as is the case in practice, then the rule is obeyed. We may apply the same reasoning to deduce that we should never connect two different ideal current sources in series.

Exercise 3.8 Draw a circuit, analogous to the one in Fig. 3.26, with two ideal current sources i_1 and i_2 in series with a resistance R_L and verify that it violates Kirchhoff's current rule. What about a circuit with the same elements connected in parallel analogous to the one in Fig. 3.27?

Thus, the element and circuit rules must be obeyed in a circuit at all instants of time. In the following section, we shall look at some more circuits and define what it means to *analyse a circuit*.

3.3 Simple Circuit Analysis

As we have been discussing all along, we arrive at circuits as a consequence of modelling several interesting phenomena around us. Once there is a circuit, the next step is to *analyse* for its behaviour so that we shall have an idea of what the original phenomenon is. Accordingly, we may define *circuit analysis* as a process of determining specific voltages across and currents through the individual elements that make up the given circuit. An opposite process of this may be called *network synthesis* wherein we choose a set of elements and make up circuits so as to achieve branch voltages and currents according to certain specifications. Synthesis is generally more difficult than analysis and it is a vast subject by itself. A thorough knowledge of analysis lays the necessary foundation for synthesis. We present the basic ideas of network synthesis in Chapter 11 after an in-depth exploration of analysis.

Let us first summarize the notation, i.e., the alphabet of the language we are going to use in this textbook. By and large the notation is uniform across several texts, but there may be slight differences here and there. At the outset, let us emphasize that we make use of SI units for all the physical quantities we come across.

3.3.1 Summary of Notation

1. *Basic variables* Throughout this subject, we consider *time* as an independent variable and reserve the symbol t (in lower case) for this. The other variables we come across are
   (a) charge at any instant of time, $q(t)$, measured in coulombs,
   (b) current at any instant of time, $i(t) (= dq/dt)$, measured in amperes (A), and
   (c) voltage at any instant of time, $v(t)$, measured in volts (V).

   Notice that we have used lowercase alphabet for the symbols. All these three variables, in general, are functions of time and hence we show the symbol t as an argument next to the variable within parentheses. If any of these is a constant with respect to time, such as the dc voltage supplied by a battery, we use uppercase alphabets, Q, I, and V, respectively, for charge, current, and voltage. There is a special class of time-varying quantities called *sinusoids* to be introduced in Chapter 6. We use script letters such as $\mathcal{V}$ and $\mathcal{I}$, read as 'script V' and 'script I', respectively, to denote the sinusoidal voltage and sinusoidal current. However, if it is clear from the context, there is no harm in using the lowercase letters for both time-varying and constant quantities, and we do so. Together with these symbols, we use subscripts, such as v_1, i_{C_2}, etc. to denote the voltage or current of a particular branch; for instance, v_1 may denote the potential of node v_1 in the circuit, and i_{C_2} may denote the current through capacitor C_2 in the circuit.

   The reciprocal of time is also a fundamental quantity, called frequency. It is denoted by the symbol f and is measured in hertz, or cycles per second. Multiplying f with 2π, we get the *radian frequency*, denoted by ω and measured in 'radians per second'. Depending upon the context, it may be convenient to work with ω rather than f. There is another symbol s, a lowercase letter, that also has the unit of frequency. This is a complex variable in contrast to the real variables f and ω, and is called the *Laplace variable*. It will be defined in a later chapter.

2. *Nodes and meshes*
   (a) *Node indices:* By definition, a node is the part where no field exists. In practice, a node is a junction of several circuit elements. Each node has its own node voltage with respect to a common ground. Accordingly, we denote a node j by the symbol v_j, to indicate the node potential v of the jth node.
   (b) *Mesh indices:* The jth mesh of a circuit is shown with the symbol m_j enclosed by a rectangular box inside the mesh. Conventionally we traverse a mesh in the clockwise direction and we indicate this with an arrow around the rectangular box. The node and mesh indices are shown in Fig. 3.28.

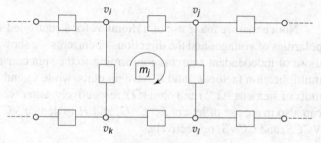

Fig. 3.28

3. *Sources*
   (a) *Independent sources:* In this book we make use of two types of independent sources: the voltage source and the current source. These have been defined in the preceding chapter but we draw the circuit symbols here once again in Figs 3.29(a) and 3.29(b). If it is a dc voltage source of, say U volts, we write 'U V' next to the circle either to its left or to its right. Likewise, if is a time-varying current source of $u(t)$ amperes, we write '$u(t)$ A' next to the circle. The current direction for a voltage source is from the negative node of the source to the positive node. For a current source the node with a higher potential is near the head of the arrow and the node with a lower potential is near the tail of the arrow, conforming to the sign convention. Generally, we use symbols such as v_{sj} and i_{sk} to denote the jth voltage source and kth current source, respectively, in a circuit.

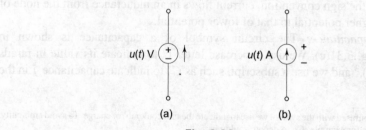

(a)   (b)

Fig. 3.29

(b) *Controlled sources:* We have defined four controlled sources in the preceding chapter. These are shown once again in Figs 3.30(a)–3.30(d).

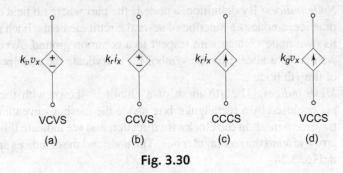

VCVS	CCVS	CCCS	VCCS
(a)	(b)	(c)	(d)

Fig. 3.30

Notice that we make use of a rhombus for a controlled source. The polarities of voltages and the directions of currents are shown similar to those of independent sources, conforming to the sign convention. The multiplication factors k_v and k_i have no units, while k_g and k_r have the units of siemens (Ω^{-1}) and ohm (Ω), respectively. Later we see that we can also make use of letters E, F, G, and H to denote VCVS, CCCS, VCCS, and CCVS, respectively.

4. *Passive elements*

(a) *Resistance:* The circuit symbol of a resistance is shown in Fig. 3.31(a). We use uppercase letter R to indicate its value in ohms (Ω), and we use a subscript, such as R_4 to indicate resistance 4 in the circuit. According to the sign convention, current flows in a resistance from the node of higher potential to that of lower potential. Sometimes, it is convenient for us to work with the reciprocal of a resistance, called conductance. Fortunately, this has a valid unit, siemens, and we use the symbol G for this. As was mentioned earlier, under the assumptions of linearity, lumpedness, and bilaterality, the words resistance and resistor may be used interchangeably. We would rather prefer to say that 'a branch offers a resistance of ...' rather than saying 'there is a resistor whose value is ...'. Similar remarks hold for the other two passive elements.

(b) *Inductance:* The circuit symbol of a inductance is shown in Fig. 3.31(b). We use uppercase letter L to indicate its value in henrys (H), and we use a subscript, such as L_j to indicate inductance j in the circuit. According to the sign convention, current flows in an inductance from the node of higher potential to that of lower potential.

(c) *Capacitance:* The circuit symbol of a capacitance is shown in Fig. 3.31(c). We use uppercase letter C to indicate its value in farads (F),[3] and we use a subscript, such as C_j to indicate capacitance j in the

---

[3] C is not to be confused with the symbol we use to indicate the unit coulomb for charge. To avoid ambiguity, we may choose, for example, Cq to denote coulombs.

circuit. According to the sign convention, current flows in a capacitance from the node of higher potential to that of lower potential.

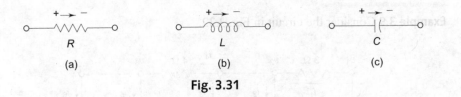

Fig. 3.31

(d) *Reactance and susceptance:* In the later chapters, we would be doing complex algebra. We prefer to see an inductance and/or a capacitance in terms of a resistance offered to the flow of current. This is defined as *reactance* of the corresponding element and we denote it with the symbol X (uppercase) with an appropriate subscript; for instance, the reactance offered by an inductance is X_L and that offered by a capacitance is X_C. We use ohm as the unit of reactance. The reciprocal of reactance, akin to conductance, is defined as *susceptance*, and we denote it with the symbol B with appropriate subscripts. The unit of B is Ω^{-1}.

(e) *Impedance and admittance:* While we do complex algebra, we need to talk about the real part and the imaginary part. While the resistance R, or the conductance G, appears as the real part, the reactance X, or the susceptance B, appears as the imaginary part. The complex number $R + jX$ is defined as *impedance* and we denote it with the uppercase symbol Z; the unit is ohm, of course. Likewise, the complex number $G + jB$ is defined as *admittance* and we use the uppercase letter Y for this. What would be the unit of Y?

(f) *Time-constant:* In Chapter 5, we shall see that the product $R \times C$ and the quotient L/R have the units of time. Both these are denoted by the Greek letter τ, read as 'tau', which serves as an invaluable tool in circuit theory as well as in general systems theory.

5. *Unknowns* In circuit analysis, we are interested in determining the node voltages and/or mesh currents. We may use the symbols v and i for these. However, sometimes, we may have to determine the voltage across a branch, or the current through a branch, which requires some more operations (add or subtract) on v and i. To avoid any confusion, we prefer to use the symbol d or y (in lowercase) with appropriate subscripts. This lowercase y should not be confused with the symbol Y for admittance.

6. *Power and energy* Power is the product of $v(t)$ and $i(t)$, and hence is a time-varying quantity in general. We use the symbol $p(t)$ to denote the instantaneous power. In the special case when the power is a constant, we resort to the uppercase letter P. Energy is the product of power and time, and we use the uppercase symbol E to denote it. Recall that the unit of power is watt (= J/s) and the unit of energy is joule.

There might be some more quantities, such as the damping factor denoted by ζ, but we prefer to enumerate them as and when they appear.

3.3.2 Examples

Example 3.9 Consider the circuit in Fig. 3.32.

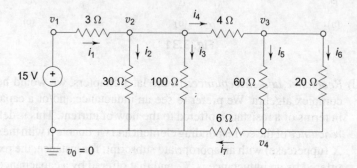

Fig. 3.32

As has been mentioned above, circuit analysis deals with determining the node voltages and mesh currents, and hence the voltages across and the currents through each of the individual elements. Though the reader would have already solved problems of this sort in an earlier physics course, we suggest that he/she follow the solution offered here with an enthusiasm to learn better alternatives.

- Identify all the five nodes, including the ground, and enumerate them as shown in Fig. 3.32. The choice of ground node is arbitrary, and we prefer the negative terminal of the source. Also, enumerate the four meshes.

- The potential of node 1 is $v_1 = 15$ V, since we have a dc voltage source connected between this node and the ground.

- Since we are interested in the voltages across and the currents through each of the branches, let us enumerate the branch currents through each of the resistances as i_1, i_2, i_3, i_4, i_5, i_6, and i_7, as shown in Fig. 3.32.

- Observe that two resistances, 60 Ω and 20 Ω, are across the same node pair v_3-v_4. Using Ohm's law, it is easy to find that

$$20 \times i_6 = v_3 - v_4 = 60 \times i_5$$

and hence

$$i_6 = 3 \times i_5$$

- Next, using KCR at node v_3, we have

$$i_4 - i_5 - i_6 = 0$$

and hence

$$i_5 = \frac{1}{4} i_4$$

Applying KCR at node v_4 gives us

$$i_5 + i_6 - i_7 = 0$$

and hence

$$i_5 + i_6 = i_7 = i_4$$

- We may replace the parallel combination with an equivalent resistance R_1 such that

$$\frac{1}{R_1} = \frac{1}{20} + \frac{1}{60}$$

from which we get

$$R_1 = 15\,\Omega$$

This is shown in Fig. 3.33(a).

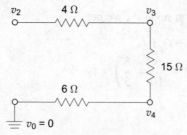

Fig. 3.33(a)

Observe that, since R_1 is in series with the $4\,\Omega$ and $6\,\Omega$ resistances, the same current i_4 flows through it. Thus, using Ohm's law, we get

$$v_3 - v_4 = 15 \times i_4$$

- Looking at the $6\,\Omega$ resistance between the nodes v_4 and v_0, we readily observe that

$$v_4 = 6 \times i_4$$

and hence,

$$v_3 = v_4 + 15 \times i_4 = 21 \times i_4$$

- Next, we shall see what is connected to the right of nodes v_2 and v_0 in Fig. 3.33(a). Clearly, there is a series combination of $4 + 15 + 6 = 25\,\Omega$ carrying a current i_4, and this is parallel to a $100\,\Omega$ resistance carrying a current i_3, and a $30\,\Omega$ resistance in parallel carrying a current i_2. Thus, we have

$$v_2 = 25 \times i_4 = 100 \times i_3 = 30 \times i_2 \text{ V}$$

Using this equation, we may deduce that

$$i_4 = 4 \times i_3$$

$$i_3 = \frac{3}{10} i_2$$

and hence

$$i_4 = \frac{6}{5} i_2$$

In terms of i_4, we may write

$$v_4 = \frac{36}{5} i_2 \quad \text{and} \quad v_3 = \frac{126}{5} i_2$$

- From Fig. 3.33(a), we also observe that

$$v_2 - v_3 = 4 \times i_4 = \frac{24}{5} i_2$$

so that we may write

$$v_2 = 30 \times i_2$$

- Applying KCR at node v_2, we get

$$i_1 = i_2 + i_3 + i_4$$

$$= \left(1 + \frac{3}{10} + \frac{6}{5} \right) \times i_2$$

$$i_2 = \frac{2}{5} i_1$$

and hence

$$v_2 = 30 \times i_2 = 12 \times i_1$$

$$i_3 = \frac{3}{10} i_2 = \frac{3}{25} i_1$$

$$i_4 = 4 \times i_3 = \frac{12}{25} i_1$$

$$i_5 = \frac{1}{4} i_4 = \frac{3}{25} i_1$$

$$i_6 = 3 \times i_5 = \frac{9}{25} i_1$$

$$i_7 = i_4 = \frac{9}{25} i_1$$

Thus, we need to find the current i_1 and all other voltages and currents are scalar multiples of i_1, i.e., all the currents and voltages are *linearly related*. This linear relationship is a consequence of using linear elements to make up the circuit.

- Let us now further reduce the circuit to that shown in Fig. 3.33(b) by combining the elements across the node pair v_2-v_0. The equivalent resistance R_2 may be found using

$$\frac{1}{R_2} = \frac{1}{30} + \frac{1}{100} + \frac{1}{25}$$

i.e., $R_2 = 12\,\Omega$.

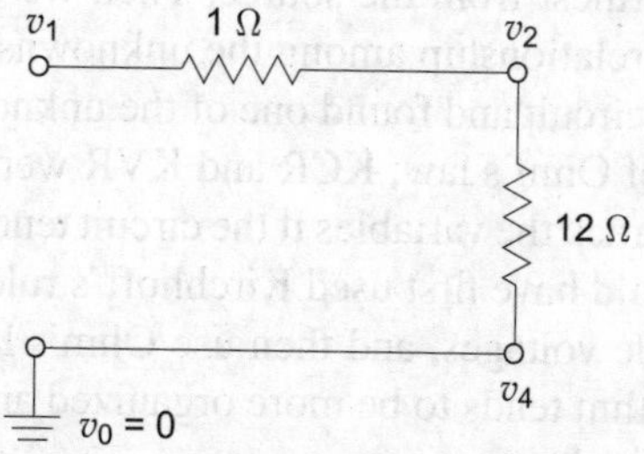

Fig. 3.33(b)

Clearly, the $3\,\Omega$ resistance and $R_2 = 12\,\Omega$ resistance are in series and the same i_1 flows through R_2. In this single-mesh equivalent circuit, if we apply KVR, we get

$$-15 + 3 \times i_1 + 12 \times i_1 = 0$$

or

$$i_1 = 1\,\text{A}$$

We see that this current i_1 is delivered by the source.

- Once we have i_1, we may solve for the rest of the currents and voltages as follows.

$$i_2 = \frac{2}{5} i_1 = 0.4\,\text{A}$$

$$i_3 = \frac{3}{10} i_2 = 0.12\,\text{A}$$

$$i_4 = 4 \times i_3 = 0.48\,\text{A}$$

$$i_5 = \frac{1}{4} i_4 = 0.12\,\text{A}$$

$$i_6 = 3 \times i_5 = 0.36\,\text{A}$$

$$i_7 = i_4 = 0.48\,\text{A}$$

$$v_2 = 12 \times i_1 = 12\,\text{V}$$

$$v_3 = \frac{126}{5} i_2 = 10.08\,\text{V}$$

$$v_4 = \frac{36}{5} i_2 = 2.88\,\text{V}$$

- The voltage across the parallel combination of $60\,\Omega$ and $20\,\Omega$ resistances in the original circuit of Fig. 3.32 is $v_3-v_4 = 7.2\,\text{V}$.

Once we have the voltages across and the currents through all the branches, the analysis is complete. In other words, we have the complete information about the behaviour of the circuit. From Fig. 3.33, we say that the dc voltage source 'sees' an equivalent network consisting of a single resistance of 15 Ω. This idea of *a source seeing an equivalent resistance* plays a crucial role in the later chapters.

Let us now analyse the algorithm we have suggested. We have started with an element that is farthest from the source. Then we moved towards the left and found an interesting relationship among the unknowns. Towards the end, we have simplified the entire circuit and found one of the unknowns, i_1. In this process, we have just made use of Ohm's law; KCR and KVR were used sparingly. Clearly, it becomes difficult to track the variables if the circuit tends to become more complex. Alternatively, we could have first used Kirchhoff's rules, say the KCR, repeatedly to determine the node voltages, and then use Ohm's law to determine the branch currents. This algorithm tends to be more organized and concise. We explain such algorithms in the next chapter.

Before proceeding to another example, let us look at a way of verifying our results. Of course, applying KCR at each of the nodes and/or applying KVR in each of the four meshes is an easy way, but we have got our solution using them. Let us think of an alternative. Since each of the passive elements in the circuit absorbs power, let us compute it and then compute the total power absorbed by the passive network. Let us use the symbols $p_1, \ldots, p_7$, respectively, to denote the power absorbed by each of the elements carrying currents $i_1, \ldots, i_7$. According to the sign convention, we put a negative sign in the following computation to denote that the power is absorbed. Thus, we have

$$p_1 = -i_1^2 \times 3 = -3\,\text{W}$$
$$p_2 = -i_2^2 \times 30 = -4.8\,\text{W}$$
$$p_3 = -i_3^2 \times 100 = -1.44\,\text{W}$$
$$p_4 = -i_4^2 \times 4 = -0.9216\,\text{W}$$
$$p_5 = -i_5^2 \times 60 = -0.864\,\text{W}$$
$$p_6 = -i_6^2 \times 20 = -2.592\,\text{W}$$
$$p_7 = -i_7^2 \times 6 = -1.3824\,\text{W}$$

and hence the total power absorbed by the passive network is

$$p_{\text{abs}} = \sum p_i = -15\,\text{W}$$

Observe that when we use symbols we need to show the negative sign along with p; however, when we say that an element has absorbed power, it is sufficient to mention the magnitude alone. In other words, saying that the passive network has *absorbed* a power of 15 W is equivalent to writing -15 W on, say, a piece of paper.

Now, where does this power come from? Obviously, the source has to *deliver* it. Clearly, since the voltage source is pumping in a current of 1 A to the rest of the network, it delivers a power of

$$p_{\text{del}} = 15\,\text{V} \times 1\,\text{A} = +15\,\text{W}$$

Thus, in any given circuit, the total power should be conserved. We call this as the *law of conservation of power* and we write it as

$$\text{Power delivered} + \text{Power absorbed} = 0 \qquad (3.19)$$

This equation has to be satisfied in any given circuit, and this provides us a way of checking our solutions.

Example 3.10 In this example, we shall derive some standard relationships. Consider the simple series circuit of Fig. 3.34.

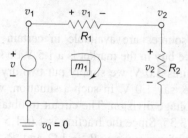

Fig. 3.34

Applying KVR around the mesh gives us

$$-v + v_1 + v_2 = 0 \tag{3.20}$$

where v_1 and v_2 are the voltages across R_1 and R_2, respectively. Since it is the same current i that passes through both the resistances in the prescribed direction, applying Ohm's law we have

$$v_1 = R_1 \times i \quad \text{and} \quad v_2 = R_2 \times i \tag{3.21}$$

Substituting Eqn (3.21) in Eqn (3.20), we get

$$i = \frac{1}{R_1 + R_2} \, v \tag{3.22}$$

$$v_1 = \frac{R_1}{R_1 + R_2} \, v \tag{3.23}$$

$$v_2 = \frac{R_2}{R_1 + R_2} \, v \tag{3.24}$$

Equation (3.22) is straightforward: the net current in the circuit is the quotient of the net voltage and the net resistance. We may generalize this and say that

$$i = \frac{1}{R_1 + R_2 + \cdots + R_N} \, v \tag{3.25}$$

i.e., in a series circuit of N resistances connected to a voltage source v, the current i flowing through all the elements may be computed using Eqn (3.25). Likewise, we may generalize Eqn (3.23) for the case of a series connection of N resistances and say that

$$v_j = \frac{R_j}{\sum\limits_{j=1}^{N} R_j} \, v \tag{3.26}$$

i.e., the voltage across jth resistance is proportional to R_j. Equation (3.26), in fact, is a rule that has to be obeyed by any series circuit having N resistances connected to a voltage source v. We call this as the *voltage division rule*, which says that *in a series circuit the voltages across each of the resistances divide in the ratio of the resistances themselves.* It is useful to rewrite Eqn (3.26) as

$$\frac{v_j}{\sum\limits_{j=1}^{N} v_j} = \frac{R_j}{\sum\limits_{j=1}^{N} R_j} \qquad (3.27)$$

Generally, voltage sources are available in certain prescribed ranges; for example, the dry cell we buy in the market is a 1.5 V dc source. If the appliance, say the walkman, requires 3.0 V, we simply put two dry cells in series. Certain appliances may require, say 1.0 V. In such a situation, we design a circuit that achieves the desired voltage division. The circuit to obtain 1.0 V out of a 1.5 V source is given in Fig. 3.35. Since the fraction is $1.0/1.5 = 2/3 = 2/(2+1)$, we can build the circuit with resistances $R_1 = 1\,\Omega$ and $R_2 = 2\,\Omega$. Notice that the ratio of the resistances is $1:2$, and hence we may choose R_1 and R_2 in several different ways, according to the availability of the components in the stock room. For instance, we may choose $0.5\,\text{k}\Omega$ and $1.0\,\text{k}\Omega$. At times, we may have to split either R_1 or R_2 or both, depending on the availability of the components.

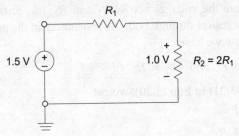

Fig. 3.35

Exercise 3.10 List all possible values of resistances available in your laboratory.

Exercise 3.11 An appliance requires 2.2 V. Build a circuit that performs the desired voltage division, using standard dry cells and standard resistances available in your laboratory.

Example 3.11 Let us now devise a circuit that performs current division. Typically, by the principle of duality, we may expect the circuit shown in Fig. 3.36.

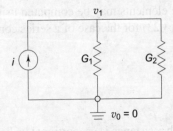

Fig. 3.36

Notice that we have used the symbol G instead of R owing to the ease in computing the parallel combination of resistances. Since this is a circuit across a pair of nodes, out of which one is designated as the ground, applying KCR at the node v_1 gives

$$i - G_1 v_1 - G_2 v_1 = 0 \qquad (3.28)$$

The parallel combination of G_1 and G_2 is simply $G_1 + G_2$, and hence, using Ohm's law, we have

$$v_1 = (G_1 + G_2) \times i \qquad (3.29)$$

and hence

$$i_1 = \frac{G_1}{G_1 + G_2} i \qquad (3.30)$$

$$i_2 = \frac{G_2}{G_1 + G_2} i \qquad (3.31)$$

i.e., the ratio of the currents is the ratio of the conductances. We may generalize Eqns (3.30) and (3.31) for the case of N resistances in parallel being driven by a current source i, and write the *current division rule* as

$$i_j = \frac{G_j}{\sum\limits_{j=1}^{N} G_1} i \qquad (3.32)$$

which may be rewritten, akin to Eqn (3.27), as

$$\frac{i_j}{\sum\limits_{j=1}^{N} i_j} = \frac{G_j}{\sum\limits_{j=1}^{N} G_j} \qquad (3.33)$$

Comparing this equation with Eqn (3.27), we see that the principle of duality is quite transparent—replace v with i and R with G. As has been mentioned in Chapter 2, current sources are more of a mathematical convenience and we exploit the duality principle throughout the textbook.

Example 3.12 Consider the circuit shown in Fig. 3.37, and compute the current passing through the 5 Ω resistance.

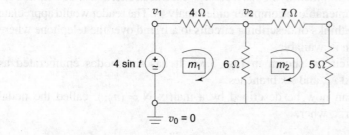

Fig. 3.37

We may quickly reduce the network across the node pair v_2-v_0 as

$$R_{eq} = (5+7) \parallel 6 = \frac{12 \times 6}{12+6} = 4\,\Omega$$

Hence, by Ohm's law, the current delivered by the ac voltage source is

$$i = \frac{4 \sin t}{4+4} = 0.5 \sin t \text{ A}$$

It is this current i that gets divided in the parallel network, whose conductances may be written as

$$G_1 = \frac{1}{6} \quad \text{and} \quad G_2 = \frac{1}{5+7} = \frac{1}{12}$$

and, according to the current division rule, the current through the 5 Ω resistance (and also through the 7 Ω resistance) is

$$\frac{\frac{1}{12}}{\frac{1}{6} + \frac{1}{12}} \, 0.5 \sin t = \frac{1}{6} \sin t \text{ A}$$

Exercise 3.1 Repeat Example 3.12 by utilizing voltage division rule.

Given a complex network containing several nodes and elements, it is often useful to identify sub-networks that are isomorphic to voltage division and/or current division circuits. Clearly, this is an extension of utilizing the series and/or parallel combination of homogeneous elements. Although we have shown voltage division and current division using resistances alone, later on we will see the generalized version of this division, which involves heterogeneous elements and complex algebra.

3.3.3   The SPICE Data Structure

So far, it is very clear that it is the nodes and the branches that make up a network, and at least one mesh enclosed by the branches makes it a circuit. While a network is being shown with graphical symbols, let us now look at an alternative description for a network that is amenable to computer-aided analysis. The reader would appreciate the idea if he/she thinks of describing circuits to a friend over the telephone where no visual medium is available.

Consider the circuit shown in Fig. 3.38. It has five nodes enumerated as v_0, v_1, v_2, v_3 and v_4, and six branches.

This circuit can now be described by a matrix $\mathbf{N} = (n_{jk})$, called the nodal connectivity matrix, where

$$n_{jk} = \begin{cases} 1 & \text{if there is a branch between nodes } j \text{ and } k \\ 0 & \text{if nodes } j \text{ and } k \text{ are not connected} \end{cases} \tag{3.34}$$

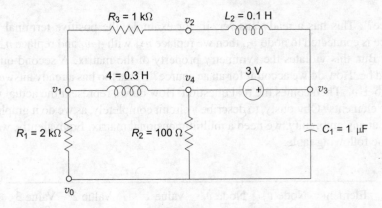

Fig. 3.38

We consider that a node is always connected to itself, and hence $n_{jj} = 1$. For the circuit in Fig. 3.38, the connectivity matrix is

$$
\mathbf{N} = \begin{pmatrix} 1 & 1 & 0 & 1 & 1 \\ 1 & 1 & 1 & 0 & 1 \\ 0 & 1 & 1 & 1 & 0 \\ 1 & 0 & 1 & 1 & 1 \\ 1 & 1 & 0 & 1 & 1 \end{pmatrix} \tag{3.35}
$$

Clearly, the connectivity matrix is always square and symmetric, with '1' on the principal diagonal. This matrix tells how elements are connected to make up a given circuit.

Exercise 3.13 Given the following matrix, draw the corresponding network.

$$
\mathbf{N} = \begin{pmatrix} 1 & 1 & 0 & 1 \\ 1 & 1 & 1 & 0 \\ 0 & 1 & 1 & 1 \\ 1 & 0 & 1 & 1 \end{pmatrix}
$$

Thus, we agree that without any graphics it is possible to describe a circuit. The connectivity matrix $\mathbf{N}$ serves the purpose. However, we need to know what exactly are the elements in the given circuit. For this, we need to modify our connectivity matrix little more. This may be done using the following way.

(i) There is a resistance R_1 between the nodes v_1 and v_0 in the circuit of Fig. 3.38. We simply replace '1' in the positions n_{01} and n_{10} with R_1.

(ii) Similarly, there is a dc voltage source v_s between nodes v_3 and v_4. Here, we replace '1' in the positions n_{34} and n_{43} with v_s, and so on.

This would immediately ring an alarm! There arise several questions. The first question, perhaps, is 'How do we account for the exact polarities of the voltage source?'. This has a ready answer; if, for example, the positive terminal of the source is connected to node v_3, then we replace n_{34} with $+ v_s$ and replace n_{43} with $- v_s$. But, this violates the symmetry property of the matrix. A second question would be 'How do we account for an ac source?'. This too has a ready answer: add 'ac' to $\pm v_s$. Then comes the real question: How do we represent the actual values of the elements? Obviously, to describe a circuit completely, as we do it graphically without any ambiguity, we need a multidimensional matrix. Equivalently, we may use the following table:

Element	Node v_j	Node v_k	Value 1	Value 2	Value 3
R1	1	0	2 K Ohm		
L1	1	4	0.3 H		
R3	1	2	1K Ohm		
L2	2	3	0.1 H		
Vs	3	4	dc	3 V	
C	3	0	1 UF		
R2	4	0	100 Ohm		

There are a few observations that could be readily made from the above table.

1. The order in which we describe the elements is immaterial; care should be taken not to omit any particular element. To this end, writing the connectivity matrix first would be of great help.
2. Since we are interested in this tabular description for computer-aided analysis, the subscripted variables, such as R_1, have been written in programming style, e.g., $R1$. Each of the elements has a standard symbol given in the following table:

Element	Symbol
Capacitor	C
Voltage-controlled voltage source (VCVS)	E
Current-controlled current source (CCCS)	F
Voltage-controlled current source (VCCS)	G
Current-controlled voltage source (CCVS)	H
Independent current source	I
Inductor	L
Resistor	R
Independent voltage source	V

We suffix each of these symbols with an appropriate subscript, e.g., R_1.

3. The element value has to be suffixed with the corresponding units according to the following table.

Units	Suffix
volts	V
amperes	A
hertz	Hz
ohm	Ohm
henry	H
farad	F
degree	Degree

Of course, we may abbreviate the number of zeros in terms of standard suffixes such as those given in the following table.

Metric prefix	Multiplying factor	Abbreviated suffix
femto	10^{-15}	F
pico	10^{-12}	P
nano	10^{-9}	N
micro	10^{-6}	U
milli	10^{-3}	M
kilo	10^{3}	K
mega	10^{6}	MEG
giga	10^{9}	G
tera	10^{12}	T

For instance, in the circuit of Fig. 3.38, the capacitor's value is one microfarad, abbreviated as 1 UF in the table.

4. We have specified three columns for the value of an element, but we have not used the second and third columns for the passive elements. In general, as we see in later chapters, the capacitors and inductors shall have *initial conditions*, which may be spelt out in the second column. For controlled sources, we need three columns for the value—the first two to specify the branch, the voltage across/current through which controls the source, and the last one to specify the multiplying factor k_v, k_i, k_g, or k_r. We may include an additional column to write our comments, much the same way we do in any other programming language.

There already exists a widely accepted computer program called *simulation program with integrated circuit emphasis*, popularly known as *SPICE*, that performs circuit analysis. This program was developed in the early 1970s at the University of California, Berkeley. Typically, there are two versions of *SPICE* available: one for the mainframes, available commercially as *HSPICE, IGSPICE*, etc. and a low-cost one for the personal computers, available as *PSPICE*. A student version is generally available on the world wide web[4] to students and teachers free of charge. In this textbook, we are just beginning to learn what are circuits, and how they can be analysed and synthesized. Accordingly, the number of components in a typical circuit is very small, say not even ten! Once we master the fundamental ideas, we can graduate into more complex circuits containing tens, hundreds, and thousands of components commercially available as *integrated circuits*, ICs. SPICE was originally developed as a program that simulates the performance of ICs (read the expansion of SPICE again), and provide a simple, cost-effective means of confirming the desired behaviour prior to building them physically in the laboratory or factory. In fact, the program can handle many more elements such as diodes, transistors, etc. and there are standard symbols for such components too. We have enumerated the symbols of only those elements we use in this book.

As such, there is no point spending more time learning about the program in this book; rather, we just use it as an *aid* in understanding the descriptions of networks initially, and then in verifying the answers to several problems, at a later stage. In this chapter, we shall just give little more introduction to the program. A tutorial is given at the end of the book in Appendix A. We suggest (Rashid 1995) and (Roberts 1997) for a programming enthusiast.

In order to use the program, we need to create a computer file, called the SPICE *input file*. This file has three parts as follows. First we need to describe the circuit to the program. This is simply done by creating a table, such as the one we have developed for the circuit in Fig. 3.38. Second, we need to specify the analysis we seek. So far, we are familiar with very simple analysis. Later on, we see that there are dc, transient, and ac analyses that could be performed on the circuits of interest in this book. We shall introduce appropriate SPICE commands in Appendix A. These commands are added to the earlier table, below the circuit description. Lastly, we obtain an output from the program. This could be a printed statement of the values of desired variables, or plots of variables as functions of time. In addition, we may also graphically visualize these plots as if we are using a cathode-ray oscilloscope (CRO). We call this table (multidimensional array) as the *data structure*, i.e., how the data of a particular circuit is organized, for SPICE.

It is strongly recommended that it is adequate, at this point, for the reader to be able to develop the first part of the SPICE input files for the circuit description. Let us look into the following examples and exercises.

Example 3.13 [Adopted from (Rashid 1995)] Consider the circuit shown in Fig. 3.39.

---

[4] Interested reader may search for this using GOOGLE. One such version, called AIMSPICE, is available with the author.

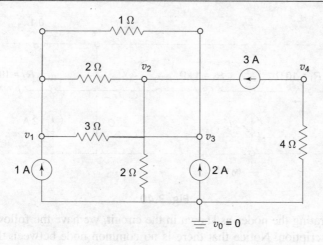

Fig. 3.39

Observe that the branches between nodes v_1 and v_3, and v_2 and v_0, do not have a common node. Likewise, node v_3 has no common node with the branch between v_2 and v_4; otherwise, nodes v_2 and v_3 are identical.

Let us write the SPICE circuit description. Enumerating the nodes as shown in the figure, we readily have the following table.

Element	Node v_j	Node v_k	Value 1	Value 2
R1	1	3	1 Ohm	
R2	1	2	2 Ohm	
R3	1	3	3 Ohm	
R4	2	0	2 Ohm	
R5	4	0	4 Ohm	
I1	0	1	dc	1 A
I2	0	3	dc	2 A
I3	4	2	dc	3 A

Observe the way we have specified the current sources. It is according to the sign convention of a passive element; node v_j is at the tail of the arrow and node v_k is at the head of the arrow. Nevertheless, this is the syntax in a SPICE program; just follow the arrow.

Example 3.14 Consider the circuit in Fig. 3.40 containing all types of controlled sources.

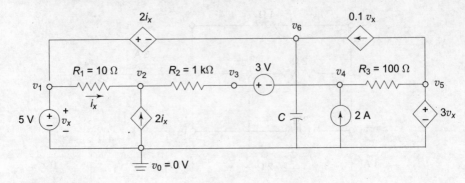

Fig. 3.40

Enumerating the nodes as shown in the circuit, we have the following SPICE circuit description. Notice that there is no common node between the branches joining the node pairs v_3-v_4 and v_6-v_0.

Element	Node v_j	Node v_k	Value 1	Value 2	Value 3	Comments
R1	1	2	10 Ohm			
R2	2	3	1K Ohm			
R3	4	5	100 Ohm			
C	6	0	1M F			; 1×10^{-3}F capr
Vx	1	0	dc	5V		
Is	0	4	dc	2A		
Vs	3	4	dc	3V		
E	5	0	1	0	3	; VCVS
F	0	2	1	2	2	; CCCS
G	5	6	1	0	0.1	; VCCS
H	1	6	1	2	2	; CCVS

Example 3.15 Consider the circuit shown in Fig. 3.41. With the nodes enumerated as shown in the figure, it is easy to write the following SPICE circuit description. Observe that the branches carrying resistances 50 Ω and 40 Ω do not have a common node; otherwise, node v_5 itself is the ground.

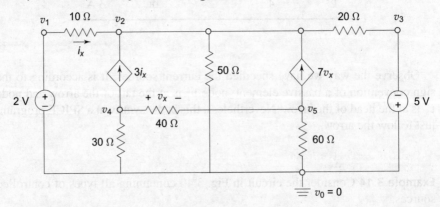

Fig. 3.41

Element	Node v_j	Node v_k	Value 1	Value 2	Value 3	Comments
R1	1	2	10 Ohm			
R2	2	3	20 Ohm			
R3	2	0	50 Ohm			
R4	4	5	40 Ohm			
R5	4	0	30 Ohm			
R6	5	0	60 Ohm			
Vs1	1	0	dc	2 V		
Vs2	0	3	dc	5 V		
F	4	2	1	2	3	; CCCS
G	5	2	4	5	7	; VCCS

Exercise 3.14 Obtain the SPICE circuit description for the circuit shown in Fig. 3.42.

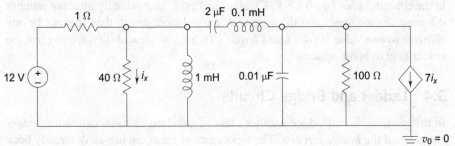

Fig. 3.42

Exercise 3.15 Draw a topologically isomorphic circuit of the one given in Fig. 3.43 and verify that both the circuits have identical SPICE circuit descriptions.

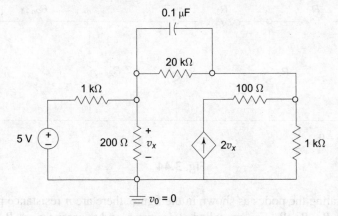

Fig. 3.43

We end this section with the following comments. First, as has been mentioned earlier, the SPICE input file has three parts—the circuit description, analysis commands, and the output. What we have seen so far is only the circuit description. As has been mentioned earlier, this is purely a graphics-free description. In certain versions of the program, there are graphical tools available wherein we may 'draw' the circuit on the computer screen itself. A detailed tutorial is presented in Appendix A. Secondly, according to the syntax of the SPICE program, we must give a title at the top of the table. This is to identify the output generated by the program. Similarly, the input file must end with **.END** line written at the end of the table. Lastly, there are certain symbols used by the SPICE program to denote different quantities; for instance, the symbol F denotes the unit *farad* of a capacitance, it denotes a controlled source (CCCS), and it also denotes the prefix *femto*. However, we need not worry about this because each of the columns of the table of SPICE input file has a specific designation and the program can easily make out what the symbol is saying in a given context. For instance, there could be a capacitance whose value is 0.1×10^{-15} F which would be represented in the column *value 1* as $0.1FF$. Clearly, the first F immediately after the number 0.1 must denote *femto* and the second F must denote *farad*; there cannot be any element whose value is '0.1 farad femto'. To improve readability, we suggest the reader to give blank spaces.

3.4 Ladder and Bridge Circuits

In this section, we introduce another class of standard circuits, called the *ladder circuits* and the *bridge circuits*. The topologies of these circuits have already been enumerated in the beginning of this chapter, along with few more topologies such as the T network, and so on. In this chapter, we shall confine ourselves to the circuits that contain only the resistances.

A typical ladder circuit is shown in Fig. 3.44.

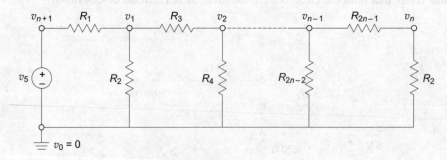

Fig. 3.44

Enumerating the nodes as shown in the figure, there are n resistance pairs; the resistances R_1, R_3, R_5, ... are called *series arms*, and the resistances R_2, R_4, R_6, ... are called *shunt arms*. A ladder circuit of this form is more specifically called an *n-stage ladder*, each of the stages being enumerated from left to right. It is obvious that the voltages and currents decrease as one moves from one stage to the next.

An ardent reader would have quickly observed that the voltage division circuit we have discussed earlier is a one-stage ladder. Clearly, these ladder circuits form the core of a class of circuits called *attenuators* whose purpose is to *attenuate* or cut down the voltages and currents in a given circuit.

Obtaining the equivalent resistance R_{eq} of the resistive network 'as seen by' the voltage source is an interesting piece of work. Let us quickly do it with a simple three-stage ladder shown in Fig. 3.45.

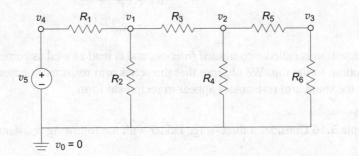

Fig. 3.45

First, there is a series combination of R_5 and R_6 at the right extreme of the circuit. This series combination is in parallel with R_4. Hence, the net resistance that gets added to R_3 is

$$R_4 \parallel (R_5 + R_6) = \cfrac{1}{\cfrac{1}{R_4} + \cfrac{1}{R_5 + R_6}}$$

Next, the resistance that is parallel to the shunt arm R_2 is then

$$R_3 + \cfrac{1}{\cfrac{1}{R_4} + \cfrac{1}{R_5 + R_6}}$$

Next, the resistance that is in series with R_1 is then

$$\cfrac{1}{\cfrac{1}{R_2} + \cfrac{1}{R_3 + \cfrac{1}{\cfrac{1}{R_4} + \cfrac{1}{R_5 + R_6}}}}$$

And, the equivalent resistance is

$$R_{eq} = R_1 + \cfrac{1}{\cfrac{1}{R_2} + \cfrac{1}{R_3 + \cfrac{1}{\cfrac{1}{R_4} + \cfrac{1}{R_5 + R_6}}}}$$

This can be generalized to the *n*-stage network as

$$R_{eq} = R_1 + \cfrac{1}{\cfrac{1}{R_2} + \cfrac{1}{R_3 + \cfrac{1}{\cfrac{1}{R_4} + \cfrac{1}{R_5 + \cfrac{1}{\cfrac{1}{R_6} + \cdots}}}}} \tag{3.36}$$

This equation is called a *continued fraction*, and is read as well as computed from the bottom to the top. We observe that the series arm resistances appear directly while the shunt arm resistances appear in reciprocal form.

Example 3.16 Consider a three-stage ladder with the following resistances:

$$R_1 = 12\,\Omega,\ R_2 = 10\,\Omega,\ R_3 = 30\,\Omega,\ R_4 = 15\,\Omega,\ R_5 = 20\,\Omega,\ \text{and } R_6 = 10\,\Omega$$

Using Eqn (3.36), we readily have

$$R_{eq} = 12 + \cfrac{1}{\cfrac{1}{10} + \cfrac{1}{30 + \cfrac{1}{\cfrac{1}{15} + \cfrac{1}{20 + \cfrac{1}{\cfrac{1}{10}}}}}} = 20\,\Omega$$

Exercise 3.16 Obtain the R_{eq} of a general four-stage ladder. If the ladder has eight identical $10\,\Omega$ resistances, determine the equivalent resistance as seen by the voltage source.

Exercise 3.17 Draw the circuit diagram of an *n*-stage ladder network driven by a current source. *Hint* Use the principle of duality.

Example 3.17 Consider the three-stage ladder of Example 3.16. If this is driven by a 10 V dc source, let us determine the node voltages. First, we identify that $v_4 = 10$ V. Since the equivalent resistance as seen by the source is $20\,\Omega$, the equivalent resistance in series with the $12\,\Omega$ resistance is $8\,\Omega$. Using voltage division, it is easy to see that

$$v_1 = \frac{8}{12 + 8}\,10 = 4\text{ V}$$

Next, the equivalent resistance across the node pair v_2-v_0 is

$$15 \parallel (20 + 10) = 10\,\Omega$$

Since this is in series with the $30\,\Omega$ resistance, once again we may use voltage division to get

$$v_2 = \frac{10}{30+10}v_1 = 1\,\text{V}$$

and

$$v_3 = \frac{10}{20+10}v_2 = \frac{1}{3}\,\text{V}$$

Let us verify our results also. The current i_j through R_j, for $j = 1$ to 6, may be computed as

$$i_6 = \frac{v_3}{10} = \frac{1}{30}\,\text{A}$$

$$i_5 = \frac{v_2 - v_3}{20} = \frac{1}{30}\,\text{A}$$

$$i_4 = \frac{v_2}{15} = \frac{1}{15}\,\text{A}$$

$$i_3 = \frac{v_1 - v_2}{30} = \frac{1}{10}\,\text{A}$$

$$i_2 = \frac{v_1}{10} = 0.4\,\text{A}$$

$$i_1 = \frac{v_s - v_1}{12} = 0.5\,\text{A}$$

Clearly,

$$i_1 = i_2 + i_3$$

$$i_3 = i_4 + i_5$$

$$i_5 = i_6 \tag{3.37}$$

and the circuit obeys KCR. Thus, we see that we may obtain a voltage division of $30 : 1$ using the three-stage ladder.

Exercise 3.18 Verify the results of Example 3.17 using KVR in each of the three meshes. Also, compute the power absorbed by each of the resistances and verify the law of conservation of power.

There is a special case of ladder circuits, called the *R–2R ladder*. This is shown in Fig. 3.46.

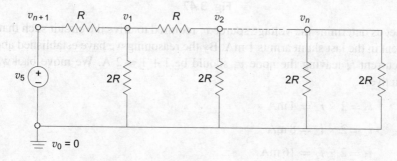

Fig. 3.46

Observe that at the extreme right, across the node pair v_n-v_0 there is *just* a parallel combination of two shunt arms, with the series arm missing in between. Clearly, the current i_n entering the node v_n is divided into two halves. Moving backwards, we may notice that the equivalent resistance across the node pair v_{n-1}-v_0 is also R and hence the current i_{n-1} entering the node v_{n-1} is also divided into two halves, one of which is i_n, i.e.,

$$i_n = \frac{1}{2} i_{n-1}$$

It is evident that if we stand at any node and look towards the right, we see an equivalent resistance of $2R \parallel (R + R) = R$. Accordingly, it is easy to generalize the current entering node v_k, $k = 2, 3, \ldots, n$ and write it as

$$i_k = \frac{1}{2} i_{k-1}, \quad k = 2, 3, \ldots, n \tag{3.38}$$

When, the successive currents get halved as we proceed from one stage to the next, why not the node voltages? It may be verified easily, using the voltage division, that

$$v_k = \frac{1}{2} v_{k-1}, \quad k = 2, 3, \ldots, n$$

with

$$v_1 = \frac{1}{2} v_{n+1}. \tag{3.39}$$

Example 3.18 Consider the four-stage R–$2R$ ladder in Fig. 3.47.

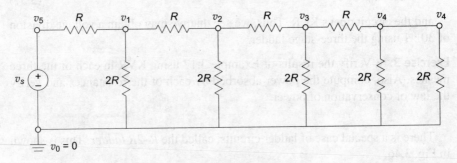

Fig. 3.47

Let us determine the voltage source v_s required to drive the circuit such that the current in the last shunt arm is 1 mA. By the reasoning we have established above, the current i_4 leaving the node v_4 should be $1 + 1 = 2$ A. We move backwards inductively:

$$i_3 = 2 \times i_4 = 4 \text{ mA}$$

$$i_2 = 2 \times i_3 = 8 \text{ mA}$$

$$i_1 = 2 \times i_2 = 16 \text{ mA}$$

The node voltage v_1 would be

$$v_1 = 2R \times i_2 = 16R \times 10^{-3}\, \text{V}$$

Hence the source voltage should be $2 \times v_1 = 32R \times 10^{-3}$ V. Since it is given that $R = 1.25\,\text{k}\Omega$, the source should be 40 V.

We may obtain a general geometric progression for the n-stage R–$2R$ ladder from the above example. If we need a current of i_{n+1} A in the last shunt arm, then the source should be

$$v_s = 2^{n-1} R\, i_{n+1}\, \text{V}$$

The R–$2R$ ladder circuits find applications in several places including digital-to-analog converters in modern audio systems.

Next, we introduce another prototype circuit called the *bridge circuit*. A typical bridge circuit is shown in Fig. 3.48.

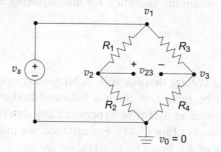

Fig. 3.48

A careful inspection of the circuit reveals that it is basically a parallel circuit across the node pair v_1-v_0. The two resistive branches, each with a series combination of two resistances, are called *bridge arms*. With the nodes enumerated as shown in the figure, we may easily observe that each of the bridge arms works like a voltage division circuit with the node voltages

$$v_2 = \frac{R_2}{R_1 + R_2} v_s \tag{3.40}$$

and

$$v_3 = \frac{R_4}{R_3 + R_4} v_s \tag{3.41}$$

so that the potential difference between the nodes v_2 and v_3 is

$$v_{23} = v_2 - v_3 = \left(\frac{R_2}{R_1 + R_2} - \frac{R_4}{R_3 + R_4} \right) v_s \tag{3.42}$$

Depending on the relative magnitudes of the resistances, we may obtain v_{23} with different polarities. For instance, suppose that

$$v_s = 4\,\text{V}, \quad R_1 = R_2 = R_3 = 1\,\text{k}\Omega, \quad \text{and} \quad R_4 = 3\,\text{k}\Omega$$

then

$$v_{23} = \left(\frac{1}{2} - \frac{3}{4} \right) 4 = -1 \,\text{V}$$

On the other hand, if $R_4 = 0.5 \,\text{k}\Omega$, then,

$$v_{23} = \left(\frac{1}{2} - \frac{0.5}{1.5} \right) 4 = +\frac{2}{3} \,\text{V}$$

Exercise 3.19 In the bridge circuit shown in Fig. 3.48, determine R_4 so that $v_{23} = +1 \,\text{V}$.

Owing to the negative sign in the parentheses in Eqn (3.40), we may suspect that under certain conditions the voltage v_{23} becomes *zero*, irrespective of v_s. It is not too difficult to obtain the condition for this; equating both the terms in the parentheses, we get

$$\frac{R_1}{R_2} = \frac{R_3}{R_4} \tag{3.43}$$

which conforms to the voltage division rule. A bridge circuit operating under such conditions is said to be *balanced*. Such a balanced bridge is generally called *Wheatstone's bridge*, and is useful in experimentally determining an unknown resistance using the ratio in Eqn (3.41). For instance, we may place an unknown resistance R_x in place of R_4. From Eqn (3.41), we see that

$$R_x = \frac{R_2 R_3}{R_1}$$

This procedure is called *null measurement technique*. Besides this application, we make use of resistive bridges in several instrumentation applications such as *resistive temperature detectors* (RTDs) and *load cells*.

3.5   Circuits with Controlled Sources

In this section, we shall look at circuits that contain controlled sources. We restrict ourselves to simpler circuits just to illustrate the idea of analysis in the presence of controlled sources. We shall straightaway proceed towards examples.

Example 3.19 Consider the circuit shown in Fig. 3.49.

The circuit has a single mesh and a current-controlled voltage source (CCVS). Since all the elements are connected in series, it is the same current i_1 that passes through all the elements. Accordingly, applying KVR around the mesh, together with Ohm's law for the resistances, we get

$$-20 + 1 \times i_1 + 25 + 2.5 \times i_1 + 1.5 \times i_1 = 0$$

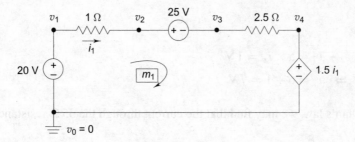

Fig. 3.49

This is a simple equation which yields

$$i_1 = -1\,\text{A}$$

Here, we observe a couple of interesting things.

1. The CCVS behaves like a simple resistance obeying Ohm's law. However, the resistance is *negative*. This is due to the polarities of the other independent sources. If we reverse the polarities of the independent sources, the CCVS offers a positive resistance. Recall that the multiplying factor k_r has the units of resistance.

2. $i_1 = -1\,\text{A}$ implies that the direction of the current is actually opposite to that shown in the circuit, i.e., antilockwise. Accordingly, the 20 V independent voltage source absorbs power along with the resistances, and this power is being supplied by the other 25 V independent voltage source and the CCVS. If the direction of the current were opposite, the CCVS would have absorbed the power.

Example 3.20 Consider the circuit shown in Fig. 3.50.

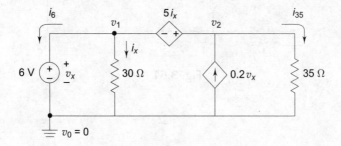

Fig. 3.50

From the circuit, $v_x = v_1 = 6\,\text{V}$, and hence

$$i_x = \frac{6\,\text{V}}{30\,\Omega} = 0.2\,\text{A}$$

Further,

$$v_2 - v_1 = 5 \times i_x = 1\,\text{V}$$
$$\Rightarrow \quad v_2 = 6 + 1 = 7\,\text{V}$$

Using Ohm's law, we may find that the current through the 35 Ω resistance is

$$i_{35} = \frac{7}{35} = 0.2\,\text{A}$$

Applying KCR at node v_2, we may easily find that a current of $1.2 - 0.2 = 1.0\,\text{A}$ leaves the node v_2 towards the left flowing through the CCVS towards v_1. Applying KCR at node v_1 suggests us that a current of

$$i_6 = 1.0 - i_x = 0.8\,\text{A}$$

leaves the node v_1 towards left and flows through the voltage source from top to bottom.

The above results may be verified using the law of conservation of power. Only the VCCS delivers a power of 8.4 W and all other elements, including the independent source, absorb power. We leave it to the reader to complete the verification.

Exercise 3.20 Analyse the circuit shown in Fig. 3.51. Verify your results using the law of conservation of power.

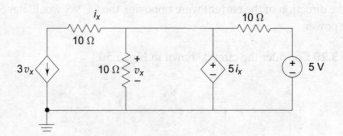

Fig. 3.51

We end this section with a note that circuits with controlled sources may be handled deftly like any other circuit using the basic tools, the element and circuit rules. We must pay attention to the definition of the variable (voltage across or current through a specific branch), including the polarity, which controls the controlled source. Any change in the controlling variable may drastically change the controlled variable. A controlled source establishes an *interdependence* between different branches of a circuit, thereby creating intriguing conditions (the negative resistance in Example 3.19, for instance) and altering the behaviour of the circuit.

In some sense, if it were not these controlled sources in circuits, circuit theory would turn out to be a routine affair.

Summary

This chapter is the fulcrum for the rest of the textbook. Having introduced circuit elements and modelling in the previous chapters, we have put several intuitive ideas on a firm mathematical and physical foundation in this chapter. We have started with the definition of a network and then we have gone further to formally define electrical circuits. Subsequently, we have defined the basic parameters, such as the nodes and meshes, of a circuit. We have summarized the notation we adopt in this text. We have given samples of circuit analysis as a forerunner to the next chapter. Towards the end of the chapter we have introduced ladder networks, bridge networks, and analysis of circuits having controlled sources. Starting with Chapter 4, we shall present the analysis in a more insightful perspective.

In this chapter we have also *developed*, rather than defined, the basic data structure to represent circuits in a computer understandable way. Our intention is to put things in a more organized perspective. We wish to make it very clear that it is worthwhile to learn circuit theory instead of gaining expertise in SPICE. We are here to *design* great many things, and analysis is the basic tool. For analysis, at least to begin with fairly simple circuits, a paper and a pen are indispensable. SPICE serves us only to quickly 'verify' our results; we do not want it to solve our problems in this introductory course on circuit theory. To this end, we shall not use SPICE in any of the following chapters. Appendix A is devoted for a tutorial and problems for the reader to learn it off-line.

Problems

3.1 Determine which of the following networks in Fig. 3.52 are circuits.

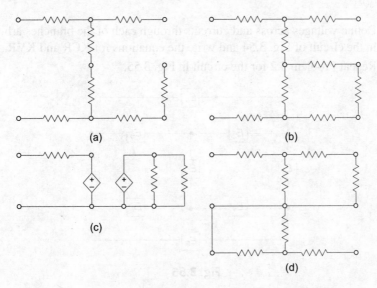

(a)

(b)

(c)

(d)

Fig. 3.52

3.2 Given the circuit in Fig. 3.53, count the number of nodes, meshes, and loops.

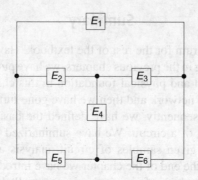

Fig. 3.53

3.3 Define voltages across and currents through each of the branches arbitrarily in the circuit of Fig. 3.53 and write the equations for KCR and KVR.

3.4 Repeat Problem 3.2 for the circuit in Fig. 3.54.

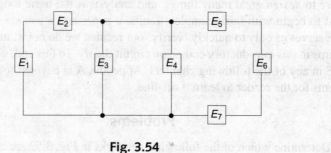

Fig. 3.54

3.5 Define voltages across and currents through each of the branches arbitrarily in the circuit of Fig. 3.54 and write the equations for KCR and KVR.

3.6 Repeat Problem 3.2 for the circuit in Fig. 3.55.

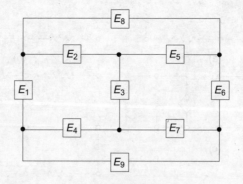

Fig. 3.55

3.7 Define voltages across and currents through each of the branches arbitrarily in the circuit of Fig. 3.55 and write the equations for KCR and KVR.

3.8 Repeat Problem 3.2 for the circuit in Fig. 3.56.

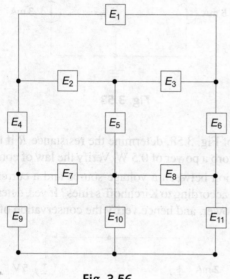

Fig. 3.56

3.9 Define voltages across and currents through each of the branches arbitrarily in the circuit of Fig. 3.56 and write the equations for KCR and KVR.

3.10 Consider the circuit shown in Fig. 3.57. It is found that the 5 V dc source delivers a power of 10 mW. Determine the resistance R. Verify the law of conservation of power.

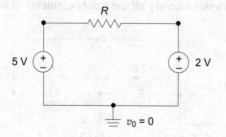

Fig. 3.57

3.11 In the circuit shown in Fig. 3.57, determine the resistance R if the 2 V dc source is found to absorb a power of 0.5 W. Verify the law of conservation of power.

3.12 Determine the resistance R in the circuit of Fig. 3.58, if the 5 mA dc current source is found to deliver 2 mW of power. Determine the power delivered/absorbed by the other current source and hence verify the conservation of power.

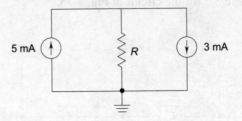

Fig. 3.58

3.13 In the circuit of Fig. 3.58, determine the resistance R if the 3 mA dc source is found to absorb a power of 0.5 W. Verify the law of conservation of power.

3.14 Is the connection between a voltage source and a current source shown in Fig. 3.59 valid according to Kirchhoff's rules? If yes, determine which source is delivering power, and hence verify the conservation of power.

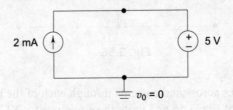

Fig. 3.59

3.15 Consider the circuit in Fig. 3.60. It is found that the element E_4, carrying a current of 3 A from right to left, absorbs a power of 6 W. Verify conservation of power and hence identify all the active elements in the circuit.

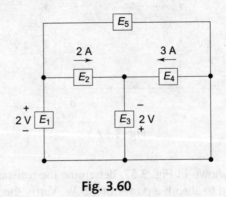

Fig. 3.60

3.16 Repeat Problem 3.15, if it is found that the extreme left is found to release a power of 6 W.

3.17 Given the currents through some of the branches in the circuit of Fig. 3.61, determine the remaining currents.

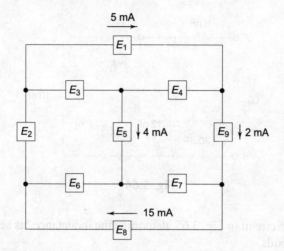

Fig. 3.61

3.18 Given the voltages across some of the branches in the circuit of Fig. 3.62, determine the remaining voltages.

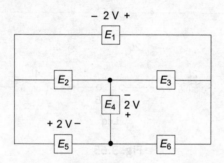

Fig. 3.62

3.19 Given the circuit in Fig. 3.63, determine the resistance 'as seen by' the port AB towards its right.

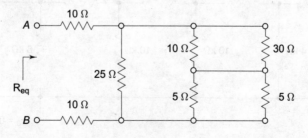

Fig. 3.63

3.20 Given the circuit in Fig. 3.64, determine the capacitance 'as seen by' the port AB towards its right.

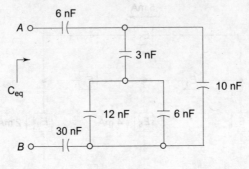

Fig. 3.64

3.21 Given the circuit in Fig. 3.65, determine the inductance 'as seen by' the port AB upwards.

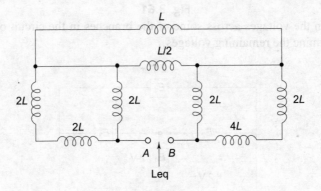

Fig. 3.65

3.22 Given the circuit in Fig. 3.66, determine the currents through all the branches of the circuit.

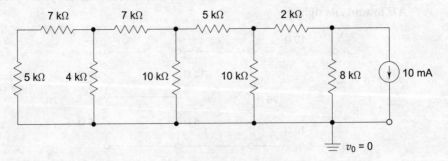

Fig. 3.66

3.23 Given the circuit in Fig. 3.67, determine the power released by the voltage source, and hence the power absorbed by all other resistances in the circuit.

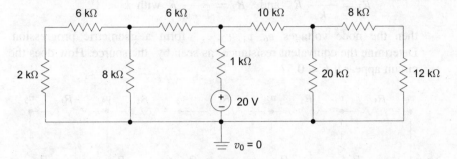

Fig. 3.67

3.24 Determine the node voltages (with respect to the ground) at the nodes v_2 and v_3 in the circuit of Fig. 3.68. What is the equivalent resistance as seen by the source? Notice that there are only four nodes in the circuit and the branches carrying $20\ \Omega$ resistances just cross each other without touching.

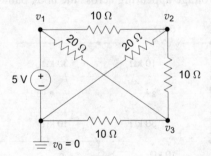

Fig. 3.68

3.25 Using voltage division and/or current division determine the current through each of the resistances in the circuit of Fig. 3.69.

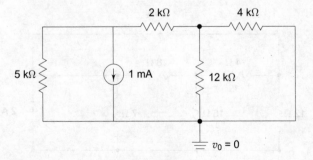

Fig. 3.69

3.26 Consider the ladder circuit in Fig. 3.70. Show that if

$$R_1 = \frac{1-k}{k} R \quad \text{and} \quad R_2 = \frac{1}{1-k} R \quad \text{with } k < 1$$

then the node voltages v_6, v_1, v_2, ... form a geometric progression. Determine the equivalent resistance 'as seen by' the source. How does the circuit appear if $k = 0.5$?

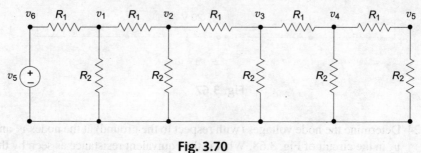

Fig. 3.70

3.27 Consider the bridge circuit of Fig. 3.71. Determine the magnitude and the polarity of the voltage appearing across the node pairs v_1-v_0 and v_2-v_3.

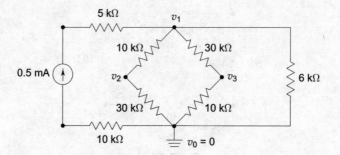

Fig. 3.71

3.28 In the circuit of Fig. 3.72, determine the power absorbed/released by the VCVS.

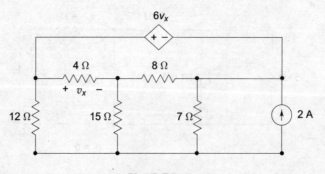

Fig. 3.72

3.29 Determine the equivalent resistance, including the VCCS, as seen by the node pair v_1-v_0 in the circuit of Fig. 3.73. Repeat the problem if v_x is defined across the $2\,k\Omega$ resistance at the bottom of the circuit across the node pair v_4-v_0.

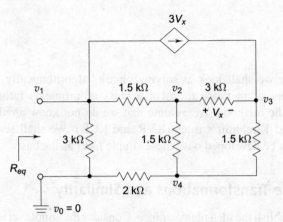

Fig. 3.73

3.30 For each of the circuits from Figs 3.63–3.73, write the SPICE circuit description.

CHAPTER 4

Analysis of Resistive Circuits

In this chapter we shall look at solving circuits algorithmically. We emphasize developing general methods to solve a class of problems rather than several problem-specific methods. We assume that we do not know anything other than Ohm's law and Kirchhoff's rules (KVR and KCR). We shall see that a host of techniques can be developed with these simple *facts* as the basis.

4.1 Source Transformations and Similarity

Let us first establish the idea of *invariance*. Consider the simple series circuit shown in Fig. 4.1.

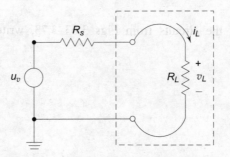

Fig. 4.1

In this figure, u_v stands for a voltage source, R_s is the source resistance in series, and R_L is the load resistance. If v_L is the voltage drop across the load then $i_L = v_L/R_L$ is the current through it. Since this is a series circuit, the same current passes through R_s, and we have from the KVR that

$$-u_v + R_s i_L + v_L = 0 \qquad (4.1)$$

Eliminating i_L, we may rewrite Eqn (4.1) as

$$-\frac{u_v}{R_s} + \left(\frac{1}{R_s} + \frac{1}{R_L}\right) v_L = 0 \qquad (4.2)$$

We quickly observe that we may use Eqn (4.2) to build the circuit shown in Fig. 4.2.

Comparing both the circuits, we notice that the load resistance is in its own place; the voltage source u_v in series with a resistance R_s is *transformed* into a current source $u_i = u_v/R_s$ in parallel with the same source resistance R_s. We leave it to the reader to verify that the voltage drop v_L across the load resistance and the current

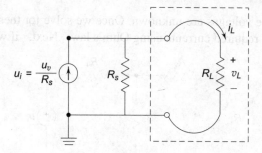

Fig. 4.2

i_L flowing through it are *invariant* under this transformation. In other words, the load resistance cannot distinguish whether it is being driven by a voltage source or a current source; it draws the same power $p_L = v_L i_L$ W. We say that the two circuits are *dual* and *equivalent* to each other. We call this transformation *similarity transformation*, more popularly known as *source transformation*.

It is interesting to derive a few important relations from the above transformation. From the series circuit in Fig. 4.1, we have

$$i_L = \frac{u_v}{R_s + R_L}$$

which can be rewritten as

$$i_L = \left(\frac{u_v}{R_s}\right) \frac{R_s}{R_s + R_L}$$

$$= u_i \frac{R_s}{R_s + R_L} \tag{4.3}$$

Applying Eqn (4.3) to the circuit in Fig. 4.2, we get the so-called *current division rule* we derived in the preceding chapter. Similarly, we get the *voltage division rule*

$$v_L = u_i \times (R_s \parallel R_L)$$

$$= u_v \frac{R_L}{R_s + R_L} \tag{4.4}$$

In the following sections we shall make use of the above ideas.

4.2 Nodal Analysis

In this section, we shall attempt to determine the currents flowing through all the resistances in the circuit shown in Fig. 4.3. We shall look at the circuit from a different perspective and develop an elegant algorithm that does the job. First, let us be assured that we know the current rule (KCR) thoroughly. Looking at the circuit from this point of view, let us quickly identify the nodes and label them v_1, v_2, v_3, and v_0. Arbitrarily we assume that v_0 is the reference node, or ground, whose potential is zero, i.e., $v_0 = 0$ V. Next, let us denote the voltage drop at each of these nodes as v_1, v_2, v_3, respectively, with respect to v_0. We know that $v_3 = u_v$ readily, but the other node voltages are unknown. Once we solve for these unknowns we may compute the required currents using Ohm's law. Next, if we apply KCR at

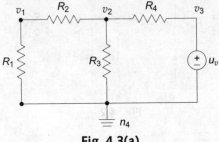

Fig. 4.3(a)

node v_1, assuming that the potential v_1 is higher than all other node potentials, we get that the algebraic sum of the currents *leaving* node v_1 is zero, i.e.,

$$\frac{v_1}{R_1} + \frac{v_1 - v_2}{R_2} = 0$$

Recall that, according to the sign convention, a current in a passive element flows from a node at a higher potential to one at a lower potential. In the above equation, we have assumed that v_1 is higher than v_2 and v_0 and accordingly assumed that the current through R_2 is positive from left to right. Later, upon solving, if we find that v_1 is lower than, say v_2, it actually means that the current through R_2 is from right to left. We are absolutely safe making these conventions.

Next, assuming that v_2 is higher than the other node potentials, let us apply KCR at node v_2 to get

$$\frac{v_2 - v_1}{R_2} + \frac{v_2}{R_3} + \frac{v_2 - v_3}{R_4} = 0$$

We need not apply KCR at node v_3 because we know its potential (with respect to the ground) and the voltage source is in series with R_4. Thus, we have two linear equations[1] in two unknowns. Since v_1 and v_2 are independent of each other, the two equations are consistent and form a linear system of equations (LSE). Let us rewrite these equations below.

$$\left(\frac{1}{R_1} + \frac{1}{R_2}\right) v_1 - \frac{1}{R_2} v_2 = 0 \tag{4.5}$$

$$-\frac{1}{R_2} v_1 + \left(\frac{1}{R_2} + \frac{1}{R_3} + \frac{1}{R_4}\right) v_2 = \frac{1}{R_4} u_v \tag{4.6}$$

Notice that, once again owing to our sign convention, the coefficient of v_1 is positive and that of v_2 is negative in Eqn (4.5). Likewise, the coefficient of v_2 is positive and that of v_1 is negative in Eqn (4.6). Moreover, the negative coefficients in each of these equations are one and the same. Our observation becomes more transparent

---

[1] Why sholud we call these 'linear equations'?

if we write the above equations in the *matrix-vector form* as

$$
\text{LSE:} \quad \left(\begin{array}{cc} \left(\dfrac{1}{R_1} + \dfrac{1}{R_2} \right) & -\dfrac{1}{R_2} \\[2ex] -\dfrac{1}{R_2} & \left(\dfrac{1}{R_2} + \dfrac{1}{R_3} + \dfrac{1}{R_4} \right) \end{array} \right) \left(\begin{array}{c} v_1 \\ v_2 \end{array} \right) = \left(\begin{array}{c} 0 \\ (u_v/R_4) \end{array} \right) \quad (4.7)
$$

or, equivalently, as

$$
\boxed{\mathcal{N}\,\bar{v} = \bar{i}} \quad (4.8)
$$

where we defined the 'vectors'

$$
\bar{v} \triangleq \left(\begin{array}{c} v_1 \\ v_2 \end{array} \right) \quad \text{and} \quad \bar{i} \triangleq \left(\begin{array}{c} 0 \\ (u_v/R_4) \end{array} \right)
$$

We assume that the reader is familiar with linear system of equations in an earlier course on mathemetics. However, we quickly give a summary here so that the reader would get an idea of writing the equations in the matrix-vector form. More details are provided in Appendix C. To make things simpler, we shall use three simultaneous equations in three unknowns to begin with. Consider the following set of equations:

$$
a_{11}x_1 + a_{12}x_2 + a_{13}x_3 = b_1
$$

$$
a_{21}x_1 + a_{22}x_2 + a_{23}x_3 = b_2
$$

$$
a_{31}x_1 + a_{32}x_2 + a_{33}x_3 = b_3
$$

Let us carefully notice the way the equations are written. On the left-hand side, the unknowns are denoted as x_j, $j = 1, 2, 3$ with a subscript j. The coefficients are written with a 'double subscript' a_{ij} with the second index j coinciding with the subscript of the unknown x_j. On the right-hand side are given the known quantities b_i, $i = 1, 2, 3$ with the subscript i which coincides with the first index of the coefficients. Each of the above three equations may be written in terms of the product of a row vector and a column vector as

$$
\begin{bmatrix} a_{i1} & a_{i2} & a_{i3} \end{bmatrix} \begin{bmatrix} x_1 \\ x_2 \\ x_3 \end{bmatrix} = b_i
$$

Since the column vector of the unknowns, denoted hereafter as $\bar{x}$, is common to all the three equations, we may nicely *pack* the three equations and obtain the matrix-vector form as

$$
\begin{bmatrix} a_{ij} \end{bmatrix} \bar{x} = \bar{b}
$$

where the vector $\bar{b}$ denotes the column vector obtained by stacking the known quantities b_i.

We summarize our observations on the matrix-vector Eqn (4.8) as follows:

1. The coefficient matrix $\mathcal{N}$ is a square matrix. We instruct the reader to prepare the circuit description table as a SPICE data structure as outlined in Chapter 3 and compare this matrix.

2. Each of the elements of $\mathcal{N}$ represents a conductance.

3. In particular, the total conductance, called *self-conductance* seen by each of the nodes has a positive sign and is placed along the principal diagonal of the matrix.

4. Each of the off-diagonal elements, $\mathcal{N}_{ij}$, $i \neq j$, represents the *trans-conductance* between the nodes i and j.

5. Clearly, $\mathcal{N}_{ij} = \mathcal{N}_{ji}$ and the matrix is symmetric. We call the matrix $\mathcal{N}$ as the *node incidence matrix*.

Obviously, a unique solution for the above LSE exists if and only if the matrix $\mathcal{N}$ has full rank. This is true as long as the node voltages are independent of each other, and this is assured here since each of the nodes has to obey Kirchhoff's current rule.

Before proceeding further, let us do the following numerical example.

Example 4.1 In the circuit of Fig. 4.3, let

$$R_1 = 1\,\Omega, \quad R_2 = 2\,\Omega, \quad R_3 = R_4 = 3\,\Omega, \quad \text{and} \quad u_v = 9\,\text{V}.$$

Substituting these values in Eqn (4.7), we obtain the solution

$$v_1 = 1\,\text{V} \quad \text{and} \quad v_2 = 3\,\text{V}$$

We notice that v_2 is higher than v_1 but lower than $v_3 = u_v$.

Exercise 4.1 Determine the currents and their directions through each of the resistances in the Example 4.1.

(***Ans.*** The source delivers 2 A. Current through R_1 is 1 A top to bottom, current through R_2 is 1 A right to left, current through R_3 is 1 A top to bottom, and current through R_4 is 1 A right to left)

At this point, an impatient reader may prefer to solve the problem in the following way:

- The voltage source u_v with the series resistance R_4 can be transformed into a current source u_v/R_4 with a parallel resistance R_4.

- We now have a current source and three parallel branches of resistances, namely, R_4, R_3, and $R_1 + R_2$. The circuit may be redrawn as shown in Fig. 4.3(b).

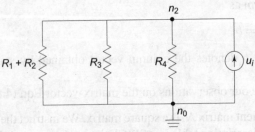

Fig. 4.3(b)

- We may now apply current division rule repeatedly and solve for the currents.

This way, he can challenge the elegance of the above algorithm.

We simply counter this argument[2] by slightly modifying the circuit with a resistance R_5 between nodes v_1 and v_3 as shown in Fig. 4.4 and challenge him to solve it using the above strategy.

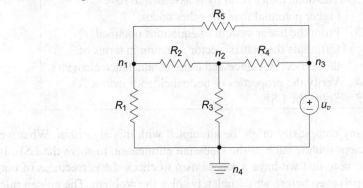

Fig. 4.4

Example 4.2 He would be baffled since there is no way to apply source transformation. We win by showing that a slight modification in Eqn (4.5) given by

$$\left(\frac{1}{R_1} + \frac{1}{R_2} + \frac{1}{R_5}\right) v_1 - \frac{1}{R_2} v_2 = \frac{u_v}{R_5} \tag{4.9}$$

would solve the problem. The modified matrix-vector equation is

LSE:

$$\begin{pmatrix} \left(\dfrac{1}{R_1} + \dfrac{1}{R_2} + \dfrac{1}{R_5}\right) & -\dfrac{1}{R_2} \\[2ex] -\dfrac{1}{R_2} & \left(\dfrac{1}{R_2} + \dfrac{1}{R_3} + \dfrac{1}{R_4}\right) \end{pmatrix} \begin{pmatrix} v_1 \\ v_2 \end{pmatrix} = \begin{pmatrix} \dfrac{u_v}{R_5} \\[2ex] \dfrac{u_v}{R_4} \end{pmatrix} \tag{4.10}$$

The elegance of our solution is its *generalization*, i.e., we have *not* offered two different algorithms to the two problems—the same algorithm works for both the circuits, with the KCR appropriately applied. Our algorithm has the following steps, independent of the circuit topology. Since our method solves for *all* the unknown node potentials at once, we call this *nodal analysis*.

---

[2] Giving 'counter examples' to disprove a claim is a great idea and has to be developed rigorously; unfortunately, this cannot be taught.

Step 1	Enumerate the nodes—0, 1, ... , —systematically. Set n_0 as the reference node and assume $v_0 = 0$. Segregate the nodes whose voltages are to be found.
Step 2	Apply KCR at each unknown node and express the currents according to passive sign convention. The node under scrutiny is assumed to have higher potential than the other nodes.
Step 3	From the linear system of equations obtained, formulate the matrix-vector equation in terms of the self-conductance and trans-conductance elements.
Step 4	Verify the properties of node incidence matrix $\mathcal{N}$.
Step 5	Solve the LSE.

Circuits of any complexity might be attempted with this algorithm. What we need is an efficient routine such as the Gaussian elimination to solve the LSE. It is interesting to note that we have a mechanism to check the correctness of our solution (in step 4) even before we completely solve the problem. The golden rule is that

verification of the results is just as important as the results themselves.

Exercise 4.2 Apply nodal analysis to determine all the branch currents in the circuit shown in Fig. 4.5.

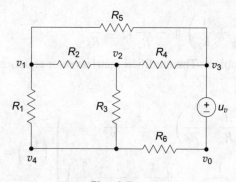

Fig. 4.5

Ans. With the nodes enumerated as shown in Fig. 4.5 and assuming that the voltage source is across n_3 and ground, the LSE to be solved is as follows:

$$\begin{pmatrix} \left(\frac{1}{R_1}+\frac{1}{R_2}+\frac{1}{R_5}\right) & -\frac{1}{R_2} & -\frac{1}{R_1} \\ -\frac{1}{R_2} & \left(\frac{1}{R_2}+\frac{1}{R_3}+\frac{1}{R_4}\right) & -\frac{1}{R_3} \\ -\frac{1}{R_1} & -\frac{1}{R_3} & \left(\frac{1}{R_1}+\frac{1}{R_3}+\frac{1}{R_6}\right) \end{pmatrix} \begin{pmatrix} v_1 \\ v_2 \\ v_4 \end{pmatrix} = \begin{pmatrix} \frac{1}{R_5} \\ \frac{1}{R_4} \\ 0 \end{pmatrix} u_v$$

Let us look at the following examples to learn more about our nodal analysis algorithm.

Example 4.3 Let us apply nodal analysis to the circuit shown in Fig. 4.6.

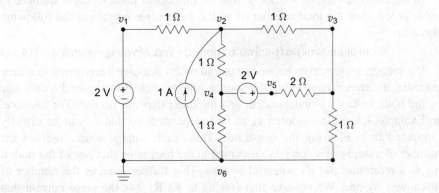

Fig. 4.6

We notice that there are six nodes, out of which one is reference. Further, we have two voltage sources and hence the following *constraints* have to be necessarily satisfied. These constraints represent the dependence of certain unknowns on others.

$$v_1 = 2\,\text{V} \quad \text{and} \quad v_5 - v_4 = 2\,\text{V}$$

Thus, we are left with only three node voltages to be determined. Applying KCR, we get the following equations:

$$\frac{v_2 - 2}{1} + \frac{v_2 - v_3}{1} + \frac{v_2 - v_4}{1} = 1$$

$$\frac{v_3 - v_2}{1} + \frac{v_3}{1} + \frac{v_3 - v_5}{2} = 0$$

$$\frac{v_4 - v_2}{1} + \frac{v_4}{1} + \frac{v_5 - v_3}{2} = 0$$

We observe that in the last equation we have both v_4 and v_5. However, we may readily eliminate v_5 using the constraint that $v_5 - v_4 = 2$ V. (Notice that we may eliminate v_4 instead.) Thus, we are left with three unknowns, namely, v_2, v_3, and v_4. The matrix-vector equation is

LSE:
$$\begin{pmatrix} 3 & -1 & -1 \\ -1 & \dfrac{5}{2} & -\dfrac{1}{2} \\ -1 & -\dfrac{1}{2} & \dfrac{5}{2} \end{pmatrix} \begin{pmatrix} v_2 \\ v_3 \\ v_4 \end{pmatrix} = \begin{pmatrix} 3 \\ 1 \\ -1 \end{pmatrix}$$

and the solution is

$$v_2 = \frac{3}{2} \text{ V}, \quad v_3 = \frac{13}{12} \text{ V}, \quad \text{and} \quad v_4 = \frac{5}{12} \text{ V}$$

In the examples above, we notice that the number of node voltages we need to solve is less than the total number of nodes. In fact, we might use the following formula:

$$(\text{No. of unknowns}) = (\text{total no. of nodes}) - (\text{no. of voltage sources}) - 1 \quad (4.11)$$

If a voltage source exists between a pair of nodes, we have a *constraint* that their potential difference must equal the source voltage itself. Consequently, only one of the node voltages is independent and the other depends on this. For instance, in Example 4.3, we considered v_4 to be independent so that v_5 can be directly computed from v_4 using the constraint. Thus, each voltage source reduces the number of unknowns by one. We assumed in the beginning that one of the nodes, n_0, is a reference and its potential is zero. This further reduces the number of unknowns by one. We observe that (owing to KCR) it is the same current that leaves the nodes on either side of the voltage source; this current is indirectly established by the surrounding network and, as such, cannot be expressed in terms of the source voltage itself. We, therefore, simply regard the source and its two nodes as a *super node*. The resulting equation shall have the voltages of both the nodes, one of which can be eliminated by making use of the constraint available to us.

Example 4.4 Let us consider the circuit shown in Fig. 4.7.

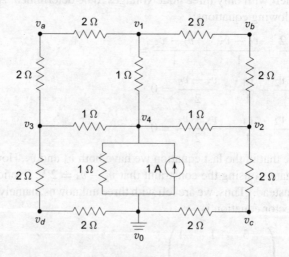

Fig. 4.7

Apparently there are nine nodes and hence eight unknowns. Is this circuit really that big? A closer inspection would reveal certain series combinations of resistances and it is necessary and sufficient to solve for only four independent voltages—corresponding to the nodes v_1, v_2, v_3, and v_4—enumerated in the figure.

The remaining four voltages can be obtained simply by applying the voltage division rule. The four equations are

$$
\text{LSE:} \quad
\begin{pmatrix}
\dfrac{3}{2} & -\dfrac{1}{4} & -\dfrac{1}{4} & -1 \\[2mm]
-\dfrac{1}{4} & \dfrac{3}{2} & 0 & -1 \\[2mm]
-\dfrac{1}{4} & 0 & \dfrac{3}{2} & -1 \\[2mm]
-1 & -1 & -1 & 4
\end{pmatrix}
\begin{pmatrix}
v_1 \\ v_2 \\ v_3 \\ v_4
\end{pmatrix}
=
\begin{pmatrix}
0 \\ 0 \\ 0 \\ 1
\end{pmatrix}
$$

After all, we are interested in solving a problem. An elegant strategy is to *reduce* the problem to a simpler one and then solve it. In this process we may use different tools available to us. However, keeping track of the reductions is equally essential.

Exercise 4.3 Complete Example 4.4 and determine all the branch voltages and branch currents. Verify your results.

(**Ans.** $v_1 = 2/3$; $v_2 = v_3 = 7/12$; $v_4 = 17/24$ V)

Exercise 4.4 Obtain the node voltages for the circuit shown in Fig. 4.8.

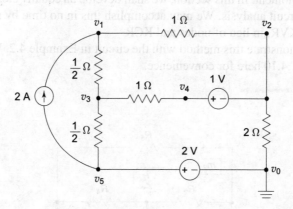

Fig. 4.8

(**Ans.** Refer to Example 4.6 and Exercise 4.8)

Exercise 4.5 Analyse the circuit shown in Fig. 4.9 for its node voltages.
(**Ans.** From the figure, $v_5 = 0$ V, $v_6 = v_1 - 3$ V, and $v_7 = v_4 + 1$ V. The remaining node voltages are $v_1 = 31/8$, $v_2 = 49/24$, $v_3 = 65/24$, and $v_4 = 5/4$)

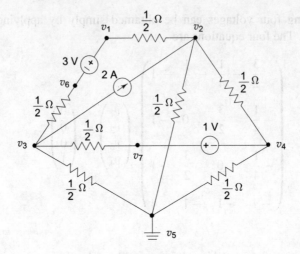

Fig. 4.9

We end this section emphasizing that it is extremely important to label the nodes appropriately before we start analysing the circuit. In writing the KCR, the node voltages have to be used in a manner that is consistent with the sign convention.

4.3 Mesh Analysis

We have noticed some sort of similarity between KCR and KVR, though it is a bit vague at the moment. In this section, we shall develop an equally elegant algorithm to perform circuit analysis. We may accomplish this in no time by making use of meshes and KVR, in lieu of nodes and KCR.

Let us demonstrate this method with the circuit in Example 4.2. We redraw the circuit as Fig. 4.10 here for convenience.

Example 4.5

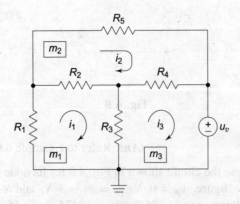

Fig. 4.10

We identify that there are three meshes in the circuit. Let us label them m_1, m_2, and m_3 as shown in Fig. 4.10. As far as a mesh is concerned, all the elements are connected in series. Hence, let us identify three currents, called *mesh currents* i_1, i_2, and i_3 flowing in their respective meshes. Further, let us arbitrarily assign the direction of these currents as clockwise, i.e., we traverse each of the meshes in the clockwise direction. In so doing, we follow the direction of the current in that particular mesh and assign the polarities across the elements. Next, we apply KVR for each of the meshes. For mesh 1, we have

$$R_1 i_1 + R_2(i_1 - i_2) + R_3(i_1 - i_3) = 0 \tag{4.12}$$

Notice that a branch of the network belongs to one or more meshes. Accordingly, the current through that branch is an algebraic sum (taking into account the directions) of the mesh currents; for instance, while applying KVR for mesh 1 in the above circuit, the current i_1 flowing in the clockwise direction is countered by the current i_2 flowing in the opposite direction through the branch R_2 and hence the net current through it is $i_1 - i_2$. Owing to the same reason, we get the current through R_3 as $i_1 - i_3$. Similarly, applying KVR for mesh 2, we get

$$R_5 i_2 + R_4(i_2 - i_3) + R_2(i_2 - i_1) = 0 \tag{4.13}$$

and for mesh 3 we get

$$R_3(i_3 - i_1) + R_4(i_3 - i_2) + u_v = 0 \tag{4.14}$$

We leave it to the reader to verify that these three equations can be arranged in the matrix-vector form as

$$\text{LSE:} \quad \begin{pmatrix} (R_1 + R_2 + R_3) & -R_2 & -R_3 \\ -R_2 & (R_2 + R_4 + R_5) & -R_4 \\ -R_3 & -R_4 & (R_3 + R_4) \end{pmatrix} \begin{pmatrix} i_1 \\ i_2 \\ i_3 \end{pmatrix} = \begin{pmatrix} 0 \\ 0 \\ u_v \end{pmatrix} \tag{4.15}$$

i.e.,

$$\boxed{M\bar{i} = \bar{v}} \tag{4.16}$$

Exercise 4.6 Verify that Eqns (4.12), (4.13), and (4.14) can be put in the form of Eqn (4.15).

One might easily note that the matrix M is *dual* to the node incidence matrix N discussed in the preceding section and hence might be called *mesh incidence matrix*. Its properties are similar to those of N. Since this method analyses the circuit in terms of its meshes, we call this *mesh analysis*. The algorithm is given below. Clearly, we get this algorithm from nodal analysis by substituting mesh for node, current for voltage, and voltage for current.

> **Step 1** Enumerate the meshes systematically. Segregate the meshes whose currents are unknown and examine the constraints.
>
> **Step 2** Apply KVR for each unknown mesh and express the voltages according to passive sign convention. The mesh currents are considered positive in the clockwise direction.
>
> **Step 3** Formulate the matrix-vector equation in terms of the self-resistance and trans-resistance terms.
>
> **Step 4** Verify the properties of mesh incidence matrix $\mathcal{M}$.
>
> **Step 5** Solve the LSE.

Exercise 4.7 Apply this algorithm to the circuit in Exercise 4.2 and verify your results.

Example 4.6 Let us analyse the circuit in Exercise 4.4 for its meshes. We redraw the circuit in Fig. 4.11 for convenience.

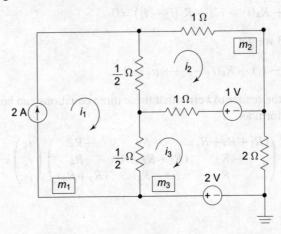

Fig. 4.11

We label the three meshes as shown in Fig. 4.11. We readily have

$$i_1 = 2\,\text{A}$$

For the other two meshes, we get

$$\frac{1}{2}(i_2 - i_1) + 1i_2 - 1 + 1(i_2 - i_3) = 0$$

$$\frac{1}{2}(i_3 - i_1) + 1(i_3 - i_2) + 1 + 2i_3 - 2 = 0$$

We put them in the matrix-vector form as follows:

$$\text{LSE:} \quad \begin{pmatrix} 2.5 & -1 \\ -1 & 3.5 \end{pmatrix} \begin{pmatrix} i_2 \\ i_3 \end{pmatrix} = \begin{pmatrix} 2 \\ 2 \end{pmatrix}$$

Exercise 4.8 Solve this LSE and compare the results with those of Exercise 4.4.

(*Ans.* $i_2 = 36/31$; $i_3 = 28/31$)

Example 4.7 Consider the circuit shown in Fig. 4.12.

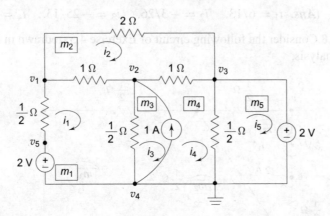

Fig. 4.12

The meshes may be numbered as shown in the figure. The currents are assumed to flow in the clockwise direction in each of the meshes. The number of unknowns may be obtained using the following formula:

$$\text{(No. of unknowns)} = \text{(total no. of meshes)} - \text{(no. of current sources)} \quad (4.17)$$

We have the constraint that

$$i_4 - i_3 = 1 \text{ A}$$

and we use this to eliminate i_3 in the following. The mesh equations are

$$-2 + \frac{1}{2}i_1 + 1(i_1 - i_2) + \frac{1}{2}(i_1 - i_3) = 0$$

$$1(i_2 - i_1) + 2i_2 + 1(i_2 - i_1) = 0$$

$$\frac{1}{2}(i_3 - i_1) + 1(i_4 - i_2) + \frac{1}{2}(i_4 - i_5) = 0$$

$$\frac{1}{2}(i_5 - i_4) + 2 = 0$$

We notice that the voltage across the current source of 1 A is equal to the voltage across the $\frac{1}{2}\,\Omega$ resistance in parallel to the source. Further, we have the constraint relating i_3 and i_4 as mentioned above and hence we have a *super mesh*, similar to the super node, and consequently, the third equation has both i_3 and i_4. We eliminate i_3 using the constraint and obtain the following matrix-vector equation:

$$\text{LSE:} \quad \begin{pmatrix} 2 & -1 & -0.5 & 0 \\ -1 & 4 & -1 & 0 \\ -0.5 & -1 & 2 & -0.5 \\ 0 & 0 & -0.5 & 0.5 \end{pmatrix} \begin{pmatrix} i_1 \\ i_2 \\ i_4 \\ i_5 \end{pmatrix} = \begin{pmatrix} 2 \\ 1 \\ -1.5 \\ -1.5 \end{pmatrix}$$

An ardent reader would notice that there is a lesser number of unknown node voltages than the mesh currents in this circuit.

Exercise 4.9 Solve the above LSE and verify the results using nodal analysis.

$$(\textbf{\textit{Ans.}}\ i_1 = 6/13, \quad i_2 = -3/26, \quad i_3 = -25/13, \quad i_4 = -64/13)$$

Example 4.8 Consider the following circuit of Exercise 4.5 (redrawn in Fig. 4.13) for mesh analysis.

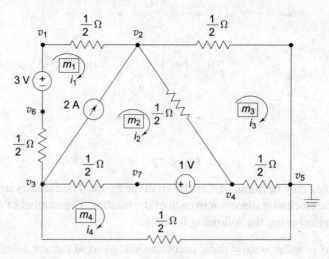

Fig. 4.13

With the currents as shown in the figure, we have the constraint that

$$i_2 - i_1 = 2\ \text{A}$$

Applying KVR results in the following matrix-vector equation:

$$
\begin{pmatrix}
2 & -\dfrac{1}{2} & -\dfrac{1}{2} \\[2mm]
-\dfrac{1}{2} & \dfrac{3}{2} & -\dfrac{1}{2} \\[2mm]
-\dfrac{1}{2} & -\dfrac{1}{2} & \dfrac{3}{2}
\end{pmatrix}
\begin{pmatrix}
i_2 \\ i_3 \\ i_4
\end{pmatrix}
=
\begin{pmatrix}
6 \\ 0 \\ -1
\end{pmatrix}
$$

Solving, we get

$$i_1 = \frac{5}{3}, \quad i_2 = \frac{11}{3}, \quad i_3 = \frac{19}{12}, \quad \text{and } i_4 = \frac{13}{12}$$

Exercise 4.10 Analyse the circuit in Example 4.3 (Fig. 4.6) using mesh analysis.

Exercise 4.11 Analyse the circuit given in Fig. 4.14 using mesh analysis and verify your results using nodal analysis.

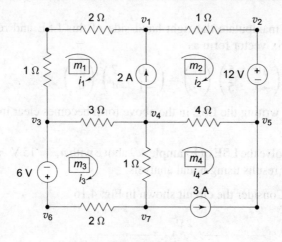

Fig. 4.14

Example 4.9 Let us attempt to analyse the circuit in Fig. 4.15.

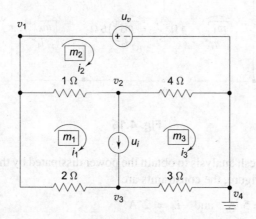

Fig. 4.15

If the meshes are labeled as shown in the figure, we have the constraint that

$$i_1 - i_3 = u_i$$

We shall eliminate i_3 using this constraint. The two equations are

$$2i_1 + 1(i_1 - i_2) + 4(i_3 - i_2) + 3i_3 = 0$$

$$u_v + 4(i_2 - i_3) + 1(i_2 - i_1) = 0$$

The resulting LSE is

$$\begin{pmatrix} 10 & -5 \\ -5 & 5 \end{pmatrix} \begin{pmatrix} i_1 \\ i_2 \end{pmatrix} = \begin{pmatrix} 7u_i \\ -1u_v - 4u_i \end{pmatrix}$$

Let us slightly manipulate the right-hand side of this LSE and rewrite it in an equivalent matrix-vector form as

$$\text{LSE:} \quad \begin{pmatrix} 10 & -5 \\ -5 & 5 \end{pmatrix} \begin{pmatrix} i_1 \\ i_2 \end{pmatrix} = \begin{pmatrix} 0 & 7 \\ -1 & -4 \end{pmatrix} \begin{pmatrix} u_v \\ u_i \end{pmatrix}$$

The purpose of writing the LSE in the above form becomes clear in the following section.

Exercise 4.12 Solve the LSE of Example 4.9 above with $u_v = 15$ V and $u_i = 5$ A and verify your results using nodal analysis.

Example 4.10 Consider the circuit shown in Fig. 4.16.

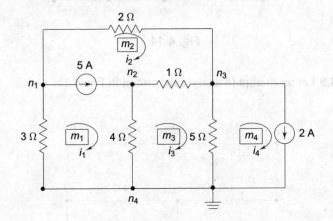

Fig. 4.16

Let us apply mesh analysis to obtain the power dissipated by the 4 Ω resistance. As shown in the figure, the constraints are

$$i_1 - i_2 = 5 \text{ A} \quad \text{and} \quad i_4 = 2 \text{ A}$$

In terms of the desired unknowns i_1 and i_3 the mesh equations are

$$\begin{pmatrix} 10 & -5 \\ -5 & 10 \end{pmatrix} \begin{pmatrix} i_1 \\ i_3 \end{pmatrix} = \begin{pmatrix} 15 \\ 5 \end{pmatrix}$$

Upon solving, we get

$$i_1 = \frac{7}{3} \quad \text{and} \quad i_3 = \frac{5}{3}$$

and the power dissipated by the 4 Ω resistance is

$$P_4 = (i_1 - i_3)^2 \times 4 = \frac{16}{9} \text{ W}$$

Exercise 4.13 Verify the result of Example 4.10 using nodal analysis.

4.3.1   Which Algorithm is Better?

An ardent reader would have observed that the circuit in Fig. 4.4 has two unknowns and hence two equations, but the same circuit in Fig. 4.10 has three equations. We need to provide a rationale for the choice of an algorithm and we do so with the following argument.

With Ohm's law and Kirchhoff's rules as our basic tools, we eye on some common features among the networks we begin to analyse. Obviously, every network has a certain number of nodes and meshes where we can deploy our tools. We devise a particular strategy depending on factors such as (i) the number of variables—the smaller the better, (ii) obtaining the results as directly as possible, (iii) objective of the analysis—one current or several voltages, and the like. For simpler circuits even repeated application of source transformations (Example 4.1) might be sufficient. We generally prefer nodal analysis for a more complex network (as in the examples and exercises above) if we are solving for a branch voltage or when the number of unknown node potentials is smaller than that of the unknown mesh currents. A similar reason is applicable for the choice of mesh analysis. As a general rule, we need to become proficient in all these algorithms and many more that follow; at times we may adopt a combination of two or more algorithms, such as a source transformation followed by mesh analysis or vice versa. In fact, it would be a good practice to solve the problem using one algorithm and verify the results using an alternative.

The elegance of an algorithm stems from its *generality*, i.e., its applicability to a broader class of complex networks rather than a particular problem. In the following sections we shall see few more elegant algorithms.

4.4   Principle of Linearity

In Chapter 2, we have emphasized that the input–output relationship of a resistor is *linear*. In this section we shall explore this much deeper.

Suppose we have the circuit shown in Fig. 4.17.

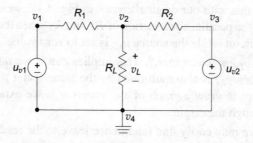

Fig. 4.17

Let us assume that we are interested in the voltage drop v_L across the resistance R_L. Applying nodal analysis we get

$$\frac{v_L - u_{v1}}{R_1} + \frac{v_L}{R_L} + \frac{v_L - u_{v2}}{R_2} = 0$$

from which we may solve for v_L as

$$v_L = \underbrace{\frac{R_2 R_L}{R_1 R_2 + R_2 R_L + R_L R_1} u_{v1}}_{v_{L_1}} + \underbrace{\frac{R_1 R_L}{R_1 R_2 + R_2 R_L + R_L R_1} u_{v2}}_{v_{L_2}} \quad (4.18)$$

Dissecting Eqn (4.18) gives us the following information.

1. The node voltage v_L has two terms v_{L_1} and v_{L_2} as shown on the right-hand side of Eqn (4.18). We shall write Eqn (4.18) as

    $$v_L = \alpha_1 u_{v1} + \alpha_2 u_{v2} \quad (4.19)$$

    where α_1 and α_2 are the *constant* coefficients of u_{v1} and u_{v2}, respectively, from Eqn (4.18).

2. Let us first rewrite $\alpha_1 u_{v1}$ in a more insightful form as

    $$v_{L_1} = \frac{\dfrac{R_2 R_L}{R_2 + R_L}}{R_1 + \dfrac{R_2 R_L}{R_2 + R_L}} u_{v1}$$

    This equation allows us to draw the circuit in Fig. 4.18.

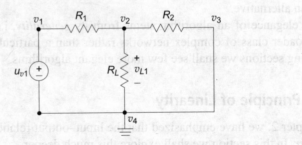

Fig. 4.18

Comparing this with our original circuit in Fig. 4.17, we see that v_{L_1} is the drop across the parallel combination of R_L and R_2 when the source u_{v1} alone drives the circuit while the source u_{v2} is set to zero volts.

3. Notice that v_{L_1} is *linear* in u_{v1}. This implies that if we multiply u_{v1} with a scalar β_1, then v_{L_1} is also multiplied by the same scalar β_1. In other words, if we attempt to draw a graph of v_{L_1} versus u_{v1}, we expect a straight line passing through the origin.

4. Similarly, we may easily find (and hence leave to the reader as an exercise) that v_{L_2} is the drop across the parallel combination of R_L and R_1 assuming that u_{v2} is the only source in the circuit; v_{L_2} is *linear* in u_{v2}.

5. Since $v_L = v_{L_1} + v_{L_2}$, the node voltage v_L may be appropriately called the *linear combination* of the source voltages u_{v1} and u_{v2}. In fact, this is what we meant in writing Eqn (4.19).

6. If u_{v1} is scaled by β_1 and u_{v2} by β_2, it is easy to observe that

    $$v_L = \beta_1 v_{L_1} + \beta_2 v_{L_2}$$

Let us now recall the results of Example 4.9. We reproduce them below for convenience.

$$\begin{pmatrix} 10 & -5 \\ -5 & 5 \end{pmatrix} \begin{pmatrix} i_1 \\ i_2 \end{pmatrix} = \begin{pmatrix} 0 & 7 \\ -1 & -4 \end{pmatrix} \begin{pmatrix} u_v \\ u_i \end{pmatrix}$$

or, equivalently,

$$\begin{pmatrix} i_1 \\ i_2 \end{pmatrix} = \begin{pmatrix} -\dfrac{1}{5} & \dfrac{3}{5} \\ -\dfrac{2}{5} & -\dfrac{1}{5} \end{pmatrix} \begin{pmatrix} u_v \\ u_i \end{pmatrix} \tag{4.20}$$

Here too, we notice that each of the currents i_1 and i_2 may be expressed in the form

$$i_1 = H_{11}\, u_v + H_{12}\, u_i$$

and

$$i_2 = H_{21}\, u_v + H_{22}\, u_i$$

i.e., each of the currents is a linear combination of the sources. Let us define *linearity* formally.

Definition 4.1(a) Homogeneity

If u_j is the only source in the circuit and d is the only output we are interested in, we write this as

$$u_j \mapsto d$$

We say u_j *causes* d. Consequently, if u_j is multiplied by a scalar β_j so is d, i.e.,

$$\beta_j u_j \mapsto \beta_j d$$

This is called *homogeneity*, or *scaling*.

Definition 4.1(b) Additivity

If there are several sources $u_1, u_2, \ldots, u_r$ in the circuit and $d_1, d_2, \ldots, d_r$ are the corresponding outputs, i.e.,

$$u_j \mapsto d_j \quad j = 1, 2, \ldots, r$$

then the output is

$$d = \sum_{j=1}^{r} d_j$$

This is called *additivity*.

Definition 4.2 Linearity

We may combine the properties of homogeneity and additivity and obtain an expression for the output as

$$\sum_{j=1}^{r} \beta_j u_j \mapsto d \tag{4.21}$$

This is called the principle of *linearity*, or *superposition*.

We may extend this principle to the case where there are multiple outputs as in the case of Example 4.9. Let us now introduce some more notation. Suppose that a circuit has r number of sources $u_1, u_2, \ldots, u_r$. We define a vector $\overline{u}$ called the *input vector*:

$$\overline{u} \triangleq \begin{bmatrix} u_1 \\ u_2 \\ \vdots \\ u_r \end{bmatrix}_{r \times 1} \tag{4.22}$$

Likewise, suppose that we are interested in p number of quantities $d_1, d_2, \ldots, d_p$, call them outputs, in the circuit. We define another vector $\overline{d}$ called the *output vector*:

$$\overline{d} \triangleq \begin{bmatrix} d_1 \\ d_2 \\ \vdots \\ d_p \end{bmatrix}_{p \times 1} \tag{4.23}$$

In general, we use symbols such as $\vec{u}$ or $\overline{u}$ to denote vectors, particularly in physics. However, in this text, we prefer to avoid such accents since we understand that a single source u is also a vector of size 1×1.

Using the principle of linearity, we get

$$d_j = \sum_{k=1}^{r} H_{jk} u_k, \quad j = 1, 2, \ldots, p \tag{4.24}$$

for some coefficients H_{jk}. In terms of the vectors u and d, we may rewrite equations of form (4.24) as

$$d = \mathcal{H} u \tag{4.25}$$

where $\mathcal{H}$ is a $p \times r$ matrix that contains the coefficients. We call this matrix *transfer function*. A formal definition is given in a later chapter. Notice that the vectors u and d need not be homogeneous, i.e., the u is generally a heterogeneous collection of voltage and current sources as in Example 4.9 and d is generally a collection of node voltages and branch currents. We say it is a *linear network* if and only if the relationship between its input u and its output d is linear as shown in Eqn (4.24). Let us do a few examples to solidify our ideas.

Example 4.11 Let us consider the circuit shown in Fig. 4.17 with

$$R_1 = R_2 = R_L = 2\,\Omega$$

and make use of Eqn (4.18). Let us assume that the source $u_{v2} = 0\,\text{V}$ initially. Consequently,

$$d = \frac{1}{3} u_{v1}$$

If we plot d versus u_{v1}, we get a straight line passing through the origin with slope 1/3 as shown in Fig. 4.19.

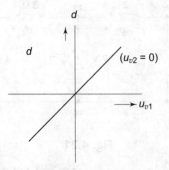

Fig. 4.19

If we introduce the source $u_{v2} = 3\,\mathrm{V}$, we get

$$d = \frac{1}{3}u_{v1} + 1$$

and the graph of d versus u_{v1} is once again a straight line, but now with an intercept at $+1$ on the d axis. If we vary u_{v2}, we get a *family of straight lines* as shown in Fig. 4.20.

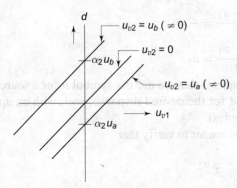

Fig. 4.20

Each of the straight lines passing through the origin represents 'homogeneity', and the intercept along the d axis represents 'additivity'. Since the slope of this family of straight lines is constant, we represent this as

$$\Delta d \propto \Delta u_{v1} \quad \text{with constant } u_{v2}$$

and

$$\Delta d \propto \Delta u_{v2} \quad \text{with constant } u_{v1}$$

Exercise 4.14 For $u_{v1} = 7\,\mathrm{V}$ and $u_{v2} = -4.5\,\mathrm{V}$ explain how you would obtain d from Fig. 4.20.

Example 4.12 Let us consider the circuit shown in Fig. 4.21.

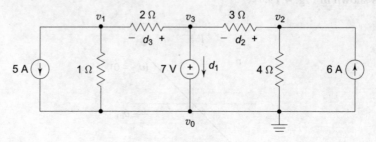

Fig. 4.21

We notice that there are three sources. Let us label them as follows:

$$u_1 = 7\,\text{V voltage source}$$

$$u_2 = 5\,\text{A current source}$$

and

$$u_3 = 6\,\text{A current source}$$

Suppose that we are interested in d_1, the current through the 7 V voltage source from top to bottom, and d_2, the voltage across the 3 Ω resistance from right to left as shown in the figure. Let us apply nodal analysis to get

$$\frac{v_1}{1} + \frac{v_1 - u_1}{2} = -u_2$$

$$\frac{v_2}{4} + \frac{v_2 - u_1}{3} = u_3$$

Recall from Chapter 3 that we use the symbol u for a source, the symbol v for node voltage, and d for the desired response, each with an appropriate subscript according to the context.

We leave it to the reader to verify that

$$v_1 = \frac{1}{3}u_1 - \frac{2}{3}u_2$$

and

$$v_2 = \frac{4}{7}u_1 + \frac{12}{7}u_3$$

From the circuit we get

$$d_1 = \frac{v_1 - u_1}{2} + \frac{v_2 - u_1}{3}$$

$$= \frac{1}{7}u_1 - \frac{1}{3}u_2 + \frac{4}{7}u_3$$

and

$$d_2 = v_2 - v_3$$

$$= \frac{5}{21}u_1 + \frac{2}{3}u_2 + \frac{12}{7}u_3$$

Observe that the influence of the source $u_3 = 6$ A is the greatest on the desired outputs d_1 and d_2.

Exercise 4.15 Obtain the transfer function in Example 4.12 with the numerical values and verify the results using mesh analysis.

$$\textbf{\textit{Ans.}} \quad \begin{pmatrix} d_1 \\ d_2 \end{pmatrix} = \frac{1}{21} \begin{pmatrix} 3 & -7 & 12 \\ 5 & 14 & 36 \end{pmatrix} \begin{pmatrix} u_1 \\ u_2 \\ u_3 \end{pmatrix}$$

So far, we have observed that we may write the response of the circuit having several sources as

$$d_j = \sum_{k=1}^{r} H_{jk} u_k, \quad j = 1, 2, \ldots, p$$

i.e., each of the p outputs is some linear combination of all the r independent sources. We will use the converse of this statement and develop yet another elegant technique. The key is the independence of the sources. We quickly count that the output d is a sum of r terms where r is the number of sources. This suggests us that we may compute partial responses due to each of the r sources using linearity and then superpose (i.e., add up) these partial responses to get the complete response. We consider Example 4.12 to illustrate this.

Example 4.13 For the circuit in Example 4.12, suppose that we are interested in d_3, the current through the 2 Ω resistance from right to left.

1. Suppose $u_1 = 6$ V voltage source alone is present in the circuit and $u_2 = u_3 = 0$, meaning that the respective branches were *open-circuited*.
2. Here the 2 Ω resistance is in series with the 1 Ω resistance, and this series combination is across u_1. We get

$$d_3 = \frac{1}{3} u_1 \text{ A from right to left}$$

due to u_1 acting alone in the circuit. Let us call this d_3^1, the superscript denoting the subscript of the source in action.
3. Suppose, now, that u_2 alone is acting in the circuit and $u_1 = u_3 = 0$, meaning that the branch in which the voltage source was present is *short-circuited* and the branch in the which the current source was present is *open-circuited*.
4. Using the current division rule we quickly get that

$$d_3^2 = \frac{1}{3} u_2 \text{ A from right to left}$$

5. Similarly, assuming that u_3 alone is acting in the circuit, we get

$$d_3^3 = 0 \cdot u_3 = 0 \text{ A}$$

6. The total current d_3 flowing through the $2\,\Omega$ resistance from right to left is thus

$$d_3 = d_3^1 + d_3^2 + d_3^3$$
$$= \frac{1}{3}u_1 + \frac{1}{3}u_2$$
$$= 4\,\text{A}$$

We verify this result from Example 4.12 as

$$d_3 = \frac{u_1 - v_1}{2}$$
$$= \frac{1}{3}u_1 + \frac{1}{3}u_2$$

Let us now clarify an important point.

- When we say a source is *set to zero* we mean that it is *not* pumping any energy into the circuit. This happens if we 'see' a short circuit (resistance $= 0\,\Omega$) in the place of a voltage source so that the voltage across the port is $0\,\text{V}$, and an open circuit (resistance $= \infty\,\Omega$) in the place of a current source so that the current through the branch is $0\,\text{A}$.

We are now ready to write the algorithm. This is essentially a 'divide and conquer' approach exploiting the linearity of the input–output relationship. Accordingly, we call this algorithm the *principle of linearity*, or equivalently, *principle of superposition*.

Step 1	Enumerate the sources systematically as $u_1, u_2, \ldots, u_r$.
Step 2	Let the desired output be the vector d. For $j = 1$ to r do steps 3 and 4.
Step 3	Assume that the source u_j is acting alone while all other sources are replaced either by short circuits or by open circuits.
Step 4	Apply source transformations, or nodal analysis, or mesh analysis and obtain the desired response d^j due to u_j.
Step 5	$d = \sum_{j=1}^{r} d^j$.

Example 4.14 Consider the circuit shown in Fig. 4.22.

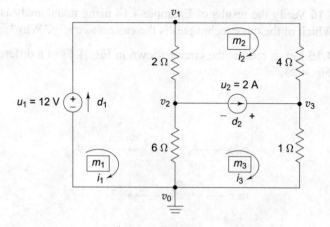

Fig. 4.22

Let us determine the power delivered (or absorbed) by the two independent sources. For this, we need d_1, the current delivered by the voltage source and d_2, the voltage across the current source, i.e., $v_3 - v_2$. We shall apply the principle of superposition.

Assuming that the circuit has the voltage source u_1 alone, we easily get

$$d_1^1 = \frac{13}{40}u_1 = 3.9\,\text{A}$$

Using voltage division rule, we get

$$d_2^1 = v_3^1 - v_2^1$$
$$= \frac{1}{1+4}u_1 - \frac{6}{6+2}u_2$$
$$= -6.6\,\text{V}$$

Next, assuming that the current source u_2 alone is acting in the circuit, we get a circuit with three meshes (including a super mesh). Assigning the currents as shown in the figure, we get

$$\begin{pmatrix} 8 & -8 \\ -8 & 13 \end{pmatrix}\begin{pmatrix} i_1^2 \\ i_3^2 \end{pmatrix} = \begin{pmatrix} -4 \\ 10 \end{pmatrix}$$

Solving for the desired quantities, we get

$$d_1^2 = 1.1 \text{ A} \quad \text{and} \quad d_2^2 = 4.6 \text{ V}$$

Thus,

$$d_1 = 5 \text{ A} \quad \text{and} \quad d_2 = -2 \text{ V}$$

and the source u_1 *delivers* a power of 60 W while the source u_2 *absorbs* a power of 4 W. In the latter case we say the power is absorbed because the potential difference across the current source is built against the sign convention.

Exercise 4.16 Verify the results of Example 4.14 using nodal analysis and mesh analysis. Which of the three techniques is the easiest to apply? Why?

Example 4.15 Let us redraw the circuit shown in Fig. 4.17 in a different way, as shown in Fig. 4.23.

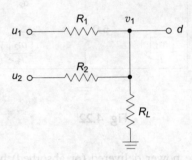

Fig. 4.23

The sources u_1 and u_2 are voltage sources with their positive terminals shown connecting to R_1 and R_2, respectively. Although it is not shown explicitly, we may understand that the negative terminals of the sources are connected to the ground. As we have already seen, the output d is the linear combination of u_1 and u_2. This topology suggests us that we might 'add' several more voltage sources in series with resistance to node n_1 and obtain the output as a linear combination of all these sources. For instance, for the case of four sources u_1, u_2, u_3, and u_4 and corresponding resistances R_1, R_2, R_3, and R_4, the output d may be readily written as

$$d = \frac{R_2 \| R_3 \| R_4 \| R_L}{R_1 + (R_2 \| R_3 \| R_4 \| R_L)} u_1$$

$$= + \frac{R_1 \| R_3 \| R_4 \| R_L}{R_2 + (R_1 \| R_3 \| R_4 \| R_L)} u_2$$

$$= + \frac{R_1 \| R_2 \| R_4 \| R_L}{R_3 + (R_1 \| R_2 \| R_4 \| R_L)} u_3$$

$$= + \frac{R_1 \| R_2 \| R_3 \| R_L}{R_4 + (R_1 \| R_2 \| R_3 \| R_L)} u_4$$

as a generalization of Eqn (4.18). We encourage the reader to focus on the nice pattern that appears in this.

The resistance R_L plays a crucial role. It cannot be $0\ \Omega$ because the output would become zero. However, it can be arbitrarily large and still we get the linear combination of the sources as an output. This topology is called *voltage summer*. We shall return back to this circuit in a later chapter on operational amplifiers.

Exercise 4.17 Using the results of Example 4.15, synthesize a network that has n voltage sources u_1, u_2, ... , u_n and an output

$$d = \frac{1}{m+n} \sum_{j=1}^{n} u_j$$

How do you include the parameter 'm' in the circuit?

(***Hint*** Use the voltage division rule repeatedly.)

Example 4.16 Consider the circuit shown in Fig. 4.24.

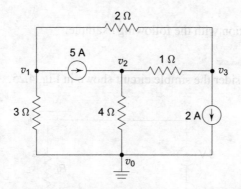

Fig. 4.24

We are interested in the power dissipated by the $4\,\Omega$ resistance. Let us label the nodes as shown in the figure. Further, let us call the 5 A source as u_1 and the 2 A source as u_2. Applying the principle of linearity we get the voltage d across the $4\,\Omega$ resistance as

$$d = d^1 + d^2$$
$$= (6) + (-4) = 2\,\text{V}$$

and a current of 0.5 A flows through it from top to bottom. The power dissipated is

$$p = \frac{(2)^2}{4} = 1\,\text{W}$$

We emphasize that the net voltage (or equivalently, the net current) through the resistance due to all the sources has to be found out first. Then we should attempt to find the power. In other words,

$$p = \frac{(d^1 + d^2)^2}{R} \neq \frac{(d^1)^2 + (d^2)^2}{R}$$

Notice the order in which we used the parentheses. The relationship between voltage and voltage, current and voltage, or current and current involves only the first degree of the quantity, whereas the relationship between voltage and power, or current and power involves second degree. Accordingly, the former relationship is *linear* and the latter relationship is *non-linear*, i.e., *not* linear; in fact, it is quadratic. Care has to be exercised in computing power dissipated using the principle of linearity. We also caution the reader to distinguish clearly, as is illustrated in the equation above, between the degree to which a quantity is raised to and the superscript.

Exercise 4.18 Verify the results of Example 4.16.

We end this section with the following example.

Example 4.17 Consider the simple circuit shown in Fig. 4.25.

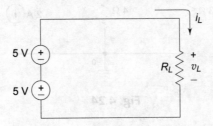

Fig. 4.25

The current i_L through the resistance is

$$i_L = \frac{1}{5}(5+5) = 2 \, \text{A}$$

Suppose the second voltage source u_{v2} is now scaled by a scalar α. We shall not assign any numerical to it, but leave it as it is. The current through the resistance is

$$i_L = \frac{1}{5}(5 + \alpha 5) = 1 + \alpha 1 \, \text{A}$$

Notice that in this latter case we are able to distinguish the response of each of the sources—the scalar α serves as an 'identification tag'.

4.5   Network Theorems

In Chapter 3 we defined a *port* as a pair of nodes across which some element is connected. In this section, we shall develop another elegant way of analysing a network by reducing its complexity drastically. We shall first take up a small circuit to introduce some terminology and to develop the intuition necessary.

Consider the simple series circuit shown in Fig. 4.26.

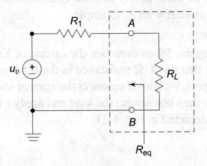

Fig. 4.26

Notice that we have intentionally marked the nodes as A and B across which the load R_L is connected. We call this pair of nodes, hereafter, as the *load port*. Looking at the rest of the circuit *from the point of view* of the load port we see a resistance R_1. We call this the *equivalent resistance* R_{eq}, *as seen by the load port*. Note that we have assumed an ideal voltage source and hence it offers no resistance in series with R_1. We may *experimentally* determine this R_{eq} doing the following.

1. Remove the source u_v and replace it with a short circuit.
2. Disconnect R_L from the load port.
3. Connect a *test voltage source* v_{test} in its place.
4. Measure the current i_{test} delivered by v_{test}
5. Since R_1 is the only resistance in this modified circuit, we get

$$R_{eq} = \frac{i_{test}}{i_{test}} = R_1$$

Next, the voltage across R_L may be obtained as

$$v_L = \frac{R_L}{R_{eq} + R_L} u_v$$

using the voltage division rule. From this equation we see that

$$u_v = v_L \mid \text{ with } R_L = \infty \ \Omega \tag{4.26}$$

We call this *open-circuit voltage* v_{oc}.

Next, the current through R_L is

$$i_L = \frac{1}{R_{eq} + R_L} v_{oc}$$

Let us define *short-circuit current* i_{sc} as

$$i_{sc} \overset{\Delta}{=} i_L \mid \text{ with } R_L = 0 \ \Omega \tag{4.27}$$

We quickly observe that

$$R_{eq} = \frac{v_{oc}}{i_{sc}} \tag{4.28}$$

We shall make use of these newly defined quantities in the following. As has been our practice, let us start with an example.

Example 4.18 Once again, let us consider the circuit of Example 4.16 shown in Fig. 4.24. We assume that the 4 Ω resistance is the load and the pair of nodes v_2 and v_4 make the load port. We first disconnect the current sources and replace them with open circuits. We then disconnect the load and apply a test voltage v_{test} across the load port. This is depicted in Fig. 4.27.

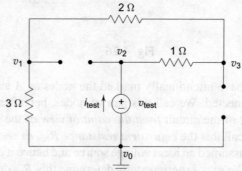

Fig. 4.27

It is a series circuit and we easily measure the current as

$$i_{\text{test}} = \frac{v_{\text{test}}}{1 + 2 + 3} \ \text{A}$$

Hence the equivalent resistance as seen by the load port is

$$R_{\text{eq}} = 6 \ \Omega$$

Starting from the load, we have a series resistance R_{eq} one end of which is connected to the node v_2. We are yet to determine the element, or elements, between the second end of R_{eq} and the ground v_0. In Example 4.16 we have seen that this load has a drop of 2 V across it. Now, let us ask the following questions.

'Does there exist a voltage source v_{eq} in series with R_{eq} and R_L, as shown in Fig. 4.28, such that the load continues to have the same potential difference of 2 V as before? If this exists, how do you derive this from the given circuit?

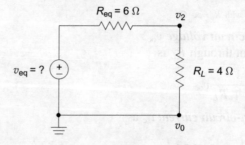

Fig. 4.28

The second question is more important, because we should not deviate from the given problem. The answer to the first question may be easily obtained by using the voltage division rule as follows:

$$\frac{R_L}{R_{eq} + R_L} v_{eq} = 2\,V$$

$$\Rightarrow \quad v_{eq} = 5\,V$$

We need to justify this because we never know v_L *a priori*. Sometimes, even R_L is not known *a priori*. Earlier, we have found that $v_{eq} = v_{oc}$.

Let us restore the independent sources in the circuit, remove R_L, and obtain v_{oc} across the load port. Applying the principle of linearity we get the simplified circuits as shown in Figs 4.29(a) and 4.29(b).

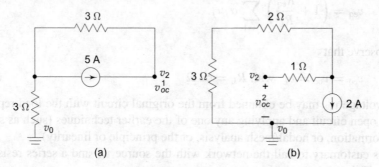

(a)   (b)

Fig. 4.29

Applying nodal analysis to each of these circuits, we get

$$v_{oc} = v_{oc}^1 + v_{oc}^2$$

$$= 15 - 10 = 5\,V$$

Exercise 4.19 Verify the results of Example 4.16.

Let us now put everything formally.

4.5.1 Thévenin's Theorem

We have an important theorem proposed by the French telegraph engineer Léon Thévenin in 1883. It is interesting to note that an earlier statement in the form of this theorem is credited to Helmholtz, and the theorem is sometimes referred to as Helmholtz–Thévenin theorem.

Theorem 4.1 *Any two-node linear network may be replaced by a voltage source equal to the open-circuit voltage between the nodes in series with the resistance as seen by a load at this port.*

Proof: Let a given circuit have r number of independent sources. Suppose that we are interested in the voltage v_L across a certain load port.

By disconnecting all the sources and putting a test source v_{test} across the load port we first determine the equivalent resistance R_{eq} as seen by the load port.

Next, by the principle of linearity we know that

$$v_L = \sum_{j=1}^{r} \alpha_j u_j \tag{4.29}$$

Now, if R_{eq} is the equivalent resistance as seen by the load, and we are interested in determining an equivalent voltage source v_{eq}, in series with R_{eq}, that would provide the same voltage v_L across the load, then we have

$$\frac{R_L}{R_{eq} + R_L} v_{eq} = \sum_{j=1}^{r} \alpha_j u_j \tag{4.30}$$

From this equation, we have

$$v_{eq} = \left(1 + \frac{R_{eq}}{R_L}\right) \sum_{j=1}^{r} \alpha_j u_j$$

We observe that

$$v_{eq} = v_{oc} = v_L \text{ with } R_L = \infty\ \Omega \tag{4.31}$$

This voltage v_{oc} may be obtained from the original circuit with the load replaced by an open circuit and applying any one of the earlier techniques (such as source transformation, or nodal/mesh analysis, or the principle of linearity).

It is customary to call the network with the source v_{oc} and a series resistance R_{eq} as *Thévenin's equivalent network*.

Q.E.D.

Let us now give an algorithm to analyse circuits using Thévenin's theorem.

Step 1	Identify the load port and remove the load. Replace all the independent voltage sources with short circuits and all the independent current with open circuits.
Step 2	Apply a test voltage source v_{test} and compute the current i_{test} it delivers to the circuit.
Step 3	The ratio v_{test} / i_{test} gives the equivalent resistance R_{eq}.
Step 4	Remove the test voltage source and retain all the independent sources in their respective positions.
Step 5	Compute the open-circuited voltage v_{oc} across the load port using any of the earlier techniques.
Step 6	Thévenin's equivalent is as shown in Fig. 4.27 with R_{eq} in series with v_{oc} computed in steps 3 and 5.

A few remarks about this algorithm are in order. First, the load is removed from its port in step 1 and is never inserted back until step 6. Second, we emphasize that the test voltage source v_{test} is kept as an algebraic figure without assigning any numerical value. This way, we find that the resulting i_{test} is strictly linear with the source and we get the ratio exactly. Some unknown errors might creep in if numerical values are assigned. However, if we employ computer programs to solve the LSE, it is preferable to test the result with several different values of v_{test}. Third, we recommend that R_{eq} be computed only this way rather than as the ratio v_{oc} / i_{sc}

since the latter method requires solving the problem (almost) twice. Moreover, in certain tricky circuits we may find that both v_{oc} and i_{sc} equal to zero and R_{eq} appears to be indeterminate while, in fact, it is not so.

Example 4.19 Let us obtain Thévenin's equivalent of the circuit shown in Fig. 4.30.

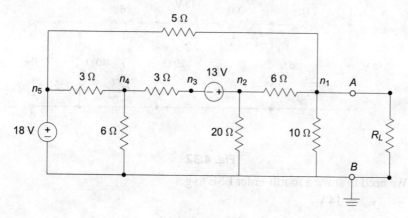

Fig. 4.30

We shall compute the equivalent resistance using the circuit in Fig. 4.31 and employing mesh analysis.

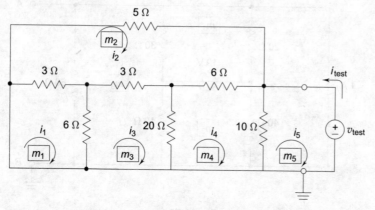

Fig. 4.31

The resulting matrix-vector equation looks formidable with fifth order. However, we encourage the readers to write their own computer programs to solve the linear system of equations using Cramer's rule and Gaussian elimination. This allows the readers to apply their prudence before accepting computer-generated results. We find that

$$i_{test} = \frac{v_{test}}{2.5}$$

and hence

$$R_{eq} = 2.5 \ \Omega$$

We shall insert the two voltage sources back in their positions and once again employ mesh analysis to find v_{oc} across the load port as shown in Fig. 4.32.

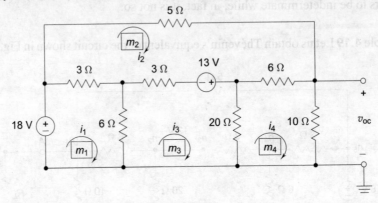

Fig. 4.32

We need to solve a fourth-order LSE to get

$$v_{oc} = 14\,\text{V}$$

Exercise 4.20 Verify the results of Example 4.18.

Example 4.20 Let us obtain Thévenin's equivalent of the circuit shown in Fig. 4.33.

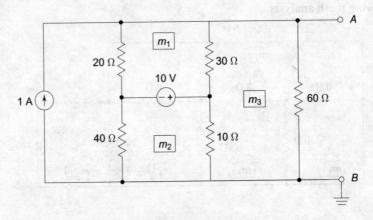

Fig. 4.33

Applying v_{test} while replacing all other independent sources with open circuits and short circuits, we get the LSE as

$$
\begin{pmatrix}
50 & 0 & -30 & 0 \\
0 & 50 & -10 & 0 \\
-30 & -10 & 100 & -60 \\
0 & 0 & -60 & 60
\end{pmatrix}
\begin{pmatrix}
i_1 \\
i_2 \\
i_3 \\
-i_{test}
\end{pmatrix}
=
\begin{pmatrix}
0 \\
0 \\
0 \\
-v_{test}
\end{pmatrix}
$$

from which we get

$$R_{eq} = \frac{v_{test}}{i_{test}} = 15\,\Omega$$

Similarly, with all the independent sources restored, applying mesh analysis gives the following LSE:

$$\begin{pmatrix} 50 & 0 & -30 \\ 0 & 50 & -10 \\ -30 & -10 & 100 \end{pmatrix} \begin{pmatrix} i_1 \\ i_2 \\ i_3 \end{pmatrix} = \begin{pmatrix} 10 \\ 50 \\ 0 \end{pmatrix}$$

From this we get

$$v_{oc} = 60 \times i_3 = 12 \text{ V}$$

Exercise 4.21 Verify the results of Example 4.19 employing nodal analysis.

Exercise 4.22 Obtain Thévenin's equivalent of the circuit shown in Fig. 4.34.

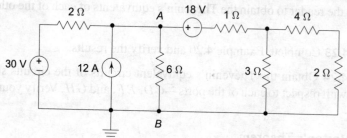

Fig. 4.34

(*Ans.* $R_{eq} = 1.2 \, \Omega$ and $v_{oc} = 39.6$ V. The load voltage $v_A = 33$ V.)

Example 4.21 Consider the two circuits shown in Figs 4.35(a) and 4.35(b).

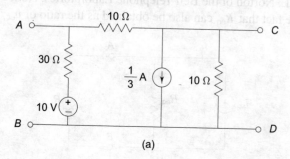

(a)

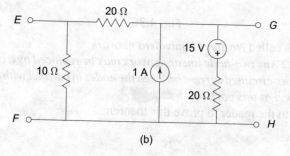

(b)

Fig. 4.35

There are four open load ports—AB, CD, EF, and GH—shown. We may obtain Thévenin's equivalent with respect to each of these ports. For instance, from the point of view of the port AB, the circuit of Fig. 4.35(a) has

$$R_{eq} = 10 + 10 \parallel 30 = 12\,\Omega \quad \text{and} \quad v_{oc} = 2\,\text{V}$$

Further, we may make a circuit with the port AB attached to the port EF and compute the power dissipated by each of the elements of the circuit in Fig. 4.35(b). In such a case, we observe that it is easier to compute with the Thévenin's equivalent of Fig. 4.35(a). We emphasize that the load port may contain a circuit itself rather than a simple single resistance R_L.

In fact, we have four different circuits with the possible port combinations. We leave it to the reader to obtain the Thévenin's equivalents of each of the other three ports.

Exercise 4.23 Complete Example 4.20 and verify the results.

Exercise 4.24 Obtain the Thévenin's equivalent circuits of the circuits shown in Fig. 4.34 with respect to each of the ports—CD, EF, and GH. Verify your results.

4.5.2 Norton's Theorem

If we have a network with a voltage source and a series resistance, we always have its dual—a current source with a parallel resistance. This was established in the beginning of this chapter. Accordingly, the circuit shown in Fig. 4.36 serves as a dual to Thévenin's equivalent network. The dual of Thévenin's theorem was proposed by E.L. Norton of the Bell Telephone Laboratories. Notice that we have made use of the fact that R_{eq} can also be obtained as the ratio of v_{oc}/i_{sc}.

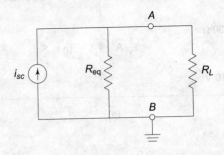

Fig. 4.36

This network is called *Norton's equivalent network*.

Theorem 4.2 *Any two-node linear network may be replaced by a current source equal to the short-circuited current between the nodes in parallel with the resistance as seen by a load at this port.*

We leave it to the reader to prove this theorem.

Exercise 4.25 Prove Norton's theorem.

The algorithms to obtain Thévenin's and Norton's equivalents are quite similar; in fact, dual to each other. Let us do Exercise 4.22 here. We leave it to the reader to redo Examples 4.18 to 4.20 and Exercise 4.24.

Example 4.22 We shall consider the circuit shown in Fig. 4.34. We shall remove the load resistance $R_L = 6\,\Omega$ and the other independent sources and apply a test voltage source[3] across the load port AB. The resulting circuit is shown in Fig. 4.37.

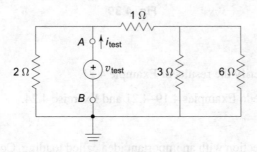

Fig. 4.37

Using source transformation and simple series/parallel reduction, we find that

$$R_{eq} = \frac{v_{test}}{i_{test}} = 1.2\,\Omega$$

With the independent sources back in their respective positions and a short-circuit across the load port, the circuit is shown in Fig. 4.38.

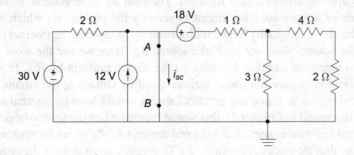

Fig. 4.38

Applying the principle of linearity, we get

$$i_{sc} = 33\,\text{A}$$

Thus, the Norton's equivalent circuit is

---

[3] Though Norton's equivalent is a dual of Thévenin's equivalent, we need not apply a 'test current source' in lieu of 'test voltage source' to determine R_{eq}. Which source is to be applied depends on the circuit at hand. Similarly, it is wise to anticipate whether it would be easier to compute v_{oc} or i_{sc} before blindly following the algorithm.

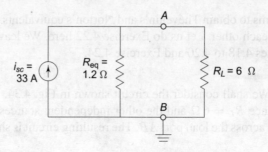

Fig. 4.39

Exercise 4.26 Verify the results of Example 4.21.

Exercise 4.27 Redo Examples 4.19–4.21 and Exercise 4.24.

We end this section with an important idea called loading. Consider Thévenin's equivalent circuit of Fig. 4.28. Recall that we have the voltage source v_{oc} and the series resistance R_{eq} *as seen by the load* R_L. Equivalently, we may also say that the load R_L is as seen by the voltage source v_{oc} with an internal resistance R_{eq}. If R_{eq} and R_L are of comparable magnitude the voltage v_{oc} is shared by the two. On the other hand, if R_L is much larger than R_{eq}, i.e.,

$$R_L \gg R_{eq}$$

then v_{oc} appears almost across R_L itself. The load R_L is, in general, an equivalent resistance of a more complex circuit as seen by the port across which we define R_L. If the above inequality holds, it is customary to say that the internal resistance R_{eq} of the source 'does not load' the source v_{oc}. Here we use the word 'load' as a verb to represent additional burden on the source putforth by R_{eq}. In principle, it is preferable to have the source deliver all of its voltage to the circuit, i.e., R_L, connected across it. Since any practical source would have an internal resistance (as was discussed in Chapter 2), this would happen when the source (together with its internal resistance) sees an infinite resistance, i.e., $R_L = \infty$. In other words, we would say that the circuit, as seen by the Thévenin's equivalent of the source, must have an infinite resistance at its input port.

We may turn around the same idea with the inequality written as

$$R_{eq} \ll R_L$$

and argue that the load R_L must see a very low resistance, ideally $0\,\Omega$, at the output port of the source to ensure that it draws maximum possible voltage from the source.

To sum up, a voltage source must ideally drive a load that offers infinite input resistance; equivalently, a load must be connected to a voltage source that offers zero output resistance. It is not difficult to make a dual statement about a current source—it does not get loaded only if $R_{eq} \gg R_L$.

This idea of loading and no-loading is very important, particularly in the electronic circuits. It is, in fact, undesirable because the output voltage of the source varies from load to load. It is the circuit designer's duty to anticipate this loading problem and allow for necessary compensation. For instance, consider the common observation we all make while travelling in a vehicle. When we start the engine with the lights on, the lights flicker. We might model this phenomenon as a current source driving two loads, a smaller R_{L1} for the lights and a larger R_{L2} for the starter motor, connected across the source. Naturally, the smaller R_{L1} would prevent much of the current flowing into the other resistance. By turning off the lights, we mean that R_{L1} is temporarily disconnected so that the source would see the starter motor alone. We shall elaborate this idea little more in a later chapter on two-port networks.

4.5.3 Maximum Power Transfer Theorem

We shall illustrate another theorem in connection with the Thévenin's and Norton's equivalents. Suppose we have a Thévenin's equivalent circuit as shown in Fig. 4.40 (Fig. 4.28 repeated again for convenience with slight modification.)

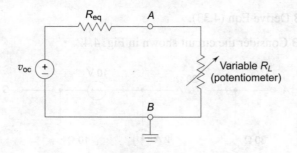

Fig. 4.40

We now ask: What is the load R_L that receives maximum power from the source? Using voltage division rule, we find that

$$i_L = \frac{1}{R_{eq} + R_L} v_{oc}$$

from which we get

$$p_L = i_L^2 R_L$$

$$= \frac{R_L}{(R_{eq} + R_L)^2} v_{oc}^2 \tag{4.32}$$

Using elementary calculus, we find that

$$p_L \text{ is maximum when } R_L = R_{eq} \tag{4.33}$$

A typical plot of p_L versus R_L is shown in Fig. 4.41.

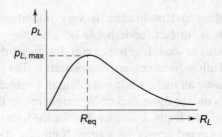

Fig. 4.41

We may compute that the maximum power is given by

$$p_{L,\,max} = \frac{v_{oc}^2}{4\,R_L} \tag{4.34}$$

In passing, we remark that this is contrary to the idea of loading, wherein we were interested in delivering maximum possible 'voltage' to the load R_L.

Exercise 4.28 Derive Eqn (4.33).

Example 4.23 Consider the circuit shown in Fig. 4.42.

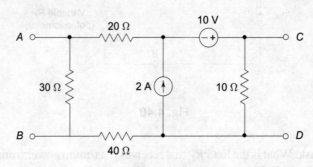

Fig. 4.42

It is easy to find that the equivalent resistances with reference to ports *AB*, *CD*, *AC,* and *BD* are 21 Ω, 9 Ω, 16 Ω, and 24 Ω, respectively. According to the maximum power transfer theorem, for instance, a load of 16 Ω receives maximum power when connected across the port *AC*.

Exercise 4.29 Verify the results of Example 4.23.

We close the section with a few remarks. In general, Thévenin's equivalent works best with series connections and Norton's equivalent (by duality) works best with parallel connections. Further, if $R_{eq} = 0\,\Omega$, then there is no choice but Thévenin's equivalent and if $R_{eq} = \infty\,\Omega$, there is no choice but Norton's equivalent. And, Thévenin's and Norton's equivalent circuits may be obtained for certain intermediate portions of any network to simplify computations significantly.

4.6 Circuits with Controlled Sources

In this chapter so far, we have discussed certain basic ideas for solving circuit problems. What we have emphasized is the *algorithm*, i.e., the elegant way of solving a problem, rather than the solution itself. For instance, given a circuit, it is possible to solve it *somehow*; but engineering is offering elegant solutions. We have illustrated the algorithms using circuits that are extremely simple. We end this chapter with an incremental complexity in the circuits—circuits with a controlled source. In later chapters we address more complex circuits and more general algorithms.

In Chapter 2 we have discussed the four types of controlled sources. These are useful in modelling amplifiers, transformers, etc. We conclude this chapter with examples of circuits that contain controlled sources. We profess that all the algorithms developed in this chapter are applicable to these circuits without any exception. We emphasize the following points before giving several examples.

1. The controlling variables v_x and i_x of the controlled sources are not known *a priori* and hence must be found using additional circuit equations.

2. While applying the principle of linearity, or while determining the equivalent resistance with reference to a load port, we *should not* replace the controlled sources with short circuits or open circuits. However, all the independent sources may be replaced as usual. This is because the constraint binding the control variable and the controlled source is lost if the controlled source is removed from the circuit. Consequently, in general, applying the principle of linearity is not worthy.

3. While determining the Thévenin's equivalent of a certain intermediate portion of any network containing a controlled source, we need to ensure that the controlling variable is also present in the same portion.

Example 4.24 Consider the circuit shown in Fig. 4.43.

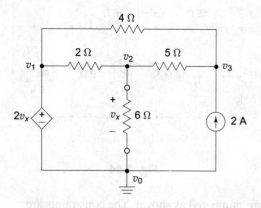

Fig. 4.43

We notice that the control variable v_x is the node voltage v_2. Hence,

$$v_1 = 2\,v_x = 2\,v_2$$

is the constraint. Applying KCR at nodes 2 and 3 would give

$$\left(\frac{1}{2} + \frac{1}{5} + \frac{1}{6}\right) v_2 - \frac{1}{2} v_1 - \frac{1}{5} v_3 = 0$$

$$\left(\frac{1}{5} + \frac{1}{4}\right) v_3 - \frac{1}{4} v_1 - \frac{1}{5} v_2 = 2$$

from which we may obtain the matrix-vector equation as follows:

$$\begin{pmatrix} -\dfrac{4}{30} & -\dfrac{1}{5} \\[2mm] -\dfrac{7}{10} & \dfrac{9}{20} \end{pmatrix} \begin{pmatrix} v_2 \\ v_3 \end{pmatrix} = \begin{pmatrix} 0 \\ 2 \end{pmatrix}$$

What is conspicuous here is that the node incidence matrix $\mathcal{N}$ no longer obeys any of the earlier properties—not all diagonal elements are positive, and it is not symmetric. So, we need to exercise care in writing the nodal and mesh equations when the circuits contain controlled sources.

Exercise 4.30 Complete Example 4.24 and verify the results using mesh analysis.

(***Ans.*** $v_2 = v_x = -2/25$ V and $v_3 = 4/75$ V)

Example 4.25 Let us now look at the circuit given in Fig. 4.44.

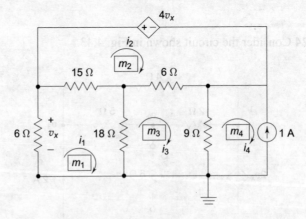

Fig. 4.44

The meshes are numbered as shown. The constraints are

$$i_4 = -1\,\text{A} \quad \text{and} \quad v_x = -6i_1$$

With these constraints in place, we get the following LSE:

$$\begin{pmatrix} 39 & -15 & 18 \\ -39 & 21 & -6 \\ -18 & -6 & 33 \end{pmatrix} \begin{pmatrix} i_1 \\ i_2 \\ i_3 \end{pmatrix} = \begin{pmatrix} 0 \\ 0 \\ -9 \end{pmatrix}$$

Exercise 4.31 Solve the LSE of Example 4.25 and verify your results using nodal analysis.

(*Ans.* $i_1 = 2/3$, $i_2 = 4/3$, $i_3 = 1/3$ A)

Exercise 4.32 Apply nodal and mesh analysis to the circuit shown in Fig. 4.45 and cross-check your results.

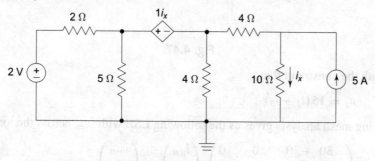

Fig. 4.45

Example 4.26 Let us obtain R_{eq} of the load port in the circuit of Fig. 4.46.

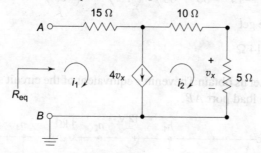

Fig. 4.46

We apply a test voltage source across the load port. If we remove the controlled source, we will not be able to study the effect of the source. So let us not disturb it. Applying mesh analysis with the constraint

$$i_1 - i_2 = 4v_x = 4 \times 5 \times i_2$$

or

$$i_1 = 21i_2$$

the super mesh equation is

$$15\,i_1 + 15\,i_2 = v_{test}$$

Since $i_{test} = i_1$, we get

$$R_{eq} = \frac{15 \times 22}{21} = 15.7143\,\Omega$$

Example 4.27. Let us obtain the R_{eq} of the load port of the circuit shown in Fig. 4.47.

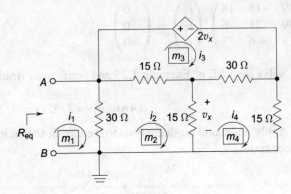

Fig. 4.47

With the constraint that

$$v_x = 15 \left(i_2 - i_3 \right)$$

applying mesh analysis gives us the following LSE with v_{test} across the load port:

$$\begin{pmatrix} 30 & -30 & 0 & 0 \\ -30 & 60 & -15 & -15 \\ 0 & 15 & 45 & -60 \\ 0 & -15 & -30 & 60 \end{pmatrix} \begin{pmatrix} i_{test} \\ i_2 \\ i_3 \\ i_4 \end{pmatrix} = \begin{pmatrix} v_{test} \\ 0 \\ 0 \\ 0 \end{pmatrix}$$

Solving this, we get

$$R_{eq} = 14 \, \Omega$$

Example 4.28 Let us obtain Thévenin's equivalent of the circuit shown in Fig. 4.48 looking into the load port AB.

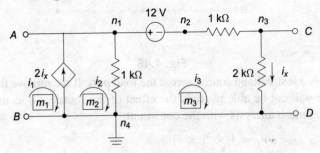

Fig. 4.48

With v_{test} and the controlled source, the constraint is

$$i_{test} - i_2 = -2 i_x$$

The mesh equations are

$$1 i_2 - 1 i_x = v_{test}$$

$$-1 i_2 + 4 i_x = 0$$

Solving, we get

$$i_2 = \frac{4}{3}\, v_{test}, \quad i_x = \frac{1}{3}\, v_{test}, \quad \text{and } i_{test} = \frac{2}{3}\, v_{test}$$

all in milliamperes if v_{test} is few volts. From this we get

$$R_{eq} = 1.5\,k\Omega$$

To find v_{oc} it is easy to apply nodal analysis with the constraint

$$v_1 - v_2 = 12\,V$$

The only super node equation we get is

$$\frac{v_{oc}}{1} + \frac{v_2}{3} = 2i_x$$

$$= 2\frac{v_2}{3}$$

and using the constraint the result is

$$v_{oc} = -6\,V$$

Exercise 4.33 Verify the results of Example 4.28.

Exercise 4.34 Obtain Norton's equivalent of the circuit in Fig. 4.48 from the point of view of the port CB.

Example 4.29 Let us obtain Norton's equivalent of the circuit in Fig. 4.49 looking into the load port AB.

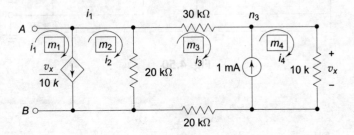

Fig. 4.49

For R_{eq} we get the following LSE applying mesh analysis:

$$\begin{pmatrix} 20 & -20 \\ -20 & 80 \end{pmatrix} \begin{pmatrix} i_2 \\ i_3 \end{pmatrix} = \begin{pmatrix} v_{test} \\ 0 \end{pmatrix}$$

with the constraint that

$$i_{test} - i_2 = \frac{v_x}{10\,k\Omega} = i_3$$

Solving, we get $i_{test} = v_{test}/12\,k\Omega$ and hence

$$R_{eq} = 12\,k\Omega$$

For computing i_{sc}, once again we find that mesh analysis is convenient. Now we have two constraints owing to the presence of an independent current source. These are

$$i_1 - i_2 = \frac{v_x}{10\,k\Omega} = i_4 \quad \text{and} \quad i_3 - i_4 = -1$$

The mesh equations reduce to

$$\begin{pmatrix} 20 & -20 \\ -20 & 80 \end{pmatrix} \begin{pmatrix} i_2 \\ i_3 \end{pmatrix} = \begin{pmatrix} 0 \\ -10 \end{pmatrix}$$

Solving, we get

$$i_2 = i_3 = -\frac{1}{6}\,\text{mA} \quad \text{and} \quad i_1 = \frac{2}{3}\,\text{mA}$$

Hence

$$i_{sc} = -i_1 = -\frac{2}{3}\,\text{mA}$$

Exercise 4.35 Find v_0/v_s in the circuit shown in Fig. 4.50.

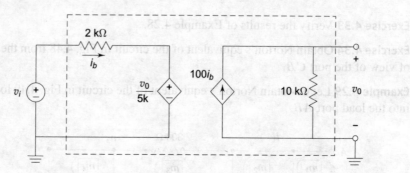

Fig. 4.50

Summary

In this chapter we have developed the essential techniques of network analysis. We began with circuits energized by *independent sources*, i.e., sources whose terminal behaviour is independent of the network connected. The theme is 'sophisticated solutions' rather than solutions *per se*. This sort of a systematic and organized way of solving the circuits naturally results in developing excellent software such as SPICE, which in turn helps us contemplate most complex circuits such as very large scale integrated (VLSI) electronic circuits. Towards the end of the chapter, we have presented examples of circuits that have *controlled sources* in addition to the independent sources and resistors. Analysis of circuits having controlled sources is also straightforward and all the techniques developed earlier, except the principle of superposition, may be applied.

We urge the reader to do the examples and exercises thoroughly before moving on to the next chapter. We encourage the reader to generate his/her own networks and apply the techniques. Further, we also recommend the reader to develop his/her own computer programs, for Gaussian eliminations and its variants, and Cramer's rule, to solve the linear system of equations. Of course, there are a large number of commercial software packages, such as MATLAB, MATRIX-X, etc. that could be used for solving the equations. Though it may not be apparent in this course, the reader must learn to appreciate the shortcomings of computer-generated results, especially in the light of round-off errors. Development of *efficient algorithms* is a subject in itself and we encourage the reader to gain some familiarity.

Problems

4.1 Use nodal analysis to determine all the node voltages and branch currents in the circuit shown in Fig. 4.51.

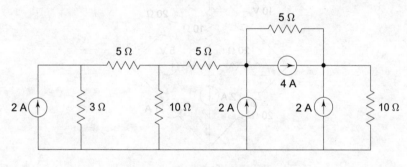

Fig. 4.51

4.2 In the circuit shown in Fig 4.52, determine the power absorbed by the 30 Ω resistance. Use nodal analysis.

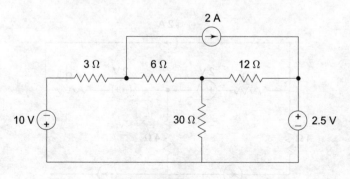

Fig. 4.52

4.3 Determine the current (and its direction) through the 4 Ω resistance in the circuit in Fig. 4.53.

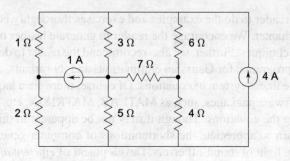

Fig. 4.53

4.4 Determine the potential differences across each of the elements of the circuit in Fig. 4.54. Use nodal analysis with supernodes around the voltage sources.

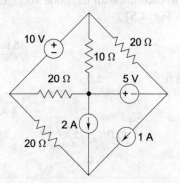

Fig. 4.54

4.5 Determine the power (delivered or absorbed) by each of the independent sources in the circuit in Fig. 4.55.

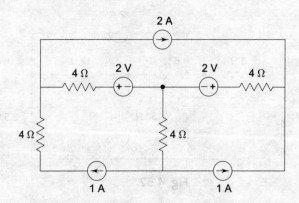

Fig. 4.55

4.6 Use mesh analysis to determine the branch currents in the circuit in Fig. 4.56.

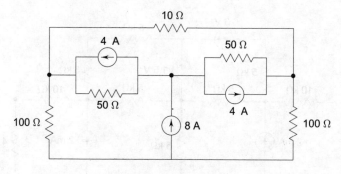

Fig. 4.56

4.7 Determine the power absorbed/delivered by each of the elements in the circuit in Fig. 4.57 and hence verify the conservation of power. Use mesh analysis.

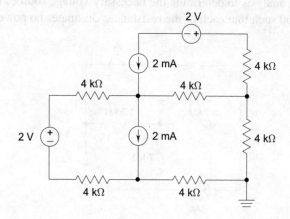

Fig. 4.57

4.8 Consider the circuit in Fig. 4.58. If it is given that the current i_1 in mesh 1 is 2 mA, determine suitable values of the resistances R_1 and R_2.

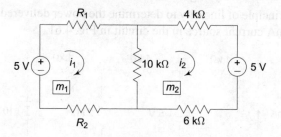

Fig. 4.58

4.9 Determine the power absorbed (or delivered) by the 1.5 V source in the circuit in Fig. 4.59.

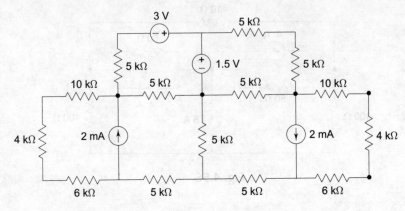

Fig. 4.59

4.10 Use mesh analysis to determine the necessary voltage sources in the circuit in Fig. 4.60 such that each of the resistances dissipates no power.

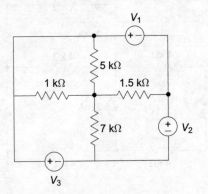

Fig. 4.60

Can you generalize the conditions under which all the passive elements in a circuit with non-zero sources dissipate no power? (***Hint*** Look at the mesh incidence matrix $\mathcal{M}$ you get for the circuit.)

4.11 Use the principle of linearity to determine the power delivered (or absorbed) by the 2 mA current source in the circuit in Fig. 4.61.

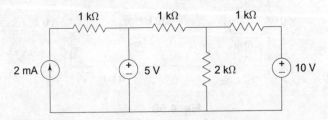

Fig. 4.61

4.12 Use the principle of linearity to determine the power dissipated by the 50 kΩ resistance in the circuit of Fig. 4.62.

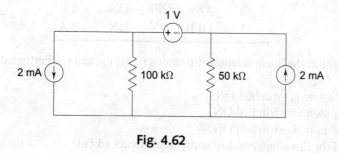

Fig. 4.62

4.13 Consider the circuit in Fig. 4.63.

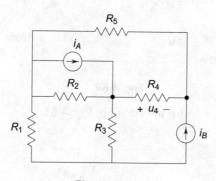

Fig. 4.63

An experiment was conducted and the following observations were made.

- With the current sources $i_A = 10$ A and $i_B = 25$ A, the voltage v_4 across the resistance R_4 is measured to be 100 V.
- With the current sources $i_A = 25$ A and $i_B = 10$ A, v_4 is measured to be -50 V.

Using the principle of linearity, obtain the general relationship among the three variables i_A, i_B, and v_4 and draw the graphs of v_4 versus i_A and i_B as a family of straight lines. Further, determine v_4 if $i_A = -10$ A and $i_b = -5$ A.

4.14 Generalizing problem 4.13, how many (different) observations are required if a circuit has n number of independent sources? Justify.

4.15 A passive network containing several resistances is energized by three independent sources—i_A, v_B, and v_C. In an experiment, while switching off one (and not *two*) of these sources at a time, the current i_x through a certain branch X is measured and the observations are recorded as follows:

S. No.	i_A	v_B	v_C	i_x
1	ON	ON	OFF	20 A
2	ON	OFF	ON	− 5 A
3	OFF	ON	ON	2 A

Determine the linear relationship among i_x, i_A, v_B, and v_C. Further, determine i_x if

- i_A alone is switched ON,
- v_B alone is switched ON,
- v_C alone is switched ON,
- all the three independent sources are switched On.

4.16 For the circuit in Fig. 4.64 determine the Thévenin's and Norton's equivalents as seen by the node pair AB.

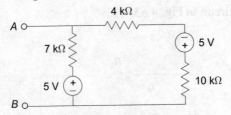

Fig. 4.64

4.17 Repeat problem 4.16 for the circuit in Fig. 4.65.

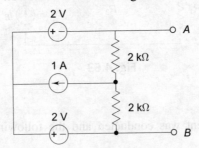

Fig. 4.65

4.18 Repeat problem 4.16 for the circuit in Fig. 4.66.

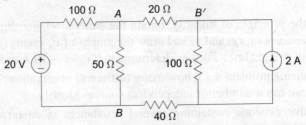

Fig. 4.66

4.19 For the circuit in Fig. 4.66, determine the power delivered to the 50 Ω resistance placed across the port AB. Verify your result using the principle of linearity.

4.20 Repeat Problem 4.18 if the port of interest is the 20 Ω resistance across the port AB'. Determine the power absorbed by this port and verify your result.

4.21 Consider the circuit in Fig. 4.67. Determine R_L that would dissipate maximum power.

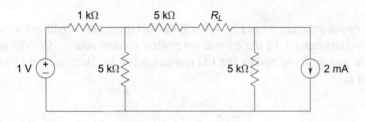

Fig. 4.67

4.22 In an experiment, it is found that a (practical) dc voltage source (with an internal source resistance R_s) delivers a power of 80 W when connected to a resistance of 20 Ω. It is also observed that the same source provides a current of 2.5 A when connected across an ammeter, i.e., when it is momentarily short-circuited. Determine the maximum power the source could deliver. What is the corresponding load resistance R_L?

4.23 Figure 4.68 shows a circuit in two stages. Determine R_L so that it would draw maximum power from the previous stage.

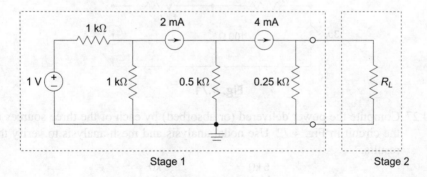

Fig. 4.68

4.24 For the circuit in Fig. 4.69, determine a condition for R in terms of the source resistance R_s and the load resistance R_L so that the source would deliver maximum power to R_L. Determine R if (i) $R_s = 10\,\text{k}\Omega$ and $R_L = 20\,\text{k}\Omega$, (ii) $R_s = 20\,\text{k}\Omega$ and $R_L = 10\,\text{k}\Omega$.

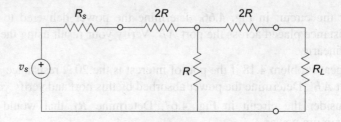

Fig. 4.69

4.25 Consider the circuit in Fig. 4.70. Show that there are two different values for the parameter k of the current-controlled voltage source (CCVS) such that the power dissipated by the 1 Ω resistance is 4 W. Determine the two values of k.

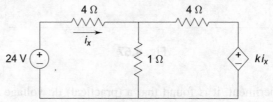

Fig. 4.70

4.26 Use nodal analysis to compute the power delivered (or absorbed) by the VCVS in the circuit in Fig. 4.71. Check your results using mesh analysis.

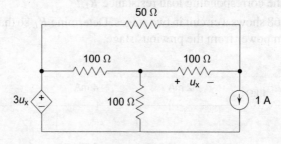

Fig. 4.71

4.27 Compute the power delivered (or absorbed) by each of the three sources in the circuit in Fig. 4.72. Use nodal analysis and mesh analysis to verify the results.

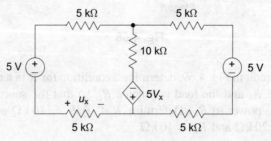

Fig. 4.72

4.28 Obtain the controlling current i_x in the circuit in Fig. 4.73 using the principle of linearity.

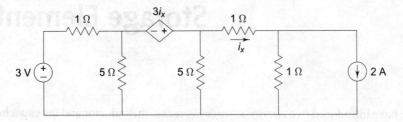

Fig. 4.73

4.29 In the circuit in Fig. 4.74, obtain the Thévenin's and Norton's equivalents as seen by (i) port AB, (ii) port AC, and (iii) port CB.

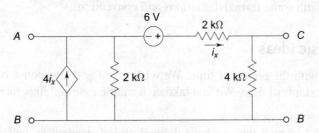

Fig. 4.74

4.30 Determine the maximum power that can be drawn by a passive load R_L when connected across the port AB in the circuit in Fig. 4.75. What is the corresponding R_L?

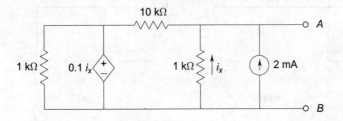

Fig. 4.75

CHAPTER 5

Transients with Energy Storage Elements

We have introduced two *energy-storage devices*—the inductor and the capacitor—in Chapter 2. However, we have not yet used them in the circuits. In this chapter we shall look at the circuits that contain one or both of these elements in addition to the resistors. The ideas presented in this chapter may at first appear to be counterintuitive. We simply remark that these are the very foundations of circuit theory and hence it is important for the reader to learn the material presented here. We begin with some formal definitions and conventions.

5.1 Basic Ideas

We begin with the notion of time. We wish to study the response of a circuit at different instants of time. We first take up a simple case and illustrate the idea.

Example 5.1. Recall that we have defined an independent dc voltage source as one which delivers a 'constant' voltage irrespective of the network it has been connected to. This may be described in terms of a graph as shown in Fig. 5.1(a).

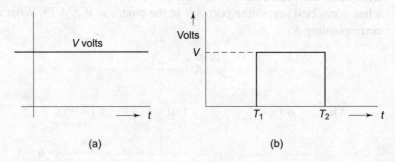

(a)   (b)

Fig. 5.1

The graph in Fig. 5.1(a) has lots of things that might not have been clearly defined. However, the message is clear: the voltage is *constant with respect to time*. If we slightly change our point of view, perhaps, it gives us a better picture. If we connect this source to an external network at time T_1 and disconnect it at a later time T_2, the network 'sees' a voltage V volts as shown in Fig. 5.1(b); there was no voltage 'before' T_1 and there would be no voltage 'after' T_2.

Let us now take a 1 Ω resistance and connect it across an independent dc voltage source of, say, 5 V. If we measure the current through the resistance, it would be

5 A 'as long as' we apply the source. The current was zero 'until' the source was connected and it would be zero 'after' we disconnect the source.

We have the following question: how do we *quantify* the words 'as long as', 'until', and 'after'? We notice that we are referring to *time*[1]. Let us formalize the convention here. Suppose we connect the source to the element now, say at 9 a.m. today morning. This refers to the 'present time'. Any time instant before this would be 'past'. Let us examine this further. First, we assume that a resistor of 1 Ω remains so forever without any change in the resistance it offers. In other words, if we seek to measure the resistance with a meter it would show 1 Ω when it was manufactured, today, tomorrow, after one year, after several years, and so on. We say that the resistor exhibits the property of *time invariance* or *shift invariance*. Under this assumption, it does not matter whether we connect the source to the circuit at 9 a.m. today or at 10 p.m. three years later. Our interest is in the response, i.e., the current flowing through the resistance *with reference to* the time at which the source is connected. Accordingly, we prefer to say that, presently, at time $t = 0$ the source is connected. It follows that $t < 0$ denotes past and $t > 0$ denotes future with reference to the present. The source may be connected for a duration of T seconds, however small or large it may be. Resistors are not the only devices to exhibit the time-invariance behaviour. By and large, several other devices including capacitors and inductors exhibit this property. Accordingly, we assume that all the elements and devices we deal with in this text exhibit this interesting property, at least over a (sufficiently large) interval of time we are observing them.

Definition 5.1 Time set $\mathcal{T}$

A time set $\mathcal{T}$ is the set of all real numbers:

$$\mathcal{T} = \{t \mid t \in \Re = (-\infty, +\infty)\} \tag{5.1}$$

Since we are interested in the past, present, and future, let us now write $\mathcal{T}$ as a union of three sets as follows:

$$\mathcal{T} = \underbrace{(-\infty, 0)}_{\text{past}} \bigcup \underbrace{[0]}_{\text{present}} \bigcup \underbrace{(0, +\infty)}_{\text{future}} \tag{5.2}$$

Since the set is continuous, it would be convenient for us to write the above union as

$$\mathcal{T} = \underbrace{(-\infty, 0^-]}_{\text{past}} \bigcup \underbrace{[0]}_{\text{present}} \bigcup \underbrace{[0^+, +\infty)}_{\text{future}} \tag{5.3}$$

where we define

Definition 5.2

$$0^- \overset{\Delta}{=} \text{an instant just before } 0$$

$$0^+ \overset{\Delta}{=} \text{an instant just after } 0 \tag{5.4}$$

---

[1] Grammatically speaking, it is 'tense'.

i.e., past ends at $t = 0^-$, present is $t = 0$, and future begins at $t = 0^+$. As noted above, the time-invariance property allows us to add (or subtract) a fixed amount of time, such as 9 a.m. today, to the present time $t = 0$ so that the past ends at $t = 9^-$ and future begins at $t = 9^+$.

5.2 Circuits with Capacitors

Let us look at the circuit in Fig. 5.2(a).

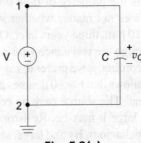

Fig. 5.2(a)

A dc voltage source is connected across a capacitance of C farads (assumed to exhibit the property of time invariance) during the interval $t \in (-\infty, 0^-]$. By KVR, the voltage across the capacitance is

$$v_C(t) = V \text{ volts}$$

We say that the capacitor is charged up to V volts, or equivalently, a *static* charge of $q = CV$ coulombs. By its basic characteristic, since the voltage across the capacitance is constant,

$$i_C(t) = C\frac{d}{dt}v_C(t) = 0$$

i.e., no current flows through the element. In other words, the capacitor acts like an 'open circuit' if it is subjected to a dc voltage.

Now assume that the source is removed at $t = 0$. The capacitor still retains the charge since it has to obey Kirchhoff's rules. We describe this property using the following equation:

$$v_C(0) = v_C(0^-) = V \text{ volts} \tag{5.5}$$

Let us next assume that at the same instance $t = 0$ (without any delay) the capacitance is connected across a conducting wire ($R = 0\,\Omega$) as shown in Fig. 5.2(b).

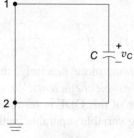

Fig. 5.2(b)

Then, again KVR has to be obeyed and the conducting wire has V volts across it *instantaneously* (i.e., at $t = 0^+$) and hence a current of ∞ A must be flowing through it. This implies that the wire dissipates a power of ∞ W and gets 'red hot' instantaneously. To have a first-hand experience of this, the reader is encouraged to buy a capacitance (for a fan motor or something like that) and request the vendor to test it.

Next, let us assume that at $t = 0$ the capacitance is connected across a resistance $R\ \Omega$ as shown in Fig. 5.2(c).

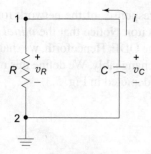

Fig. 5.2(c)

Again, by KVR, the resistance has $v_R = V$ volts across it instantaneously and hence a current of V/R A flows through it in the direction (according to the sign convention for passive elements) shown in the figure. Since this is a series circuit, the same current flows through the capacitance, therefore, the voltage $v_C(t)$ across the capacitance is no longer constant, with respect to time. Let us now determine how this voltage varies with time. Formally speaking, we have $v_C(0^+) = V$ volts to begin with and we are interested in $v_C(t)$ for all times $t \geq 0$. Since

$$i(t) = -C\frac{dv_C}{dt}$$

is the current (why is there a negative sign?) coming out of the capacitance and causing a voltage drop of

$$v_R(t) = R \times i(t) = v_C(t)$$

volts across the resistance, we have

$$-RC\frac{d}{dt}v_C(t) = v_C(t) \tag{5.6}$$

which we shall rewrite as

$$\frac{d}{dt}\, v_C(t) + \frac{1}{RC}\, v_C(t) = 0 \tag{5.7}$$

This is a *differential equation*; more precisely, this is a *'first-order' ordinary differential equation with constant coefficients*. We shall abbreviate this, hereafter, as ODE. The right-hand side of this ODE is zero and we call this a homogeneous ODE. We may solve it using variable separable method as follows:

$$\frac{dv_C(t)}{v_C(t)} = -\frac{dt}{RC}$$

integrating both sides,

$$\ln v_C(t) = -\frac{t}{RC} \quad \forall\, t \geq 0$$

or, equivalently,

$$v_C(t \geq 0) = v_C(0^+)\, e^{-t/RC}\ \text{V}$$

$$= V e^{-t/RC}\ \text{V} \tag{5.8}$$

We emphasize that this 'response' of the network for $t \geq 0$ is solely due to the *initial charge* on the capacitor. Notice that the *initial voltage* $v_C(0^+)$ serves as an *initial condition* to solve the ODE. Henceforth, we shall use the words initial charge and initial condition interchangeably. We define this response as *natural response*. This 'function of time' is depicted in Fig. 5.3.

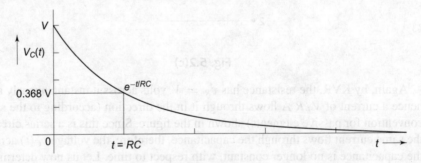

Fig. 5.3

We observe from the figure that

1. the natural response $v_C(t)$ indeed varies with time as we have guessed at the beginning,
2. the capacitance *discharges exponentially*[2] through the resistance. As time passes, i.e., as $t \to \infty$ the capacitance discharges to 0 V completely and thereafter there is no current in the circuit, and

---

[2] To get a 'feel' of what an exponential decay is, just recall how you relish a spoon of ice cream in your mouth: it melts on your tongue exponentially! The number e has a fascinating history. Interested reader has a wealth of literature on the number e available. Numerically, $e = 2.71828182845904$ upto 16 significant digits. The search engine company GOOGLE intended to raise exactly US $ 2.718,281,828 through its initial public offer recently. Interesting, isn't it?

3. the direction of the current does not change in the interval $t \in [0^+, \infty)$.

We would bring the following important point to the reader's immediate attention: *The basic property of an element, such as the resistance (or the capacitance, or the inductance) offered by an element, does not change with time. This is what we mean by time invariance and this is what results in an ODE with constant coefficients such as 1/RC. Nevertheless, the voltage across the element and the current through it may vary with time. In other words, time invariance is the fundamental property of an element, and not that of voltage and current.*

The consequence of time invariance may be *shown* as follows. Suppose the capacitor, initially charged to V volts, is connected across the resistor at a time $t = T$, instead of at $t = 0$, then the response in Eqn (5.8) may be written as

$$v_C(t \geq T) = V e^{-(t-T)/RC} \text{ V}$$

We shall come back to expressions of this sort towards the end of the chapter.

It is easy to observe that if $R = 0 \ \Omega$ in Eqn (5.8), $v_C(t)$ drops to zero instantaneously, i.e., the discharge is *rapid* and we go back to Fig. 5.2(b). Put in appropriate symbols, we have that

$$v_C(0^+) = 0 \text{ V}$$

However, in practice, there is a physical limit to such extreme events as infinite current flowing in the circuit. In other words, the capacitor cannot discharge (neither can it charge) rapidly, and postulate the following *voltage continuity rule*:

$$\boxed{v_C(0^+) = v_C(0) = v_C(0^-)} \tag{5.9}$$

We have some more observations to be examined in detail.

1. Looking at the exponential in Eqn (5.8) we observe that the product $R \times C$ has the units of time, i.e.,

$$RC = \frac{\Omega \text{ coulomb}}{\text{volt}} = \frac{\text{coulomb}}{\text{ampere}} = \text{second}$$

Let us then define

Definition 5.3 Time-constant τ

$$\boxed{\tau \stackrel{\Delta}{=} RC \text{ seconds}} \tag{5.10}$$

2. The smaller this parameter, called the time-constant (read as 'tau'), the faster is the discharge.

3. The capacitance acts as an *exponentially decaying voltage source* for all time $t \geq 0$ with the polarity intact. This is consistent with our sign convention for sources.

4. As was summarized in Chapter 3, we denote constant voltages and constant currents with uppercase letters such as 'V' and 'I'. We reserve the lowercase letters 'v' and 'i' for voltages and currents that vary with time. Of course, without loss of generality, we may use the lowercase alphabet itself, provided there is no ambiguity in the context. We use appropriate subscripts for clarity.

5.2.1   Circuits with Switches

Let us now formalize our ideas little further. All the time[3] we are talking about connecting and disconnecting instantaneously. This may be accomplished by employing switches in the circuits. We shall illustrate this in the following example.

Example 5.2 In the circuit shown in Fig. 5.4, a switch connects the capacitance C to the dc voltage source, but disconnects the resistances, in the past (i.e., for $t < 0$). The same switch connects the resistances in the present and future (i.e., for $t \geq 0$) and disconnects the source. We depict such a switch as a lever with an arrow in the appropriate direction and with a time index as shown in Fig. 5.4. We shall study this circuit in detail to understand the idea of time-constant.

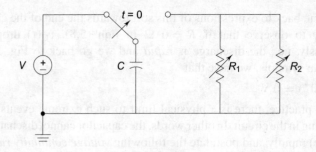

Fig. 5.4

The capacitance would be charged to

$$v_C(0^+) = V \text{ volts} = v_C(0^-)$$

at the instant it is connected across the potentiometers. If R_{eq} denotes the parallel combination of the resistances, 'as seen by the capacitance for $t \geq 0$', we have

$$v_C(t \geq 0) = V e^{-t/R_{eq}C} \text{ volts}$$

Let us take a simple practical device as an analogy to sharpen our intuition. The circuit in Fig. 5.4 could be a model of the LPG stove we use in our kitchens—the LPG cylinder is like a capacitance which is 'refilled' periodically and the potentiometers represent the variable burners of the stove.

Case A If we do not use the stove at all, it is equivalent to saying that both the resistances are open circuits and hence $R_{eq} = \infty = \tau$. In such a situation,

$$v_C(t \geq 0) = V \text{ volts}$$

and the cylinder never requires refilling. This happens because the term $\tau = \infty$ sits in the denominator of the exponential and hence $e^{-t/\tau} = 1$.

---

[3] pun intended!

Case B If we prefer to use the stove in the 'sim' mode, i.e., with high resistances and hence larger time-constant, it takes a longer time for the capacitance to discharge, i.e., we may use the cylinder for a longer period.

Case C If we prefer to use the stove in the 'high' mode, i.e., with low resistances and hence smaller time-constant, it takes a very short time for the capacitance to discharge.

We show the three cases in Fig. 5.5. Observe the scales on the *x* axis and *y* axis carefully. On the *y* axis, we have normalized the magnitude to 1 unit.

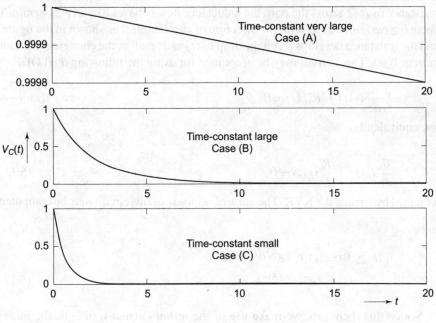

Fig. 5.5

In a way, the time-constant is an index of *inertia* in electrical systems.

5.3 Circuits with Inductors

Let us now quickly explore a circuit having an inductor and a resistor as shown in Fig. 5.6.

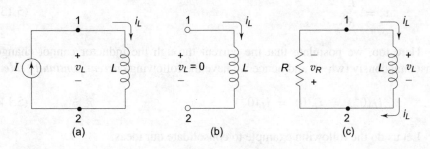

Fig. 5.6

Notice that this figure is drawn as a *dual* to Fig. 5.2. Recall that we had the following analogy in Chapter 2:

capacitor : voltage :: inductor : current

If an independent current source is connected across an inductor of L H it charges up to I A, i.e., $i_L(0^-) = I$ A. Further, since this current is constant, the voltage across the inductance is $v_L(t < 0) = 0$ V, i.e., the inductor behaves like a *short circuit* for dc voltages. If we disconnect the source at $t = 0$, the inductor still retains this current since it has to obey Kirchhoff's rules. At this instant, if we connect a resistance of R Ω across the port, the inductance now behaves like an *exponentially decaying current source*, with the direction of current intact as shown in the figure, and the resistance develops a voltage drop $v_R(t)$ as a result of the changing inductor current $i_L(t)$. This current may be accounted for using the following dual ODE:

$$ L \frac{d}{dt} i_L(t) + R i_L(t) = 0 $$

or, equivalently,

$$ \frac{d}{dt} i_L(t) + \frac{R}{L} i_L(t) = 0 \tag{5.11} $$

obtained by writing the KVR. The *natural response* of this circuit may be computed

to be

$$ i_L(t \geq 0) = i_L(0^+) e^{-t/(L/R)} \text{ A} $$
$$ = I e^{-t/(L/R)} \text{ A} \tag{5.12} $$

Notice that, here too, we make use of the *initial current* $i_L(0^+)$ as the *initial condition* to solve the ODE.

Exercise 5.1 Verify Eqns (5.11) and (5.12).

In this circuit, we may identify that the time-constant is

$$ \tau = \frac{L}{R} \tag{5.13} $$

Here too, we postulate that the current through the inductor cannot change instantaneously (why?) and hence we have the following *current continuity rule*:

$$ i_L(0^+) = i_L(0) = i_L(0^-) \tag{5.14} $$

Let us do the following example to consolidate our ideas.

Example 5.3 Consider the circuit shown in Fig. 5.7.

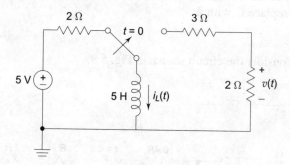

Fig. 5.7

Notice that the 2 Ω resistance in series with the voltage source in the past is mandatory (Why?). We may quickly compute that

$$i_L(0^+) = \frac{5\,\text{V}}{2\,\Omega} = 2.5\,\text{A}$$

$$i_L(t \geq 0) = 2.5\,e^{-t/\tau}\,\text{A}$$

and

$$v(t \geq 0) = -5e^{-t/\tau}\,\text{V}$$

where, for $t \geq 0$,

$$R_{\text{eq}} = 3 + 2 = 5\,\Omega$$

and hence

$$\tau = \frac{5\,\text{H}}{5\,\Omega} = 1\,\text{s}$$

We place emphasis on the polarity of the voltage developed across the inductance. As long as it is getting charged, with a constant current, it is like a short circuit with no voltage developed across it. However, when it is discharging, the direction of current does not change. Accordingly, to obey Kirchhoff's rules, a voltage has to develop according to the sign convention for current sources. It is in this sense we say that the inductor behaves like an exponentially decaying current source.

In passing, we just comment that the inductance of 5 H could be, for instance, a series combination of five thousand 1-mH inductors in series. Likewise, the resistance of 5 Ω could be an equivalent resistance offered by a large network of resistors. The simple *RL* (or the *RC*) circuit is just a series combination of equivalent resistance and equivalent inductance (or capacitance) in the sense of Thèvenin.

We also note that the expression $\tau = L/R$ for the time-constant of the RL circuit is a dual to that of the RC circuit—replace the resistance R with the conductance $G = 1/R$ and replace C with L.

Example 5.4 Consider the circuit shown in Fig. 5.8.

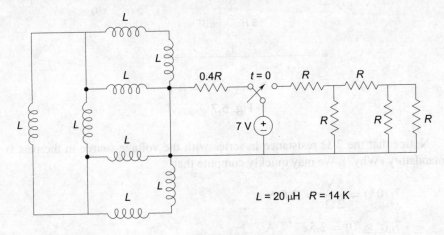

Fig. 5.8

This circuit appears to be quite messy at the outset, but can be reduced to the simple RL circuit of Fig. 5.6(c). We have the following series-parallel connections of the inductors:

$$[(L + L)\|L + L + (L + L)\|L]\,\|L = \left(\frac{2L}{3} + L + \frac{2L}{3}\right)\|L = \frac{7L}{10}$$

with the following series-parallel connections of the resistors, at $t \geq 0$:

$$[(R \| R) + R)\|R] + R + 0.4R = 2R$$

Thus, we get

$$L_{eq} = 14\ \mu H$$

$$R_{eq} = 28\ k\Omega$$

and

$$\tau = 0.5\ ms$$

and the initial current flowing into the bank of inductors, according to the sign convention, is

$$i_L(0^-) = \frac{7\,V}{0.4 \times 14\,k\Omega} = 1.25\,mA$$

For $t \geq 0$, the current *leaving* the bank of resistors would be

$$i_L(0^+) = -1.25\,mA$$

and this current decays with a time-constant of 0.5 ms, i.e.,

$$i_L(t \geq 0) = -1.25\,e^{-2000t}\ \text{mA}$$

Exercise 5.2 Determine $v(t \geq 0)$ in the circuit of Fig. 5.9.

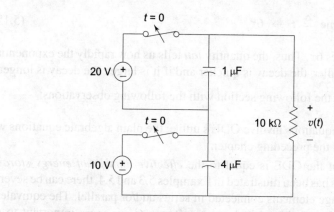

Fig. 5.9

(*Ans.* $C_{eq} = 4/5\ \mu$ F, $\tau = 8$ ms, and $v(0^+) = 30$ V)

5.4 More About the Time-constant

Let us now examine the exponential decay in some more detail. We shall consider the function

$$f(t) = Pe^{-t/\tau}$$

At the instant $t = 0$, $f(t = 0) = P$. However, we may quickly compute that

$$f(t = \tau) = 0.368P$$

i.e., at the time instant $t = \tau$, the exponential *falls to 36.8% of its initial value.* Looking at the response from this perspective, we define the time-constant in a more general way as follows.

Definition 5.4 Time-constant

A time constant is an interval in which the exponential falls to 36.8% of its initial value.

Let us now see what happens further. We may compute that

$$f(t = 2\tau) = 0.135P$$
$$f(t = 3\tau) = 0.05P$$
$$f(t = 4\tau) = 0.02P$$
$$f(t = 5\tau) = 0.007P$$
$$f(t = 6\tau) = 0.002P$$

We observe that in an interval of about 'six time-constants' the exponential falls to about 0.2% of its initial value. For all practical purposes we consider that $f(t)$ has decayed to zero and, accordingly, we define

Definition 5.5 Infinite time

$$\text{infinite time} \triangleq t \to 6\tau \tag{5.15}$$

i.e., as t approaches 6τ. Thus, the quantity *tau* tells us how rapidly the exponential decays; if it is smaller, the decay is quicker and if it is larger, the decay is longer.

We move on to the following section with the following observations:

1. The circuit equations involve ODE's unlike the plain algebraic equations we have seen in the preceding chapter.

2. The order of the ODE is equal to the *effective number of energy storage elements*. As has been illustrated in Examples 5.3 and 5.4, there can be several energy-storage elements connected in series and/or parallel. The equivalent inductance L_{eq} or the equivalent capacitance C_{eq} is more important to us rather than the connections. It is customary to call the circuit as *first order* if it has effectively only one energy-storage element. We shall pursue *second order* circuits formally in the following section.

3. As long as the passive elements—the resistance, capacitance, and inductance—exhibit the property of time-invariance, we shall have ODE's with *constant coefficients*.

4. For the first-order circuits the time-constant τ is an extremely important parameter. To compute this we need the equivalent inductance or capacitance 'as seen by the equivalent resistance' for $t \geq 0$. In turn, the equivalent resistance may be computed 'as seen by the equivalent inductance or capacitance' for $t \geq 0$. This was illustrated in Example 5.4. The circuit may be reduced to just a pair of nodes for $t \geq 0$. This is in tune with Thévenin's theorem.

5. There is an exciting result in store for us as a consequence of the linear relationship between voltage and current in our circuits. For the RC circuit we have obtained the ODE in Eqn (5.7) in terms of the voltage $v_C(t)$ across the capacitor. Since the voltage $v_R(t)$ across the resistor is identical to $v_C(t)$ for $t \geq 0$, we may obtain the following ODE to solve for $v_R(t)$:

$$\frac{d}{dt} v_R(t) + \frac{1}{RC} v_R(t) = 0$$

The solution is $v_R(0^+) e^{-t/\tau}$. Observe that this solution is of the same form as $v_C(t)$, but the initial condition is $v_R(0^+)$. For this particular circuit it so happens that $v_R(0^+) = v_C(0^+)$ and we verify that $v_R(t) = v_C(t)$ for all $t \geq 0$. We emphasize that the voltage continuity rule applies to the capacitor alone and hence we need to determine $v_C(0^+)$ first and then determine $v_R(0^+)$ in terms of $v_C(0^+)$.

Next, the current $i(t)$ in the circuit for $t \geq 0$ is simply $v_R(t)/R = v_C(t)/R$. Accordingly, we may obtain the following ODE to solve for $i(t)$:

$$\frac{d}{dt}i(t) + \frac{1}{RC}i(t) = 0$$

The solution is $i(0^+)e^{-t/\tau}$. Observe, once again, that this solution is of the same form as that of $v_C(t)$, but the initial condition is $i(0^+)$, which has to be determined from $v_C(0^+)$.

Thus, the response of any other element in the *linear circuit* at any instant of time is *linearly related* to that of the energy-storage element, at that particular instant of time. In other words, the voltage across and the current through any branch in a linear circuit decay exponentially *at the same rate*. In fact, this is illustrated in Example 5.3. This prompts us to compute the response $y(t)$ in any branch, be it the current through or the voltage across, of the circuit as

$$y(t) = y(0^+)e^{-t/\tau}, \tag{5.16}$$

where

$$\tau = \begin{cases} R_{eq}C_{eq} & \text{for } RC \text{ circuits} \\ \dfrac{L_{eq}}{R_{eq}} & \text{for } RL \text{ circuits} \end{cases}$$

This equation suggests us the following simple algorithm to compute the response:

Step 1	Determine the time-constant τ of the circuit.
Step 2	Determine the initial voltage (or initial current) $y(0^+)$ in the branch of interest.
Step 3	The response $y(t)$ at any instant of time is given by $y(t) = y(0^+)e^{-t/\tau}$.

Example 5.5. Consider the circuit shown in Fig. 5.10(a).

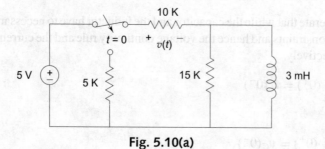

Fig. 5.10(a)

Looking at the circuit we may quickly compute

$$i_L(0^-) = \frac{5\,\text{V}}{10\,\text{k}\Omega} = 0.5\,\text{mA}, \quad v(0^-) = 5\,\text{V}$$

in the past. In the present, we have

$$i_L(0^+) = 0.5\,\text{mA}$$

flowing in the same old direction, i.e., from top to bottom. The circuit for $t \geq 0$ is shown in Fig. 5.10(b).

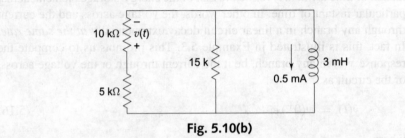

Fig. 5.10(b)

From the circuit in Fig. 5.10(b), it is clear that the inductor acts like a 0.5 mA current source, and according to the current division rule, a current of 0.25 mA flows through the 10 kΩ resistance, and hence

$$v(0^+) = 2.5\,\text{V}$$

The equivalent resistance 'as seen by' the inductance is

$$15\,\text{k}\Omega \parallel (5\,\text{k}\Omega + 10\,\text{k}\Omega) = 7.5\,\text{k}\Omega$$

The time-constant is

$$\tau = \frac{L}{R_{eq}} = 0.4\,\mu\text{s}$$

Hence,

$$v(t \geq 0) = 2.5\,e^{-2.5\times10^6 t}\,\text{V}$$

We reiterate that while the capacitor and the inductor have to necessarily obey the physical constraints and hence the voltage continuity rule and the current continuity rule, respectively,

$$i_L(0^+) = i_L(0^-)$$

and

$$v_C(0^+) = v_C(0^-)$$

there is no such restriction on the resistor since the voltage and current are *not* related by a differential. Further, the inductor voltage $v_L(t)$ and the current $i_C(t)$ through the capacitor too need not obey the continuity rules.

Exercise 5.3 In the circuit of Fig. 5.10(b) above, determine the voltage $v_L(t)$ across the inductor for $t \geq 0$. (***Ans.*** $3.75\, e^{-t/\tau}$ V)

5.5 Circuits with an Inductor and a Capacitor

Let us now consider the circuit shown in Fig. 5.11 with two independent energy-storage elements connected in *series* along with a resistance for all time $t \geq 0$, i.e., when both the switches are thrown to position 2.

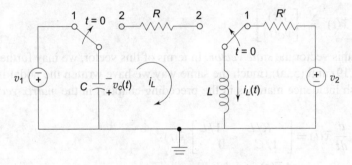

Fig. 5.11

This circuit is more popularly called *series RLC circuit* since all the three passive elements are in series for $t \geq 0$. We have shown two independent sources charging the two energy-storage elements for $t < 0$ for convenience. In practice, one source is sufficient and this is shown in the examples later in this chapter. One might immediately expect that this circuit would be described by a *second-order ODE* (with constant coefficients, of course) since there are effectively two energy-storage elements. There is much more in store[4] for us.

First let us observe the polarity of the potential difference and the direction of current. Applying KVR we get the following *homogeneous* equation:

$$v_C(t) + Ri(t) + L\frac{d}{dt}i(t) = 0 \qquad (5.17)$$

If we wish to solve for the current $i(t)$, we need the voltage $v_C(t)$ across the capacitance so that

$$i(t) = C\frac{d}{dt}v_C(t) \qquad (5.18)$$

---
[4] pun unintentional!

and the voltage $v_C(t)$, in turn, depends on the the current $i(t)$ according to Eqn (5.17). Observe that we have a *pair of ODE's* and we need the solution to one of them to obtain the solution of the other. Accordingly, we say that these equations are *coupled ODE's*. Since $i(t)$ and $v_C(t)$ give us the 'status' of the circuit at any instant of time, we call them *state variables*. We realize that these two quantities are *independent variables* in the sense that they describe two independent energy-storage elements. The word 'state' here refers to the current status, and has to be interpreted just the same way as the words statement, state-of-the-art, etc.

Before proceeding to solve these equations, let us rewrite them as

$$\frac{d}{dt} i(t) = -\frac{R}{L} i(t) - \frac{1}{L} v_C(t) \qquad\qquad (5.19)$$

$$\frac{d}{dt} v_C(t) = \frac{1}{C} i(t) \qquad\qquad (5.20)$$

Let us now define a vector $\bar{x}(t)$ as

$$\bar{x}(t) \triangleq \begin{bmatrix} i(t) \\ v_C(t) \end{bmatrix} \qquad\qquad (5.21)$$

We call this vector the *state vector*. In terms of this vector, we may further rewrite Eqns (5.19) and (5.20) (much the same way we have written the nodal incidence and mesh incidence matrices in the preceding chapter) in the *matrix-vector form* as

$$\frac{d}{dt} \bar{x}(t) = \begin{bmatrix} -R/L & -1/L \\ 1/C & 0 \end{bmatrix} \bar{x}(t)$$

$$= A \bar{x}(t) \qquad\qquad (5.22)$$

where A represents the 2×2 matrix called the *circuit matrix*. The solution to this matrix-vector differential equation (what is the order?) is given in terms of the eigenvalues of the matrix A. We shall pursue this approach in more detail in Chapter 9. There is an exciting subject called 'modern control theory', generally taught in later semesters, which emphasizes equations of this kind and several ways of obtaining their solutions. For the time being, we shall look at an interesting insight provided by Eqn (5.22). Suppose we are interested in the current $i(t)$ in the circuit. This may be simply written as the matrix-vector equation

$$i(t) = \begin{bmatrix} 1 & 0 \end{bmatrix} \bar{x}(t)$$

Similarly, if we are interested in the voltage $v_C(t)$, we may write it as

$$v_C(t) = \begin{bmatrix} 0 & 1 \end{bmatrix} \bar{x}(t)$$

Further, if we are interested in the voltage $v_R = Ri(t)$ across the resistance, we may write it as

$$v_R(t) = \begin{bmatrix} R & 0 \end{bmatrix} \bar{x}(t)$$

Notice that we have simply *scaled* the current $i(t)$, i.e., multiplied with a scalar R. Still further, if we are interested in the voltage $v_L(t)$ across the inductance we may write it, using KVR, as

$$v_L(t) = -v_C(t) - Ri(t)$$
$$= \begin{bmatrix} -R & -1 \end{bmatrix} \overline{x}(t)$$

Notice that we were able to express this voltage $v_L(t)$ as a *linear combination* of the state variables without having to differentiate the current $i(t)$. Finally, if we are simultaneously interested in all the three voltages—v_C, v_R, v_L—we may write them as

$$\overline{y}(t) \triangleq \begin{bmatrix} v_C(t) \\ v_R(t) \\ v_L(t) \end{bmatrix} = \begin{bmatrix} 0 & 1 \\ R & 0 \\ -R & -1 \end{bmatrix} \overline{x}$$

In general, whatever may be the quantity we are interested in, we may write it as

$$\overline{y}(t) = E\,\overline{x}(t) \tag{5.23}$$

where the *output vector* $\overline{y}(t)$ denotes the quantities of interest to us. In general, the dimension of the circuit matrix is $n \times n$, i.e., it is a square matrix. Similarly, the dimension of $\overline{y}$ could be $m \times 1$ when there are m quantities of interest to us to be computed, and hence the dimension of the E matrix is $m \times n$. We emphasize that *linear algebra* plays an important role in all the manipulations we make here. We recommend the readers to learn it rigorously so that they would appreciate more abstract courses such as control systems and signal processing. We shall provide a basis to understand the linear algebra in Chapter 9.

Let us now turn our attention to solving for the two independent variables $i(t)$ and $v_C(t)$. Equations (5.19) and (5.20) may be combined and written as

$$\frac{d^2}{dt^2} i(t) + \frac{R}{L}\frac{d}{dt} i(t) + \frac{1}{LC} i(t) = 0 \tag{5.24}$$

or, equivalently, as

$$\frac{d^2}{dt^2} v_C(t) + \frac{R}{L}\frac{d}{dt} v_C(t) + \frac{1}{LC} v_C(t) = 0 \tag{5.25}$$

Notice the 'duality' in the two equations. The reader is strongly advised not to make nonsensical conclusions such as $v_C(t) = i(t)$. We shall see how the story unfolds beautifully before us.

Exercise 5.4 Derive Eqns (5.24) and (5.25) from Eqns (5.19) and (5.20).

Thus, we got the second-order ODE our intuition told us at the beginning of this section. We solve this in the following, rather popular, way. Let us choose to solve Eqn (5.24) for $i(t)$. First, to match the physical dimensions we observe that L/R and $\sqrt{LC}$ should have the units of time; the former is already familiar to us, as the time-constant, from the RL circuit. Accordingly, their reciprocals would have the units of frequency. In the light of this observation, we shall define two parameters:

natural frequency

$$\omega_n \triangleq \frac{1}{\sqrt{LC}} \qquad (5.26)$$

damping factor

$$\zeta \triangleq \frac{R}{2}\sqrt{\frac{C}{L}} \qquad (5.27)$$

so that we may rewrite Eqn (5.24) as

$$\frac{d^2}{dt^2} i(t) + 2\zeta\omega_n \frac{d}{dt} i(t) + \omega_n^2 i(t) = 0 \qquad (5.28)$$

Next, we observe that the equation has three terms—$i(t)$ itself, its first derivative, and its second derivative—all of which have to be homogeneous. What would be an appropriate function $i(t)$ satisfying this condition? A moment's thought would suggest us either a sinusoid or an exponential. We prefer the latter for two reasons—(i) in this chapter we have seen the response of a first-order circuit to be an exponential, and (ii) a sinusoid may also be written as an exponential using Euler's formula.

Let us *guess* that the solution is e^{st}, where s is some complex number. Why should it be complex? We shall discover the answer as we go through. Substituting our guess in Eqn (5.28), we get

$$\left(s^2 + 2\zeta\omega_n s + \omega_n^2\right) i(t) = 0$$

Since $i(t) \neq 0$, we have

$$s^2 + 2\zeta\omega_n s + \omega_n^2 = 0 \qquad (5.29)$$

i.e., the complex number s in our guess should satisfy this equation. Observe that s has the units of frequency. We attribute nepers per second as the units to distinguish this from the popular notion of frequency measured either in 'cycles per second (Hz)' or 'radians per second'. We call this equation the *characteristic equation* for an obvious reason we shall see just now. We notice that this characteristic equation is quadratic and hence has two roots—s_1 and s_2. This suggests that, in fact, there are two solutions—$e^{s_1 t}$ and $e^{s_2 t}$—to our ODE in Eqn (5.28). It is yet another interesting fact that a *linear combination* of these two solutions is a solution since our ODE is linear. Recall elementary lessons in calculus—the derivative and the integral are linear operators, for example,

$$\frac{d}{dt}\left[px(t) + qy(t)\right] = p\frac{d}{dt}x(t) + q\frac{d}{dt}y(t)$$

Thus, the *general solution* to our ODE is

$$i(t) = A\,e^{s_1 t} + B\,e^{s_2 t} \qquad (5.30)$$

Notice that it is the roots s_1 and s_2 that *characterize* how the response $i(t)$ looks like, and this is the reason we have given the name 'characteristic equation' to Eqn (5.29).

Exercise 5.5 Verify that substituting the general solution in Eqn (5.30) into Eqn (5.28) also results in the same characteristic equation in Eqn (5.29).

We call this a 'general' solution because still we have to solve for the coefficients A and B. These may be solved using the initial conditions

$$\boxed{i(0^+)}$$

and

$$\boxed{\frac{d}{dt}i(t)|_{t=0^+}}$$

Further, if we were solving Eqn (5.25) for $v_C(t)$ we would get the same *general* solution with a different set of coefficients A and B since the initial conditions $v_C(0^+)$ and $d/dt\, v_C(t)|_{t=0^+}$ are different in this case.

While we are solving for the inductor current $i(t)$ the initial condition $i(0^+)$ is straightforward—it is the current to which the inductor is initially charged to. However, the second initial condition $d/dt\,[i(t)]$ at $t = 0^+$ might cause some concern since the other initial charge in the circuit we know is the voltage $v_C(0^+)$ across the capacitor. We shall resolve this using the KVR around the mesh. Since

$$v_C(t) + v_R(t) + v_L(t) = 0 \;\; \forall\, t$$

it follows that, since $v_L(t) = L\, di_L(t)/dt$,

$$\frac{d}{dt}i(t)|_{t=0^+} = -\frac{Ri(0^+) + v_C(0^+)}{L}$$

In a similar fashion, while solving for $v_C(t)$, we might determine $d/dt\,[v_C(t)]$ at $t = 0^+$ using the initial current in the inductor. We emphasize that writing the circuit equations initially in terms of the state variables helps us identify the initial conditions directly.

Let us dig a little deeper. The roots of the characteristic equation are

$$s_{1,2} = -\zeta\omega_n \pm \omega_n\sqrt{\zeta^2 - 1} \tag{5.31}$$

Looking at the discriminant we have the following possibilities enumerated in Secs 5.5.1–5.5.4.

5.5.1 Case (i): $\zeta > 1$

In this case, the two roots are

$$s_1 = -\zeta\omega_n + \omega_n\sqrt{\zeta^2 - 1}$$

$$s_2 = -\zeta\omega_n - \omega_n\sqrt{\zeta^2 - 1} \tag{5.32}$$

i.e., the roots are *(negative) real and distinct* and the response is a sum of two exponentials, one of which decaying slightly faster than the other one. The constants A and B may be obtained as follows. Given the initial conditions $i(0^+)$ and $d/dt\,[i(t)]$ at $t = 0^+$, we notice that

$$A + B = i(0^+) \tag{5.33}$$

$$s_1 A + s_2 B = \frac{d}{dt} i(t)|_{t=0^+}$$

$$= -\frac{Ri(0^+) + v_C(0^+)}{L} \tag{5.34}$$

Equations (5.33) and (5.34) may be put in the following matrix-vector form:

$$\begin{pmatrix} 1 & 1 \\ s_1 & s_2 \end{pmatrix} \begin{pmatrix} A \\ B \end{pmatrix} = \begin{pmatrix} 1 & 0 \\ -\dfrac{R}{L} & -\dfrac{1}{L} \end{pmatrix} \begin{pmatrix} i(0^+) \\ v_C(0^+) \end{pmatrix}$$

In passing, we just mention that matrices such as the coefficient matrix on the left hand side are called *Vandermonde matrices*. It turns out the matrix A we have defined earlier in Eqn (5.22) has the same characteristic equation and hence the eigenvalues are s_1 and s_2. The eigenvectors corresponding to these eigenvalues appear as the columns of the Vandermonde matrix. Interested reader may verify this. We may solve this pair of linear equations to obtain the constants A and B. From the equations, we immediately observe that each of the constants A and B is a *linear combination* of the initial conditions $i(0^+)$ and $v_C(0^+)$. Further, we may notice that if s_1 and s_2 are complex conjugates so are the constants A and B.

Exercise 5.6 Verify that the constants A and B turn out to be complex conjugates if s_1 and s_2 are complex conjugates.

The general shape of the response is shown in Fig. 5.12.

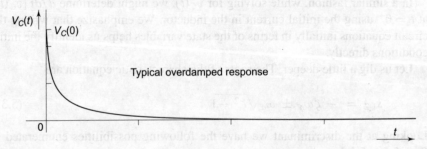

Typical overdamped response

Fig. 5.12

Example 5.6 Let us consider the series RLC circuit shown in Fig. 5.11 with

$$R = 5\,\Omega, \quad L = 0.2\,\text{H}, \quad C = 50\,\text{mF}$$

and

$$i(0^+) = 0\,\text{A}, \quad v_C(0^+) = 12\,\text{V}$$

In this case,

$$\zeta = \frac{5}{2}\sqrt{\frac{50 \times 10^{-3}}{0.2}} = 1.25$$

and

$$\omega_n = \frac{1}{\sqrt{0.2 \times 50 \times 10^{-3}}} = 10\,\text{rad/s}$$

and

$$s_1 = 10\,(-1.25 + \sqrt{1.25^2 - 1}) = -5$$
$$s_2 = 10\,(-1.25 - \sqrt{1.25^2 - 1}) = -20$$

We may use Eqns (5.33) and (5.34) and determine the constants A and B as

$$\begin{pmatrix} 1 & 1 \\ -5 & -20 \end{pmatrix} \begin{pmatrix} A \\ B \end{pmatrix} = \begin{pmatrix} 1 & 0 \\ -25 & -5 \end{pmatrix} \begin{pmatrix} 0 \\ 12 \end{pmatrix}$$

or

$$\begin{pmatrix} A \\ B \end{pmatrix} = \begin{pmatrix} -4 \\ 4 \end{pmatrix}$$

The response may be written as

$$i(t) = -4\,e^{-5t} + 4\,e^{-20t}\,\text{A}$$

This response is plotted in Fig. 5.13.

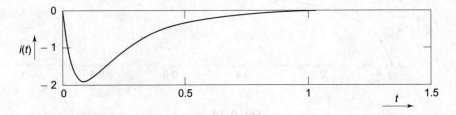

Fig. 5.13

The voltage drop across the resistance may be readily found to be

$$v_R(t) = Ri(t) = -20\,e^{-5t} + 20\,e^{-20t}\quad\text{V}$$

The voltage drop across the capacitance may be computed to be

$$v_C(t) = -L\frac{d}{dt}i(t) - v_R(t)$$

$$= 16\,e^{-5t} - 4\,e^{-20t} \text{ V}$$

Notice that we have got $v_C(0^+) = 12$ V.

We recommend the following general algorithm for the analysis of series *RLC* circuits:

Step 1	Determine the ζ, ω_n and the constants A and B.
Step 2	Determine the current through the inductor:
	$i(t) = A\,e^{S_1 t} + B\,e^{S_2 t}$
Step 3	Deduce the voltages $v_R(t)$ and $v_C(t)$
	using $i(t)$ and $di(t)/dt$.
Step 4	Verify the initial conditions and
	the KVR around the loop:
	$v_R(t) + v_L(t) + v_C(t) = 0 \ \forall t$
	to ensure the correctness of the solution.

Example 5.7 Let us consider the same circuit as in Example 5.6, but with a different set of initial conditions:

$$i(0^+) = -4\,A$$

and

$$v_C(0^+) = 10\,V$$

In this case, the response may be verified to be

$$i(t) = -2\left(e^{-5t} + e^{-20t}\right) \text{ A}$$

This is shown in Fig. 5.14.

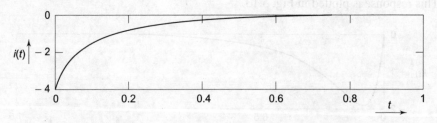

Fig. 5.14

Notice that the exact shape of the response depends on the constants A and B, which in turn depend upon the initial conditions.

Exercise 5.7 Verify the results of Example 5.7 by computing $v_R(t)$ and $v_C(t)$.

Exercise 5.8 Obtain the response of the circuit shown in Fig. 5.11 for the following parameters: $R = 0.5 \, \text{k}\Omega$, $L = 1 \, \text{mH}$, $C = 0.4 \, \mu\text{F}$ with $i(0^+) = -2 \, \text{A}$ and $v_C(0^+) = -5 \, \text{V}$.

Exercise 5.9 A series RLC circuit has the following natural response:

$$i(t) = 7 e^{-5t} - 3 e^{-t} \, \text{A}$$

If $R = 6 \, \Omega$, find the inductance L and the capacitance C. (**Hint** Express ζ and ω_n in terms of s_1 and s_2.)

(**Ans.** $L = 1 \, \text{H}$ and $C = 0.2 \, \text{F}$.)

5.5.2 Case (ii): $\zeta = 1$

In this case, the roots are

$$s_{1,2} = s = -\omega_n \tag{5.35}$$

i.e., the roots are *(negative) real, but identical*. If this is the case [5], e^{st} and $d/ds \, e^{st} = t e^{st}$ are the two solutions and hence the linear combination

$$i(t) = (At + B) e^{-\omega_n t} \tag{5.36}$$

is the solution to our ODE. This appears to be a slightly tricky function. However, we can tame it using L'Hospital rule—for very small values of t (i.e., $t \to 0$), the term At dominates since the exponential is close to $e^0 = 1$; for arbitrarily large values of t (i.e., $t \to \infty$), the exponential which is now close to $e^{-\infty} = 0$ dominates. The constants A and B may be evaluated using $i(0^+)$ and $d/dt \, [i(t)]$ at $t = 0^+$ as follows:

$$B = i(0^+), \tag{5.37}$$

$$A + sB = -\frac{R i(0^+) + v_C(0^+)}{L} \tag{5.38}$$

---

[5] Suppose we have a polynomial $f(s)$ that has two identical roots at $s = s_0$. We may write this polynomial as

$$f(s) = (s - s_0) g(s)$$

If we differentiate both sides with respect to s and evaluate it at $s = s_0$, we have

$$\tfrac{d}{ds} f(s)|_{s=s_0} = g(s)|_{s=s_0}$$

Since the *multiplicity* of the root s_0 is 2, it follows that s_0 is a root of $g(s)$ as well and hence

$$\tfrac{d}{ds} f(s)|_{s=s_0} = 0$$

We may apply this result to Eqn (5.28) to get

$$(s + \omega_n)^2 e^{st} = 0$$

suggesting that $i(t) = e^{-\omega_n t}$ is a solution, as expected. Further, once again differentiating both sides with respect to s and evaluating at $s = -\omega_n$ gives us

$$(s + \omega_n)^2 (t e^{st})|_{s=-\omega_n} = 0$$

and it follows that

$$i(t) = t e^{-\omega_n t}$$

is also a solution.

Here the matrix-vector form appears as follows:

$$\begin{pmatrix} 0 & 1 \\ 1 & s \end{pmatrix} \begin{pmatrix} A \\ B \end{pmatrix} = \begin{pmatrix} 1 & 0 \\ -\frac{R}{L} & -\frac{1}{L} \end{pmatrix} \begin{pmatrix} i(0^+) \\ v_C(0^+) \end{pmatrix}$$

Here, comparing with the matrix-vector form for case (i), we notice that the first column of the coefficient matrix is *also* differentiated with respect to s. This is also a Vandermonde matrix.

We shall look at the following examples for more accuracy in the shape of the response with respect to time. We shall make use of the same algorithm as before for computing the responses. However, notice that we employ Eqns (5.37) and (5.38) in the present case to compute the constants A and B in step 1.

Example 5.8 Consider the series RLC circuit shown in Fig. 5.11 with the following parameters:

$$R = 4\,\Omega, \quad L = 0.2\,\text{H}, \quad C = 50\,\text{mF}$$

and

$$i(0^+) = 0\,\text{A}, \quad v_C(0^+) = 12\,\text{V}$$

It is easy to compute that

$$\zeta = 1, \quad \omega_n = 10\,\text{rad/s}$$

and hence

$$A = -60 \quad \text{and} \quad B = 0$$

The current $i(t)$ in the loop may be computed to be

$$i(t) = -60\,t\,e^{-10t}\;\text{A}$$

This is plotted in Fig. 5.15.

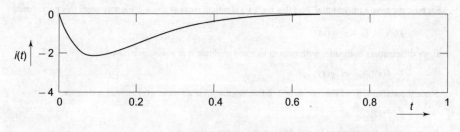

Fig. 5.15

Observe from the figure that initially the response looks like a straight line— $-60t$ with respect to the time axis, with a negative slope. Thereafter, the response takes an exponential shape and approaches 0 as $t \to \infty$, despite the presence of the multiplying factor t.

Exercise 5.10 Verify the results of Example 5.8 by computing $v_R(t)$ and $v_C(t)$ and verifying the *KVR* around the loop.

Example 5.9 Consider the same series RLC circuit as in Example 5.8 with the following initial conditions: $i(0^+) = 3$ A, and $v_C(0^+) = -8$ V.

In this case, we may compute that

$$i(t) = (10t + 3) e^{-10t} \text{ A}$$

This is plotted in Fig. 5.16(a). Observe that the response has two terms, one of which is a purely decaying exponential. It is this term that, more or less, dictates the response in this case.

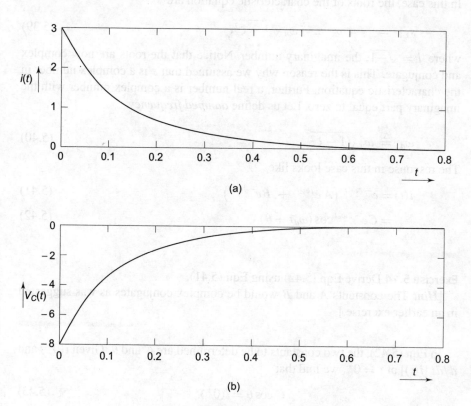

(a)

(b)

Fig. 5.16

Let us determine the voltage $v_C(t)$ across the capacitor:

$$v_C(t) = -Ri(t) - L \frac{d}{dt} i(t)$$

$$= -(20t + 8) e^{-10t} \text{ V}$$

A plot of this voltage is given in Fig. 5.16(b).

Exercise 5.11 Sketch the response $v_R(t)$ in Example 5.9. What do you observe?

Exercise 5.12 Sketch the response of the series RLC circuit with the following parameters: $R = 5\,\Omega$, $L = 0.5$ H, $C = 0.125$ F, with $i(0^+) = 1$ A and $v_C(0^+) = 5$ V.

Exercise 5.13 Find the time t_p at which the response $i(t)$ of the series RLC circuit with $\zeta = 1$ attains its peak amplitude. Also, determine the percentage of the stored energy still left in the circuit. [**Hint** Differentiate $i(t)$ with respect to time and equate it to 0.]

5.5.3   Case (iii): $\zeta < 1$

In this case, the roots of the characteristic equation are

$$s_{1,2} = -\zeta \omega_n \pm j \omega_n \sqrt{1 - \zeta^2}, \tag{5.39}$$

where $j = \sqrt{-1}$, the imaginary number. Notice that the roots are now complex and conjugate. This is the reason why we assumed that s is a complex number in the characteristic equation. Further, a real number is a complex number with the imaginary part equal to zero. Let us define *damped frequency*,

$$\omega_d \overset{\Delta}{=} \omega_n \sqrt{1 - \zeta^2} \tag{5.40}$$

The response in this case looks like

$$i(t) = e^{-\zeta \omega_n t} \left(A e^{j \omega_d t} + B e^{-j \omega_d t} \right) \tag{5.41}$$

$$= C e^{-\zeta \omega_n t} \cos(\omega_d t + \theta) \tag{5.42}$$

Exercise 5.14 Derive Eqn (5.42) using Eqn (5.41).
[**Hint** The constants A and B would be complex conjugates as was suggested in an earlier exercise.]

In Eqn (5.42), the two constants to be determined are C and θ. Given $i(0^+)$ and $d/dt [i(t)]$ at $t = 0^+$, we find that

$$C \cos \theta = i(0^+) \tag{5.43}$$

$$C \omega_n \left(\zeta \cos \theta + \sqrt{1 - \zeta^2} \sin \theta \right) = -\frac{d}{dt} i(t)|_{t=0^+} \tag{5.44}$$

Exercise 5.15 Verify Eqn (5.43) and Eqn (5.44). Obtain the Vandermonde matrix in this case.

We recommend the reader to first compute the constants A and B using the Vandermonde matrix, just as we did for the case $\zeta > 1$. These constants turn out to be complex conjugates of the form

$$\boxed{A = p + jq}$$

and

$$\boxed{B = p - jq = A^*}$$

Now, it would be easy to see that

$$C = 2\sqrt{p^2 + q^2}$$

and

$$\theta = \tan^{-1}\frac{q}{p}$$

Notice that there are some oscillations in the response. Where do they come from? We shall first look at the following examples and then see if we get an answer to the question.

Example 5.10 Consider the series RLC circuit shown in Fig. 5.11 with the following parameters:

$$R = 2.4\,\Omega, \quad L = 0.2\,\mathrm{H}, \quad C = 50\,\mathrm{mF}$$

and

$$i(0^+) = 4\,\mathrm{A}, \quad v_C(0^+) = 3.2\,\mathrm{V}$$

Here we may quickly compute that

$$\zeta = 0.6$$

and

$$s_1 = -6 + j8, \quad s_2 = -6 - j8$$

We may use Eqns (5.33) and (5.34) to compute the constants A and B as

$$A = 2 + j2.5 = B^*$$

Notice that we have got the constants to be complex conjugates. The current $i(t)$ may be written, using Eqn (5.44), as

$$i(t) = e^{-6t}\,(4\cos 8t - 5\sin 8t)$$

$$= \sqrt{41}\,e^{-6t}\cos\left(8t + \tan^{-1}\frac{5}{4}\right)$$

making use of trigonometric identities. The response $i(t)$ is shown in Fig. 5.17.

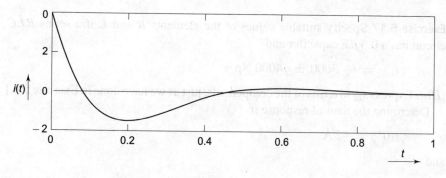

Fig. 5.17

Exercise 5.16 Verify the results of Example 5.10 by computing $v_C(t)$ and $v_R(t)$.

Example 5.11 Consider the series RLC circuit in Fig. 5.11 with the following parameters:

$$R = 10\,\Omega, \quad L = 5\,\text{H}, \quad C = 0.05\,\text{F}$$

and

$$i(0^+) = 2\,\text{A}, \quad v_C(0^+) = -3\,\text{V}$$

We get the following:

$$\zeta = 0.5$$
$$\omega_n = 2\,\text{rad/s}$$
$$\omega_d = \sqrt{3}\,\text{rad/s}$$
$$s_{1,2} = -1 \pm j\sqrt{3}$$

and using Eqn (5.44)

$$i(t) = 2.157\,e^{-t}\cos(\sqrt{3}\,t + 22°)$$

Notice that the initial conditions are satisfied. This response is plotted in Fig. 5.18.

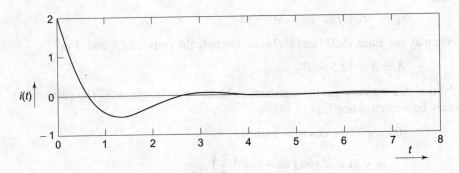

Fig. 5.18

Exercise 5.17 Specify suitable values of the elements R and L if a series RLC circuit has a $0.5\,\mu\text{F}$ capacitor and

$$s_{1,2} = -3000 \pm j4000\,\text{Np/s}$$

[**Hint** Express s_{12} in terms of the components, just as we have done in Exercise 5.9.] Determine the natural response if

$$i(0^+) = -2\,\text{A}$$

and

$$v_C(0^+) = 7\,\text{V}$$

5.5.4 Case (iv): $\zeta = 0$

In this case, the roots are

$$s_{1,2} = \pm j\omega_n \qquad (5.45)$$

and the response looks like

$$i(t) = A \cos (\omega_n t + \theta) \qquad (5.46)$$

i.e., the response is oscillatory. This is rather surprising given that the circuit is *a priori* energized by dc voltages alone. Let us look at an example to gain better physical insight.

Example 5.12 First let us identify that the circuit does not have a resistance since $\zeta = 0$. Thus, we see the following LC circuit in Fig. 5.19 for $t \geq 0$.

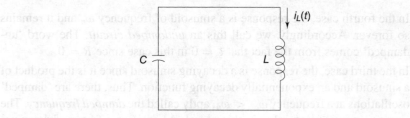

Fig. 5.19

Let us further assume that the capacitor is initially not charged, i.e., $v_C(0^+) = 0$ V. It is easy to see that the response will be

$$i(t) = i(0^+) \cos \omega_n t$$

i.e., the current through the inductor, and hence through the capacitor, 'oscillates' between $+i(0^+)$ and $-i(0^+)$ with a frequency $\omega_n/2\pi$ cycles/s. In particular, the inductor discharges through the capacitor in one half of the cycle and the capacitor returns this charge to the inductor in the next half of the cycle. We shall confirm that this is indeed so. The voltage developed across the capacitor is

$$v_C(t) = -v_L(t) = -L\frac{d}{dt} i(t) = i(0^+)L\omega_n \sin \omega_n t$$

Clearly, the power is conserved. The instantaneous energies of L and C for $t \geq 0$ are given by

$$w_L(t) = \frac{1}{2} Li^2(t) = \frac{1}{4} Li^2(0) (1 + \cos 2\omega_n t)$$

$$w_C(t) = \frac{1}{2} Cv_C^2(t) = \frac{1}{4} Li^2(0) (1 - \cos 2\omega_n t)$$

where we have used the trigonometric identity

$$\cos 2\omega_n t = 2\cos^2 \omega_n t - 1 = 1 - 2\sin^2 \omega_n t$$

and the definition $\omega_n = 1/\sqrt{LC}$.

We observe that both energies alternate between 0 and $0.5Li^2(0)$, but when one of these reaches its maximum, the other reaches its minimum, and vice versa. This suggests that the energy is transferred back and forth between the inductor and the capacitor. It is in this sense that we call ω_n *natural frequency*. The LC circuit is also called an *oscillator*.

It is easy to extend the above idea to the case when the capacitor is also initially charged and we leave it as an exercise to the reader.

Exercise 5.18 Suppose that the capacitor in Fig. 5.19 is charged to $v_C(0^+)$ initially. Verify that the energy flows back and forth between L and C in this case also.

Let us now fix names for each of the four cases we have discussed. We shall begin with the fourth one.

1. In the fourth case, the response is a sinusoid of frequency ω_n and it remains so forever. Accordingly, we call this an *undamped circuit*. The word 'undamped' comes from the fact that $\zeta = 0$ in this case since $R = 0$.

2. In the third case, the response is a decaying sinusoid since it is the product of a sinusoid and an exponentially decaying function. Thus, there are 'damped' oscillations at a frequency $\omega_d < \omega_n$, aptly called the *damped frequency*. The quantity ζ comes into play in this case and serves as a factor of damping, sitting in the decaying exponent. Accordingly, we call this the *damping factor*. Since the oscillations are not completely damped, we call the circuit an *underdamped circuit*. This damping stems from the energy *loss* in the resistance.

3. Extending this further, we may call the circuit in the first case an *overdamped circuit*. There could have been oscillations but the resistance, and hence ζ is too much and the oscillations are overly damped.

4. The second case, which sits between the first and the third, is called the *critically damped* circuit.

Let us now look at another possible arrangement of the three elements.

5.5.5 The Parallel *RLC* Circuit

If we modify the circuit in Fig. 5.11 slightly, we get the *parallel RLC circuit* as shown in Fig. 5.20.

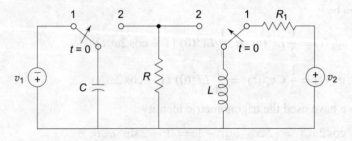

Fig. 5.20

Here, by duality with the series RLC circuit, applying KCR at $t \geq 0$, we get

$$\frac{d^2}{dt^2} i(t) + \frac{1}{RC} \frac{d}{dt} i(t) + \frac{1}{LC} i(t) = 0 \tag{5.47}$$

or, equivalently, as

$$\frac{d^2}{dt^2} v_C(t) + \frac{1}{RC} \frac{d}{dt} v_C(t) + \frac{1}{LC} v_C(t) = 0 \tag{5.48}$$

Here too, we will define two quantities—natural frequency and damping factor—as follows:

$$\omega_n \overset{\Delta}{=} \frac{1}{\sqrt{LC}} \tag{5.49}$$

$$\zeta \overset{\Delta}{=} \frac{1}{2R} \sqrt{\frac{L}{C}} \tag{5.50}$$

Once again, notice the duality with the series RLC circuit—replacing R with $G = 1/R$, L with C, and C with L we get the above definitions. Observe that the factor of '2' does not have a dual. It is important to note the difference:

$$\zeta_{\text{series}} = \frac{R}{2} \sqrt{\frac{C}{L}}$$

and

$$\zeta_{\text{parallel}} = \frac{1}{2R} \sqrt{\frac{L}{C}} = \frac{1}{2} \frac{1}{R \sqrt{C/L}}$$

We may rewrite Eqn (5.47) as

$$\frac{d^2}{dt^2} i(t) + 2\zeta \omega_n \frac{d}{dt} i(t) + \omega_n^2 i(t) = 0 \tag{5.51}$$

Rest of the story is routine. We shall have four different cases—the overdamped, the critically damped, the underdamped, and the undamped circuits; we may ignore the fourth case since it would be identical to the circuit shown in Fig. 5.19 with $R = \infty$. We shall look at the following examples to see what goes on. We follow the same algorithm with slight modifications:

Step 1	Determine the ζ, ω_n and the constants A and B or, equivalently, C and θ for the underdamped case.
Step 2	Determine the voltage across the port: $v_C(t) = A e^{s_1 t} + B e^{s_2 t}$
Step 3	Deduce the currents $i_R(t)$ and $i_L(t)$ using $v_C(t)$ and $dv_C(t)/dt$.
Step 4	Verify the initial conditions and the KCR: $i_R(t) + i_L(t) + i_C(t) = 0 \ \forall t$ to ensure the correctness of the solution.

Example 5.13 In the parallel RLC circuit shown in Fig. 5.20, let

$$R = 150\,\Omega, \quad L = 5\,\text{mH}, \quad C = 20\,\text{nF}$$

and

$$i_L(0^+) = 12\,\text{mA}, \quad v_C(0^+) = 0\,\text{V}$$

For this circuit, we have

$$\zeta = \frac{1}{2}\frac{1}{150}\sqrt{\frac{5\times 10^{-3}}{20\times 10^{-9}}} = \frac{5}{3}$$

$$\omega_n = \frac{1}{\sqrt{5\times 10^{-3}\times 20\times 10^{-9}}} = 100\,\text{krad/s}$$

$$s_1 = -300\,\text{kNp/s}$$

$$s_2 = -33.333\,\text{kNp/s}$$

Let us find the node voltage $v_C(t)$. One of the initial conditions is directly available to us. The other may be found using KCR as follows:

$$C\frac{d}{dt}v_C(t) + \frac{1}{R}v_C(t) + i_L(t) = 0$$

$$\Rightarrow \frac{d}{dt}v_C(t)|_{t=0} = -\frac{1}{C}\left(i_L(0) + \frac{v_C(0^+)}{R}\right) \tag{5.52}$$

We can solve for the constants A and B using the Vandermonde matrix as

$$\begin{pmatrix} 1 & 1 \\ -300\times 10^3 & -33.333\times 10^3 \end{pmatrix}\begin{pmatrix} A \\ B \end{pmatrix} = \begin{pmatrix} 0 & 1 \\ -\frac{1}{C} & -\frac{1}{RC} \end{pmatrix}\begin{pmatrix} i_L(0) \\ v_C(0^+) \end{pmatrix}$$

or

$$\begin{pmatrix} A \\ B \end{pmatrix} = \begin{pmatrix} 9/4 \\ -9/4 \end{pmatrix}$$

Thus, the node voltage is

$$v_C(t) = \frac{9}{4}\left(e^{-300,000t} - e^{-33,333t}\right)\,\text{V}$$

Exercise 5.19 Verify the results of Example 5.13 by computing the currents through the inductor and the resistor and verifying the KCR. Sketch, to a reasonable scale, each of these currents and the voltage $v_C(t)$ as a function of time.

Example 5.14 In the parallel RLC circuit shown in Example 5.13, let $R = 250\,\Omega$. For this case, we may easily compute that

$$\zeta = 1, \quad s = -\omega_n, \quad A = -6\times 10^5$$

and

$$B = 0$$

and the node voltage is

$$v_C(t) = -6 \times 10^5 \, e^{-100,000t} \text{ V}$$

The currents through each of the elements may be computed as follows:

$$i_C(t) = -12 \, (1 - 100,000t) \, e^{-100,000t} \text{ mA}$$

$$i_R(t) = -2400 \, e^{-100,000t} \text{ A}$$

$$i_L(t) = \left(12 \times 10^{-3} + 1200t \right) e^{-100,000t} \text{ A}$$

Exercise 5.20 Verify the results of Example 5.14. Sketch each of the currents and the node voltage as a function of time.

Example 5.15 If we choose the resistance to be $500 \, \Omega$, we get the underdamped case with

$$\zeta = 0.5$$
$$\omega_d = 50 \times \sqrt{3} \text{ krad/s}$$
$$A = j \, 2\sqrt{3}$$
$$B = -j \, 2\sqrt{3}$$

The node voltage may be computed to be

$$v_C(t) = -4\sqrt{3} \, e^{-50,000t} \, \sin(\omega_d t) \text{ V}$$

Exercise 5.21 Verify the results of Example 5.15 and sketch the time functions with a reasonable scale.

In all these examples, we suggest the reader to plot the responses for a better understanding.

Exercise 5.22 Specify suitable elements for a parallel *RLC* circuit so that $s_1 = -2000$ and $s_2 = -5000$ with the constraint that the current in mA through the resistance and the voltage in volts across it at $t = 0$ are numerically equal.

5.5.6   Retrospection

Having done quite a lot of things so far in this chapter, it is time that we look back once to fix certain ideas permanently in our minds.

First, we started with a circuit that has one energy-storage element. This was described by a differential equation, which we solved religiously to obtain the circuit's response. We observed that (i) the ODE was a first-order equation because there was only one energy-storage element effectively, (ii) it was homogeneous because there was no external independent source given to the circuit, and (iii) the coefficients were constant because the passive elements are (assumed to be) time-invariant. We preferred to call this circuit itself a first-order circuit. The response of the circuit is a decaying exponential, the rate of which, called the time-constant, depends on the choice of the elements.

Next, we could easily extend most of the ideas from the first-order circuit to the second-order circuit—there were two energy-storage elements, the equation was a second-order ODE with constant coefficients. However, we solved this second-order ODE rather intuitively by guessing that the response might be an exponential in this case also. Fortunately, the guess turned out to be an *almost* correct one; in fact, the solution was a linear combination of two exponentials. We should have expected this because we saw, in terms of the state variables, that a second-order circuit actually has *two* first-order differential equations [Eqns (5.17) and (5.18)].

We shall now look at more subtle issues. Look out for the italicized words in the following:

1. The *first*-order circuit has
   - *one* initial condition,
   - *one* (decaying) exponential and, hence,
   - *one* constant to be determined, which was found to be directly proportional to the initial condition itself, and
   - the response is governed by *one* parameter called the time-constant.

     Instead of solving mathematically the first-order ODE, had we gone guessing we would have obtained the same result. But then, we would not have had a basis to claim that the response in the second-order case too could be exponential in nature. In the second-order circuit, we eschewed the mathematical intricacies but guessed an answer that was later corrected with an appropriate reasoning. Consequently,

2. the *second*-order circuit has
   - *two* initial conditions, and the response is a *linear combination* of
   - *two* (decaying) exponentials and hence
   - *two* constants to be determined [Each of the constants was found to be a *linear combination* of the initial conditions $i(0^+)$ and $v_C(0^+)$.],
   - its response governed by *two* parameters, namely, the damping factor ζ and the natural frequency ω_n.

Oddly enough, however, there were *four* different cases arising in the second-order circuit.

We have a valid extension of the above observation—an nth-order homogeneous circuit shall have n exponentials governed by n parameters with n constants to be obtained using the n initial conditions. In particular, all the resistive circuits, considered in Chapter 4, are called *zeroth-order circuits*. They are described by simple linear algebraic equations in lieu of differential equations. There were no initial conditions and no exponentials. It is customary to call the zeroth-order circuits *static circuits* and nth-order ($n \geq 1$) circuits *dynamic circuits*. It is customary to call the dynamic circuits that are *not* driven by any source for $t \geq 0$ as *source-free circuits*, or *homogeneous circuits*, or *autonomous circuits*.

The response of these first- and second-order source-free circuits for $t \geq 0$ is solely due to the stored energy for $t < 0$. We call this 'natural response' since we are not forcing it using external sources for $t \geq 0$. Since the response invariably consists of *decaying* exponentials, we see that the response decays to zero very quickly in about six time-constants. In this sense, we also call the response *transient response*. The word transient refers to something that lasts only a short time. While the parameter time-constant is straightforward in first-order circuits, it is indirect in the second-order circuits. For the overdamped case we consider the time-constant as

$$\tau_{over} \approx \max\left(\frac{1}{s_1}, \frac{1}{s_2}\right)$$

since one of the exponents decays faster than the other one, it is sufficient to consider the response as approximately equal to the longer decaying exponential and hence the above expression. For the critically damped case,

$$\tau_{critical} \approx \frac{1}{\omega_n}$$

and for the underdamped case,

$$\tau_{under} \approx \frac{1}{\zeta\omega_n}$$

Typically, all the electrical networks will have time-constants in the range of a few microseconds to a few milliseconds.

Next, we shall move on to general RLC circuits.

5.5.7 General *RLC* Circuits

In this section we shall look at some general source-free circuits. We shall learn a few things with the following examples.

Example 5.16 Consider the circuit shown in Fig. 5.21.

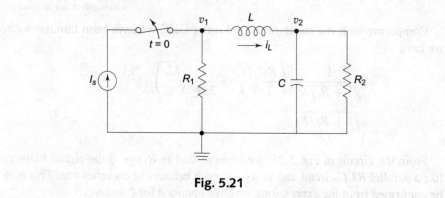

Fig. 5.21

For $t \geq 0$, this is neither a series RLC circuit nor a parallel one. Let us determine the voltage $v_C(t)$ across the capacitor.

First, assuming that the source has been driving the circuit for $t < 0$, we have

$$i_L(0^-) = \frac{R_1}{R_1 + R_2} I_s$$

and

$$v_C(0^-) = R_2 i_L(0^-)$$

since the inductor behaves like a short circuit and the capacitor behaves like an open circuit.

Next, let us apply nodal analysis to the circuit for $t \geq 0$. We get the following node equations:

$$\frac{v_1(t)}{R_1} + i_L(t) = 0 \qquad (5.53)$$

$$\frac{v_2(t)}{R_2} + C \frac{d}{dt} v_2(t) - i_L(t) = 0 \qquad (5.54)$$

and we have a constraint

$$L \frac{d}{dt} i_L(t) = v_1(t) - v_2(t) \qquad (5.55)$$

The variable i_L can be eliminated using Eqn (5.54) in Eqn (5.55) to get

$$C \frac{d^2}{dt^2} v_2(t) + \frac{1}{R_2} \frac{d}{dt} v_2(t) = \frac{v_1(t) - v_2(t)}{L}$$

$$= -\frac{R_1 i_L(t) - v_2(t)}{L}$$

where we made use of Eqn (5.53) to eliminate v_1 altogether. Further, eliminating i_L once again and rearranging the terms, we get

$$\frac{d^2}{dt^2} v_2(t) + \left(\frac{1}{R_2 C} + \frac{R_1}{L} \right) \frac{d}{dt} v_2(t) + \left(\frac{1}{LC} + \frac{R_1/R_2}{LC} \right) v_2(t) = 0 \qquad (5.56)$$

Exercise 5.23 Derive Eqn (5.56).

Comparing with the standard second-order ODE we have been familiar with, we have

$$\zeta = \frac{1}{\sqrt{1 + R_1/R_2}} \left(\frac{R_1}{2} \sqrt{\frac{C}{L}} + \frac{1}{2 R_2} \sqrt{\frac{L}{C}} \right)$$

$$\omega_n = \sqrt{\frac{1 + R_1/R_2}{LC}}$$

From the circuit in Fig. 5.21, we observe that as $R_1 \to 0$ the circuit behaves like a parallel RLC circuit and as $R_2 \to \infty$ it behaves like a series one. This may be confirmed from the expressions we have obtained for ζ and ω_n.

Exercise 5.24 Repeat Example 5.16 for the circuit in Fig. 5.21 with the positions of inductor and capacitor interchanged with each other.

Exercise 5.25 Repeat Example 5.16 and Exercise 5.24 for the following circuit parameters:

$$I_s = 2 \, A, \quad R_1 = 18 \, \Omega, \quad R_2 = 6 \, \Omega, \quad L = 1 \, H, \quad C = \frac{1}{25} \, F$$

Example 5.17 Let us now consider the following circuit in Fig. 5.22.

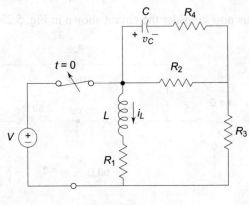

Fig. 5.22

The initial conditions are

$$i_L(0^+) = \frac{V}{R_1}$$

and

$$v_C(0^+) = \frac{R_2}{R_2 + R_3} V$$

Applying mesh analysis with the current directions as shown, we get

$$C \frac{d}{dt} v_C(t) = -i_C(t)$$

$$(R_2 + R_4)C \frac{d}{dt} v_C(t) + v_C(t) + R_2 i_L(t) = 0$$

$$L \frac{d}{dt} i_L(t) + (R_1 + R_2 + R_3)i_L(t) + R_2 C \frac{d}{dt} v_C(t) = 0$$

To simplify the expressions, let us assume

$$R_1 = R_2 = R_3 = R_4 = R$$

Rearranging the above equations, we get the following state equations:

$$\frac{d}{dt} i_L(t) = -\frac{5R}{2L} i_L(t) + \frac{1}{2L} v_C(t)$$

$$\frac{d}{dt} v_C(t) = -\frac{1}{2C} i_L(t) - \frac{1}{2RC} v_C(t)$$

We emphasize that writing these state equations helps us in using the correct initial conditions—$i_L(0)$ and $di_L/dt|_{t=0}$ or $v_C(0)$ and $dv_C/dt|_{t=0}$. Further manipulations would give us the desired second-order ODE as

$$\frac{d^2}{dt^2} i_L(t) + \frac{1}{2}\left(\frac{5R}{L} + \frac{1}{RC}\right) \frac{d}{dt} i_L(t) + \frac{3}{2LC} i_L(t) = 0$$

Exercise 5.26 In Example 5.17 determine the values, if they exist, of R and L so that the circuit is (i) overdamped, (ii) critically damped, and (iii) underdamped.

Example 5.18 Let us now consider the circuit shown in Fig. 5.23.

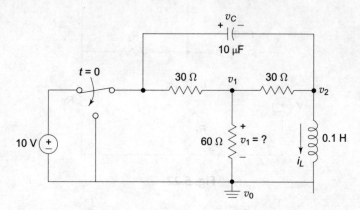

Fig. 5.23

The initial conditions may be obtained by inspection as follows:

$$i_L(0^+) = \frac{2}{15} \, A, \quad v_C(0^+) = 10 \, V$$

For $t \geq 0$, we have the following nodal analysis.

Applying KCR at node 1 gives

$$v_1(t) \left(\frac{1}{30} + \frac{1}{60} + \frac{1}{30} \right) - \frac{v_2(t)}{30} = 0$$

or, equivalently, $v_1(t) = 0.4 \, v_2(t)$. At node 2, we shall first identify that

$$v_2(t) = -v_C(t) = L \frac{d}{dt} i_L(t)$$

and hence

$$v_1(t) = 0.04 \frac{d}{dt} i_L(t)$$

Applying KCR at node 2 gives us

$$\frac{v_2(t) - v_1(t)}{30} + i_L(t) + C \frac{d}{dt} v_2(t) = 0$$

Differentiation once on either side and eliminating v_1 and v_2 we get

$$10^{-6} \frac{d^2}{dt^2} i_L(t) + 0.002 \frac{d}{dt} i_L(t) + i_L(t) = 0$$

or, equivalently,

$$\frac{d^2}{dt^2} i_L(t) + 2000 \frac{d}{dt} i_L(t) + 10^6 i_L(t) = 0 \tag{5.57}$$

From Eqn (5.57) we find that

$$\zeta = 1$$

and

$$\omega_n = 1000 \, \text{rad/s}$$

The initial conditions are

$$i_L(0^+) = \frac{2}{15} \, \text{A}$$

and

$$\frac{d}{dt} i_L|_{t=0^+} = -\frac{v_C(0^+)}{L} = -100$$

Since this is a critically damped circuit, we have

$$i_L(t) = \left(\frac{100}{3} t + \frac{2}{15}\right) e^{-1000t} \, \text{A}$$

From this, we get the desired voltage v_1 as

$$v_1(t) = 0.04 \frac{d}{dt} i_L(t) = -4 \left(\frac{10^3}{3} t + 1\right) e^{-1000t}$$

Exercise 5.27 Derive Eqn (5.57) and verify the results of Example (5.18).

Example 5.19 Consider the circuit shown in Fig. 5.24 with a controlled current source.

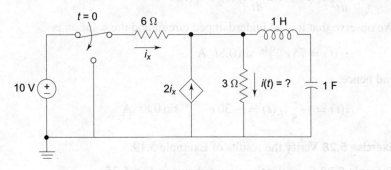

Fig. 5.24

For $t < 0$, we have $i_L(0^-) = 0$ since the capacitor acts like an open circuit and hence no current flows through the inductor. Applying Kirchhoff's rules, we also have

$$i(t) = 3i_x(t)$$

and

$$6i_x(t) + 3i(t) = 10 \, \text{V}$$

Solving these equations, we get

$$i_x(0^-) = \frac{10}{15} \, \text{A}, \quad i(0^-) = 3i_x = 2 \, \text{A}$$

and

$$v_C(0^-) = 3i(0^-) = 6 \, \text{V}$$

Thus, the initial conditions are

$$i_L(0^+) = 0\,\text{A}$$

and

$$v_C(0^+) = 6\,\text{V}$$

For $t \geq 0$, the circuit is slightly trickier. First we have

$$-3\,i(t) + \frac{d}{dt}\,i_L(t) + v_C(t) = 0$$

in the mesh containing the inductor and the capacitor, and from this we deduce that

$$\frac{d}{dt}\,i_L(t)\big|_{t=0^+} = \frac{3\,i_1(0^+) - v_C(0^+)}{L} = -6\,\text{A/s}$$

Further, we also notice that

$$3\,i(t) = -6\,i_x(t) = \frac{d}{dt}\,i_L(t) + v_C(t)$$

and

$$i_L(t) = 3\,i_x(t) - i(t)$$

Eliminating i and i_x we get the following ODE:

$$\frac{d^2}{dt^2}\,i_L(t) + 1.2\,\frac{d}{dt}\,i_L(t) + i_L(t) = 0$$

We observe that it is an underdamped circuit and the solution is

$$i_L(t) = 75\,e^{-0.6t}\,\sin 0.8t \quad \text{A}$$

and hence

$$i(t) = -\frac{2}{5}\,i_L(t) = -30\,e^{-0.6t}\,\sin 0.8t \quad \text{A}$$

Exercise 5.28 Verify the results of Example 5.19.

Example 5.20 Consider the circuit shown in Fig. 5.25.

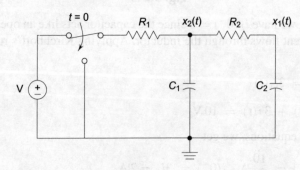

Fig. 5.25

Notice that there is no inductor, but there are *two* independent capacitors. These are independent in the sense that we cannot combine them either in series or in parallel. Hence it is a second-order circuit. Let us, for simplicity, assume that $R_1 = R_2 = R$ and $C_1 = C_2 = C$. For the purpose of illustration, let us resort to the notion of state variables. Let us define two variables $x_1(t)$ and $x_2(t)$ as the voltages across the two capacitors as shown in the figure. Applying nodal analysis, we have the following first-order ODE's:

$$\frac{d}{dt} x_1 = -\frac{1}{RC} (x_1 - x_2)$$

$$\frac{d}{dt} x_2 = \frac{1}{RC} (x_1 - 2 x_2)$$

Differentiating the first ODE and eliminating x_2, we have the second-order ODE as

$$\frac{d^2}{dt^2} x_1(t) + \frac{3}{RC} \frac{d}{dt} x_1(t) + \frac{1}{(RC)^2} x_1(t) = 0$$

The initial conditions would be

$$x_1(0^+) = x_2(0^+) = V$$

and

$$\frac{d}{dt} x_1(t)|_{t=0^+} = -\frac{1}{RC} (x_1 - x_2)|_{t=0^+} = 0$$

Exercise 5.29. Verify the ODE in Example 5.20.

Exercise 5.30. Repeat Example 5.20 with non-identical resistors R_1 and R_2 and non-identical capacitors C_1 and C_2. Verify that the circuit is *never* underdamped.

We conclude this section with the following observations:

- As long as the effective number of energy-storage elements is 2, we have a second-order circuit.
- We might obtain an equivalent resistive circuit for $t < 0$ with inductors replaced with short circuits and capacitors replaced with open circuits to determine the initial conditions.
- As long as the ODE governing the circuit is second order, we can always compare it with the standard ODE to obtain the parameters ζ and ω_n.
- Once ζ and ω_n are known, we might identify one of the four possible cases discussed earlier in connection with the series and parallel RLC circuits and compute the constants A and B and hence the response.
- Since we are dealing with linear circuits, the voltage across or the current through in any branch of the circuit shall have the same general form

$$y(t) = A e^{s_1 t} + B e^{s_2 t}$$

i.e., a *linear combination* of the voltage across the capacitor and the current through the inductor. It would be systematic to first obtain the mesh equations (or the node equations) in terms of the inductor current and/or the capacitor voltage. Once $i_L(t)$ and $v_C(t)$ are determined, any other voltage or current may be determined as a linear combination of these two.

- The two energy-storage elements of a second-order circuit could be, trivially, a capacitor and an inductor, or two capacitors, or two inductors. In any case, we shall have two *state variables* telling us the status of the circuit at any given time and the response in any branch of the circuit is simply a linear combination of the state variables. However, we must exercise care in obtaining the *correct* initial conditions.

- Independent of the circuit topology—whether it is a series RLC circuit, or a parallel one, or something similar to the circuit shown in Example 5.20— it would be advantageous to obtain the second-order ODE in terms of the variable of interest, the voltage or the current $y(t)$ that is asked for, and proceed ahead determining ω_n, ζ, and appropriate initial conditions $y(0)$ and $dy(t)/dt \, |_{t=0^+}$ in terms of the circuit's actual initial conditions. We discuss the rationale behind this and present an algorithm in a later section.

We shall next study circuits driven by dc sources for all times $t \in \mathcal{T}$.

5.6 The Unit Step Forcing Function

So far in this chapter we have been connecting dc sources in the past, removing them in the present, and studying the circuit in the future. Let us now look at the behaviour of the circuits when we connect the sources in the present. First, we shall formalize our notion of connection and removal.

Suppose we wish to apply a dc voltage source of V volts at $t = 0$ across a port of an arbitrary passive network. This may be shown in Fig. 5.26(a) with a switch.

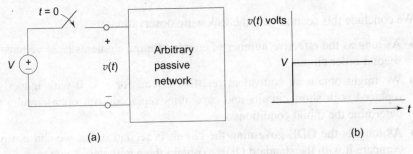

(a)   (b)

Fig. 5.26

Alternatively, we may depict the voltage $v(t)$ across the port as a function of time in Fig. 5.26(b). This representation suggests us that the voltage $v(t)$ may be expressed as

$$v(t) = \begin{cases} V \text{ volts} & \text{for } t \geq 0 \\ 0 & \text{for } t < 0 \end{cases} \tag{5.58}$$

Observe that the function appears like a 'step', as in a flight of stairs, at $t = 0$. It is customary to call such a function a *step function*. Hereafter, we prefer to say 'applying a step forcing function' to 'connecting a source' at $t = 0$. Of course, we may wish to apply the step forcing function at any time t_0, either before $t = 0$ or after that. Accordingly, we define a general step function as follows:

$$v(t - t_0) = \begin{cases} V \text{ volts} & \text{for } t \geq t_0 \\ 0 & \text{for } t < t_0 \end{cases} \tag{5.59}$$

However, since we assume that the passive elements exhibit the property of time-invariance, we need not worry about t_0.

Further, we have been dealing with linear circuits. By definition 4.1 of Chapter 4, linearity comprises homogeneity (or scaling) and additivity. Let us now make use of the homogeneity property and say that

- if an arbitrary source of 1 V, or 1 A, causes a response $y(t)$ in an arbitrary passive network, then a source of α V, or α A causes a response of $\alpha \times y(t)$ in the same network.

In other words, it is sufficient for us to study the circuit behaviour for sources that deliver voltages, or currents, of one 'unit' in magnitude. In the light of this observation, we shall slightly modify the step function in Eqn (5.58) and define a *unit step function*, without reference to any particular source, as follows:

$$q(t) \overset{\triangle}{=} \begin{cases} 1 & \text{for } t \geq 0 \\ 0 & \text{for } t < 0 \end{cases} \tag{5.60}$$

A dc voltage source of V volts may be represented symbolically as

$$v(t) = Vq(t) \text{ V } \forall t \tag{5.61}$$

Likewise, a dc current source of I A may be represented as

$$i(t) = Iq(t) \text{ A } \forall t \tag{5.62}$$

We very often use these step forcing functions when we are operating switches in various gadgets such as torch lights, computers, automobiles, etc. The nature of these forcing functions, or forces, as the general name suggests, need not be always electrical; they could be mechanical, pneumatic, and so on. For example, when we play an audio system, we press the PLAY button, which in turn supplies a step dc voltage to the amplifier from that instant onwards. Simultaneously, a step mechanical force is exerted on the button to hold it in the pressed position. The opposite happens when we press the STOP button. As yet another example, we have semiconductor diodes, popularly called the *p-n* junction diodes, wherein the density of the impurities changes abruptly from one side of the diode to the other. We call them 'step-graded' diodes.

Exercise 5.31 Brainstorm on various step signals you see around in the day-to-day activities.

Put briefly, a step function suggests that there was nothing in the past, but *suddenly* there is a force in the present and is *constant* in the future, forever.

In reality, however, *sudden* changes in magnitude at $t = 0$ may not be possible due to inertia. Nevertheless, in most cases the 'switching delay' is negligibly small relative to the time span in which the circuit is studied and hence our mathematics makes sense. In fact, what is interesting to us is the *discontinuity* at $t = 0$; it is not so exciting otherwise since a capacitor is an open circuit and an inductor is a short circuit when subjected to dc voltages and/or currents. The discontinuity might cause some ambiguity—what is the magnitude at $t = 0$? We remove this ambiguity by *defining* the magnitude to be one unit, as in Eqn (5.60), at $t = 0$; notice that the 'equality sign' is included in the upper part of the definition in Eqn (5.60).

There is yet another practical convenience in studying step functions. Generally speaking, any arbitrary function of time, hereafter referred to as a waveform, may be expressed as a linear combination of several steps. We shall take a quick look at the following examples to illustrate the idea.

5.6.1 Waveform Synthesis

Example 5.21 Suppose we have a 'pulse' function as shown in Fig. 5.27.

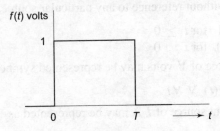

Fig. 5.27

This is yet another practical signal; for instance, it represents the voltage supplied to a torch light by keeping the switch 'on' for a duration of T seconds. This signal can be represented as an algebraic sum of two step functions:

$$f(t) = q(t) - q(t - T)$$

It is not too difficult for the reader to visualize now that $-q(t - T)$ is also a step signal that is zero until $t = T$ and changes from 0 to -1 at $t = T$ and remains -1 thereafter. Thus, during the interval $0 \leq t \leq T$, $f(t) = q(t) = 1$ and during the interval $T \leq t < \infty$ $f(t) = q(t) - q(t - T) = 1 - 1 = 0$.

We might generalize this to describe a pulse voltage of magnitude V volts of duration $T_2 - T_1$, such as the one shown in Fig. 5.1(b), as

$$f(t) = V \times [q(t - T_1) - q(t - T_2)]$$

We might further generalize this to describe a 'pulse train' shown in Fig. 5.28 as

$$f(t) = V \times [q(t - T_1) - q(t - T_2) + q(t - T_3) - q(t - T_4) + q(t - T_5) - \cdots]$$

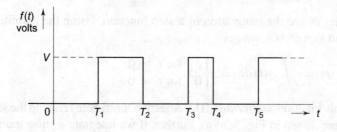

Fig. 5.28

This pulse train is a useful signal in, for example, representing a binary number in a computer.

Next, let us see what we get if we differentiate a step function. Owing to the discontinuity at $t = 0$, we get

$$\delta(t) = \frac{d}{dt} q(t) \triangleq \begin{cases} \infty & \text{if } t = 0 \\ 0 & \text{if } t \neq 0 \end{cases}$$

i.e., a 'spike' at $t = 0$ as shown in Fig. 5.29(a).

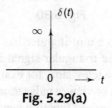

Fig. 5.29(a)

Notice that this is like a very narrow pulse of width ϵ and magnitude $1/\epsilon$. It is customary to call this signal a *unit impulse* (also called as Dirac's delta function in mathematical physics.) The adjective 'unit' here refers to the unit area ($= \epsilon \times 1/\epsilon$) of the narrow pulse. Since differentiation is a linear operation, differentiating a pulse train would give us an 'impulse train' in which the positive going and negative going impulses alternate. This is shown in Fig. 5.29(b). For a convenient pictorial representation of the unit impulse, it is customary to show the 'arrow' to be one unit long. Accordingly, in Fig. 5.29(b) we show the 'amplitude' of the spikes to be $\pm V$ volts.

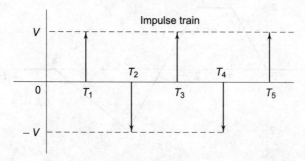

Fig. 5.29(b)

Next, let us see the integration of a step function. Using the definition of step function in Eqn (5.60), we get

$$r(t) = \int q(t)\,dt \triangleq \begin{cases} t & \text{for } t \geq 0 \\ 0 & \text{for } t < 0 \end{cases}$$

We call this function a *unit ramp*. The adjective 'unit' here refers to the slope of the straight line shown in Fig. 5.30(a). Further, if we integrate a pulse train, we get a 'triangular signal' as shown in Fig. 5.30(b) with positive going and negative going ramps alternating.

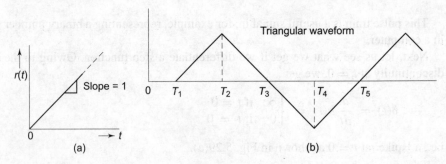

(a)   (b)

Fig. 5.30

The response of a circuit to a unit impulse has important consequences as we shall explore in Chapter 9. The triangular signal is another very practical signal found in the sweep circuits in a television, for example.

From an experimental point of view, a pulse train can be very easily synthesized and is commercially available on any function generator (also called an oscillator) we use in the laboratories. Owing to the properties of linearity and time-invariance of the circuits we are interested in, we may *experimentally* study the step response of a circuit by subjecting it to a pulse train. Then we might differentiate the step response to get the impulse response or integrate it to obtain the ramp response.

We conclude this subsection with the following example.

Example 5.22 Let us look at the waveform $f(t)$ in Fig. 5.31.

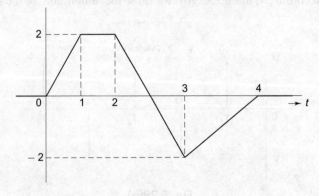

Fig. 5.31

Intuitively, we feel that it is made up of steps and ramps. Let us dissect it and write it precisely. First, there is a ramp in the interval $0 \leq t \leq 1$ with slope 2. Put in symbols, we have

$$f(t) = 2\,r(t) \text{ for } 0 \leq t \leq 1 \text{ s}$$

Beyond $t = 1$ the ramp does not exist, but the waveform is constant (i.e., slope = 0) with magnitude equal to 2. This can be achieved by *subtracting* another ramp of the same slope, but starting at $t = 1$ so that

$$f(t) = 2\,r(t) - 2\,r(t-1) \quad \text{for } 0 \leq t \leq 2 \text{ s}$$

Observe that, during this interval, $f(t) = 2t - 2(t-1) = 2$ and hence our synthesis is justified. Notice that the *net slope* of the waveform in this interval is zero, as is desired.

Next, for $2 \leq t \leq 3$, we have another ramp with slope -4, but the earlier constant is missing. This can be achieved by first subtracting a ramp of slope 4 at $t = 2$.

$$f(t) = 2\,r(t) - 2\,r(t-1) - 4\,r(t-2) \quad \text{for } 0 \leq t \leq 3 \text{ s}$$

But this would add a constant 8 to the earlier constant 2. This can be nullified by subtracting a step of magnitude 10 at $t = 2$. Thus,

$$f(t) = 2\,r(t) - 2\,r(t-1) - 4\,r(t-2) - 10\,q(t-2) \text{ for } 0 \leq t \leq 3 \text{ s}$$

For the interval $3 \leq t \leq 4$, there is a ramp of slope 2 and this can be achieved by adding a ramp of slope 6 at $t = 3$. Hence,

$$f(t) = 2r(t) - 2r(t-1) - 4r(t-2) - 10q(t-2) + 6r(t-3) \text{ for } 0 \leq t \leq 4$$

Since this would add a constant of -18, we need to add a step of magnitude 18 at $t = 3$. Beyond $t = 4$, we need a constant of magnitude 0. This can be achieved by first subtracting a ramp of slope 2 at $t = 4$ and then subtracting a step of magnitude 8. This finally gives us the desired waveform as

$$
\begin{aligned}
f(t \geq 0) = & \; 2\,r(t) - 2\,r(t-1) \\
& - 4\,r(t-2) - 10\,q(t-2) \\
& + 6\,r(t-3) + 18\,q(t-3) \\
& - 2\,r(t-4) - 8\,q(t-4)
\end{aligned}
$$

In this example, we have deliberately introduced the notion of *shifting* a function to the right along the time axis. This should not be surprising to the reader since translations and rotations (perhaps more popularly, coordinate transformations) are usually taught at the high-school level. Using the linearity and time-invariance properties of our circuits, it is trivial to study the response of any circuit subjected to a waveform like this.

Exercise 5.32. Verify the results of Example 5.22.

5.7   Step Response of *RC*, *RL*, and *RLC* Circuits

In this section we shall study the behaviour of circuits containing energy-storage elements subjected to a step (sudden) change in voltage/current sources.

5.7.1   First-order Circuits

Now let us turn our attention to studying the first-order circuits with a unit step voltage applied at $t = 0$. Consider the *RC* circuit shown in Fig. 5.32.

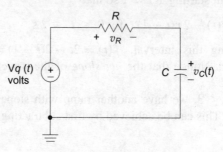

Fig. 5.32

Observe that we have applied a step forcing function at $t = 0$. Let us proceed with an assumption that the capacitor is not initially charged, i.e., $v_C(0^+) = 0$. We shall relax this assumption after a while. Let us try our intuition first and see what it says. Since the initial capacitor voltage is zero it acts like a short circuit, the drop across the resistor is V volts according to KVR. Further, since a dc voltage is being applied, after a sufficiently long time we expect the capacitor to behave like an open circuit with V volts across it, depriving the resistance of its voltage drop. Thus, the voltage across the capacitor changes and writing KVR formally around the loop gives us the following first-order ODE:

$$Ri(t) + v_C(t) = V$$

or

$$\frac{d}{dt} v_C(t) + \frac{1}{RC} v_C(t) = \frac{1}{RC} V \quad \forall \, t \geq 0 \tag{5.63}$$

Here, we need to guess a function $v_C(t)$ such that it derivative and the (scaled) function itself are homogeneous and add to a constant. With our earlier experience and little more reasoning, let us guess that

$$v_C(t) = A \, e^{st} + B$$

We need to include a constant B since the right-hand side of the ODE is a constant. Apparently, there is something inconsistent—if a first-order ODE has only one initial condition, then how is it that there are two unknown constants A and B? We shall discover an answer to this question as we proceed. Substituting our proposed solution into Eqn (5.62), we get

$$A \left(s + \frac{1}{RC} \right) e^{st} + \frac{B}{RC} = \frac{V}{RC}$$

Clearly,

$$A \left(s + \frac{1}{RC} \right) e^{st} = 0$$

and

$$B = V$$

We can easily rule out the possibility[6] of $A = 0$ since $v_C(t)$ is changing with time. Another impossibility is $t = -\infty$ since we are talking about $t \geq 0$. Yet another impossibility is $s = -\infty$, which we will consider in a moment. The only possibility is $s = -1/RC$. This appears to match our intuition since in the source-free RC circuit earlier the response was of the form $e^{-t/RC}$. Further, if $R = 0$ then $s = -\infty$ and, physically, the capacitor gets charged to V volts instantaneously which should be ruled out owing to the voltage continuity rule of any capacitor. Thus, we have $B = V$ and $s = -1/RC$. We are now left with determining A in the solution. This can be easily done by looking at the initial condition

$$v_C(0) = A + B = 0$$

which implies that $A = -B = -V$. Thus, the solution to the ODE in Eqn (5.63) is

$$v_C(t \geq 0) = -Ve^{-t/\tau} + V \tag{5.64}$$

with $v_C(0) = 0$. Observe that

$$v_C(\infty) = V$$

i.e., the capacitor behaves like an open circuit only as the time $t \to \infty$. But then, we have defined earlier 'infinite time' as about 6τ. Thus, our intuition appears to be on the right track. Not only the initial value, but also the *final value* is needed to obtain the solution to the ODE. In general, the initial and final values are collectively called the boundary values or *boundary conditions*. We will formally prove our intuition is indeed correct after we discuss some more basic ideas.

Suppose that we have started with an initially charged capacitor. In this case too, the capacitor voltage changes and the same ODE in Eqn (5.63) is valid. However, it is easy to verify that the solution in Eqn (5.64) is modified to

$$v_C(t) = v_C(0^+) e^{-t/\tau} + V \left(1 - e^{-t/\tau} \right) \tag{5.65}$$

$$= \left[v_C(0^+) - V \right] e^{-t/\tau} + V \tag{5.66}$$

We call this the *complete response*. Notice that Eqn (5.64) is now a special case, with $v_C(0^+) = 0$, of Eqn (5.66).

Exercise 5.33 Verify that Eqn (5.66) gives the complete response.

---

[6] 'It is an old maxim of mine that when you have excluded the impossible, whatever remains, however improbable, must be truth.'

—Sherlock Holmes

Let us now prove that our guess for the solution of Eqn (5.63) is indeed correct.

Proof. To *prove* our intuition formally, we use the method of integrating factors and obtain the solution of the ODE. Multiplying both sides of the ODE with the integrating factor $e^{t/\tau}$, we get

$$e^{t/\tau} \frac{d}{dt} v_C(t) + \frac{1}{\tau} e^{t/\tau} v_C(t) = \frac{V}{\tau} e^{t/\tau}$$

Now, the left-hand side may be identified as

$$\frac{d}{dt} [e^{t/\tau} v_C(t)]$$

Integrating on both sides we get

$$e^{t/\tau} v_C(t) = V e^{t/\tau} + \text{some constant } A$$

Dividing both sides by the integrating factor, applying the initial condition, and observing that as $v_C(t \to \infty) = V$, we get

$$v_C(t) = [v_C(0^+) - V] e^{-t/\tau} + V$$

Q.E.D.

Since $v_C(\infty) = V$, we may rewrite $v_C(t)$ as

$$v_C(t \geq 0) = [v_C(0^+) - v_C(\infty)] e^{-t/\tau} + v_C(\infty) \qquad (5.67)$$

We have a few things to observe. The response $v_C(t)$ may be written as in Eqn (5.65) wherein we identify that the first term is the natural response due to the initial condition. The second term is due to the forcing function V and we call it the *forced response*. Thus, we have the following *superposition* of partial responses:

complete response = natural response + forced response

This is a routine way of describing the solutions to differential equations. We observe that a portion of the forced response also has a decaying exponential and hence is a transient.

If we look at the response as in Eqn (5.66), we have the following superposition of partial responses:

complete response = transient response + steady-state response

where we have called that portion of the response that remains *forever*, i.e., as $t \to \infty$, the *steady-state response*. This later description is more useful to us. We emphasize that the natural response is *included* in the transient response.

Owing to the linearity of the circuits, we might naturally extend our ideas to say that the voltage across or the current through any branch of the circuit, except the source voltage, may be expressed in the form of Eqn (5.66), i.e.,

$$y(t \geq 0) = [y(0^+) - y(\infty)] e^{-t/\tau} + y(\infty) \qquad (5.68)$$

Notice that if $y(\infty) = 0$, we get the source-free response. Thus, a forced circuit is more general in the sense that it includes a source-free circuit. And, Eqn (5.68) suggests us the following algorithm to compute the complete response.

Step 1	Determine the time-constant τ of the circuit.
Step 2	Determine the initial voltage (or initial current), $y(0^+)$ in the branch of interest.
Step 3	Determine the final value $y(\infty)$ as $t \to \infty$ in the branch of interest.
Step 4	The response $y(t)$ at any instant of time $t \geq 0$ is given by $y(t) = [y(0^+) - y(\infty)] e^{-t/\tau} + y(\infty)$.

Although we have started with an *RC* circuit in this section, we might end up with the same conclusions for an *RL* circuit also. We leave it as an exercise to the reader.

Exercise 5.34 Verify that Eqn (5.68) holds good to compute the response in an *RL* circuit.

Let us now look at a few examples.

Example 5.23 Consider the circuit in Fig. 5.33. We wish to determine the voltage $y(t)$ across the 1 Ω resistance for all times $t \geq 0$.

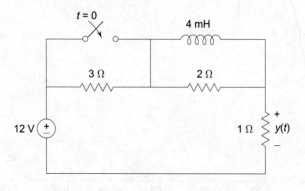

Fig. 5.33

First, let us identify the circuit in the past to determine the initial conditions. Since the voltage supplied is dc, the inductor behaves like a short circuit with $i_L(0^-) = 12/(3 + 1) = 3$ A through it. The 2 Ω resistor should be treated like a short circuit since it is in parallel with the inductor. We may quickly compute that

$$y(0^-) = \frac{1}{3 + 1} 12 = 3 \text{ V}$$

At $t = 0^+$, we have the dc current source, the inductor, along with the dc voltage source. Since the switch is closed, the $3\,\Omega$ resistor should be treated like a short circuit. If we apply the principle of superposition we obtain $y(0^+)$ as

$$y(0^+) = \frac{1}{1+2} \times 12\,\text{V} + \frac{2}{2+1} \times 3\,\text{A} \times 1\,\Omega = 6\,\text{V}$$

Recall that it is only the inductor current that obeys the continuity rule and hence $y(0^+) \neq y(0^-)$.

As $t \to \infty$, the inductor behaves like a short circuit once again and hence $y(\infty) = 12\,\text{V}$. Further, for $t \geq 0$, the inductor 'sees' a parallel combination of $1\,\Omega$ and $2\,\Omega$ resistors and hence the time-constant is

$$\tau = \frac{4\,\text{mH}}{2 \parallel 1\,\Omega} = 6\,\text{ms}$$

Thus, the complete response is

$$y(t \geq 0) = -6e^{-t/6\times 10^{-3}} + 12\,\text{V}$$

and this is plotted in Fig. 5.34.

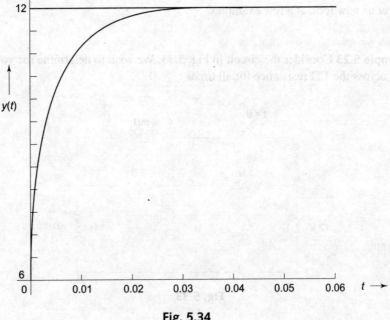

Fig. 5.34

In this example, effectively, we need four equivalent circuits: for $t < 0$, at $t = 0^+$ one at $t = \infty$, and the source-free circuit for $t \geq 0$ to determine R_{eq} and hence the time-constant.

Exercise 5.35 Draw the four equivalent circuits suggested in Example 5.23 above and verify the results.

Example 5.24 Let us consider the circuit in Fig. 5.35.

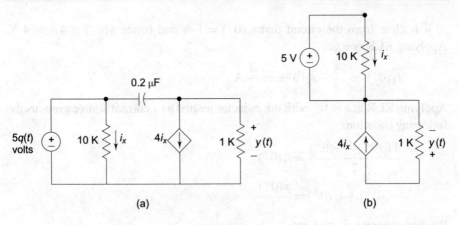

(a)   (b)

Fig. 5.35

Let us assume that the capacitor is initially discharged, i.e., $v_C(0^+) = 0\,V$. At $t = 0^+$, since the capacitor behaves like a short circuit, all the 5 V of the voltage source appears across the $1\,k\Omega$ resistor. Thus,

$$y(0^+) = 5\,V$$

Next, as $t \to \infty$, the capacitor behaves like an open circuit and the equivalent circuit may be drawn as shown in Fig. 5.35(b). From this, we may readily obtain

$$i_x = 0.5\,mA$$

and hence

$$y(\infty) = -4 \times i_x \times 1\,k\Omega = -2\,V$$

The source-free circuit for $t \geq 0$ simply consists of the capacitor and the $1\,k\Omega$ resistor and hence $\tau = 0.2\,ms$. Thus, the complete response is

$$y(t \geq 0) = 7\,e^{-5000t} - 2\,V$$

Exercise 5.36 Sketch the voltage $y(t)$ in Example 5.24 to a reasonable scale.

Example 5.25 Consider the circuit in Fig. 5.36.

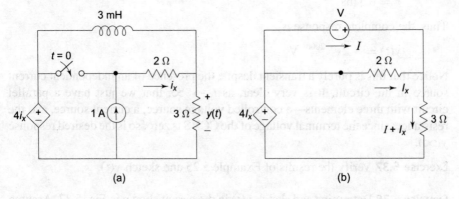

(a)   (b)

Fig. 5.36

It is clear from the circuit that $i_x(0^-) = 1$ A and hence $y(0^-) = 4 i_x = 4$ V. Applying KCR, we get

$$i_L(0^-) = \frac{4}{3} - i_x(0^-) = \frac{1}{3} \text{ A}$$

Applying KCR at $t = 0^+$ with the inductor treated as a current source gives us the following equations:

$$\frac{4 i_x(0^+) - y(0^+)}{2} = i_x(0^+)$$

$$i_x(0^+) + i_L(0^+) = \frac{y(0^+)}{3}$$

We may eliminate i_x and get

$$y(0^+) = -2 \text{ V}$$

As $t \to \infty$, the inductor is a short circuit and we should treat the $2\,\Omega$ resistor also as a short circuit. This gives us $i_x(\infty) = 0$ and hence

$$y(\infty) = 4 i_x(\infty) = 0 \text{ V}$$

Finding R_{eq} for $t \geq 0$ is slightly tricky in this case owing to the presence of a controlled source. The equivalent source-free circuit is shown in Fig. 5.36(b). Since we are interested in the resistance as seen by the inductor, let us remove the inductor and place, instead, a test voltage of V volts. Then, using nodal analysis we shall have

$$I_x = -\frac{V}{2} \text{ A}$$

and

$$I + I_x = \frac{V + 4I_x}{3}$$

Eliminating I_x gives us $I = V/6$ and hence

$$R_{eq} = 6\,\Omega$$

and

$$\tau = 0.5 \text{ ms}$$

Thus, the complete response is

$$y(t) = -2 e^{-2000t} \text{ V}$$

Notice that this is purely a transient despite the presence of an independent current source in the circuit. It is very clear, as $t \to \infty$, that we just have a parallel circuit with three elements—a controlled voltage source, a current source, and the resistance; since the terminal voltage of the CCVS is zero, so is the desired response $y(\infty)$.

Exercise 5.37 Verify the results of Example 5.25 and sketch $y(t)$.

Exercise 5.38 Determine and sketch $v(t)$ in the circuit shown in Fig. 5.37. Assume that the inductor has no initial stored energy.

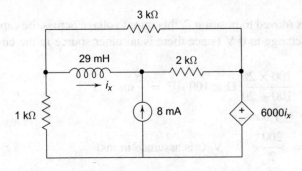

Fig. 5.37

(*Ans.* $v(t \geq 0) = \frac{336}{29} + \left(16 - \frac{336}{29}\right) e^{-t/\tau}$ V, with $\tau = 4\,\mu$s. Take care of the controlled source while computing the equivalent resistance as seen by the inductance)

Exercise 5.39 Determine and sketch $i(t)$ in the circuit shown in Fig. 5.38.

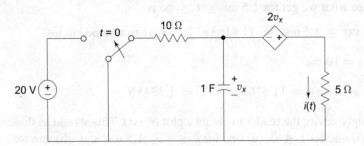

Fig. 5.38

(*Ans.* $i(0^+) = 0$, $i(\infty) = 12/7$, and $\tau = 10/3$)

Exercise 5.40 Consider the circuit shown in Fig. 5.39(a).

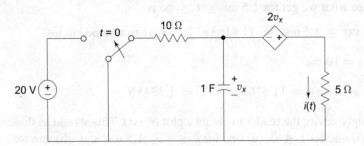

Fig. 5.39(a)

The switch is thrown to position 2 at $t = 0$ and then to position 3 at $t = 1.5$ ms. Determine and sketch $v(-2\,\text{ms} \leq t \leq 10\,\text{ms})$.

[*Ans.* Assuming that the switch is in position 1 for a very long time,

$$v(-2\,\text{ms}) = v(0^-) = v(0^+) = \frac{100}{100 + 20 + 20}\,40 = \frac{200}{7}\,\text{V}$$

If the switch is moved to position 2, this initial voltage across the capacitor has to eventually discharge to 0 V (since there is no other source in the circuit), with a time-constant

$$\tau_1 = \frac{100 \times 20}{100 + 20} \, \Omega \times 100 \, \mu F = \frac{5}{3} ms$$

i.e., for $0 \leq t \leq 1.5$ ms,

$$v(t) = \frac{200}{7} e^{-3t/5} \, V \quad (t \text{ is assumed in ms})$$

At $t = 1.5$ ms, when the switch is thrown to position 3,

$$v(1.5 \, ms) = \frac{200}{7} e^{-3 \times 1.5/5} = 11.6163 \, V$$

This voltage now decays to zero with a time-constant

$$\tau_2 = \frac{100 \times 80}{100 + 80} \, \Omega \times 100 \, \mu F = \frac{40}{9} \, ms$$

and hence what we get for $1.5 \, ms \leq t \leq \infty$ is

$$v(t \geq 1.5 \, ms) = 11.6163 \, e^{-9t/40} \, V \quad (t \text{ is assumed in ms})$$

And, at $t = 10$ ms,

$$v(10 \, ms) = 11.6163 \, e^{-90/40} = 1.2243 \, V$$

We strongly advise the reader to obtain a plot of $v(t)$. This should be done in three pieces: (i) one for $t \leq 0$, (ii) one for $0 \leq t \leq 1.5$ ms, and (iii) one for $t \geq 1.5$ ms. In the first part $v(t)$ is a constant, in the second part $v(t)$ decays exponentially with the time-constant τ_1, and in the third part it decays with the time-constant τ_2. The plot of $v(t)$ (not to scale) is shown in Fig. 5.39(b).]

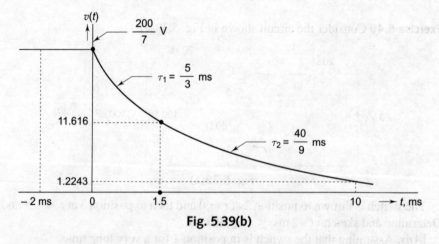

Fig. 5.39(b)

Exercise 5.41 Determine and sketch the currents through the inductors for $t \geq 0$ in the circuit of Fig. 5.40. Give a physical interpretation to justify your answer.

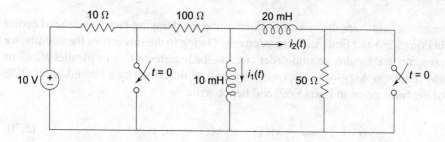

Fig. 5.40

(*Ans.* Here, the trick is the parallel combination of inductances. $i_2(t) = 0$ A for all times. $i_1(0^+) = 1/3$ A, $i_1(\infty) = 0$ A, and $\tau = 1/3$ ms)

5.7.2 Second-order Circuits

Let us quickly look at the series RLC circuit in Fig. 5.41 where the capacitor is charged initially to $v_C(0^-)$, the inductor is charged initially to $i_L(0^-)$, and a dc voltage source is connected to it for all $t \geq 0$.

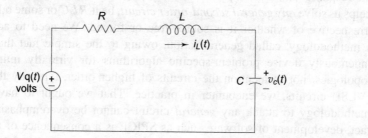

Fig. 5.41

For all $t \geq 0$, applying KVR, we get the following two first order ODE's:

$$R i_L(t) + L \frac{d}{dt} i_L(t) + v_C(t) = V$$

$$C \frac{d}{dt} v_C(t) = i_L(t)$$

Using these two equations we get the following standard second-order ODE's:

$$\frac{d^2}{dt^2} i_L(t) + 2\zeta \omega_n \frac{d}{dt} i_L(t) + \omega_n^2 i_L(t) = \omega_n^2 V \tag{5.69}$$

or, equivalently,

$$\frac{d^2}{dt^2} v_C(t) + 2\zeta \omega_n \frac{d}{dt} v_C + \omega_n^2 v_C(t) = 0$$

where ω_n and ζ are the natural frequency and the damping factor as defined earlier in Eqn (5.26) and Eqn (5.27), respectively. Owing to the linearity of the circuits, we presume that *for any* second-order circuit—be it series RLC, or parallel RLC, or any other topology—the response $y(t)$ shall be described by a second-order ODE of the form given in Eqn (5.69) and hence write

$$\frac{d^2}{dt^2} y(t) + 2\zeta\,\omega_n\,\frac{d}{dt}\,y(t) + \omega_n^2\,y(t) \; = \; \omega_n^2\,y(\infty) \tag{5.70}$$

We shall extend the ideas from the forced first-order circuits to write the complete response of the second-order circuits as

$$y(t) = \underbrace{Pe^{s_1 t} + Qe^{s_2 t}}_{\text{natural response}} + \underbrace{Re^{s_1 t} + Se^{s_2 t} + T}_{\text{forced response}}$$

$$= \underbrace{Ae^{s_1 t} + Be^{s_2 t}}_{\text{transient response}} + \underbrace{y(\infty)}_{\text{steady-state response}} \tag{5.71}$$

Here again we might expect the four cases—overdamped, critically damped, underdamped, and undamped. We shall employ the following algorithm to compute the response $y(t)$. We emphasize that this algorithm is the most general one that helps us solve *any general second-order circuit*, be it RLC or some other topology, irrespective of whether it is source-free or forced. We need to advocate such a methodology, called generalization, owing to the simple fact that we cannot ingeniously devise problem-specific algorithms for virtually infinite possible topologies, not to mention the circuits of higher order, typically the integrated (VLSI) circuits, we encounter in practice. That we ought to have a uniform methodology to attack any *general* circuit cannot be overemphasized. And, in fact, development of software such as SPICE is a consequence of this sort of a generalization. We assure that the reader would enjoy the fruits of this exercise in Chapter 9.

Step 1	Determine the circuit's initial conditions: $i_L(0^+)$, $\frac{di_L(t)}{dt}\big|_{t=0^+}$, $v_C(0^+)$, and $\frac{dv_C(t)}{dt}\big|_{t=0^+}$ and hence the boundary values $y(0^+)$, $\frac{dy(t)}{dt}\big|_{t=0^+}$ and $y(\infty)$ of the desired variable.
Step 2	Apply an appropriate technique, e.g., nodal analysis, to obtain the second-order ODE in $y(t)$.
Step 3	Determine ω_n, ζ, s_1, s_2, and the constants A and B
Step 4	The complete response is given by $y(t) = A\,e^{s_1 t} + B\,e^{s_2 t} + y(\infty).$

Let us look at the following examples.

Example 5.26 Consider the second-order circuit in Fig. 5.42.

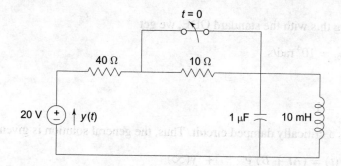

Fig. 5.42

Looking at the circuit for $t < 0$, we observe that the circuit's initial conditions are

$$i_L(0^+) = \frac{20}{40} = 0.5 \text{ A}, \quad v_C(0^+) = 0 \text{ V}$$

Next we need to obtain the boundary conditions. This is easy, using the initial conditions obtained above:

$$y(0^+) = \frac{20 - v_C(0^+)}{50}$$

$$= 0.4 \text{ A}$$

$$\Rightarrow \quad i_C(0^+) = C \frac{d}{dt} v_C(t)|_{t=0^+} = 0.4 - 0.5 = -0.1 \text{ A}$$

$$\frac{d}{dt} y(t)|_{t=0^+} = \frac{\frac{d}{dt} 20 - \frac{d}{dt} v_C(t)|_{t=0^+}}{50}$$

$$= \frac{0 - \frac{1}{1 \mu F} i_C(0^+)}{50}$$

$$= \frac{0.1}{1 \times 10^{-6} \times 50} = 2 \times 10^3$$

Further, the final value $y(\infty)$ can be obtained from the circuit directly as

$$y(t \rightarrow \infty) = 0.4 \text{ A}$$

Since we need to compute $y(t)$ delivered by the dc source, let us obtain the ODE in terms of $y(t)$ as follows. Applying mesh analysis for $t \geq 0$, we get

$$50 y(t) + v_C(t) = 20$$

$$v_C(t) = 10^2 \frac{d}{dt} i_L(t)$$

with

$$y(t) = 10^{-6} \frac{d}{dt} v_C(t) + i_L(t)$$

Juggling around with the three equations above, we get the following ODE:

$$\frac{d^2}{dt^2} y(t) + 2 \times 10^4 \frac{d}{dt} y(t) + 10^8 y(t) = 2 \times 10^3$$

Comparing this with the standard ODE, we get

$$\omega_n = 10^4 \text{ rad/s}$$

and

$$\zeta = 1$$

i.e., this is a critically damped circuit. Thus, the general solution is given by

$$y(t) = (At + B)\, e^{-\omega_n t} + y(\infty)$$

Unlike the source-free circuit, we need to be cautious about the initial conditions, owing to the presence of a *constant* $y(\infty)$. We shall show it formally as

$$y(0^+) = B + y(\infty)$$

$$\frac{d}{dt} y(t)|_{t=0^+} = A - \omega_n B$$

Putting this in the matrix-vector form, we have

$$\begin{pmatrix} 0 & 1 \\ 1 & -\omega_n \end{pmatrix} \begin{pmatrix} A \\ B \end{pmatrix} = \begin{pmatrix} y(0^+) - y(\infty) \\ \frac{d}{dt} y(t)|_{t=0^+} \end{pmatrix}$$

Notice that on the right-hand side we have $y(0^+) - y(\infty)$ instead of $y(0^+)$. Solving for A and B, we get $A = 2000$ and $B = 0$, and hence

$$y(t) = 2000\, t\, e^{-10,000t}$$

Exercise 5.42 Verify the results of Example 5.26 by obtaining the ODE for the capacitor voltage and then applying the KVR. Sketch $y(t)$. Is there any change in ω_n and ζ in the ODE? Comment.

(***Ans.*** There will not be any change in ω_n and ζ.)

Example 5.27 Consider the second-order circuit in Fig. 5.43. Let us determine the current $y(t)$ through the $2\,\Omega$ resistor.

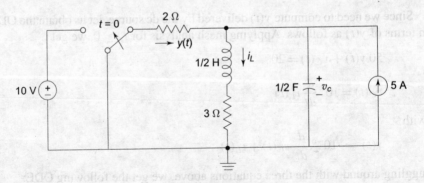

Fig. 5.43

First, let us determine the circuit's initial conditions. These may be obtained by inspection as follows. Since the 5 A current source is driving the circuit, the inductor is already a short circuit. Therefore, according to the current division rule,

$$i_L(0^+) = \frac{2}{2+3} 5 = 2 \text{ A}$$

and hence

$$v_C(0^+) = 2 \times 3 = 6 \text{ V}$$

Next, we might determine the boundary conditions as

$$y(0^+) = \frac{10 - v_C(0^+)}{2}$$

$$= 2 \text{ A}$$

$$\frac{d}{dt} y(t)|_{t=0^+} = -\frac{1}{2} \frac{d}{dt} v_C(t)|_{t=0^+}$$

$$= -i_C(0^+)$$

$$= -\left[y(0^+) - i_L(0^+) - 5 \right]$$

$$= -5 \text{ A/s}$$

$$y(t \to \infty) = \frac{10 \text{ V}}{2+3} - \frac{3}{3+2} 5 \text{ A}$$

$$= -1 \text{ A}$$

Then, applying nodal analysis, we get

$$y(t) = \frac{10 - v_C(t)}{2}$$

$$= i_L(t) + \frac{1}{2} \frac{d}{dt} v_C(t) - 5$$

with

$$\frac{1}{2} \frac{d}{dt} i_L(t) + 3 i_L(t) = v_C(t)$$

We might eliminate $i_L(t)$ and $v_C(t)$ to get the ODE as

$$\frac{d^2}{dt^2} y(t) + 7 \frac{d}{dt} y(t) + 10 y(t) = -10$$

We observe that it is an overdamped circuit since the discriminant $\sqrt{7^2 - 4 \times 1 \times 10} > 0$. Accordingly, the general response is

$$y(t) = A e^{-2t} + B e^{-5t} + y(\infty)$$

Using the boundary conditions we get $A = 10/3$ and $B = -1/3$ so that

$$y(t) = \frac{1}{3} e^{-2t} - \frac{1}{3} e^{-5t} - 1$$

Exercise 5.43 Verify the results of Example 5.27 by computing $i_L(t)$ and then applying KCR. Sketch $y(t)$.

Example 5.28 Let us consider the bridge circuit in Fig. 5.44.

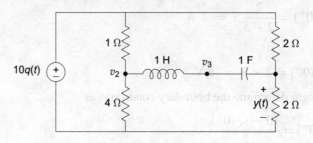

Fig. 5.44

It is evident (why?) that the circuit's initial conditions are all zero. Since we need to determine the voltage y, let us resort to nodal analysis so that we get

$$\frac{v_2 - 10}{1} + \frac{v_2}{4} + i_L = 0$$

$$\frac{y - 10}{2} + \frac{y}{2} - i_L = 0$$

Further, we observe that

$$1\frac{d}{dt} i_L(t) = v_2(t) - v_3(t)$$

$$1\frac{d}{dt} (v_3 - y) = i_L(t)$$

Let us determine the boundary conditions. Since the inductor–capacitor branch behaves like an open circuit at $t = 0^+$ and as $t \to \infty$, by simple voltage division we have

$$y(0^+) = y(\infty) = \frac{2}{2 + 2} 10 = 5 \text{ V}$$

We may eliminate all the variables except y using the above equations and obtain the ODE,

$$\frac{d^2}{dt^2} y(t) + \frac{9}{5} \frac{d}{dt} y(t) + y = 5$$

We recongnize that this circuit is an underdamped one with $\omega_n = 1$, $\zeta = 0.9$, and $\omega_d = \sqrt{0.19}$. Solving for the constants A and B, we get

$$A = -B = -j\frac{3}{2\omega_d}$$

Thus, the complete response is

$$y(t) = 6.8825\, e^{-0.9t} \sin 0.436t + 5 \text{ V}$$

Exercise 5.44 Use mesh analysis to compute the current through the bottom right $2\,\Omega$ resistor and verify the results of Example 5.28. Sketch the results. Examine what happens if all the resistors in the circuit are identical.

Exercise 5.45 Consider the circuit in Fig. 5.45.

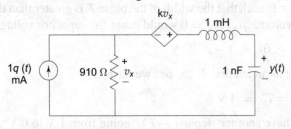

Fig. 5.45

Determine the coefficient k of the voltage controlled voltage source so that the circuit is underdamped.

Exercise 5.46 Prove that the circuit in Fig. 5.46 is never underdamped.

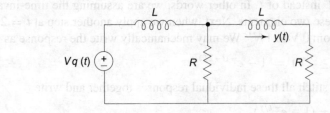

Fig. 5.46

5.7.3 Pulse Train Response

Recall that a pulse train may be *built* using step signals. Let us take up one such train for $t = 0$ described as

$$p(t) = q(t) - q(t-T) + q(t-2T) - q(t-3T) + \cdots \qquad (5.72)$$

For the sake of simplicity, we have considered a 'periodic' pulse train with period $2T$ seconds as shown in Fig. 5.47.

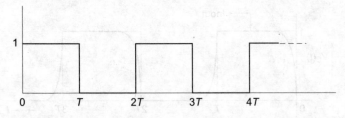

Fig. 5.47

We have also assumed that this pulse train has a non-negative amplitude for all $t \geq 0$ and the maximum amplitude is normalized to unity. Let us apply this signal to the following circuits.

Example 5.29 Suppose that the pulse train in Eqn (5.72) is applied to the first-order RC circuit in Fig. 5.32 with no initial capacitor voltage. Let us see what the capacitor voltage would be at any later time. To begin with, let us assume that the time-constant τ is such that the width of the pulse T is greater than the infinite time 6τ. The step voltage $q(t)$ at $t = 0$ would cause the capacitor voltage to be

$$v_C(t \geq 0) = 1 - e^{-t/\tau} \, \text{V}$$

Since we have assumed that $T \gg 6\tau$, we have

$$v_C(t = T) = 1 \, \text{V}$$

At $t = T$ we have another step $q(t - T)$, 'going from 1 V to 0 V', applied to the circuit. For this, we have

$$v_C(t \geq T) = -\left(1 - e^{-(t-T)/\tau}\right) \, \text{V}$$

under the assumption that the capacitance of the capacitor and the resistance of the resistor do not change in the interval $t \geq T$; notice that the exponent has now the term $t - T$ instead of t. In other words, we are assuming the time-invariance property of these two elements. Next, when we apply another step at $t = 2T$, this time 'going from 0 V to 1 V'. We may mechanically write the response as

$$v_C(t \geq 2T) = 1 - e^{-(t-2T)/\tau}$$

Now, we may stitch all these individual responses together and write

$$
\begin{aligned}
v_C(t) = &(1 - e^{-t/\tau}) \\
&-(1 - e^{-(t-T)/\tau}) \\
&+(1 - e^{-(t-2T)/\tau}) \\
&-(1 - e^{-(t-3T)/\tau}) \\
&+ \cdots \\
= &\sum_{k=0}^{\infty} (-1)^k \left(1 - e^{-(t-kT)/\tau}\right)
\end{aligned}
\tag{5.73}
$$

This response is plotted in Fig. 5.48.

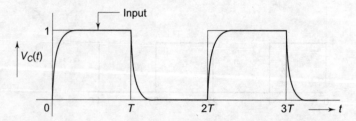

Fig. 5.48

In the example above, we assumed that the half-period T of the pulse train is much larger than the infinite time 6τ of the circuit. There are two more possibilities—(i) $T \sim \tau$, i.e., T is comparable to the time-constant, say, $T = 3\tau$, and (ii) $T \ll \tau$.

In the former case, we do not allow the transients to completely die out and the response alternates between v_{low} and v_{high} according to the following expressions:

$$v_{\text{low}} = v_{\text{high}} \, e^{-t/\tau}$$

$$v_{\text{high}} = (v_{\text{low}} - 1) \, e^{-t/\tau} + 1$$

where we take the maximum amplitude, and hence $v_C(\infty)$, of the pulse train to be equal to 1 V. v_{high} is the amplitude reached by the response at $t = T$, $3T$ etc., and v_{low} is the amplitude reached by the response at $t = 2T$, $4T$, etc.

In the latter case, since a single time-constant itself is much larger than the half-period of the pulse train, we allow only the initial portions of the transients. And, these portions are, almost, straight line segments. Hence, the output $v_C(t)$ approximates a *triangular* waveform with a very small amplitude. Since, we have learnt that a triangular waveform may be obtained by 'integrating' a pulse train, the first-order RC circuit of Fig. 5.32 with $\tau \gg T$ is called *integrator*.

Exercise 5.47 Repeat Example 5.29 for an arbitrary pulse train:

$$p(t) = q(t) - q(t - T_1) + q(t - T_2) - q(t - T_3) + \cdots$$

Exercise 5.48 Repeat Example 5.29 if $T \equiv 2\tau$.

Let us discuss what is going on in some more details.

Example 5.30 In this example we shall see what happens to an underdamped second-order response when the circuit is subjected to a periodic pulse train given in Eqn (5.72). Let us consider the voltage $v_C(t)$ across the capacitor in the series RLC circuit of Fig. 5.41.

If we assume that the initial conditions are zero, i.e.,

$$v_C(0^+) = i_L(0^+) = 0$$

then a unit step $q(t)$ at $t = 0$ gives us

$$v_C(t \geq 0) = 1 - \frac{\omega_n}{\omega_d} e^{-\zeta \omega_n t} \cos(\omega_d t + \theta)$$

and a unit step $q(t - kT)$, $k = 0, 1, 2, \ldots$, gives us

$$v_C(t \geq kT) = 1 - \frac{\omega_n}{\omega_d} e^{-\zeta \omega_n (t - kT)} \cos[\omega_d(t - kT) + \theta]$$

Thus, the response due to the pulse train of Eqn (5.72) may be written as

$$v_C(t \geq kT) = \sum_{k=0}^{\infty} (-1)^k \left(1 - \frac{\omega_n}{\omega_d} e^{-\zeta\omega_n(t-kT)} \cos\left[\omega_d(t - kT) + \theta\right] \right)$$

and this is shown in Fig. 5.49.

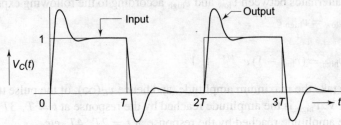

Fig. 5.49

Exercise 5.49 Obtain the response of a second-order underdamped circuit to the waveform discussed in Example 5.22. (**Hint** A ramp signal is obtained by integrating a step signal.)

Exercise 5.50 Obtain the response of a second-order critically damped circuit to the waveform shown in Fig. 5.50 under the assumptions (i) $\tau \ll T$, (ii) $\tau \sim T$, and (iii) $\tau > T$.

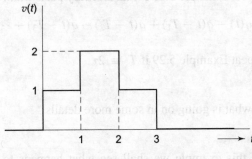

Fig. 5.50

The concept of switching and step response of energy-storage elements is an extremely important one. However, it is impossible to demonstrate these ideas in a laboratory since we cannot generate step signals. We would rather tap on the time-invariance property of the elements and apply a pulse train to the circuit. We take care that the 'half-period' T of the pulse train is much larger than the time-constant of the circuit; more specifically, $T \gg 6\tau$. On a storage oscilloscope, for instance, the waveform can be captured, to an appropriate scale, as step response.

Summary

This chapter has been quite exhaustive and we have uncovered *several* basic concepts. While the energy-storage elements behave like open circuits or short circuits for dc voltages, their response to switching is amazing. The circuits have been described by ordinary differential equations rather than algebraic equations, such as those we have seen until Chapter 4. It is rather exciting to note that, under certain circumstances, the response of a second-order circuit is sinusoidal, at least initially, although the circuit is not driven by any sinusoidal sources. We have learnt to circumvent arduous procedures to solve the differential equations by looking at the simplest possible first-order case carefully, and making a reasonable guess for the second-order equations thereafter. All through, the idea of linearity has been the undercurrent. For the second-order circuits we have presented a *systematic* approach to determine any quantity that is of interest to us. We might extend this approach to circuits of any order. For the sake of laboratory demonstrations, we have also presented the pulse train response of circuits exploiting the time-invariance property of the passive elements.

We conclude this chapter with a word of caution. Once we are familiar with the so-called *Laplace transform* and the results of Chapter 9, we might feel that we have wasted a lot of time on this chapter. It is not so; circuits and their theory, *per se*, are very much down-to-earth and we need to envisage the 'physics' behind the behaviour of each of the elements, particularly the energy-storage elements. The results of Chapter 9 have a mathematical, rather than physical, flavour and are geared to extrapolate and generalize the ideas of circuit analysis to circuit synthesis and henceforth to the so-called *system theory*.

Problems

5.1 Consider the circuit in Fig. 5.51. Determine $i_L(t)$ and plot it.

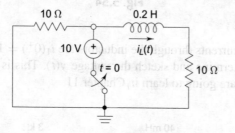

Fig. 5.51

5.2 Consider the circuit in Fig. 5.52. Determine and plot the inductor current $i_L(t)$ and $y(t)$ for $t \geq 0$.

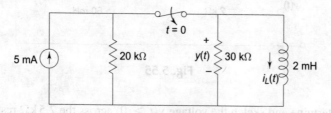

Fig. 5.52

5.3 Consider the circuit in Fig. 5.53. Determine and plot $y(t)$ for $t \geq 0$.

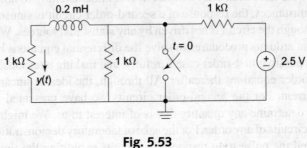

Fig. 5.53

5.4 Determine the current through and the voltage across the capacitor in the circuit of Fig. 5.54.

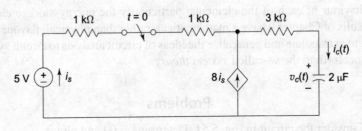

Fig. 5.54

5.5 The initial currents through the inductors are $i_1(0^+) = 10$ mA and $i_2(0^+) = 15$ mA. Determine and sketch the voltage $y(t)$. This is one of the standard circuits we are going to learn in Chapter 11.

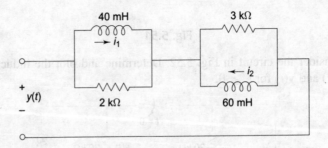

Fig. 5.55

5.6 Determine and sketch the voltage $y(t \geq 0)$ across the 7.5 kΩ resistor in the circuit of Fig. 5.56.

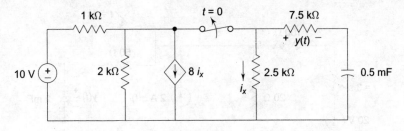

Fig. 5.56

5.7 Obtain a sketch of the voltage $v_C(t \geq 0)$ across the capacitance in the circuit of Fig. 5.57.

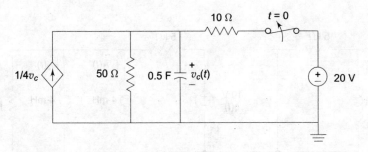

Fig. 5.57

5.8 For all times $-\infty \leq t \leq +\infty$, plot the voltage $y(t)$ across the capacitor in the circuit of Fig. 5.58.

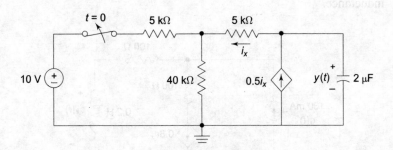

Fig. 5.58

5.9 In the time interval $-5 \leq t \leq +5$ s, sketch the voltage $y(t)$ across the 2 mF capacitor in the circuit of Fig. 5.59.

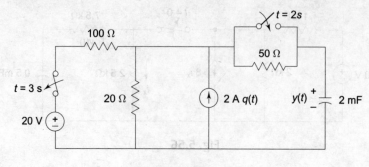

Fig. 5.59

5.10 Determine and sketch the inductor currents $i_1(t)$ and $i_2(t)$ in the circuit of Fig. 5.60.

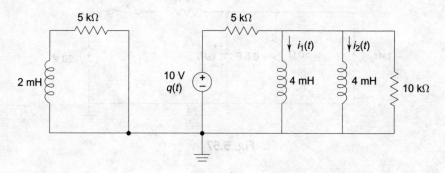

Fig. 5.60

5.11 In the circuit in Fig. 5.61, compute the voltage $v_x(t \geq 0)$ across the inductance.

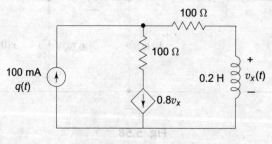

Fig. 5.61

5.12 In the following first-order circuit shown in Fig. 5.62, determine the voltages $v_1(t)$ and $v_2(t)$ across each of the capacitances connected in series.

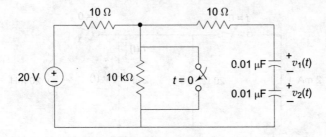

Fig. 5.62

5.13 The switch in the circuit of Fig. 5.63 is in position 1 for a long time. At $t = 0$, it is thrown to position 2, and back to position 1 at $t = 1$ ms, then to position 2 at $t = 2$ ms, and so on. Compute and sketch the voltage $v_C(t)$ across the capacitor for the time interval $-\infty \leq t \leq 10$ ms.

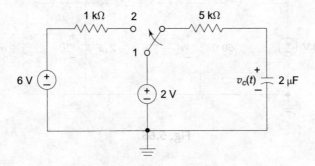

Fig. 5.63

5.14 Compute the damping factor ζ and the natural frequency ω_n and hence determine the voltage $v_C(t \geq 0)$ across the capacitor in the circuit of Fig. 5.64.

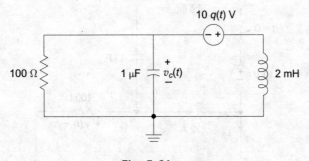

Fig. 5.64

5.15 Repeat problem 5.14 for the circuit of Fig. 5.65.

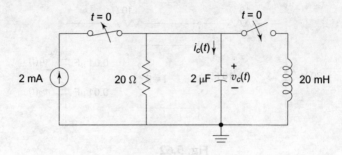

Fig. 5.65

5.16 Repeat Problem 5.14 for the circuit of Fig. 5.66.

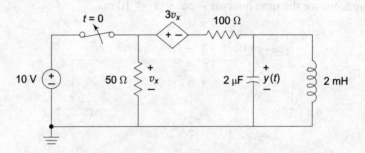

Fig. 5.66

5.17 Change the capacitance in the circuit of Fig. 5.66 so that its response becomes critically damped.

5.18 Determine and sketch the voltage $y(t)$ across the resistance in the circuit of Fig. 5.67.

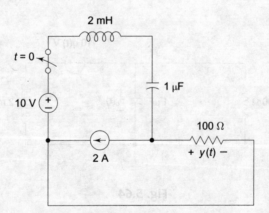

Fig. 5.67

5.19 In the circuit of Fig. 5.68, determine the energy stored in (i) the inductance and (ii) the capacitance at $t = 0.1$, 0.2, and 0.5 s.

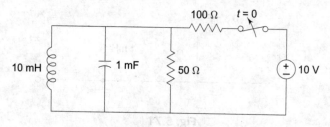

Fig. 5.68

5.20 Determine and sketch the voltage $y(t \geq 0)$ across the capacitance in the circuit of Fig. 5.69.

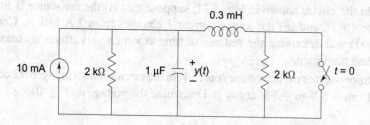

Fig. 5.69

5.21 A series RLC circuit, with $R = 20\ \Omega$, $L = 2$ mH, and $C = 10\ \mu$F is driven by a pulse train whose amplitude varies between 0 V and 10 V with a period $T = 2$ ms. Compute and plot the mesh current.

5.22 Obtain an expression for the voltage $y(t \geq 0)$ in the circuit of Fig. 5.70.

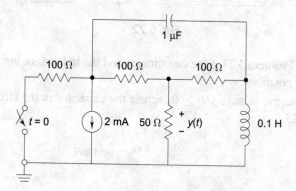

Fig. 5.70

5.23 Repeat Problem 5.22 with the inductance and the capacitance interchanged in their positions.

5.24 In the circuit shown in Fig. 5.71, the source current I changes from -1 A to $+1$ A at $t = 0$. Determine and sketch the voltage $y(t)$ during the interval $-\infty \leq t \leq +\infty$.

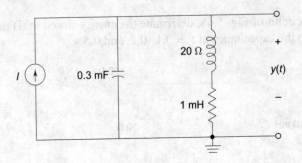

Fig. 5.71

5.25 Repeat Problem 5.24 with the capacitance and the inductance interchanged in their positions.

5.26 In the circuit shown in Fig. 5.71, suppose that (i) the resistance is lowered to 10 Ω, and (ii) the source current I changes from 2 A to 1 A. Compute $y(t)$ and determine the instants of time at which $y(t)$ attains its maximum and minimum.

5.27 Suppose that the dc voltage source V in the circuit shown in Fig. 5.72 changes from -5 V to $+5$ V at $t = 0$. Determine the voltage $v_0(t \geq 0)$.

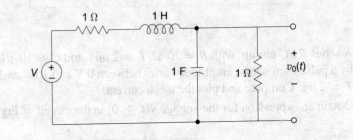

Fig. 5.72

5.28 Repeat Problem 5.27 if the capacitance and the inductance are interchanged in their positions.

5.29 Compute the voltage $y(t \geq 0)$ across the capacitor in the circuit shown in Fig. 5.73.

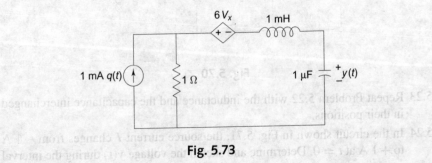

Fig. 5.73

5.30 Compute the voltage across the capacitor and the current through the inductor in the circuit shown in Fig. 5.74 for $t \geq 0$. Compare your results with those of Example 5.28.

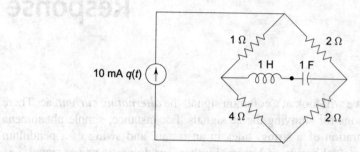

10 mA $q(t)$

1 Ω 2 Ω

1 H 1 F

4 Ω 2 Ω

Fig. 5.74

CHAPTER 6

Sinusoidal Sources and Response

In this chapter we will look at a common signal, the *alternating current,* ac. There are several reasons for studying the ac signals. For instance, simple phenomena such as the vibration of a string, tides in an ocean, and swing of a pendulum suggest the sinusoidal beauty of Nature. Further, our domestic power signal, i.e., 220 V at 50 cycle is best known for its efficiency in generation, transmission, and distribution. Most interesting of all, mathematically this provides us an elegant tool to generalize our network analysis and we will focus on this aspect in this chapter. Before we dig deeper, let us remind ourselves of certain popular convention. In contrast to the ac signal, the signal that is constant with respect to time is called *direct current*, dc. Clearly, the frequency of this signal is zero cycles per second. Though the 'c' in ac and dc stands for current, by 'dc' and 'ac' we traditionally refer to the frequency—zero or non-zero—rather than the current itself. Accordingly, it is common to use phrases such as 'ac voltage', 'dc current', etc. It does not make any sense to read these as alternating current voltage or direct current current. They denote, respectively, a voltage that varies sinusoidally with a specific frequency and a current that does not change with time.

6.1 Sinusoids

When we say a 'sinusoid' what comes to our mind is the expression $\sin \theta$ from the simple harmonic motion we studied in basic physics earlier. In this text, we are interested in voltages and/or currents. Accordingly, we define a *sinusoidal source*

$$u(t) = U_m \cos(\omega t + \theta) \tag{6.1}$$

at any instant of time t. Here we identify the following parameters:

- U_m, the maximum amplitude of excursion of the sinusoid, in 'volts' for voltage source and in 'amperes' for current source;
- t, the instance of time in 'seconds';
- ω, the *frequency* in 'rad/s';
- θ, the phase shift in 'degrees'.

These are depicted in Fig. 6.1. We also show the sinusoid for three different values of θ. Throughout this book we prefer 'radians per second' as the unit of frequency instead of hertz and 'degrees' as the unit of phase shift instead of radians. However, we will remember to make the expression $\omega t + \theta$ meaningful by converting the units appropriately.

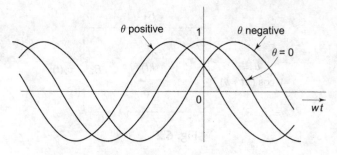

Fig. 6.1

One might immediately ask the importance of *co*sinusoid in our expression. We defer a detailed explanation for this for a while. Instead, we study the following circuits to learn some basics. Meanwhile, we suggest the reader to ponder over this question: What is $u(t)$ if we set $\omega = \theta = 0$?

Example 6.1 Consider the sinusoid

$$u(t) = 220\sqrt{3}\,\cos(1000\pi t + 210°)$$

Here,

$$U_m = 220\sqrt{3}, \quad \omega = 1000\,\pi \text{ rad/s},$$

$$f = \omega/2\pi = 500\,\text{Hz}, \quad T = \frac{1}{f} = 2\,\text{ms}, \quad \theta = 210°$$

Further,

$$u(0) = -110$$

and

$$u(0.5\,\text{ms}) = 110\sqrt{3}$$

and so on. The angle $\theta = +210°$ indicates that the basic sinusoid $U_m \cos(\omega t)$ is *shifted to the left* on the ωt axis by $210°$. For instance, $U_m \cos(\omega t)$ attains its maximum U_m at $\omega t = 0$, i.e., at $t = 0$ s, while $U \cos(\omega t + 210°)$ attains its maximum at $\omega t + 210° = 0$, or equivalently at $t = -7/6$ ms. If the phase angle were $-210°$, $u(t)$ attains its maximum at $t = +7/6$ ms since the sinusoid would be *shifted to the right* by $210°$ in this case. Notice that this shifting left or shifting right is *relative* to our reference time $t = 0$. We encourage the reader to look at a few arbitrary sinusoids of his/her choice.

Example 6.2 Consider the simple series circuit in Fig. 6.2.

Looking at a loop we are tempted to apply KVR, assuming a current $i(t)$ flows in the loop, and the result is

$$(R_1 + R_2)i(t) = u(t) = V_m \cos(\omega t + \theta)$$

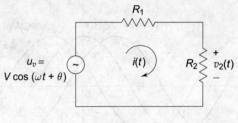

Fig. 6.2

Recall that the voltage and the current are directly proportional *at any instant of time* in a resistance, according to Ohm's law. Thus, we observe that the current is also a *sinusoid of the same frequency and phase shift* as that of the input voltage source, i.e.,

$$i(t) = I_m \cos(\omega t + \theta)$$

where

$$I_m = \frac{V_m}{R_1 + R_2}$$

If we look at the voltage across R_2, we observe that

$$v_2(t) = \frac{R_2}{R_1 + R_2} v(t)$$

i.e., the voltage division rule of Chapter 3 holds without any modification.

Example 6.3 Consider the simple series circuit in Fig. 6.3. Assume that the inductor does not have any initial current flowing through it.

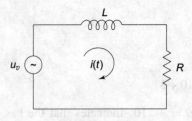

Fig. 6.3

Applying KVR, we get

$$L \frac{d}{dt} i(t) + Ri(t) = u_v(t)$$

Again we have ended up with an *ordinary differential equation with constant coefficients*, or ODE for short. We need to solve this ODE to obtain the current $i(t)$. We observe that on the right-hand side is a sinusoid and on the left-hand side is a linear combination of the variable $i(t)$ and its first derivative[1]. Since the derivative

---

[1] Recall that differentiation and integration are linear operations.

of a sinusoid is another sinusoid, we might suspect that $i(t)$ is also a sinusoid of the same frequency, perhaps with a different amplitude and a phase *difference* with the source. As was suggested in Chapter 3, we advocate the following general methodology:

- assume something with a reason;
- work out the problem;
- justify the assumption.

If the assumption is not justified, we make another reasonable assumption and repeat the three steps. Usually, we have only two alternatives to assume and the assumption is either justified or contradicted in the third step. If it is justified, it is okay; if contradicted, we assume the alternative. This methodology becomes more and more transparent as we proceed.

Let us now 'assume' that

$$i(t) = I_m \cos(\omega t + \phi)$$

where I_m and ϕ are to be determined. Substituting this in the ODE, we get

$$- \omega L I_m \sin(\omega t + \phi) + R I_m (\omega t + \phi) = V_m (\omega t + \theta)$$

We may now put the left-hand side in the form of a standard trigonometric identity:

$$\cos A \, \cos B - \sin A \, \sin B = \cos(A + B)$$

We need to do some manipulations to get

$$\sin A = \frac{\omega L}{\sqrt{R^2 + \omega^2 L^2}}, \quad \cos A = \frac{R}{\sqrt{R^2 + \omega^2 L^2}}$$

so that

$$A = \tan^{-1} \frac{\omega L}{R}$$

and

$$B = \omega t + \phi$$

The result is

$$I_m \cos\left(\omega t + \phi + \tan^{-1} \frac{\omega L}{R}\right) = \frac{V_m}{\sqrt{R^2 + \omega^2 L^2}} \cos(\omega t + \theta)$$

Comparing the left-hand side with the right-hand side, we finally get

$$I_m = \frac{V_m}{\sqrt{R^2 + \omega^2 L^2}}$$

and

$$\phi = \theta - \tan^{-1} \frac{\omega L}{R}$$

Thus, we have justified our assumption and the current is indeed a *sinusoid of the same frequency* as that of the source, but with a different amplitude and phase shift. In particular, we notice the following.

- The current $i(t)$ 'lags' behind the source in phase, i.e., the current sinusoid is delayed by an angle $\tan^{-1}(\omega L/R)$ relative to the voltage source.
- The amplitude and the phase shift of the current are *dependent on the frequency* in addition to V_m, L, and R.

We remark that the solution is rather tedious, involving some *non-intuitive* and mechanical manipulations.

In general, if the response is $d(t) = D\cos(\omega t + \phi)$ due to a source $u(t)$, we write

$$D = D(\omega) \quad \text{and} \quad \phi = \phi(\omega) \tag{6.2}$$

It may be verified that the voltage v_R across the resistance is $Ri(t)$ and, unlike the previous example, the voltage division rule is not applicable here.

Exercise 6.1 Verify the results of Example 6.2 with the input voltage $u_v(t) = V_m \sin(\omega t + \theta)$.

(**Ans.** No change from the results of Example 6.2.)

Example 6.4 Let us now consider the simple series circuit in Fig. 6.4. Here we are interested in the voltage $v_C(t)$ across the capacitance. We assume that the capacitor is initially discharged.

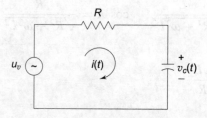

Fig. 6.4

Applying KVR once again, we get

$$i(t) = C\frac{d}{dt}v_C(t)$$

and

$$RC\frac{d}{dt}v_C(t) + v_C(t) = u_v(t)$$

We may assume $v_C(t)$ is a sinusoid and solve the ODE to get

$$v_C(t) = \frac{V_m}{\sqrt{1 + (\omega RC)^2}}\cos(\omega t + \theta - \tan^{-1}\omega RC)$$

We notice that the amplitude and the phase shift depend upon the frequency ω and the voltage v_C 'lags' behind the source by an angle $\tan^{-1} \omega RC$.

Exercise 6.2 Verify the results of Example 6.3. Redo the problem with $u_v(t) = V_m \sin(\omega t + \theta)$.

(**Ans.** No change in magnitude and phase angle.)

Let us summarize our observations from the examples of this section.

1. The input is a sinusoidal source of frequency ω and phase shift θ.
2. The output is *also* a sinusoid of the *same frequency* ω, but with a different amplitude and phase shift, each of which are dependent on ω.
3. If we look at the *phase difference* between the output and the input, i.e., $\psi = \phi - \theta$, we say that
   - the output *leads* the input if $\psi > 0°$,
   - the output *lags* behind the input if $\psi < 0°$, and
   - the output is *in phase* with the input if $\psi = 0°$.

Exercise 6.3 Obtain the voltage across the inductance and the resistance in Example 6.2.

Exercise 6.4 Obtain the current through the capacitance and the voltage across the resistance in Example 6.3.

We shall digress briefly from the mainstream and make a note of the following important idea.

6.1.1   Transients

For the ODE in Example 6.3 we guessed that the response could be a sinusoid and went on to justify that. Apparently, we are correct. But with the experience from Chapter 5, particularly with the source-free circuits, we have also missed something. Let us look at this carefully. Suppose we augment our initial guess and say that

$$y(t) = A e^{st} + B \cos(\omega t + \theta)$$

we still find it a valid guess and it turns out that $s = -1/\tau$. In other words, even if the source is purely sinusoidal, the response shall have a *transient* in addition to the steady state. We strongly advise the reader to be aware of this not-so-trivial observation.

However, for the sake of clarity, the focus of this chapter is only on the sinusoidal steady-state part of the response and we 'ignore' the transients. We shall come back to the transients again in Chapter 9.

6.2 Linearity and Euler's Identity

Let us quickly recapitulate some elementary ideas of complex numbers before we proceed.

6.2.1 Complex Numbers

Complex numbers have a real part and an imaginary part with $j = \sqrt{-1}$ acting as an identification tag. We use script letters such as $\mathcal{A}$ to denote these complex numbers. A complex number can be written as

$$\mathcal{A} = a_1 + ja_2$$

in *rectangular form* or as

$$\mathcal{A} = \sqrt{a_1^2 + a_2^2} \; \angle \; \tan^{-1} \frac{a_2}{a_1}$$

in *polar form*. Alternatively, we may also express this polar form as

$$\mathcal{A} = \sqrt{a_1^2 + a_2^2} \; e^{j\,\tan^{-1}(a_2/a_1)}$$

using Euler's identity. We show this in Argand's plane in Fig. 6.5

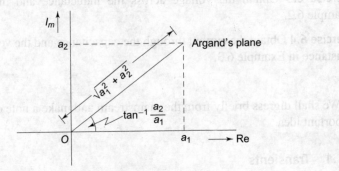

Fig. 6.5

If we multiply a complex number by a real scalar α, we get

$$\alpha\mathcal{A} = \alpha\, a_1 + j\alpha\, a_2$$

$$= \alpha \sqrt{a_1^2 + a_2^2} \; e^{j\,\tan^{-1}(a_2/a_1)}$$

We notice that only the magnitude is scaled, but the angle remains the same. However, if the scalar is the imaginary number $j = \sqrt{-1}$, we see that

$$j\mathcal{A} = j\, a_1 - a_2$$

$$= \sqrt{a_1^2 + a_2^2} \; e^{j\,\tan^{-1}(a_1/-a_2)}$$

$$= \sqrt{a_1^2 + a_2^2} \; e^{j[90° + \tan^{-1}(a_2/a_1)]}$$

i.e., the magnitude remains unaltered but the angle is now shifted by $+90°$. It is easy to see that if we divide by j, the angle will be shifted by $-90°$.

Two complex numbers A and B may be added or subtracted, using the familiar parallelogram law, and it is easy to do this when these numbers are expressed in rectangular form. For instance,

$$A \pm B = (a_1 \pm b_1) + j(a_2 \pm b_2)$$

i.e., the real and imaginary parts are added pairwise. It is not as convenient to do this addition in polar form. However, we may express the result in polar form if the context demands so.

Likewise, two complex numbers may be multiplied or divided, and it is easy to do this in polar form. For instance,

$$A \cdot B = \sqrt{a_1^2 + a_2^2}\, e^{j\,\tan^{-1}(a_2/a_1)} \cdot \sqrt{b_1^2 + b_2^2}\, e^{j\,\tan^{-1}(b_2/b_1)}$$

$$= \sqrt{(a_1^2 + a_2^2)(b_1^2 + b_2^2)}\, e^{j(\tan^{-1}(a_2/a_1)} + \tan^{-1}(b_2/b_1)}$$

i.e., we take the product of the magnitudes and add the angles. Likewise, we perform the division in polar form by taking the quotient of the magnitudes and subtracting the angles. Clearly, these two operations are difficult to perform using rectangular coordinates; we may express the end result in the rectangular form, if the context demands so.

We need to have a closer look at the angle part of the complex numbers expressed in polar form. Consider the following complex numbers:

$$A_1 = a_1 + ja_2$$
$$A_2 = -a_1 + ja_2$$
$$A_3 = -a_1 - ja_2$$
$$A_4 = a_1 - ja_2$$

All the four numbers have the same magnitude

$$|A_i| = \sqrt{a_1^2 + a_2^2}, \quad i = 1,\, 2,\, 3,\, 4$$

However, the angles are different, i.e.,

$\angle A_1 = \tan^{-1}\left(\dfrac{a_2}{a_1}\right)$ is in the first quadrant of Fig. 6.5.

$\angle A_2 = \tan^{-1}\left(\dfrac{a_2}{-a_1}\right)$ is in the second quadrant

$\angle A_3 = \tan^{-1}\left(\dfrac{-a_2}{-a_1}\right)$ is in the third quadrant

$\angle A_4 = \tan^{-1}\left(\dfrac{-a_2}{a_1}\right)$ is in the fourth quadrant

It is slightly tricky to compute these angles since, for instance, we might mistake

$$\angle A_1 = \angle A_3$$

and

$$\angle A_2 = \angle A_4$$

using hand-held calculators. We have to be cautious to see whether a negative sign is appearing in the numerator as in $\mathcal{A}_3$ and $\mathcal{A}_4$, or in the denominator as in $\mathcal{A}_2$ and $\mathcal{A}_3$, or in both as in $\mathcal{A}_3$, or in none as in $\mathcal{A}_1$.

Let us now redo Example 6.3 using a voltage source

$$u_v(t) = V_m \sin(\omega t + \theta)$$

The reader has the answer already (cf. Exercise 6.2):

$$v_C(t) = \frac{V}{\sqrt{1+(\omega RC)^2}} \sin(\omega t + \theta - \tan^{-1} \omega RC)$$

In other words, if the source is 'cos($\cdot$)' the response is 'cos($\cdot$)' and if the source is 'sin($\cdot$)' the response is 'sin($\cdot$)' without any change in the respective arguments. Further, we may observe that if the amplitude of a source is multiplied with a scalar α, then the amplitude of the corresponding response is also multiplied by the *same scalar* α. In short, we observe that the circuits containing energy-storage elements are also linear and the principle of linearity holds good.

Let us now recollect Example 4.17 and assume that the circuit shown in Example 6.3 has two sources as shown in Fig. 6.6.

Example 6.5 Consider the series RC circuit with two sources as shown in Fig. 6.6.

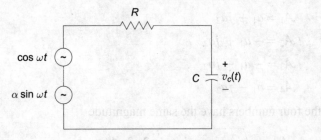

Fig. 6.6

The response shall have two parts—one due to 'cos($\cdot$)' and the other due to 'α sin($\cdot$)'. We write it as

$$v_C(t) = V_C \cos(\omega t + \phi)$$
$$+ \ \alpha V_C \sin(\omega t + \phi)$$

What happens if we choose $\alpha = j$, the imaginary number $\sqrt{-1}$? The number j also serves as the 'identification tag' and we may rewrite the above equation with α replaced by j, i.e.,

$$v_C(t) = \underbrace{V_C \cos(\omega t + \phi)}_{v_{\text{real}}} + \underbrace{jV_C \sin(\omega t + \phi)}_{v_{\text{imag}}} \qquad (6.3)$$

We observe that, thanks to Euler's identity, the two sources may be written as

$$u_v(t) = V_m e^{j(\omega t + \theta)}$$

and the response can be written as

$$v_C(t) = V_C e^{j(\omega t + \phi)}$$

We observe that the source and the response are *complex functions*. Let us plug in the actual parameters ω, R, and C to get

$$v_C(t) = \frac{V}{\sqrt{1 + (\omega RC)^2}} e^{j(\omega t + \theta - \tan^{-1}\omega RC)}$$

$$= \frac{1}{\sqrt{1 + (\omega RC)^2}} e^{-j(\tan^{-1}\omega RC)} u_v(t)$$

We may write the above equation in the rectangular form as

$$v_C(t) = \frac{1}{1 + j\omega RC} u_v(t) \tag{6.4}$$

i.e., the output is proportional to the input *at every instant of time* and the proportionality constant is a complex number.

Exercise 6.5 Verify Eqn (6.4).

It appears that the complex numbers really make things complex. However, we assure that we are in the process of simplifying things. In fact, it was the need for more freedom in formal calculations that brought about the use of complex numbers. For instance, if we are interested in the response due to $V_m \cos(\omega t + \theta)$, i.e., the real part of the complex source $V_m e^{j(\omega t + \theta)}$, we make use of the identification tag to extract the real part of $v_C(t)$. For instance, in Example 6.5,

$$v_{C\text{real}}(t) = \text{Real part of } \frac{V}{1 + j\omega RC} [\cos(\omega t + \theta) + j \sin(\omega t + \theta)]$$

$$= \frac{V}{\sqrt{1 + (\omega RC)^2}} \cos(\omega t + \theta - \tan^{-1}\omega RC) \tag{6.5}$$

Imaginary[2] sources never exist physically, but we use them in the analysis (i) to simplify the otherwise laborious differential equations, and (ii) to generalize the techniques developed in Chapter 4.

Exercise 6.6 Verify Eqn (6.5).

We end this section with a simple but important observation. We may rewrite Eqn (6.4) as

$$v_C(t) = \frac{\dfrac{1}{j\omega C}}{R + \dfrac{1}{j\omega C}} u_v(t) \tag{6.6}$$

---

[2] It is a common experience of many eminent personalities that if we add 'imagination' to the reality, life is indeed far more simpler!

which suggests that the voltage division rule has come back in disguise in terms of complex algebra. We shall elaborate on this in the following section.

Exercise 6.7 Redo Examples 6.1 and 6.2 using the complex voltage source and express the outputs in the form of Eqn (6.6).

Exercise 6.8 Redo Exercises 6.3 and 6.4 using the complex voltage source and express the outputs in the form of Eqn (6.6).

6.2.2 Phasors

Let us put everything we have discussed so far in a formal perspective. The source (whether it is a voltage source or a current source) is a complex function in the frequency ω, i.e.,

$$u(t) = U_m e^{j(\omega t + \theta)}, \quad U_m, \omega, t, \theta \in \Re$$

It has both a real part and an imaginary part with 'j' serving as an identification tag. By the principle of linearity, the response of the circuit (be it voltage or current) is also complex, i.e.,

$$d(t) = D e^{j(\omega t + \phi)}, \quad D, \phi \in \Re$$

In terms of the phase difference $\psi = \phi - \theta$, we may easily rewrite the response as

$$d(t) = \frac{D}{U_m} e^{j\psi} u(t) \tag{6.7}$$

We observe the following.

- The source and hence the response are both sinusoids. We need not explicitly mention them again and again.
- The frequency of the response is the same as that of the source.
- When frequency ω is available, time t is automatically available. Therefore, one of them is sufficient. At present we prefer ω.
- The source and the response have different *real* amplitudes U_m and D, respectively, and different *real* phase shifts θ and ϕ, respectively.

Let us then use a shorthand notation to represent the complex source as

$$\mathcal{U}(\omega) \triangleq U_m e^{j\theta} \tag{6.8}$$

and represent the response as

$$\mathcal{D}(\omega) \triangleq D e^{j\phi} \tag{6.9}$$

We call the script letters $\mathcal{U}$ and $\mathcal{D}$ *phasors* since they embed the magnitude and phase information. Observe that these phasors are actually complex numbers, expressed in polar form, given the frequency and the phase shift. We suppress the term $e^{j\omega t}$ in both the definitions since it is understood that we are speaking about sinusoids of a specific frequency ω. Hereafter, we use both the representations (with and without $e^{j\omega t}$) of the polar form interchangeably.

6.3 Behaviour of Elements with ac Signals

Let us look at our basic elements and their response to our complex source in Eqn (6.8).

Resistance If a voltage source $\mathcal{V}_m(\omega)$ is applied across a resistance R, we see that the current through it is

$$\mathcal{I}_m(\omega) = \frac{V_m}{R} e^{j\theta} = \frac{V_m}{R} \quad \forall \, \omega, \theta$$

using Ohm's law. We emphasize that the resistance R is independent of frequency, i.e., the element offers the same resistance at all frequencies[3]. We shall rewrite the above expression as

$$\mathcal{V}_m = R\mathcal{I}_m \quad \forall \omega, \theta \tag{6.10}$$

Observe that, since R is a real number, the phase difference $\psi = 0°$. Thus, the voltage and current in a resistance are *in phase*.

Inductance If a current source $\mathcal{I}_m(\omega)$ is applied across an inductance L, we see that the voltage drop across it is

$$\mathcal{V} = j\omega L \times I\, e^{j\theta}$$
$$= j\omega L \mathcal{I} \,\forall\, \omega, \,\theta \tag{6.11}$$

using the differential relationship $v = L di/dt$. Apparently, the phase shift of the voltage ϕ is missing. But, if we observe carefully, we see that the proportionality constant has a 'j' which can be written as $e^{j90°}$ using Euler's identity. Remember that differentiating a sinusoid results in a phase shift of 90°; we are not deviating from the basic facts. The reader is advised to obtain plots of $\cos \omega t$, and $\cos(\omega t \pm 90°)$ and see the effects of differentiation and phase shifts. Thus, the phase shift of the output is

$$\phi = \theta + 90°$$

and the phase difference is $\psi = 90°$ and the *voltage leads the current by 90° in an inductance.*

Looking at *Ohm's law like* Eqn (6.11) we notice that the voltage and current in an inductor are directly proportional, and the proportionality constant $j\omega L$ has the units of ohms. Further, this proportionality constant varies directly with the frequency ω. In other words, the 'resistance-like' quantity offered by the inductance varies directly with frequency. For small frequencies, close to dc, the inductor acts like a short circuit (which we have already seen in Chapter 5), and for high frequencies it acts like an open circuit.

---

[3] The phrase 'at any instant of time' simply gets transformed into 'at any frequency'.

Capacitance If we apply a voltage source $\mathcal{V}_m(\omega)$ across a capacitance C, we see that the current through it is

$$\mathcal{I} = j\omega C \times V e^{j\theta}$$

$$= \omega C V e^{j(\theta+90°)} \; \forall \; \omega, \theta \qquad (6.12)$$

Thus the phase shift of the output is

$$\phi = \theta + 90°$$

and the phase difference is $\psi = 90°$ and the *current leads the voltage by 90° in a capacitance*. We may write Eqn (6.12) as

$$\frac{\mathcal{V}}{\mathcal{I}} = \frac{1}{j\omega C} \; \Omega$$

Here too, the frequency dependent proportionality constant $1/j\omega C$ has the units of ohms. However, this varies inversely with the frequency. In other words, the capacitance acts like an open circuit at low frequencies and like a short circuit at high frequencies.

Exercise 6.9 Derive the $\mathcal{V}$-$\mathcal{I}$ relationship of Eqn (6.11) for inductance and that of Eqn (6.12) for capacitance.

Exercise 6.10 Justify that the units of the products ωL and $1/\omega C$ are ohms. (***Hint*** What is a time-constant?)

In all the three cases above, we are concerned with the *phase difference* between the input and the output, rather than the absolute phase shift. Accordingly, we may assume that the phase shift of source $\theta = 0°$ without loss of generality and we do so hereafter. These elements are said to possess a property called *shift invariance*, i.e, if the input has a phase shift of θ, the output has a *corresponding* phase shift of $\phi + \theta$ and the phase difference is always ϕ. As has been mentioned earlier in this chapter, it is not difficult to see that the phase shift measured in degrees is directly related to the 'time shift' in seconds. A shift of $\theta°$ along the ωt axis is the same as a shift of $\theta/2\pi f$ seconds along the t axis. Since the frequency f is inversely related to the period T, a phase shift of 90° of a sinusoid, for instance, corresponds to a time shift of 1/4 period in seconds. This is a consequence of the time-invariance property we have discussed in the preceding chapter. We shall see later that it is not only these elements, but the networks themselves also are shift invariant.

Let us define the following:

$$X_R \triangleq R \; \Omega$$

$$X_L \triangleq j\,\omega L \; \Omega \qquad (6.13)$$

$$X_C \triangleq \frac{1}{j\omega C} = -j\,\frac{1}{\omega C} \; \Omega$$

We call X_L *inductive reactance* and X_C *capacitive reactance*. We use the word 'reactance', to distinguish it from 'resistance', which has the same units. We also refer to the inductance and the capacitance as *reactive elements*. In terms of these quantities and the phasor voltage and current, we may write the *extended Ohm's law* as

$$\mathcal{V} = X\mathcal{I} \tag{6.14}$$

In terms of the quantity X, we shall fix the following notation:

- The voltage and the current in X_R are *in phase*,
- the voltage *leads* the current in X_L by $90°$, and
- the voltage *lags* behind the current in X_C by $90°$.

We can summarize this section as follows.

1. Any *real* source may be written as $\mathcal{U}_m(\omega)$ (with $\theta = 0°$) in the complex polar form, adding a (fictitious) imaginary source of the same amplitude and frequency. If $\omega = 0$, we have a dc source, otherwise it is an ac source.
2. If the element we are scrutinizing is a resistor, then the magnitude of the input $\mathcal{U}_m$ is *scaled* by R, i.e., if the source is a current source, the voltage drop is obtained by multiplying the source magnitude by R; if the source is a voltage source, the current is obtained by multiplying the source magnitude by $1/R$.
3. Given frequency ω, we obtain X for the reactive element under scrutiny, and scale the source magnitude by X to obtain the response magnitude. Since X contains $\pm j$, the phase shift $\phi \, (= \psi)$ is already there.
4. For easy manipulations, in general, we represent the sources and the signals, i.e., currents and voltages, in the polar form, and the elements, i.e., R, L, and C, in the rectangular form.

In passing, we note that a phase shift of $\pm 90°$ is achieved by multiplying with the imaginary number $\pm j$. Further, differentiation (integration) with respect to time can be achieved by multiplying (dividing) by $j\omega$. We shall come back to these ideas in a more general form later.

Thus, any source, whether ac or dc, is represented by a circle containing appropriate symbols for voltage or current, and labelled $\mathcal{U}_m(\omega)$, and any passive element is represented by a small rectangular box with an appropriate label, R or X. This is shown in Fig. 6.7. However, when there is no ambiguity in the context, we may use the time functions such as $v(t)$ and $i(t)$ themselves instead of the script letters. The quantities that are important to us are the magnitude and the phase angle of the source, the resistance and the reactance as a function of the given frequency, followed by the complex algebra.

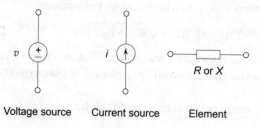

Voltage source   Current source   Element

Fig. 6.7

Once this is done, the analysis is similar to what we have done with dc sources and resistances alone. A fundamental difference is that it is real algebra in the case of dc sources and resistances, and it is complex algebra in the present case. Since the set of complex numbers $\mathcal{C}$ contains the set of real numbers $\mathfrak{R}$, we have not lost our roots; rather, we have built up the *generalization* firmly. The KCR and the KVR are intact since they are independent of elements. The electrical-mathematical genius Charles Proteus Steinmetz (1865–1923) was behind this ingenious algebra. Instead of dealing with the sinusoid functions of time, it is easy to achieve the goals of ac circuit analysis with phasors and complex algebra. In the following section we will present several examples that would highlight this idea.

6.4 Impedance and Admittance

Let us now redo our earlier examples.

Example 6.6 (cf. Example 6.3)

In tune with Fig. 6.7, we redraw the circuit as shown in Fig. 6.8.

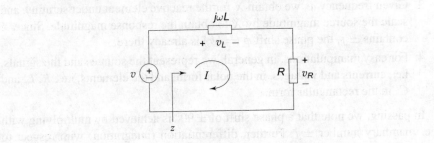

Fig. 6.8

Since this is a series circuit, the current through all the three elements is the same. However, it is the complex current we have to account for. Applying KVR, we get

$$-\mathcal{V} + j\omega L \mathcal{I} + R\mathcal{I} = 0$$

Since both R and $j\omega L$ have the same units, we may add them. Notice that one of them is a real number and the other is an imaginary one; however, both of them are *complex numbers*, one with zero imaginary part and the other with zero real part. Thus, the source 'sees' an *equivalent complex resistance*

$$\mathcal{Z}(j\omega) = R + j\omega L \ \Omega = \sqrt{R^2 + (\omega L)^2} \, e^{j \tan^{-1}(\omega L/R)} \ \Omega$$

We call this phasor as *impedance* and we use $\mathcal{Z}$ as its symbol. This impedance is a function of ω. Notice that the word 'impedance' is synonymous to 'resistance' in a literary sense.

Let the total voltage pumped into the circuit be

$$\mathcal{V} = V_m \, e^{j\omega t} \ \text{V}$$

Owing to the linearity, we may write the *generalized Ohm's law* in terms of the phasors as

$$\mathcal{V} = \mathcal{Z}\mathcal{I}$$

and obtain

$$\mathcal{I} = \frac{\mathcal{V}}{\mathcal{Z}}$$

$$= \frac{V_m}{\sqrt{R^2 + (\omega L)^2}} e^{-j\tan^{-1}(\omega L/R)}$$

Except for the complex algebra everything else is just the same as in the case of resistive circuits with dc sources, the focus of Chapter 4.

Observe that we need not really resolve this into real and imaginary parts. We know that the response to a sinusoid is always a sinusoid. Hence, we need to determine only the magnitude and phase difference. Recall that in our earlier method also, we assumed a sinusoidal form and determined the magnitude and phase difference. These two quantities are readily available in the phasor for the response. For simplicity, we may leave the phasor as it is, either in the rectangular form or in the polar form.

We may also obtain the voltage division rule in similar lines. For instance, if we are interested in the voltage across the inductor, we get it readily as

$$\mathcal{V}_L = j\omega L \mathcal{I}$$

$$= \frac{j\omega L}{R + j\omega L}\mathcal{V}$$

$$= \frac{\omega L V}{\sqrt{R^2 + (\omega L)^2}} e^{j[90° - \tan^{-1}(\omega L/R)]}$$

Notice the change in phase difference.

We reiterate that it is easy to perform the addition (or subtraction) of complex numbers in rectangular form and perform the multiplication (or division) in the polar form. However, the end result may be expressed in either of the forms.

Example 6.7 (cf. Example 6.4)

For the series *RC* circuit of Example 6.4, we may readily obtain

$$\mathcal{Z} = R + \frac{1}{j\omega C}$$

$$\mathcal{I} = \frac{\mathcal{V}}{\mathcal{Z}}$$

$$\mathcal{V}_C = \frac{\dfrac{1}{j\omega C}}{R + \dfrac{1}{j\omega C}}$$

Exercise 6.11 Obtain the magnitudes and phase shifts in Example 6.7.

We observe that in a series circuit the impedances can be added just as we add the resistors, i.e.,

$$\mathcal{Z}_{\text{series}} = \mathcal{Z}_1 + \mathcal{Z}_2 + \cdots + \mathcal{Z}_n \qquad (6.15)$$

We may expect to do the same thing in a parallel circuit also. However, we need to extend our definitions as follows. The inverse of reactance X is called *susceptance* and is denoted as B, i.e.,

$$B_L = \frac{1}{j\omega L}$$

and

$$B_C = j\omega C$$

The units, as may be expected identical to conductance, are siemens Ω^{-1}. Further, we may 'add' a conductance G to a susceptance B to get the phasor *admittance*[4] denoted by $\mathcal{Y}$, i.e.,

$$\mathcal{Y}(j\omega) = G + jB\ \Omega^{-1}$$

And, admittances can be added in parallel to obtain the equivalent admittance as

$$\mathcal{Y}_{\text{parallel}} = \mathcal{Y}_1 + \mathcal{Y}_2 + \cdots + \mathcal{Y}_n \qquad (6.16)$$

We shall illustrate this in the following example. Having done with algebraic quantities, we shall look at numericals now.

Example 6.8 Consider the parallel RLC circuit shown in Fig. 6.9.

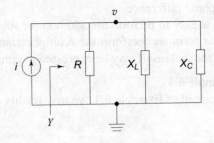

Fig. 6.9

The admittance of the circuit as seen by the current source $\mathcal{I}$ is

$$\mathcal{Y} = G + j\omega C + \frac{1}{j\omega L}$$

$$= 2 + j1 - j2\ \Omega^{-1}$$

$$= \sqrt{5}\,e^{j\,\tan^{-1}(-1/2)}$$

---

[4] synonymous with conductance

We may use the current division rule to obtain the individual branch currents. For instance,

$$\mathcal{I}_C = \frac{j\omega C}{\mathcal{Y}} \mathcal{I}$$

$$= \frac{1}{\sqrt{5}} e^{j[90° - \tan^{-1}(-1/2)]}$$

and the voltage across the capacitance may be obtained using the extended Ohm's law.

We also note another interesting issue in this example. For the given frequency and the component values, we have

$$\mathcal{Z} = \frac{1}{\mathcal{Y}} = \frac{1}{\sqrt{5}} e^{j \tan^{-1}(1/2)}$$

i.e., the voltage (across the three components) *leads* the total current $\mathcal{I}$ by an angle $\tan^{-1}(1/2)$. We say that 'the current source $\mathcal{I}$' sees an *inductive load*. At first, it appears to be counterintuitive. However, we need to keep in mind that at this frequency, the inductive reactance dominates the capacitive reactance in the sense that *relatively* more current flows through the inductance than through the capacitance.

Let us look at the circuit when the source has a frequency $\omega = 10\sqrt{2}$ rad/s. Here,

$$X_L = j\frac{1}{\sqrt{2}} = -X_C$$

i.e., both the reactances nullify each other and the net impedance is only the resistance $R = 0.5\ \Omega$. We say it is a *resistive load*. Now we may suspect that at a higher frequency the source might see a *capacitive load*, and it is in fact true.

Let us now summarize our observations above in a graph of $|\mathcal{Z}|$ versus ω as shown in Fig. 6.10.

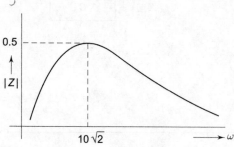

Fig. 6.10

The graph reads as follows.

- At low frequencies, the capacitance has a very high (near infinite) reactance and the inductance has a very low (near zero) reactance. The net impedance $\mathcal{Z}$ is low at low frequencies.

- At $\omega = 10\sqrt{2}$ rad/s, the impedance is purely resistive. It is interesting to note that $|\mathcal{Z}|$ is the highest at this frequency.
- At high frequencies, the inductance has a very high (near infinite) reactance and the capacitance has a very low (near zero) reactance. The net impedance $\mathcal{Z}$ is, once again, low at high frequencies.

We shall furnish elaborate details about these frequency functions in the following chapter.

Exercise 6.12 Verify the results of Example 6.8.

Exercise 6.13 Repeat Example 6.8 for a series RLC circuit driven by a voltage source. Assume the same component values. Obtain the plot (similar to that of Fig. 6.10) of $\mathcal{Z}$ versus ω.

6.5 Network Analysis Using Phasors

In this section we shall revisit the techniques of Chapter 4, namely, the nodal analysis, the principle of superposition, Thévenin's and Norton's theorems, and the maximum power transfer theorem.

6.5.1 Nodal Analysis

Example 6.9 Let us consider the circuit in Fig. 6.11.

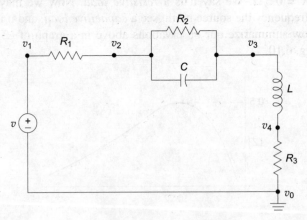

Fig. 6.11

There are five nodes including the ground and the source. For the three unknown node voltages labelled in the figure, we may obtain the following matrix-vector equation:

$$
\begin{pmatrix}
\dfrac{1}{R_1}+\dfrac{1}{R_2}+j\omega C & -\left(\dfrac{1}{R_2}+j\omega C\right) & 0 \\[2mm]
-\left(\dfrac{1}{R_2}+j\omega C\right) & \dfrac{1}{R_2}+\dfrac{1}{j\omega L}+j\omega C & -\dfrac{1}{j\omega L} \\[2mm]
0 & -\dfrac{1}{j\omega L} & \dfrac{1}{R_3}+\dfrac{1}{j\omega L}
\end{pmatrix}
\begin{pmatrix} v_2 \\ v_3 \\ v_4 \end{pmatrix}
=
\begin{pmatrix} \dfrac{v}{R_1} \\ 0 \\ 0 \end{pmatrix}
$$

We have the node incidence matrix $\mathcal{N}$ again with its properties, such as symmetry, intact. However, this time the elements are complex numbers. Though we have shown the time functions $v(t)$, $v_2(t)$, etc., the solution to this LSE is actually the corresponding phasors. As was mentioned earlier, as long as the context does not pose any ambiguity, we are free to use a uniform notation.

Example 6.10 Let us now look at the circuit in Fig. 6.12 with the following components: $v = 5\cos(20\,t)\,\text{V}$, $L_1 = 0.5\,\text{H}$, $L_2 = 0.25\,\text{H}$, $L_3 = 0.2\,\text{H}$, $C_1 = 0.25\,\text{F}$, $C_2 = 0.5\,\text{F}$, $C_3 = 0.1\,\text{F}$, and $R = 10\,\Omega$.

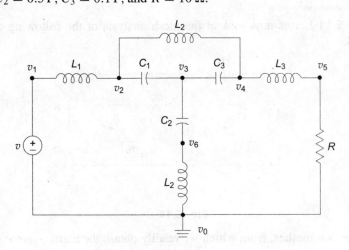

Fig. 6.12

We leave it to the reader to verify that the LSE is

$$
\begin{pmatrix}
j4.7 & -j5 & j0.2 & 0 & 0 \\
-j5 & j17 & -j2 & 0 & -j10 \\
j0.2 & -j2 & j1.55 & j0.25 & 0 \\
0 & 0 & j0.25 & 0.1-j0.25 & 0 \\
0 & -j10 & 0 & 0 & j9.8
\end{pmatrix}
\begin{pmatrix} v_2 \\ v_3 \\ v_4 \\ v_5 \\ v_6 \end{pmatrix}
=
\begin{pmatrix} -j0.5 \\ 0 \\ 0 \\ 0 \\ 0 \end{pmatrix}
$$

Apparently mesh analysis appears to be more tangible in the above two examples. However, from a practical point of view, nodal analysis is preferred to mesh analysis since we need not break the circuit to insert an ammeter. And, this is the price we pay for solving larger LSE.

Exercise 6.14 Solve Example 6.10.

Exercise 6.15 Obtain the node voltages of the circuit shown in Fig. 6.13.

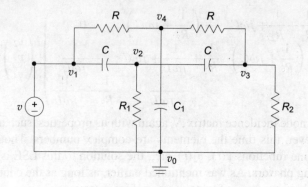

Fig. 6.13

6.5.2 Mesh Analysis

Example 6.11 Let us now look at the mesh analysis of the following circuit in Fig. 6.14.

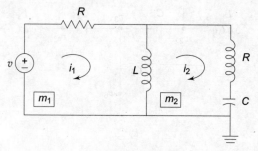

Fig. 6.14

We see two meshes, from which we readily obtain the matrix-vector equation as

$$\begin{pmatrix} R + j\omega L & -j\omega L \\ -j\omega L & R + j\omega L + \dfrac{1}{j\omega C} \end{pmatrix} \begin{pmatrix} i_1 \\ i_2 \end{pmatrix} = \begin{pmatrix} v \\ 0 \end{pmatrix}$$

Example 6.12 Consider the circuit shown in Fig. 6.15.

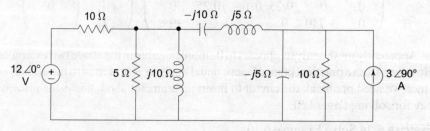

Fig. 6.15

We will not straightaway apply mesh analysis. We will first simplify the circuit by reducing the two parallel combinations and the series combination. In particular, we define

$$R = 10\,\Omega$$
$$\mathcal{Z}_1 = 5 \parallel j10 = 4 + j2\,\Omega$$
$$\mathcal{Z}_2 = -j10 + j5 = -j5\,\Omega$$
$$\mathcal{Z}_3 = -j5 \parallel 10 = 2 - j4\,\Omega$$

The simplified circuit is shown in Fig. 6.16.

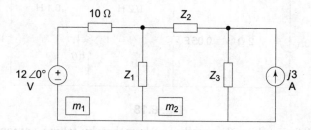

Fig. 6.16

In terms of these quantities we may write the mesh equations to get

$$\begin{pmatrix} 14 + j2 & -4 - j2 \\ -4 - j2 & 18 - j1 \end{pmatrix} \begin{pmatrix} i_1 \\ i_2 \end{pmatrix} = \begin{pmatrix} 12 \\ 0 \end{pmatrix}$$

Exercise 6.16 Solve Example 6.12 and verify the results using nodal analysis.

Exercise 6.17 Apply mesh analysis to the circuit shown in Fig. 6.17.

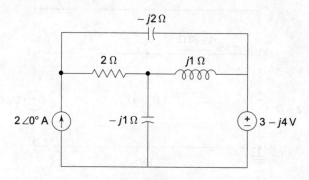

Fig. 6.17

6.5.3 Principle of Linearity

In this section we will look at a couple of examples to illustrate the principle of linearity. However, we would remark at the outset that all the sources acting in a circuit shall have the same frequency. In general, a circuit may have different sources with different frequencies. We shall study this idea in the following chapter. Meanwhile, we encourage the reader to ponder over this.

Example 6.13 Let us determine the current $i(t)$ through the 1 Ω resistance in the circuit shown in Fig. 6.18 using the principle of superposition.

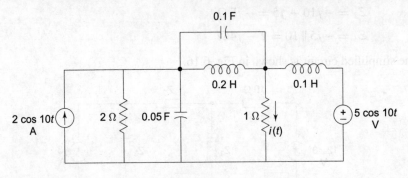

Fig. 6.18

We shall first reduce the number of branches by taking appropriate parallel combinations. Then we shall apply source transformation and obtain the desired result. The result is shown in Figs 6.19(a) and 6.19(b).

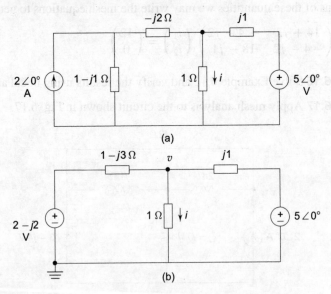

Fig. 6.19

Assuming that the left-hand-side current source alone is acting, we get

$$i^1 = \frac{6 + j10}{17} \text{ A}$$

Similarly, assuming that the other voltage source alone is acting, we get

$$i^2 = \frac{35 - j55}{17} \text{ A}$$

Thus the current through the 1 Ω resistance due to both the sources is

$$i = i^1 + i^2 = \frac{41 - j45}{17} = 3.581 \angle 312.34°$$

Exercise 6.18 Complete Example 6.13 and verify the results using nodal analysis and mesh analysis.

Example 6.14 Let us redo Example 6.12 to determine the voltage across $\mathcal{Z}_2 = -j5$ Ω with + ve on the left.

Using the same definitions $\mathcal{Z}_1$, $\mathcal{Z}_2$, and $\mathcal{Z}_3$ as in Example 6.12, we may apply source transformation to obtain a single mesh circuit with two voltage sources

$$\mathcal{V}_1 = \frac{24(2 + j1)}{14 + j2} \text{ V}$$

and

$$\mathcal{V}_2 = 12 + j6 \text{ V}$$

and three impedances $R \parallel \mathcal{Z}_1$, $\mathcal{Z}_2$, and $\mathcal{Z}_3$. This is shown in Fig. 6.20.

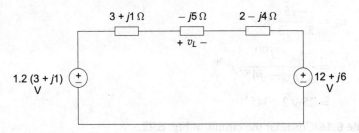

Fig. 6.20

We may simply apply the voltage division rule to get

$$v(t) = 5.1275 \angle 177.75°$$

Exercise 6.19 Verify the results of Example 6.14 using the results of Example 6.12.

We remark, once again, that for certain networks such as those in the above examples the principle of linearity appears to be inferior to nodal analysis. Nevertheless, we strongly recommend that each of the techniques is equally important irrespective of the circuit under consideration and hence to be mastered thoroughly.

6.5.4 Thévenin's and Norton's Theorems

Example 6.15 Let us now determine Thévenin's equivalent of the load port AB of the circuit in Fig. 6.21.

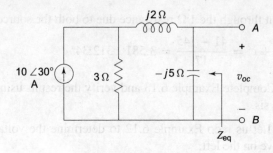

Fig. 6.21

Clearly, the load port 'sees'

$$Z_{eq} = (3 + j2) \, \| -j5 = \frac{25 - j5}{6} \, \Omega$$

i.e., at the given frequency, the equivalent impedance is capacitive. We encourage the reader to apply a test voltage source v_{test} and measure the current i_{test} to verify the equivalent impedance given above. Further, using source transformation and voltage division we get

$$v_{oc} = \frac{-j5}{3 - j3} \, 30 \, e^{j30°}$$

$$= \frac{5 \, e^{-j90°}}{3 \sqrt{2} \, e^{-j45°}} \, e^{j30°}$$

$$= 25 \sqrt{2} \, \angle 345°$$

Example 6.16 Consider the circuit in Fig. 6.22.

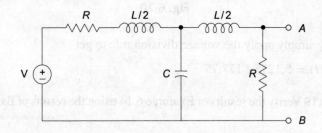

Fig. 6.22

It may be easily verified that

$$Z_{eq} = \left\{ \left[\left(R + j\omega \frac{L}{2} \right) \, \middle\| \, \frac{1}{j\omega C} \right] + j\omega \frac{L}{2} \right\} \| R$$

The open-circuit voltage v_{oc} may be obtained using nodal or mesh analysis.

Exercise 6.20 Solve Example 6.16 completely. Obtain Norton's equivalents of the circuits in Examples 6.15 and 6.16. Do these problems independently to check your results.

Exercise 6.21 Obtain Thévenin's equivalent and verify your results for the circuit shown in Fig. 6.23.

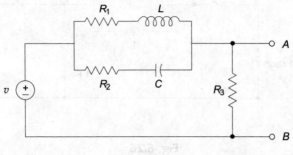

Fig. 6.23

Example 6.17 Consider the circuit shown in Fig. 6.24.

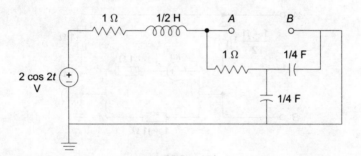

Fig. 6.24

Though the circuit appears tricky, tracing the nodes gives us a simplified circuit as shown in Fig. 6.25.

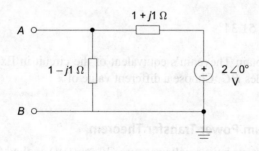

Fig. 6.25

From this we easily get $v_{oc} = 1 - j1$ and $\mathcal{Z}_{eq} = 1\,\Omega$.

Exercise 6.22 Verify the results of Example 6.17.

Example 6.18 Let us look at the circuit shown in Fig. 6.26 to determine Thévenin's equivalent with respect to the nodes AB. Assume $k = -1$.

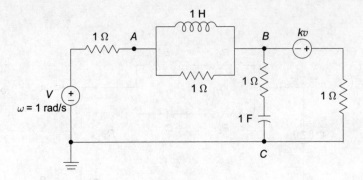

Fig. 6.26

We shall redraw the circuit as shown in Fig. 6.27 with reference to the load port.

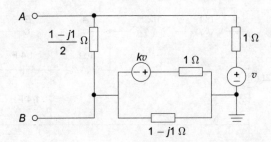

Fig. 6.27

This facilitates us to obtain

$$Z_{eq} = \frac{7 + j4}{15} \ \Omega$$

and

$$v_{oc} = 1 \angle 51.34° \tag{6.17}$$

Exercise 6.23 Obtain Thévenin's equivalent of the circuit in Example 6.18 with respect to the nodes BC. Choose a different value of k.

6.5.5   Maximum Power Transfer Theorem

Let us assume that we have a voltage source $V_m \cos(\omega t)$ applied to a resistance R. The power dissipated by this resistance is

$$p(t) = \frac{V_m^2 \cos^2(\omega t)}{R} = \frac{V_m^2}{2R} [1 + \cos(2\omega t)] \ \text{W}$$

i.e., the power absorbed by the resistance is also a sinusoid with twice the frequency of the source. However, we also have a constant along with the sinusoid. This is shown in Fig. 6.28, wherein we may observe that the sinusoid is symmetric about the constant $V_m^2/2R$. The physical significance of this figure follows.

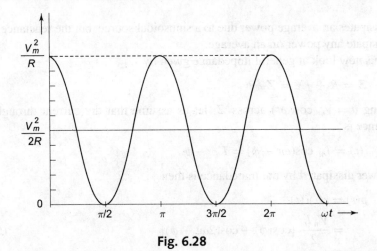

Fig. 6.28

If we look at the *average power* dissipated by the resistance over a cycle 0 to 2π, we get

$$P_{\text{average}} = \frac{1}{2\pi} \int_0^{2\pi} p(t)\, d\omega t$$

$$= \frac{V^2}{2R} \text{ W}$$

It is this average power that is of utmost interest to us from a practical standpoint. In Fig. 6.28, we observe that the sinusoid is actually 'lifted up' by this average value.

Now let us apply the same sinusoid across a capacitance C. The current passing through it is

$$i(t) = C\frac{dv}{dt} = -\omega C V \sin(\omega t)$$

and the power dissipated is

$$p(t) = -\omega C V^2 \sin(\omega t)\cos(\omega t) = -\frac{\omega C V^2}{2} \sin(2\omega t)$$

i.e., the sinusoid is not lifted up by any constant as it was done in the case of resistance. And the average power dissipated over a cycle is simply

$$P_{\text{average}} = \frac{1}{2\pi} \int_0^{2\pi} p(t)\, d\omega t$$

$$= 0 \text{ W}$$

Surprisingly, the capacitance does not dissipate any average power. In fact, it absorbs power during one half-cycle from the source and delivers the same amount of power back to the source during the other half-cycle. It is easy to verify that an inductance also does not dissipate any average power. Thus, it is only the resistance that dissipates an average power due to a sinusoidal source, but the reactance does not dissipate any power on an average.

Let us now look at general impedance given by

$$\mathcal{Z} = R + jX = Z \angle \phi$$

Applying $v = V_m \cos(\omega t)$ across $\mathcal{Z}$, let us assume that the current through the impedance is

$$i(t) = I_m \cos(\omega t - \phi) = I \angle - \phi$$

The power dissipated by the impedance is then

$$p(t) = v(t)i(t)$$

$$= \frac{V_m I_m}{2} [\cos(\phi) + \cos(2\omega t - \phi)] \tag{6.18}$$

Observe that we have a constant term, $\cos(\phi)$, in addition to the sinusoid, and the average power is

$$P_{\text{average}} = \frac{V_m I_m}{2} \cos \phi \tag{6.19}$$

The factor $\cos \phi$ is called *power factor* and we elaborate in a later chapter on power systems. For now, we shall interpret this Eqn (6.19) in a slightly different way. There are surprising results in store for us.

First, we observe that the average power is maximum when $\phi = 0°$. And this happens when the impedance $\mathcal{Z}$ is a pure resistance R.

Second, let us see what the following expression evaluates to:

$$\sqrt{\frac{1}{2\pi} \int_0^{2\pi} U_m^2 \cos^2(\omega t) \, d\omega t} \tag{6.20}$$

As we read this expression from left to right, we call it *root of mean of square of sinusoid*, briefly called *root mean square* (rms) of the sinusoid. We encourage the reader to verify that this evaluates to

$$U_{\text{rms}} \triangleq \frac{U_m}{\sqrt{2}} \tag{6.21}$$

Using Eqn (6.21) we may write Eqn (6.19) as

$$P_{\text{average}} = V_{\text{rms}} I_{\text{rms}} \cos \phi$$

Next we observe that, since ϕ is the phase angle of $\mathcal{Z}$,

$$R = |\mathcal{Z}| \cos \phi$$

and

$$X = |\mathcal{Z}| \sin \phi$$

Accordingly,

$$P_{\text{average}} = V_{\text{rms}} I_{\text{rms}} \frac{R}{|\mathcal{Z}|}$$

Further we observe, owing to the generalized Ohm's law, that

$$\frac{V_{\text{rms}}}{|Z|} = I_{\text{rms}}$$

Thus, we have

$$P_{\text{average}} = I_{\text{rms}}^2 R = \frac{V_{\text{rms}}^2}{R}, \tag{6.22}$$

a neat expression for the average power dissipated by the impedance. This is very much similar to the power expression we had for purely resistive circuits. From this point of view we also call the rms value of a sinusoid as the *effective value* of the sinusoid since it represents the *effective* dc current that would cause the resistor to dissipate the same amount of power as the original sinusoid. Further, Eqn (6.22) ratifies our earlier observation that power, if any, is dissipated (over a cycle on an average) only in a resistance and not in a reactance.

We shall use the above observation to look at the ac version of the maximum power transfer theorem. Consider once again Thévenin's equivalent with a source and equivalent impedance given by

$$v_{\text{oc}}(t) = V_{\text{oc}} \cos(\omega t + \theta)$$

and

$$Z_{\text{eq}} = R_{\text{eq}} + jX_{\text{eq}}$$

The question is: What load $Z_L = R_L + jX_L$ absorbs the maximum average power from the source?

Since the total impedance seen by the source is

$$Z = Z_{\text{eq}} + Z_L = \left(R_{\text{eq}} + R_L\right) + j\left(X_{\text{eq}} + X_L\right) = |Z| \angle \phi$$

we may assume that the source V_{oc} delivers an rms current

$$I_{\text{rms}} = \frac{1}{\sqrt{2}} \frac{V_{\text{oc}}}{|Z|} \angle \theta - \phi$$

and the average power dissipated by the load is

$$P_{L,\,\text{ave}} = I_{\text{rms}}^2 R_L$$

$$= \frac{V_{\text{oc}}^2}{2\left[(R_{\text{eq}} + R_L)^2 + (X_{\text{eq}} + X_L)^2\right]} R_L \tag{6.23}$$

As has been observed earlier, the average power dissipated is maximum when the impedance is a pure resistance. This demands that

$$X_L = -X_{\text{eq}}$$

With this condition plugged into Eqn (6.23), we get

$$P_{L,\,\text{ave}} = \frac{V_{\text{oc}}^2}{2\left(R_{\text{eq}} + R_L\right)^2} R_L$$

This is the same old expression we have seen in Chapter 4 and this becomes maximum if

$$R_L = R_{\text{eq}}$$

Thus, we have that

$$R_L = R_{\text{eq}} \quad \text{and} \quad X_L = -X_{\text{eq}}$$

or, equivalently,

$$Z_L = Z_{\text{eq}}^* \quad \text{(where the superscript } * \text{ denotes complex conjugate)} \quad (6.24)$$

for maximum average power transfer. We call this load the *matching load*. The theorem says that when a matching load is put at the load port the net impedance in the circuit is purely resistive, equal to $2R_{\text{eq}}$, and the current would be *in phase* with v_{oc}.

To summarize, we observe that impedance dissipates power that oscillates with twice the frequency of the source; i.e, at some instant of time it attains its maximum, at some other instant it is at its minimum, and at yet another instant it is zero in half-a-cycle of the input signal. Naturally, we are then interested in the net power dissipated in a cycle (of the input signal) and this turns out to be the average power we have defined earlier. We found that the reactances do not dissipate any power, on an average, but would facilitate transfer of energy back and forth from the source. The theorem states that the average power dissipated is maximum when the impedance is a pure resistance, without any reactance. This is equivalent to saying that $\cos \phi = 1$ and hence *unity power factor* is an important quantity for us. We shall elaborate on this in a later chapter on power systems. For now, we shall look at the following examples.

Example 6.19 Consider the circuit in Fig. 6.29.

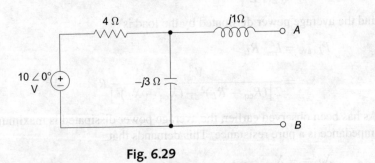

Fig. 6.29

We find that Thévenin's equivalent impedance is

$$Z_{\text{eq}} = \frac{36 - j23}{25}$$

and hence the load that absorbs the maximum power is

$$Z_L = Z_{eq}^* = \frac{36 + j23}{25}$$

The open-circuit voltage is

$$V_{oc} = \frac{-j3}{4 - j3} \, 10 \angle 0°$$

$$= 3.6 - j4.8 \text{ V}$$

$$= 6 \angle -53.13°$$

and the rms current is

$$I_{rms} = \frac{6}{\sqrt{2}(Z_{eq} + Z_L)} = 1.473 \text{ A}$$

and the average power dissipated is

$$P_{max} = I_{rms}^2 \, R_L$$

$$= 3.125 \text{ W}$$

We suggest the following algorithm to the reader.

Step 1	Obtain Z_{eq}.
Step 2	$Z_L = Z_{eq}^*$.
Step 3	Obtain the phasor $V_{oc} = V_{oc} \angle \theta$.
Step 4	Determine the phasor $\mathcal{I} = I \angle \theta$, and hence the rms magnitude $I_{rms} = I/\sqrt{2}$.
Step 5	$P_{max} = I_{rms}^2 \, R_L$.

Exercise 6.24 Repeat Example 6.19 with the inductance and the capacitance interchanged.

Example 6.20 Let us find the load that receives the maximum power in the circuit shown in Fig. 6.30.

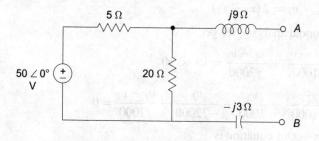

Fig. 6.30

It is easy to see that

$$V_{oc} = \frac{20}{20 + 5} \, 50 \angle 0° = 40 \angle 0° \text{ V}$$

and

$$Z_{eq} = 20 \parallel 5 + j9 - j3 = 4 + j6 \ \Omega$$

Accordingly,

$$Z_L = 4 - j6 \ \Omega$$

and

$$P_{max} = 50 \ \text{W}$$

6.5.6 Circuits with Controlled Sources

Example 6.21 Consider the circuit with a controlled source shown in Fig. 6.31.

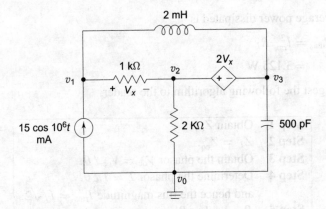

Fig. 6.31

Here the frequency is $\omega = 10^6$ rad/s and hence

$$X_L = j2000 \ \Omega \quad \text{and} \quad X_C = -j2000 \ \Omega$$

We have, as might be expected, a super node with the constraint

$$v_2 - v_3 = 2(v_1 - v_2)$$

Applying nodal analysis, we get

$$\frac{v_1 - v_2}{1000} + \frac{v_1 - v_3}{j2000} = 15 \times 10^{-3}$$

$$\frac{v_2 - v_1}{1000} + \frac{v_2}{1000} + \frac{v_3}{-j2000} + \frac{v_3 - v_1}{j2000} = 0$$

The matrix-vector equation is

$$\begin{pmatrix} 1 - j\dfrac{3}{2} & -1 + j\dfrac{3}{2} \\ -1 - j\dfrac{1}{2} & \dfrac{3}{2} \end{pmatrix} \begin{pmatrix} v_1 \\ v_2 \end{pmatrix} = \begin{pmatrix} 15 \\ 0 \end{pmatrix}$$

Exercise 6.26 Solve the LSE in Example 6.21 and verify the results using mesh analysis.

Example 6.22 Let us determine the control variable i_x in the circuit shown in Fig. 6.32.

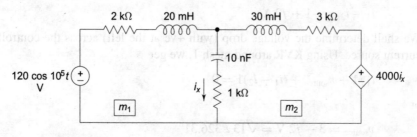

Fig. 6.32

Let us apply mesh analysis. For the given frequency $\omega = 10^5$ rad/s, we have

$$X_{L1} = j2000\,\Omega, \quad X_{L2} = j3000\,\Omega, \quad X_C = -j1000\,\Omega$$

The constraint is

$$i_x = i_1 - i_2$$

The LSE is as follows:

$$10^3 \begin{pmatrix} 3+j1 & -1+j1 \\ 3+j1 & j2 \end{pmatrix} \begin{pmatrix} i_1 \\ i_2 \end{pmatrix} = \begin{pmatrix} 120 \\ 0 \end{pmatrix}$$

Solving this, we get

$$i_x = 108 - j48\,\text{mA}$$

Exercise 6.27 Verify the results of Example 6.22 using nodal analysis.

Example 6.23 Consider Fig. 6.33.

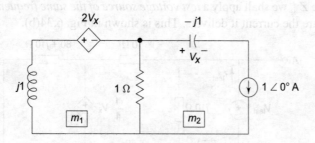

Fig. 6.33

Here we have

$$i_2 = 1\angle 0^\circ \quad \Rightarrow \quad v_x = -j1\,\text{V}$$

and

$$i_1 = -2\,v_x = j2\,\text{A}$$

The current through the $1\,\Omega$ resistance is

$$i_1 - i_2 = -1 + j2\,\text{A}$$

We shall determine the voltage drop (with +ve at the left) across the controlled current source. Using KVR around mesh 1, we get

$$i_1 j1 + v_{\text{source}} + (i_1 - i_2)1 = 0$$

or

$$v_{\text{source}} = 3 - j2\,\text{V} = \sqrt{13}\,\angle\,326.31°$$

We see that the controlled source offers an impedance

$$Z_{\text{source}} = 1 + j\frac{3}{2}\,\Omega$$

i.e., an inductive impedance at the given frequency.

Exercise 6.28 Verify the results of Example 6.23.

Example 6.24 Consider the circuit shown in Fig. 6.34(a).

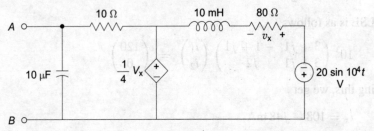

Fig. 6.34(a)

We shall first simplify the circuit and obtain the reactances offered by the components at the given frequency $\omega = 10^4$ rad/s. To determine the equivalent impedance Z_{eq} we shall apply a *test voltage source of the same frequency 10^4 rad/s* and measure the current it delivers. This is shown in Fig. 6.34(b).

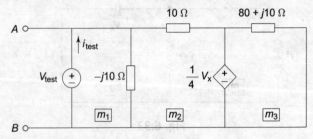

Fig. 6.34(b)

We have the constraint

$$v_x = -80 \times i_3$$

Applying mesh analysis, noticing a super mesh, we get

$$\begin{pmatrix} -j10 & j10 & 0 \\ j10 & 10 - j10 & -20 \\ 0 & 0 & 100 + j10 \end{pmatrix} \begin{pmatrix} i_{\text{test}} \\ i_2 \\ i_3 \end{pmatrix} = \begin{pmatrix} v_{\text{test}} \\ 0 \\ 0 \end{pmatrix}$$

from which we obtain

$$Z_{\text{eq}} = 5 - j5 \ \Omega$$

Now, we may apply nodal analysis to obtain

$$v_{\text{oc}} = 28 \angle 40° \text{ V}$$

Exercise 6.29 Verify the results of Example 6.24. Why is the test voltage source required to be of the same frequency as that of the independent voltage source acting in the network?

Example 6.25 Consider the network shown in Fig. 6.35.

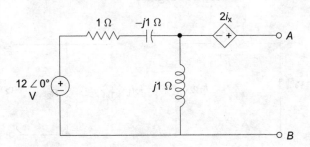

Fig. 6.35

Let us first suppress the independent voltage source and apply a test voltage source across AB. With the constraint that

$$i_x = i_1$$

we apply mesh analysis to get

$$\begin{pmatrix} 1 & -j1 \\ -2 - j1 & j1 \end{pmatrix} \begin{pmatrix} i_1 \\ i_2 \end{pmatrix} = \begin{pmatrix} 0 \\ v_{\text{test}} \end{pmatrix}$$

from which we get

$$Z_{\text{eq}} = \frac{v_{\text{test}}}{-i_2} = 1 - j1 \ \Omega$$

Since we have a controlled source connected to the load port AB, let us determine the short-circuit current i_{sc} rather than v_{oc}. The simplified figure is shown in Fig. 6.36.

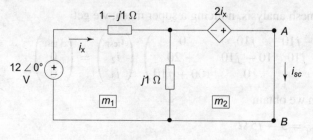

Fig. 6.36

It is easy to apply mesh analysis once again and obtain

$$i_{sc} = 18.97 \angle 71.57°$$

We encourage the reader to verify the results of this example using nodal analysis.

Exercise 6.30 Obtain Thévenin's equivalent of the load port of the circuit having two dependent sources as shown in Fig. 6.37.

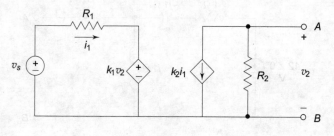

Fig. 6.37

Summary

In this chapter, we were basically concerned with the most common voltage signal, the sinusoid. There are three quantities of importance to us—the maximum amplitude U_m, frequency ω, and phase shift ψ. What is most interesting, and obviously intuitive, to us is that the response (i.e., the voltage across or the current through) of the circuit elements is also a sinusoid of the same frequency. However, the amplitude and the phase shifts are *functions* of the frequency ω. We have also observed that given a frequency, we may define a 'resistance-like' parameter, called the reactance for inductors and capacitors. In terms of these reactances we have extended Ohm's law. Once we match the units of dimensions of the physical quantities, we may simply add them and the result is impedance. We had no difficulty in further generalizing Ohm's law to impedances, i.e., combinations of circuit elements. All through, the idea of linearity played a crucial role. After learning some basics, we have avoided using the trigonometric functions; instead, we opted for complex exponentials owing to their connections to the trigonometric functions through Euler's formula. We learnt that the techniques developed in

Chapter 4 are readily extendable to general circuits, provided we are willing to do complex algebra.

The results of this chapter are to be thought of as 'generalizations' over simpler circuits of Chapter 4, which contained dc sources and resistances alone. Of course, we preferred not to include the results of Chapter 5. We just mentioned that the responses of circuit elements also have transients, even though they are energized by smooth sinusoids. Later on, in Chapter 9, we shall further generalize the results of this chapter and include the transients. We advise the reader to think of this generalization as a hierarchy. We suggest the following picture. Imagine a three-floor building. Standing on the ground floor your vision is rather restricted to a smaller neighborhood. Going up, from the first floor the vision is better, which obviously includes what you see from the ground floor. And going further up, the sight is *even better* from the second floor. We will also discuss this concept briefly in Chapter 14 on general linear systems.

Problems

6.1 Obtain the differential equation and compute the current $i_R(t)$ through the resistance in the circuit shown in Fig. 6.38.

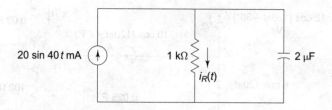

20 sin 40t mA   1 kΩ $i_R(t)$ 2 µF

Fig. 6.38

6.2 Use KVR to determine the voltage $v_R(t)$ across the resistance in the circuit shown in Fig. 6.39. Use appropriate trigonometric identities.

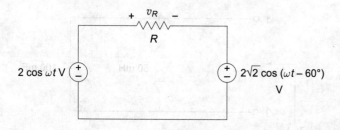

$+$ v_R $-$

R

2 cos ωt V $2\sqrt{2} \cos(\omega t - 60°)$ V

Fig. 6.39

6.3 (a) If the susceptance of a capacitor is measured to be $j4.7$ mΩ^{-1} at a frequency of $\omega = 100\pi$ rad/s, determine its reactance at $\omega = 20{,}000\pi$ rad/s.

(b) If the reactance of an inductor is measured to be $j20$ Ω at a frequency of $\omega = 100\pi$ rad/s, determine its susceptance at $\omega = 20{,}000\pi$ rad/s.

6.4 Determine the voltage $v_R(t)$ across the 10 Ω resistance in the circuit shown in Fig. 6.40.

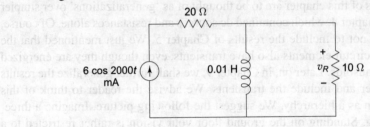

Fig. 6.40

6.5 Determine the current $i_L(t)$ through the inductance in the circuit shown in Fig. 6.41.

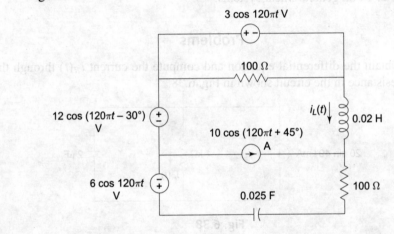

Fig. 6.41

6.6 Determine the impedance $\mathcal{Z}(\omega = 250\,\text{rad/s})$, as seen by the port 1-1' of the circuit in Fig. 6.42.

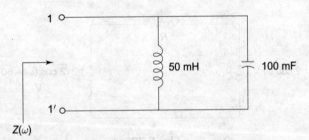

Fig. 6.42

6.7 Compute the impedance $\mathcal{Z}(\omega = 250\,\text{rad/s})$, as seen by port 1-1' of the circuit in Fig. 6.43. How about the impedance at a frequency of 600 rad/s? Determine the frequency at which the impedance is purely real. Is there any frequency at which the impedance becomes purely imaginary?

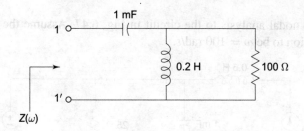

Fig. 6.43

6.8 In the circuit shown in Fig. 6.44, determine the capacitance C at which the impedance as seen by port 1-1' is real, if the operating frequency is 500 rad/s.

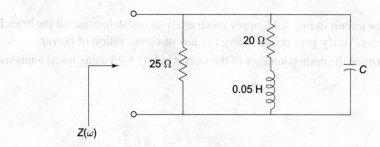

Fig. 6.44

6.9 Use voltage division rule to determine the voltage $v_C(t)$ across the capacitance in the circuit shown in Fig. 6.45.

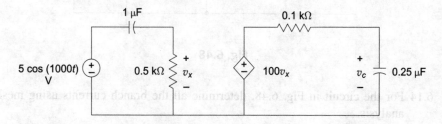

Fig. 6.45

6.10 Use current division rule to determine the voltage $v_C(t)$ across the capacitance in the circuit shown in Fig. 6.46.

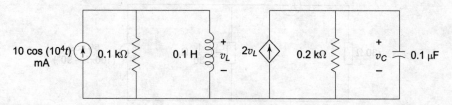

Fig. 6.46

6.11 Apply nodal analysis to the circuit in Fig. 6.47. Assume the frequency of operation to be $\omega = 100$ rad/s.

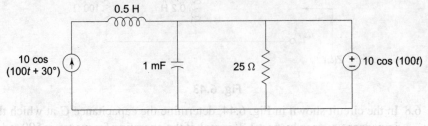

Fig. 6.47

6.12 For the circuit in Fig. 6.47, apply mesh analysis and determine all the branch currents. Verify your results using the law of conservation of power.

6.13 Determine the node potentials of the circuit in Fig. 6.48 using nodal analysis.

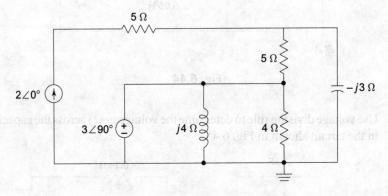

Fig. 6.48

6.14 For the circuit in Fig. 6.48, determine all the branch currents using mesh analysis.

6.15 Use the principle of linearity and determine the current delivered by the voltage source in the circuit shown in Fig. 6.49.

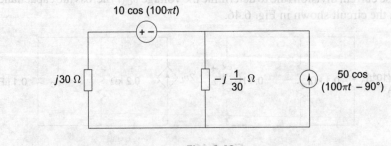

Fig. 6.49

6.16 Use the principle of linearity and determine the voltage across the capacitance in the circuit shown in Fig. 6.47.

6.17 Use the principle of linearity and determine the current delivered by the voltage source in the circuit shown in Fig. 6.48.

6.18 Consider the circuit shown in Fig. 6.50.

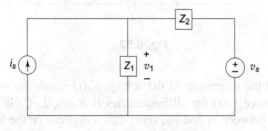

Fig. 6.50

The following observations were made in an experiment.

If $v_s(t) =$	and $i_s(t) =$	then $v_1(t) =$
0	$10\cos(2\pi t)$	$20\cos(2\pi t + 45°)$
$10\cos(2\pi t + 90°)$	0	$5\cos(2\pi t + 135°)$

Determine $v_1(t)$ if $v_s(t) = 20\cos(2\pi t)$ and $i_s(t) = 5\cos(2\pi t + 45°)$.

6.19 Use the principle of linearity and determine the current delivered by the voltage source in the circuit shown in Fig. 6.51.

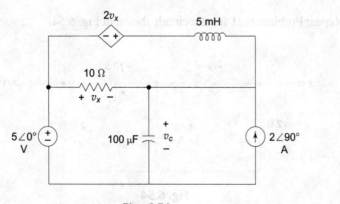

Fig. 6.51

6.20 Consider the circuit in Fig. 6.52.

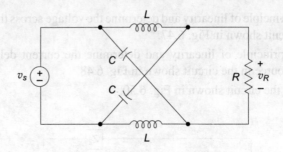

Fig. 6.52

Verify that the magnitude of the voltage $v_R(t)$ equals the magnitude of the voltage source $v_s(t)$ for all frequencies if $R = \sqrt{L/C}$. If ψ is the phase difference between v_s and v_R, verify that it depends on the frequency of the source, and obtain an expression for ψ as a function of the frequency ω. Plot $\psi(\omega)$ for $0 \le \omega \le \infty$.

6.21 For the circuit in Fig. 6.53, obtain Thévenin's and Norton's equivalents as seen by port 1-1'.

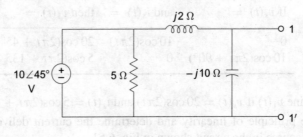

Fig. 6.53

6.22 Repeat Problem 6.21 for the circuit shown in Fig. 6.54.

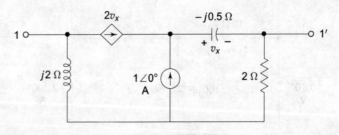

Fig. 6.54

6.23 Repeat Problem 6.21 for the circuit shown in Fig. 6.55. Verify that an impedance of $\mathcal{Z}_{eq}^*$ connected across port 1-1' would dissipate maximum average power.

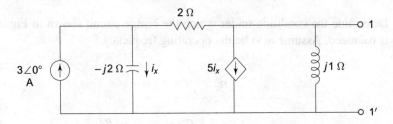

Fig. 6.55

6.24 Repeat Problem 6.23 for the circuit shown in Fig. 6.56. Assume that the frequency of operation is $\omega = 10,000$ rad/s.

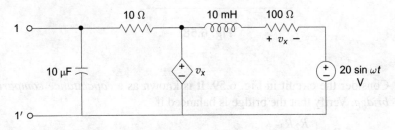

Fig. 6.56

6.25 Consider the circuit in Fig. 6.57.

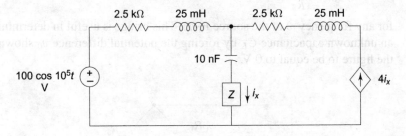

Fig. 6.57

Compute current i_x through element Z if the element is
   (a) a $2\,k\Omega$ resistance,
   (b) a $12\,nF$ capacitance,
   (c) a $22\,mH$ inductance,
   (d) a series connection of the above three elements,
   (e) a parallel connection of the above three elements.

6.26 An active ac circuit delivers a voltage $v_Z = 2 \cos(1000t - 30°)$ to a resistance of $1\,k\Omega$ connected across port 1-1'. The same circuit is found to deliver a voltage $v_Z = 3 \cos(1000t - 135°)$ to a capacitance of $1\,\mu F$ connected across port 1-1'. Obtain Thévenin's equivalent of the ac circuit as seen by port 1-1'. Also, if an inductance of 1 H is connected across port 1-1', compute the current through the inductance.

6.27 Determine the condition under which the bridge circuit shown in Fig. 6.58 is balanced. Assume ω to be the operating frequency.

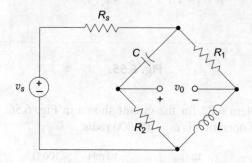

Fig. 6.58

6.28 Consider the circuit in Fig. 6.59. It is known as a *capacitance comparison bridge*. Verify that the bridge is balanced if

$$R_L = \frac{R_2 R_3}{R_1}$$

and

$$C_L = \frac{R_1 C_2}{R_3}$$

for any frequency ω of the source $v_S(t)$. This bridge is useful in determining an unknown capacitance C_L by forcing the potential difference v_0 shown in the figure to be equal to 0 V.

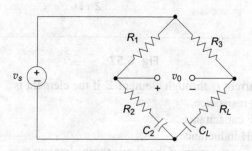

Fig. 6.59

6.29 The circuit shown in Fig. 6.60 is known as *Maxwell's bridge* used to determine unknown inductances. Obtain the expressions for R_L and L_L when the bridge is balanced.

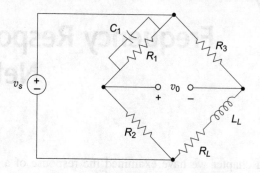

Fig. 6.60

6.30 The circuit in Fig. 6.61 is known as *Hay bridge*, also used to determine unknown inductances. However, unlike Maxwell's bridge, verify that the expressions for R_L and L_L depend upon the operating frequency ω of the source v_S.

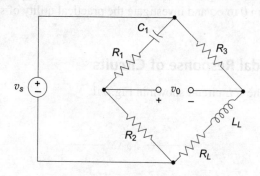

Fig. 6.61

If at a frequency $\omega = 200\pi$ krad/s the bridge is balanced with $R_1 = 25.33\,\Omega$, $C_1 = 100\,\text{pF}$, and $R_2 = R_3 = 1\,\text{k}\Omega$ determine R_L and L_L.

CHAPTER 7

Frequency Response of Networks

In the preceding chapter we have examined the response of a passive network to a sinusoidal excitation. We assumed that the source, once given to us, has a fixed frequency ω of oscillations. We observed that the 'impedance' and hence the magnitude and the phase shift of the voltages and currents are dependent on this source frequency ω. Further, we generalized the analysis techniques by adopting complex algebra. In this chapter we shall look at the response of networks to sinusoidal excitations of variable frequency. In other words, we shall summarize grossly the behaviour of a given network as we 'sweep' the frequency of oscillations of the source from 0 to ∞ and investigate the practical utility of such a network.

7.1 Sinusoidal Response of Circuits

Let us consider the RL circuit shown in Fig. 7.1.

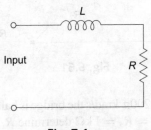

Fig. 7.1

Here, we would look at a few interesting things. First, the excitation can be either an independent voltage source or an independent current source. We will call this 'input'. Second, the response we are interested in can also be either a current or a voltage, for instance, across resistance R. We will call this 'output'. Thus, we may view the network as an *operator* that acts on an input signal and produces an output signal. And, we notice that there are four possible *input–output pairs*:

- voltage–voltage
- voltage–current
- current–voltage
- current–current

For example, if the input is a current source $i_S(t)$ and we are interested in the voltage $v_L(t)$ across the inductance, we have

$$v_L = X_L i_S = j\omega L i_S$$

using the phasor notation. Here the reactance $X_L = j\omega L$ serves as an operator. And, of course, this is a complex number dependent on the frequency ω of the source. For the above four possibilities, we will generalize the *operation* as

$$\text{output} = H(j\omega) \times \text{input}$$

We call this complex operator $H(j\omega)$ as the *transfer function*. Observe that in this equation, the initial inductor current has no role to play; we simply assume it to be zero. Recall that we have introduced the transfer function earlier in Chapter 4 in Eqn (4.25). There the circuits were purely resistive and hence we did not consider any initial stored energy. More formally, we define the transfer function as follows.

Definition 7.1 Transfer function

$$H(j\omega) \triangleq \frac{\text{output phasor}}{\text{input phasor}} \quad \text{with the reactive elements initially relaxed.}$$

Clearly, the transfer function $H(j\omega)$ is a complex function, i.e., it is a function of a complex variable $j\omega$. Observe that we have called the variable as '$j\omega$' instead of ω; we have a lot of convenience in doing so. The notion of an operator or transfer function is akin to 'mechanical advantage' of simple machines we have learnt in our elementary science. Recall that we have defined mechanical advantage as the ratio of load to effort. Here, what we call output is the voltage across/current through the load element, and the input is the effort or force we apply to the circuit. The aim of this chapter is to study the transfer functions[1] of circuits as the frequency ω is varied from 0 to ∞. The transfer function may be dimensionless, or may have the units of impedance or admittance depending upon the input–output pair. We give a further generalization to the idea of transfer function later in Chapter 9. We also use the term 'gain' interchangeably with 'transfer function' hereafter.

Example 7.1 Let us consider the reactance in the above RL circuit as the transfer function, i.e.,

$$H(j\omega) = j\omega L$$

Since this is complex, we split it into two parts—the magnitude $|H(j\omega)|$ and phase $\angle H(j\omega)$. We see that

$$|H(j\omega)| = L\omega$$

$$\angle H(j\omega) = \tan^{-1}\frac{\omega L}{0} = 90°$$

---

[1] perhaps *electrical advantage*!

It is more informative[2] to look at the graphs of these functions of ω, and it is easy too. We readily get a straight line passing through the origin with slope L for the magnitude function. Likewise, we get a horizontal line parallel to the ω-axis for the phase function. This is shown in Fig. 7.2. Notice that since the frequency is always positive, we show the first quadrants in each of the graphs.

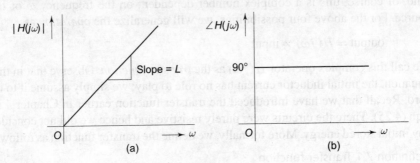

Fig. 7.2

The graphs give us the following information.

- Since $H(j\omega)$ is a ratio, the magnitude graph tells us that the output *relative to* the input is always greater in magnitude. Further this magnitude increases linearly with frequency. In this context, the term 'gain' is justified.

- The output, i.e., the voltage developed across the inductance, leads the input current by $90°$ at all frequencies.

- In other words, if the input is

$$i_S(t) = \cos(\omega t) \text{ A}$$

then the output is

$$v_L(t) = \omega L \cos(\omega t + 90°) \text{ V}$$

as might be expected.

- Thus, both the graphs put together are required to investigate the circuit's behaviour. Accordingly, we use the term *frequency response* to denote both the graphs.

Example 7.2 Suppose in the same RL circuit we consider an input voltage source $v_S(t)$ with the voltage across the resistance $v_R(t)$ as the output, the transfer function, using the voltage division rule, is

$$H(j\omega) = \frac{R}{R + j\omega L}$$

The frequency response of this transfer function is shown in Fig. 7.3. Practically, we obtain this by evaluating point by point the above function for several values of ω from 0 to a very large value, say several Grad/s, and drawing a smooth curve passing through all these points. We suggest the reader to do this exercise. We elaborate an *analytical method* of obtaining this in the following sections.

---

[2] After all, a picture is worth one thousand words!

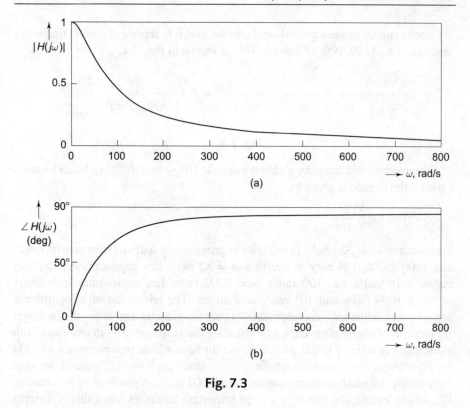

Fig. 7.3

Exercise 7.1 Verify the frequency response in Example 7.2. Draw the graphs as outlined earlier.

Exercise 7.2 Obtain the frequency response if the transfer function is

$$H(j\omega) = \frac{j\omega L}{R + j\omega L}$$

$$\left(\textbf{Ans. } |H(j\omega)| = \frac{\omega L}{\sqrt{R^2 + \omega^2 L^2}} \text{ and } \angle H(j\omega) = 90° - \tan^{-1} \frac{\omega L}{R} \right)$$

7.2 Magnitude and Phase

Obtaining the frequency response *practically* as mentioned above is not a simple thing as the reader would have experienced doing the above exercises. The reason is that the horizontal ω-axis simply extends to infinity. To avoid this difficulty, it is customary to consider a 'logarithmic axis' instead of the linear axis. This would save our space. To sketch the plots the reader may purchase a *semilog* graph paper available at stationery shops. However, preparing his/her own logarithmic scale on an ordinary graph paper will help the reader better appreciate the convenience of this scale. The reader is advised to prepare at least one such graph sheet immediately.

We need to make *equally spaced marks* on the x-axis to represent *decade* frequency intervals, i.e., 1, 10, 100, and so on. This is shown in Fig. 7.4.

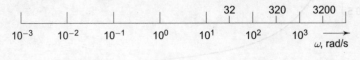

Fig. 7.4

Given a specific frequency within the decade $10^k \le \omega \le 10^{k+1}$ rad/s, its location l within the decade is given by

$$l = \log_{10} \frac{\omega}{10^k}$$

For instance, $\omega = 320$ rad/s ($k = 2$) lies approximately halfway between 100 rad/s and 1000 rad/s. It is easy to see that $\omega = 32$ rad/s lies approximately halfway between 10 rad/s and 100 rad/s; $\omega = 3200$ rad/s lies approximately halfway between 1000 rad/s and 10^4 rad/s, and so on. The advantage of a logarithmic scale is very apparent—it compresses higher frequencies and expands the lower frequencies, thereby allowing us to visualize the response at both extremes with a comparable level of detail. However, we do have a little inconvenience too. On the logarithmic scale we cannot have '$\omega = 0$' since $\log 0 = -\infty$. Instead, we start with arbitrarily small frequencies, say $\omega = 0.001$ rad/s, depending on the context. We simply emphasize that $\omega = 1$ is an important landmark since this represents '0' on the logarithmic scale.

We further define one more quantity, the *decibel*.

Definition 7.2 Decibel

The decibel (dB in short) value of a transfer function $H(j\omega)$ is defined as

$$|H(j\omega)| \ \text{dB} \triangleq 20 \log_{10} |H(j\omega)|$$

We notice that *unity magnitude* corresponds to 0 dB, *gain* corresponds to positive dB, and *attenuation* (or loss) corresponds to negative dB. Further, a magnitude of 0.707 corresponds to -3dB, a magnitude of 2 corresponds to $+6$dB, and so on. Converting the dB gain to ordinary values is trivial. By making use of a logarithmic scale on the magnitude axis preserves the linearity of the function. However, we notice a shift as explained in the following example.

Example 7.3 In Example 7.1, we have

$$|H(j\omega)| = 20 \log_{10} \omega L = 20 \log_{10} \omega + 20 \log_{10} L$$

With $\log_{10} \omega$ on the x-axis, we observe that the above equation is of the form $y = mx + c$. The intercept on the frequency axis may be found by putting $\omega = 1/L$ since $|H(j1/L)| = 0$ dB. We call this frequency $\omega = 1/L$ as the *corner frequency* ω_c. Let us not worry about the intercept on the magnitude axis at this moment. The graph is shown in Fig. 7.5(a).

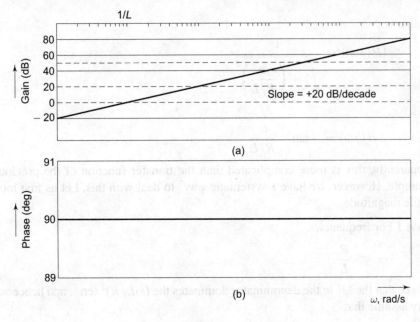

Fig. 7.5

It is apparent from the magnitude plot that for every tenfold increase in ω or equivalently for a change by a decade on the log ω-axis, i.e., from $\omega = 1$ to $\omega = 10$, from $\omega = 10$ to $\omega = 100$, etc., the slope changes by $+20$ dB. Accordingly, we say that the slope is *+20 dB per decade*. Since we do not have a point $\omega = 0$ rad/s as in Fig. 7.2, our origin is

$$\omega = \frac{1}{L} \text{ rad/s and } |H(j\omega)| = 1 = 0 \text{ dB}$$

so that the intersection of these two points is $(0, 0)$ when both x- and y-axes are logarithmic, and we have both first and fourth quadrants on the graph.

Coming to the 'phase' part of the transfer function, we have earlier noticed that the phase shift is $+90°$ irrespective of the frequency. Hence, whether the frequency axis is linear or logarithmic, the graph is a straight line parallel to the ω-axis as shown in Fig. 7.5(b). This is because the transfer function is purely imaginary. We have interesting results regarding the phase shift in the following example. In what follows, a logarithm is, by default, always to base 10 even though we might not explicitly show it in many places.

Example 7.4 Let us consider Example 7.2 once again. We will rewrite the transfer function in a more insightful form as

$$H(j\omega) = \frac{1}{1 + j\dfrac{\omega}{R/L}}$$

so that

$$|H(j\omega)| = \frac{1}{\sqrt{1 + \left(\dfrac{\omega}{R/L}\right)^2}}$$

and

$$\angle H(j\omega) = -\tan^{-1} \frac{\omega}{R/L}$$

Apparently, this is more complicated than the transfer function of the previous example. However, we have a systematic way[3] to deal with this. Let us first look at the magnitude.

Case 1 For frequencies

$$\omega \ll \frac{R}{L}$$

we see that the '1' in the denominator dominates the $(\omega L/R)^2$ term, and hence we may assume that

$$|H(j\omega)| \approx 0 \text{ dB } \forall \omega \ll \frac{R}{L}$$

i.e., the graph grazes the x-axis until $\omega = R/L$. Clearly, the slope of this straight line is 0 dB per decade.

Case 2 At $\omega = R/L$,

$$\left| H\left(j\frac{R}{L} \right) \right| = \frac{1}{\sqrt{2}} = -3 \text{ dB}$$

We call this frequency as *corner frequency* ω_c. The significance of the word 'corner' becomes clear in a moment.

Case 3 For frequencies

$$\omega \gg \frac{R}{L}$$

the '1' in the denominator can be neglected and

$$|H(j\omega)| \approx -20\log\omega - 20\log\frac{L}{R}$$

This is a straight line with slope -20 *dB per decade* with an intercept at the corner frequency on the x-axis.

Now we can stitch these three cases together. The straight lines in case 1 and case 3 are called *asymptotes* since the frequencies under consideration are far away on either side of the corner frequency. These asymptotes turn out to be tangents to the actual curve. We notice that the slope changes from 0 dB/decade to -20 dB/decade at the *corner* frequency. The actual graphs almost graze these asymptotes except in

---

[3] There is always a method in madness!

the neighbourhood of the corner frequency. The magnitude at the corner frequency is −3 dB. Thus, the maximum deviation of the actual graph from the asymptotes occurs at the corner frequency and it is just 3 dB. Accordingly, we recommend the following four steps to obtain the magnitude graph.

Step 1	Identify the corner frequency and mark it on the x-axis.
Step 2	Draw the asymptotes with appropriate slopes.
Step 3	Mark the 3 dB point at the corner frequency.
Step 4	Starting with the asymptote at low frequencies, draw a smooth curve almost grazing the asymptotes and passing through the 3 dB point in step 3.

The graph is shown in Fig. 7.6(a).

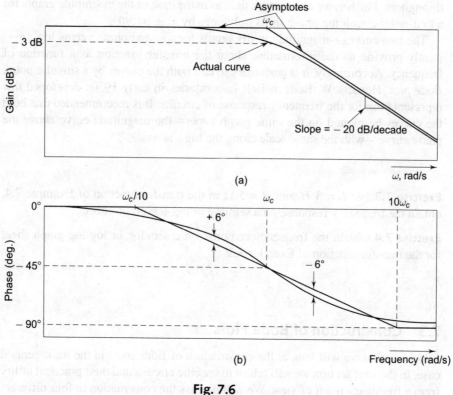

Fig. 7.6

Here we just remark that since it is a first-order circuit, it has just one corner frequency and the slope of the asymptotes changes just once, i.e., from 0 dB to −20 dB. We will come back to this issue after a while.

Let us now turn our attention to the phase shift. Here too we have three different cases as in the magnitude graph. For very low frequencies the shift is almost 0°; for very high frequencies the shift is almost −90°. Accordingly, we draw the *asymptotes* for the phase graph. At the corner frequency the shift is −45°. Thus, we can draw the phase graph almost grazing the asymptotes at the extreme frequencies but passing through $-45°$ at the corner frequency. This is shown in Fig. 7.6(b). Notice that we have not taken the logarithm of the phase angle. It is not necessary to do so (Why?). In the neighbourhood of the corner frequency ω_c, in particular for the frequencies

$$\frac{\omega_c}{10} \leq \omega \leq 10\omega_c$$

i.e., for frequencies within a decade of the corner frequency on either side, the curve is almost linear and hence is first approximated by a straight line passing through $-45°$ at ω_c. The maximum deviation is about $\pm 6°$. Using this deviation we then draw a smooth curve between the asymptotes.

We remark that the phase shift in this example varies 'non-linearly' between $0°$ and $-90°$ with frequency, unlike the previous example where it was constant throughout. Further, we also remark that, as in the case of the magnitude graph, for a first-order circuit the phase shifts only once by *at most* $90°$.

The two curves—magnitude in dB versus $\log_{10} \omega$ and phase versus $\log_{10} \omega$—jointly provide us the information about the transfer function as a function of frequency. Accordingly, it is customary to call both the curves by a singular noun, *Bode plot*. Hendrik W. Bode at Bell Laboratories in early 1930s developed this representation for the frequency response of circuits. It is recommended that both the curves be plotted on the same graph paper—the magnitude curve above the phase curve—with the same scale along the $\log_{10} \omega$-axis.

Exercise 7.3 For $L = 3\ H$ and $R = 5\ \Omega$ in the transfer function of Example 7.4, obtain the frequency response on a semilog or log-log graph sheet.

Exercise 7.4 Obtain the frequency response on a semilog or log-log graph sheet for the transfer function of Exercise 7.2.

7.3   Construction of Bode Plots

In this section we will look at the construction of Bode plots in the most general case. In the next section we will return to specific circuits and their practical utility from a frequency point of view. We will discuss the construction in four different cases as follows.

7.3.1   Zeroth-order Transfer Function

Let

$$H(j\omega) = K \tag{7.1}$$

where K is a constant. This transfer function is independent of frequency. Therefore the magnitude curve is a horizontal straight line for all frequencies. If K is positive, the phase shift is $0°$ and if K is negative it is $180°$ for all frequencies. The Bode plot is shown in Fig. 7.7.

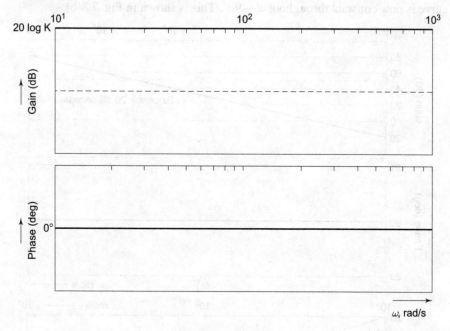

Fig. 7.7

7.3.2 First-order Transfer Function

Let the transfer function be

$$H(j\omega) = j\omega \tag{7.2}$$

i.e.,

$$|H(j\omega)|dB = 20 \log \omega$$

$$\angle H(j\omega) = \tan^{-1}\frac{\omega}{0} = 90° \tag{7.3}$$

The Bode plot of this transfer function with a corner frequency at $\omega = 1$ rad/s is shown in Fig. 7.8.

The expression for the magnitude is of the form $y = mx$, where both y and x are in the logarithmic scale. Accordingly, the magnitude curve is a straight line with slope $m = +20$ dB/decade. We say that the magnitude curve *rolls up* at 20 dB/decade. At the corner frequency the magnitude is 0 dB. Thus, the magnitude curve is a straight line with slope 20 dB/decade with an intercept of 0 dB at $\omega_c = 1$ rad/s. The phase curve, however, is constant at $+90°$ for all frequencies.

If, however,

$$H(j\omega) = \frac{1}{j\omega} \tag{7.4}$$

it is easy to see that the magnitude curve *rolls down* or *rolls off* at the rate of -20 dB/decade, once again with an intercept of 0 dB at $\omega_c = 1$ rad/s. The phase curve is now constant throughout at $-90°$. This is shown in Fig. 7.8(b).

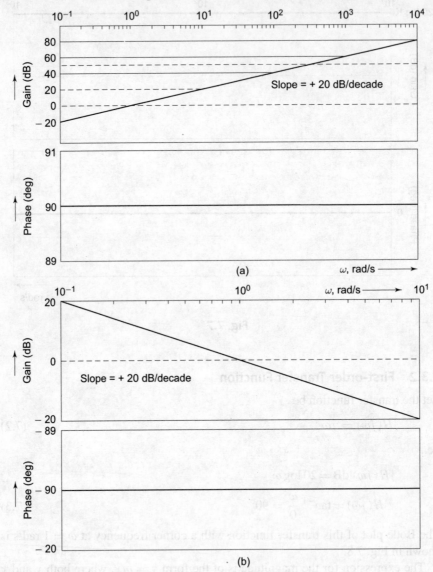

Fig. 7.8

Exercise 7.5 If $H(j\omega) = (j\omega)^2$ obtain the Bode plot. Generalize this result for the case $H(j\omega) = (j\omega)^{\pm m}$, where $m \geq 1$.

(***Hint*** Be cautious with the phase curve.)

(***Ans.*** magnitude = $\omega^{\pm m}$ and phase angle = $\pm m \times 90°$)

Let us now look at a more general first-order transfer function as follows. Suppose

$$H(j\omega) = 1 + \frac{j\omega}{\alpha} \tag{7.5}$$

so that

$$|H(j\omega)| \, \text{dB} = 20 \log \left[\sqrt{1 + \left(\frac{\omega}{\alpha}\right)^2} \right]$$

$$\angle H(j\omega) = \tan^{-1} \frac{\omega}{\alpha} \tag{7.6}$$

Let us first examine the term

$$20 \log \left[\sqrt{1 + \left(\frac{\omega}{\alpha}\right)^2} \right]$$

in the expression for the magnitude. If $\omega \ll \alpha$, then

$$20 \log \left[\sqrt{1 + \left(\frac{\omega}{\alpha}\right)^2} \right] \approx 20 \log 1 = 0 \, \text{dB} \tag{7.7}$$

i.e., the magnitude is almost 0 dB, or the magnitude curve is a straight line with 0 slope and grazing the frequency axis, for frequencies much less than the *corner frequency* α. If $\omega = \alpha$, the corner frequency, then

$$20 \log \left[\sqrt{1 + \left(\frac{\omega}{\alpha}\right)^2} \right] = 20 \log \sqrt{2} \approx 3 \, \text{dB} \tag{7.8}$$

Finally, if $\omega \gg \alpha$, then

$$20 \log \left[\sqrt{1 + \left(\frac{\omega}{\alpha}\right)^2} \right] \approx 20 \log \frac{\omega}{\alpha} \tag{7.9}$$

and the magnitude curve rolls up at +20 dB/decade with an intercept of 0 dB at $\omega = \alpha$. The straight lines we obtain for the two cases $\omega \ll \alpha$ and $\omega \gg \alpha$ are the *asymptotes*. The actual magnitude curve will be very close to these asymptotes passing through the 3 dB point at the corner frequency. This is shown in Fig. 7.9(a). The maximum deviation of the actual magnitude curve from the asymptote is 3 dB occurring at the corner frequency.

The phase curve is an inverse tangent function of ω. For $\omega \le 0.1\alpha$, the phase is almost 0° and for $\omega \ge 10\alpha$, the phase is almost 90°. In the range $0.1\alpha \le \omega \le 10\alpha$, the phase curve is almost a straight line passing through 45° at $\omega = \alpha$. This is shown in Fig. 7.9(b). Again the maximum deviation of the actual phase curve from the straight line approximation is about $\pm 6°$.

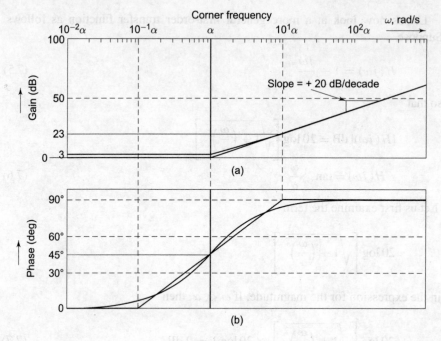

Fig. 7.9

Exercise 7.6 Obtain the Bode plot of

$$H(j\omega) = \frac{1}{1 + \dfrac{j\omega}{\alpha}}, \quad \alpha > 0$$

and generalize the results to obtain the Bode plot of

$$H(j\omega) = \left(1 + \frac{j\omega}{\alpha}\right)^{\pm m}, \quad \alpha > 0, \quad m \geq 1$$

$$\left(\text{Ans. } |H(j\omega)| = \left(\frac{1}{\sqrt{1 + \dfrac{\omega^2}{\alpha^2}}}\right)^{\pm m} \text{ and } \angle H(j\omega) = \pm m \times \tan^{-1}\frac{\omega}{\alpha}\right)$$

At this point let us consider an example and learn how to construct the Bode plots in a general case.

Example 7.5 Let

$$H(j\omega) = \frac{j6\omega}{2 + j\omega}$$

Notice that this transfer function is a combination of the three cases discussed so far, i.e., there is a constant factor independent of frequency, there is a factor with corner frequency at $\omega = 1$ rad/s, and there is a factor similar to $1 + j\omega/\alpha$ in the denominator. Let us then rewrite $H(j\omega)$ as

$$H(j\omega) = j\omega \frac{1}{1 + \dfrac{j\omega}{2}} \frac{6}{2}$$

Observe the way we have split the factors—the constant has been taken to the end. Further, the corner frequencies have been put in an *ascending* order. The purpose of splitting $H(j\omega)$ in this fashion becomes clear if we look at the expressions for the magnitude and the phase.

$$|H(j\omega)|\text{dB} = |H_1(j\omega)| + |H_2(j\omega)| + |H_3(j\omega)|$$

$$= 20\log\omega + 20\log\frac{1}{\sqrt{1 + (\dfrac{\omega}{2})^2}} + 20\log\frac{6}{2}$$

$$\angle H(j\omega) = \angle H_1(j\omega) + \angle H_2(j\omega) + \angle H_3(j\omega)$$

$$= 90° - \arctan\frac{\omega}{2} + 0°$$

By way of logarithms we have nicely transformed the multiplication '·' of terms in $H(j\omega)$ to simple addition '+' in $|H(j\omega)|$ dB; $\angle H(j\omega)$ is already in the addition form. Now the idea is clear:

There may be any number of factors in the transfer function $H(j\omega)$. All we need to do is to simply ADD algebraically, at every frequency, the Bode plots of the constituting factors.

Figure 7.10(a) shows the individual magnitude asymptotes drawn to the same frequency scale. Figure 7.10(b) shows the *addition* of the individual asymptotes shown in Fig. 7.10(a). For instance, at $\omega = 1$ rad/s

$$|H(j1)| = 0 + 0 + 20 \ \log 3 = 9.54 \text{ dB}$$

Similarly, at $\omega = 1.5$ rad/s

$$|H(j1.5)| = 3.52 + 0 + 20 \ \log 3 = 13.06 \text{ dB}$$

In fact, until $\omega = 2$ rad/s, $|H_2|$ is 0 dB.

At $\omega = 2$ rad/s, the magnitude is $6 + 0 + 9.54 = 15.54$ dB. For $\omega > 2$ rad/s, the first term $|H_1|$ rolls up at $+20$ dB/decade and the second term $|H_2|$ rolls off at -20 dB/decade. Therefore, the sum of these two factors is a constant 15.54 dB $\forall \omega > 2$ rad/s. This is where the slopes of the curves are needed. The actual magnitude curve may be obtained from the asymptotic curve by making a simple error correction at the corner frequency. At the corner frequency 2 rad/s, the magnitude of $|H_2|$ is -3 dB. Therefore, it would be sufficient to mark a point 3 dB below 15.54 dB on the

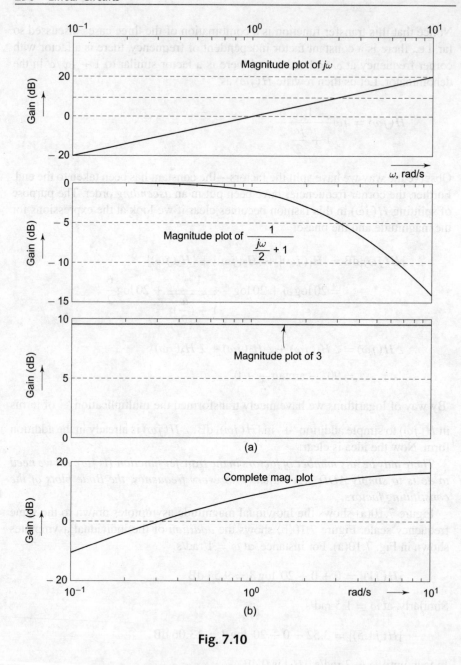

Fig. 7.10

magnitude asymptote and draw a smooth curve along the asymptotes and passing through this point. As was mentioned earlier, the maximum deviation is only 3 dB and that too occurs only at the corner frequency. In general, we may consider the errors as negligible at all other frequencies. In other words, asymptotes themselves are reasonably good Bode plots. With little more practice, the reader would find it easier to add the asymptotes using the slopes for all simpler functions of $j\omega$ and then simply shift the result by the constant arising out of the *last* term.

Figure 7.11 shows the complete Bode plot of the transfer function. We leave it as an exercise to the reader to study the phase curve. Here too, compared to large phase angles, the deviation of $\pm 6°$ may be considered negligible.

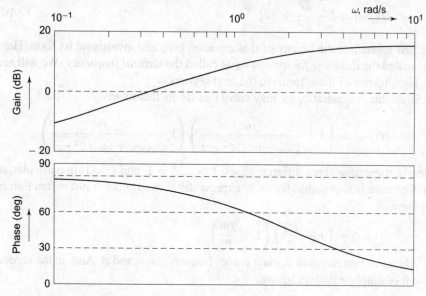

Fig. 7.11

Exercise 7.7 Verify the results of Example 7.5 by obtaining a table as follows for a wide range of frequencies from 0.1 rad/s to a few krad/s.

S. No.	ω	$|H_1|$	$|H_2|$	$|H_3|$	$|H|$	$\angle H_1$	$\angle H_2$	$\angle H$

Exercise 7.8 Obtain the Bode plot of

$$H(j\omega) = \frac{7(1 + j\omega)(3 + j\omega)^2}{(j\omega)^3(2 + j\omega)(4 + j\omega)}$$

(**Hint** $H(j\omega) = \dfrac{1}{(j\omega)^3}(1 + j\omega)\dfrac{1}{1 + \dfrac{j\omega}{2}}\left(1 + \dfrac{j\omega}{3}\right)^2\dfrac{1}{1 + \dfrac{j\omega}{4}}\dfrac{7 \times 3^2}{2 \times 4}$.

The corner frequencies should be arranged in an ascending order. The asymptotic Bode plot for the magnitude starts at -60 dB/decade at low frequencies, changes its slope to -40 dB/decade at 1 rad/s, changes again to -60 dB/decade at 2 rad/s, changes to -20 dB/decade at 3 rad/s (since there are two factors in the numerator at this frequency), and finally changes to -40 dB/decade at 4 rad/s. Similarly, the phase curve starts at $-270°$, increases between 1 rad/s and 2 rad/s, decreases again between 2 rad/s and 3 rad/s, increases, rather drastically, between 3 rad/s and 4 rad/s, and finally reaches $-180°$ asymptotically for high frequencies.)

7.3.3 Second-order Transfer Function

Consider a second-order transfer function of the form

$$H(j\omega) = 1 - \frac{\omega^2}{\omega_n^2} + j2\zeta\frac{\omega}{\omega_n} \tag{7.10}$$

We just appreciate the beauty of this equation here and investigate its form. Here ζ is called the *damping factor* and ω_n is called the *natural frequency*. We will see the significance of these terms in the next section.

Since this is quadratic, we may solve this for its factors as

$$H(j\omega) = \left(1 + \frac{j\omega}{\zeta\omega_n - \omega_n\sqrt{\zeta^2 - 1}}\right)\left(1 + \frac{j\omega}{\zeta\omega_n + \omega_n\sqrt{\zeta^2 - 1}}\right)$$

Clearly, there arise three different cases: $\zeta > 1$, $\zeta = 1$, and $\zeta < 1$. In particular, in the first case it is possible for us to express the second-order transfer function in the form of

$$H(j\omega) = \left(1 + \frac{j\omega}{\alpha}\right)\left(1 + \frac{j\omega}{\beta}\right)$$

i.e., there are two *real and distinct* corner frequencies, α and β. And, in the second case it is simply a perfect square,

$$H(j\omega) = \left(1 + \frac{j\omega}{\alpha}\right)^2$$

with two *real and identical* corner frequencies at α. These two cases boil down to the transfer functions discussed in Section 7.3.2. Thus, we need to learn only about the last case, where $0 \leq \zeta < 1$ since we cannot obtain *real* corner frequencies.

The magnitude and phase functions can be written routinely as

$$|H(j\omega)| = 20\log\left[\sqrt{\left(1 - \frac{\omega^2}{\omega_n^2}\right)^2 + \left(2\zeta\frac{\omega}{\omega_n}\right)^2}\right]$$

$$\angle H(j\omega) = \tan^{-1}\frac{2\zeta\frac{\omega}{\omega_n}}{1 - \omega^2/\omega_n^2} \tag{7.11}$$

For the magnitude curve, we will look at the asymptotic cases. First, if $\omega \ll \omega_n$, then

$$|H(j\omega)| \approx 0 \text{ dB} \tag{7.12}$$

Secondly, at the corner frequency $\omega = \omega_n$,

$$|H| = 20\log 2\zeta \text{ dB} \tag{7.13}$$

i.e., the magnitude is a function of the damping factor alone. If ζ is very small, the magnitude is very large with a negative sign; if $\zeta = 0.5$, the magnitude is 0 dB; and if ζ is close to 1, the magnitude is close to 6 dB. Thus, the magnitude at the corner frequency can be anywhere in the range $-\infty < |H| < 6$ dB. Lastly, if $\omega \gg \omega_n$, then

$$|H| \approx 40\log\frac{\omega}{\omega_n} \tag{7.14}$$

This is because, the squared term ω^4/ω_n^4 dominates 1 as well as $4\zeta^2\omega^2/\omega_n^2$. Thus, for frequencies greater than ω_n, the magnitude curve rolls up at $+40$ dB/decade. The reader should have expected that at high frequencies a single first-order term contributes $+20$ dB/decade, hence a pair of first-order terms in the numerator contribute $+40$ dB/decade. The asymptotes are shown in Fig. 7.12(a). Here we have given the actual magnitude curve at the corner frequency for several values of ζ. For clarity we have chosen $\omega_n = 4$.

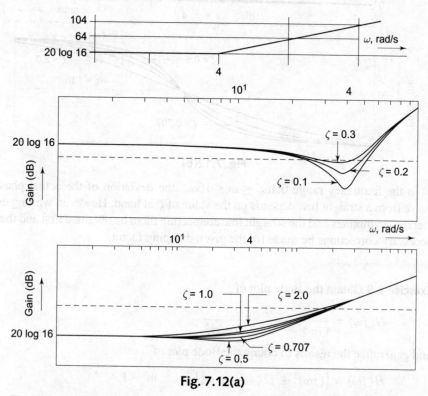

Fig. 7.12(a)

The phase curve is, once again, an inverse tangent function of ω. For small values of ω, i.e., $\omega \leq 0.1\omega_n$, the phase is almost $0°$. In the range $0.1\omega_n \leq \omega \leq 10\omega_n$, the phase depends on the exact value of the damping factor ζ. The only exception is at $\omega = \omega_n$, where the phase is exactly $90°$. And, for large values of ω, i.e., $\omega \geq 10\omega_n$, the phase is almost $180°$. This is because the argument

$$\frac{2\zeta\dfrac{\omega}{\omega_n}}{1-(\omega/\omega_n)^2} \to 0 \quad \text{as} \quad \frac{\omega}{\omega_n} \to \infty$$

Further, the denominator is always negative, suggesting us that the phase approaches the negative real axis from above. The phase curve for several values of ζ is given in Fig. 7.12(b). Here too, as one might readily anticipate, the two first-order factors contribute $2 \times 90 = 180°$ asymptotically.

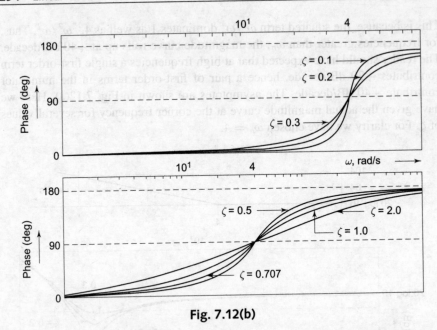

Fig. 7.12(b)

In the frequency range $0.1\omega_n \leq \omega \leq 10\omega_n$, the deviation of the actual phase curve from a straight line depends on the value of ζ at hand. However, we suggest that the asymptotes and the straight line connecting them be obtained first and then necessary corrections be made for the given damping factor.

Exercise 7.9 Obtain the Bode plot of

$$H(j\omega) = \frac{1}{(j\omega)^2 + j2\zeta\omega_n\omega + \omega_n^2}$$

and generalize the results to obtain the Bode plot of

$$H(j\omega) = \left[(j\omega)^2 + j2\zeta\omega_n\omega + \omega_n^2 \right]^{\pm m}, \quad m > 1$$

7.3.4 General Transfer Function

Let us look at the following example to solidify our ideas above.

Example 7.6
Let

$$H(j\omega) = \frac{50(j\omega + 2)}{j\omega \left((j\omega)^2 + j4\omega + 100 \right)}$$

This may be factored as

$$H(j\omega) = \frac{1}{j\omega}(j\omega + 2)\frac{1}{\left(\frac{j\omega}{10} \right)^2 + 2 \times 0.2 \left(\frac{j\omega}{10} \right) + 1} \times \frac{50 \times 2}{100}$$

where we have identified the quadratic factor in the denominator to have $\zeta = 0.2$ and $\omega_n = 10$. The complete Bode plot is shown in Fig. 7.13.

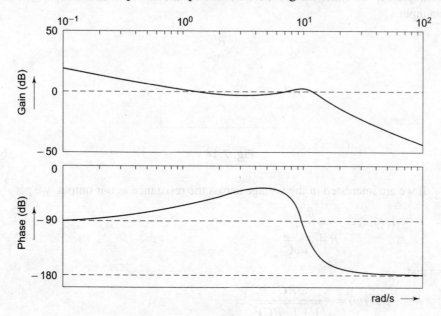

Fig. 7.13

We end this section with a general algorithm. Refer to the examples we have solved so far while studying this algorithm.

Step 1	Given an arbitrary $H(j\omega)$, factorize and rewrite this in terms of the standard first-order and second-order factors. Identify the corner frequencies.
Step 2	Rearrange the terms in step 1 in the ascending order of the corner frequencies. Take the constant term to the end.
Step 3	For each of the terms in step 2, obtain the individual Bode plots.
Step 4	Add 'algebraically' each of the Bode plots in Step 3 above.

Exercise 7.10 Obtain the Bode plot of

$$L(j\omega) = \frac{10}{(j\omega + 2)\left[(j\omega)^2 + j8\omega + 25\right]}$$

(**Ans.** Corner frequencies: $\omega = 2$ and $\omega = 5$. For the second-order part, $\zeta = 0.032$)

Now we will look at some more circuits and discuss their frequency response. We strongly recommend the reader to study the examples and exercises in this section thoroughly before moving on to the next section or next chapter.

Example 7.7 Consider the RC network shown in Fig. 7.14 with voltage $v_S(t)$ as an input.

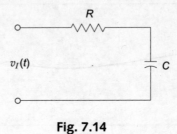

Fig. 7.14

If we are interested in the voltage across the resistance as our output, we get

$$H(j\omega) = \frac{R}{R + \dfrac{1}{j\omega C}}$$

so that

$$|H(j\omega)| = \frac{\omega RC}{\sqrt{1 + (\omega RC)^2}}$$

$$\angle H(j\omega) = 90° - \tan^{-1}\omega RC$$

The corner frequency is $1/RC$ and the Bode plot is shown in Fig. 7.15.

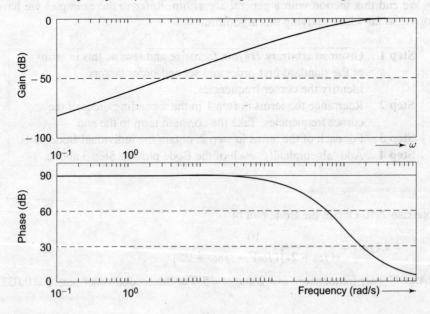

Fig. 7.15

Let us do some numericals here. Assuming $R = 1$ kΩ and $C = 1$ μF, we have the corner frequency at 1 krad/s. The Bode plot says that

- For input sinusoids of very low frequencies, close to 0 rad/s, the 'attenuation' is quite large, i.e., no output voltage is available across the resistance. This might be expected because for very low frequencies, say dc, the capacitor acts like an open circuit and hence no current flows through the resistance.
- For a sinusoid of frequency ω equal to the corner frequency 1 krad/s, the 'attenuation' is only 3 dB, i.e., $|H(j\omega)| = -3$ dB.
- For sinusoids of frequencies greater than or equal to the corner frequency, almost all the input voltage is available across resistance R since the capacitance tends to behave like a short circuit (when compared to resistance R) at these frequencies.
- The phase plot says that there is a phase lead of about 90° at low frequencies, of 45° at the corner frequency, and the output is almost in phase with the input at higher frequencies.
- Since more current flows in the circuit at higher frequencies, the average power dissipated by the resistance is maximum in the high-frequency limit and

$$P_{max} = \frac{1}{2}\frac{V_m^2}{R}$$

where V_m is the peak amplitude of the input sinusoidal voltage. The coefficient $1/2$ vanishes from the picture if we are interested in the r.m.s. quantities. At the corner frequency, we notice that

$$P_{average}(\omega = \omega_c) = \frac{1}{2}P_{max}$$

since the magnitude of the output is -3 dB relative to the magnitude of the input at the corner frequency. Accordingly, we also call the corner frequency as the *half power frequency*.

We can 'control' the behaviour of this circuit by choosing different values of R and C and hence the corner frequency. Since the lower (than the corner frequency) frequency signals are severely attenuated and higher frequency signals are 'passed through', we may call this as a *high pass transfer function*. We will provide more details in the next section.

Exercise 7.11 In Example 7.7 with (i) $RC = 1$ ms, (ii) $RC = 10$ ms, and (iii) $RC = 0.2$ ms, obtain the output voltage if the input is

(a) $3\cos(t + 30°)$
(b) $7\cos(10t + 45°)$
(c) $5\cos(100t - 45°)$
(d) $4\cos(1000t + 120°)$
(e) $10\cos(10,000t)$
(f) $\cos(10^6 t + 90°)$

Calculate the average power dissipated by the resistance in each of the 18 cases above.

(*Ans.* output $= H(j\omega) \times$ input

with the input preferably put in the polar form for easy multiplication. For instance, with $RC = 1$,

$$H(j\omega) = \frac{\omega}{\sqrt{1 + \omega^2}} \angle 90° - \tan^{-1}\omega$$

and if the input is $3\cos(t + 30°)$,

$$\text{output} = \frac{3\omega}{\sqrt{1 + \omega^2}} \angle 120° - \tan^{-1}\omega|_{\omega=1} = \frac{3}{\sqrt{2}} \angle 75°)$$

Exercise 7.12 Repeat Example 7.7 if the output is the voltage across the capacitance. Convince yourself that we may call this as a 'low pass transfer function'.

Example 7.8 Consider the RLC circuit shown in Fig. 7.16.

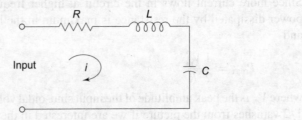

Fig. 7.16

If we are interested in the current flowing in the circuit as the output,

$$H(j\omega) = \frac{\mathcal{I}(j\omega)}{\mathcal{V}(j\omega)} = \mathcal{Y} = \frac{1}{R + j\omega L + \dfrac{1}{j\omega C}}$$

$$= \frac{j\omega C}{1 - \omega^2 LC + j\omega RC}$$

Comparing with the standard second-order transfer function given in Eqn (7.10), we get

$$\omega_n = \frac{1}{\sqrt{LC}}, \quad \zeta = \frac{R}{2}\sqrt{\frac{C}{L}}$$

We remark that a first-order circuit has just one open parameter—the corner frequency, whereas a second-order circuit has two open parameters—the natural frequency and the damping factor. Extending this, we expect that for a general nth-order circuit we will have n number of open parameters.

In terms of these two parameters, the transfer function can be written as

$$H(j\omega) = \frac{j\dfrac{\omega}{\omega_n}}{1 - (\omega/\omega_n)^2 + j2\zeta(\omega/\omega_n)} \frac{2\zeta}{R} \tag{7.15}$$

A typical Bode plot is shown in Fig. 7.17.

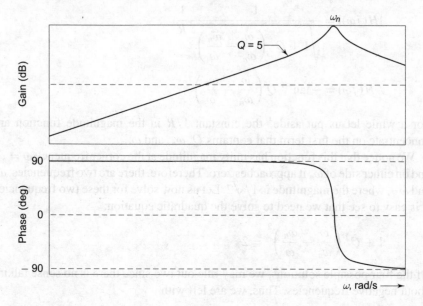

Fig. 7.17

We notice that the current is maximum in magnitude at the corner frequency $\omega_c = \omega_n$ and is in phase with the input voltage. This was discussed in the context of a parallel RLC circuit in Example 6.8 of the previous chapter. We notice that the Bode plot here is similar (except for the scales adopted) to Fig. 6.10. In fact, in Exercise 6.13 we have asked the reader to obtain the Bode plot of the impedance function of a series RLC circuit. Continuing the argument, we see that at this particular frequency ω_n the circuit exhibits a purely resistive behaviour. For frequencies $\omega < \omega_n$, it has a capacitive behaviour and for frequencies above ω_n it has an inductive behaviour. We call the circuit behaviour at ω_n as *unity power factor resonance* since the magnitude of the current is maximum and the phase shift is $0°$ and hence the average power delivered to the circuit is maximum. Even though the individual reactances X_L and X_C are not zero, they cancel each other at resonance.

We will define a new quantity now before proceeding further.

Definition 7.3 Selectivity factor

$$Q \triangleq \frac{1}{2\zeta}$$

This 'selectivity factor' allows us to speak about the frequency response in a more intuitive way as follows. We first simplify Eqn (7.15) by dividing the numerator and denominator by $j\omega/\omega_n$ to get

$$H(j\omega) = \frac{1}{R\left[1 + jQ\left(\dfrac{\omega}{\omega_n} - \dfrac{\omega_n}{\omega}\right)\right]} \tag{7.16}$$

It follows that

$$|H(j\omega)| = \frac{1}{\sqrt{1 + Q^2 \left(\dfrac{\omega}{\omega_n} - \dfrac{\omega_n}{\omega}\right)^2}} \frac{1}{R}$$

$$\angle H(j\omega) = -\tan^{-1} Q \left(\frac{\omega}{\omega_n} - \frac{\omega_n}{\omega}\right)$$

For a while let us put aside[4] the constant $1/R$ in the magnitude function and concentrate on the first term that contains Q, ω_n, and ω.

We notice that this function has unity magnitude at the corner frequency $\omega = \omega_n$ and on either side of ω_n it approaches zero. Therefore, there are two frequencies, ω_L and ω_H, where the magnitude is $1/\sqrt{2}$. Let us now solve for these two frequencies. It is easy to see that we need to solve the quadratic equation:

$$1 + Q^2 \left(\frac{\omega}{\omega_n} - \frac{\omega_n}{\omega}\right)^2 = 2$$

Of the four possible solutions, we may rule out two since there is no sense talking about negative frequencies. Thus, we are left with

$$\omega_L = \omega_n \left(\sqrt{1 + \frac{1}{(2Q)^2}} - \frac{1}{2Q}\right)$$

and

$$\omega_H = \omega_n \left(\sqrt{1 + \frac{1}{(2Q)^2}} + \frac{1}{2Q}\right) \tag{7.17}$$

Extending our ideas on power from Example 7.7, these two frequencies may be called as *half power frequencies*. We leave it to the reader to verify that the three frequencies – ω_L, ω_n, and ω_H are in geometric progression. We now define another quantity.

Definition 7.4 Bandwidth BW

$$\text{BW} \triangleq \omega_H - \omega_L$$

From Eqn (7.17), we have

$$\text{BW} \triangleq \frac{\omega_n}{Q} \tag{7.18}$$

It is clear that the larger the Q, the narrower the bandwidth BW. In other words, the two half power frequencies get closer to the corner frequencies. In fact, asymptotically, as $Q \to \infty$, BW $\to 0$. The roll-off rate on either side of the corner frequency is -40 dB/decade. This implies that a circuit with very high Q 'selects' the natural frequency more than any other frequency. Hence the name *selectivity factor* for Q. We give the Bode plot of the transfer function in Eqn (7.16) for different values of Q in Fig. 7.18.

---

[4] We can always add $20 \log(1/R)$ later.

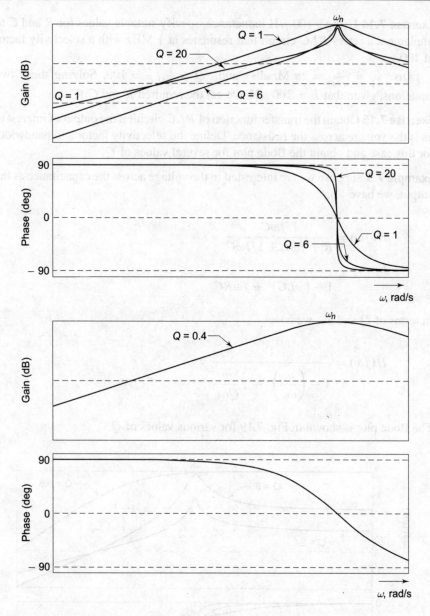

Fig. 7.18

We will have yet another interpretation for Q after a couple of examples. We will come back to the bandwidth issue in the next section.

Exercise 7.13 A resistance of 20 Ω, an inductance of 0.25 H, and a capacitance are connected to the household supply of 220 V at 50 Hz. Obtain the value of capacitance that will maximize the power transfer. Compare with the case in which the capacitance is replaced with a conducting wire.

Exercise 7.14 Using a 100-μH inductance, specify suitable values for R and C to implement a series RLC circuit that resonates at 1 MHz with a selectivity factor of 100.

(*Ans.* $\omega_n = \frac{1}{\sqrt{LC}} = 2\pi$ Mrad/s and $Q = \frac{1}{R}\sqrt{\frac{L}{C}} = 100$. Solving these two equations, given that $L = 100$ μH, we get the required R and C.)

Exercise 7.15 Obtain the transfer function of RLC circuit if the output of interest to us is the voltage across the resistance. Define the selectivity factor and bandwidth for this case and obtain the Bode plot for several values of Q.

Example 7.9 Suppose we are interested in the voltage across the capacitance as the output, we have

$$H(j\omega) = \frac{1/j\omega C}{R + j\omega L + 1/j\omega C}$$

$$= \frac{1}{1 - (\omega LC)^2 + j\omega RC}$$

In terms of Q and ω_n, we have

$$H(j\omega) = \frac{1}{1 - \left(\dfrac{\omega}{\omega_n}\right)^2 + j\dfrac{\omega}{Q\omega_n}}$$

The Bode plot is shown in Fig. 7.19 for various values of Q.

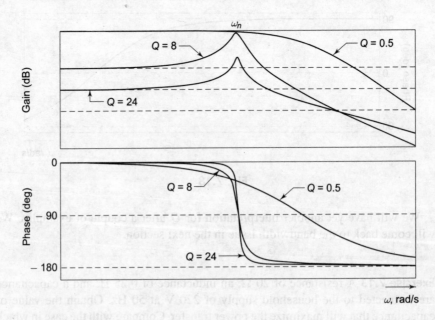

Fig. 7.19

Here too, the selectivity factor plays a major role. For instance, if $Q = 0.5$, we have

$$H(j\omega) = \frac{1}{\left(1 + j\dfrac{\omega}{\omega_n}\right)^2}$$

The asymptotes meet at the corner frequency ω_n as ever. But the actual response is -6 dB at this frequency. If $Q > 0.5$, the magnitude plot exhibits peaking at the corner frequency. As Q becomes larger, the magnitude 'blows up' to infinity. Once again Q serves as a selectivity factor. The phase shift at the corner frequency is $-90°$ and confirms that this circuit basically exhibits resonance at the corner frequency and hence the capacitance voltage has a $90°$ lagging with reference to the input voltage.

However, for $Q < 0.5$, we see that the second-order transfer function is factored into two first-order factors of the form

$$H(j\omega) = \frac{1}{\left(1 + j\dfrac{\omega}{\omega_1}\right)\left(1 + j\dfrac{\omega}{\omega_2}\right)}$$

Hence there are two corner frequencies. In other words, Q does not come into picture as it is, but the two open parameters are the two corner frequencies. Accordingly, the Bode plot has a slightly different shape, as is shown in Fig. 7.20.

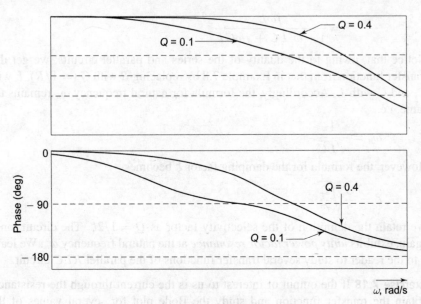

Fig. 7.20

Exercise 7.16 Repeat Example 7.9 if $R = 12\,\Omega$, $L = 0.3\,\text{H}$, and $C = \frac{1}{3}\,\text{mF}$.

(*Ans.* $\omega_n = 100$ and $\zeta = 0.2$)

Exercise 7.17 Suppose we are interested in the voltage across the inductance in a series RLC circuit. Obtain the transfer function and study the Bode plot for several values of the selectivity factor.

Example 7.10 Consider the parallel RLC circuit driven by a current source as shown in Fig. 7.21.

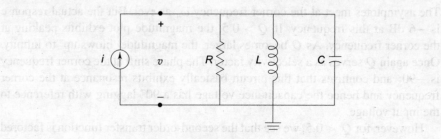

Fig. 7.21

Suppose we are interested in the voltage across the parallel combination. Then we have

$$H(j\omega) = \mathcal{Z}(j\omega) = \cfrac{1}{G + j\omega C + \cfrac{1}{j\omega L}}$$

$$= \frac{j\omega L}{1 - \omega^2 LC + j\omega GL}$$

Notice that, owing to the duality of the series and parallel circuits, we get this transfer function from that in Example 7.8 by replacing R with $G\ (= 1/R)$, L with C, and C with L. Accordingly, the formula for natural frequency ω_n remains the same, i.e.,

$$\omega_n = \frac{1}{\sqrt{LC}}$$

However, the formula for the damping factor ζ becomes

$$\zeta = \frac{G}{2}\sqrt{\frac{L}{C}}$$

We retain the definition of the selectivity factor as $Q = 1/2\zeta$. The circuit, once again, exhibits *unity power factor resonance* at the natural frequency ω_n. We leave it to the reader to study several transfer functions of the parallel RLC circuit.

Exercise 7.18 If the output of interest to us is the current through the resistance, obtain the transfer function and study the Bode plot for several values of the selectivity factor.

Exercise 7.19 Obtain the expressions for half power frequencies and bandwidth for a parallel RLC circuit.

Exercise 7.20 If the output of interest to us is the current through the inductance, obtain the transfer function and study the Bode plot for several values of the selectivity factor.

Exercise 7.21 If the output of interest to us is the current through the capacitance, obtain the transfer function and study the Bode plot for several values of the selectivity factor.

Example 7.11 Let us look at the selectivity factor Q from a different angle in this example. Consider a parallel RLC circuit once again. As was mentioned in the previous example, the resonant frequency is ω_n and the parallel combination of the inductance and capacitance acts like an open circuit. Effectively, the source 'sees' the resistance alone at resonance. Hence the voltage across the node pair is in phase with the input current. Since this is a parallel circuit, the voltage across the capacitance is

$$v_C(t) = Ri(t)$$

Using the relationship between voltage, power, and energy, it is easy to derive that the instantaneous 'energy' stored in the capacitance is given by

$$w_C(t) = \frac{1}{2}Cv_C^2(t) \tag{7.19}$$

At resonant frequency this evaluates to

$$\tilde{w}_C(t) = \frac{1}{2}R^2CI_m^2\cos^2(\omega_n t)$$

$$= \frac{QR}{\omega_n}I_{rms}^2\cos^2(\omega_n t) \tag{7.20}$$

using the relationship among R, C, Q, and ω_n. Since the phase shift is zero at this frequency, the energy stored in the capacitance is maximum at this frequency according to Eqn (7.20).

Exercise 7.22 Derive Eqns (7.19) and (7.20).

Similarly, we have

$$v_L(t) = Ri(t) \quad \text{and} \quad i_L(t) = \frac{R}{j\omega L}i(t)$$

By using duality, we may expect that

$$\tilde{w}_L(t) = \frac{QR}{\omega_n}I_{rms}^2\sin^2(\omega_n t) \tag{7.21}$$

is the maximum energy stored in the inductance at the resonant frequency.

Exercise 7.23 Derive Eqn (7.21).

The total energy stored in the circuit at ω_n is

$$\tilde{w}_C(t) + \tilde{w}_L(t) = \frac{QR}{\omega_n}I_{rms}^2 \tag{7.22}$$

Surprisingly, this maximum stored energy is independent of time. We call this $\tilde{w}$. What this tells us is that at resonance there is no transfer of energy between the source and the parallel combination of L and C; there is an *internal* exchange of energy between these two elements. The source caters only to the energy dissipated by the resistance.

Let us now look at the *energy dissipated by the circuit per second* at resonance. This is the maximum average power:

$$P_{\max} = RI_{\text{rms}}^2 \text{ W}$$

If we divide this by the *number of cycles per second*, we get the 'energy dissipated per cycle', i.e.,

$$\tilde{w}_R = \frac{2\pi}{\omega_n} RI_{\text{rms}}^2$$

$$= 2\pi \frac{\tilde{w}}{Q}$$

We can now have a fresh expression for the selectivity factor as

$$Q \triangleq 2\pi \times \frac{\text{(maximum stored energy)}}{\text{(energy dissipated per cycle)}} \quad (7.23)$$

We thus define the selectivity factor Q in two different ways: (i) using the bandwidth and corner frequency, and (ii) using the energy stored and dissipated. Notice that these two expressions are independent of the particular circuit under consideration; it could be a series RLC circuit or a parallel one. However, the basic definition comes from the damping factor ζ, which, in turn, depends on the actual circuit elements and their topology.

Let us now do a simple numerical exercise to get a physical insight. Suppose the current source is a sinusoid of maximum amplitude 1 mA and the circuit parameters are

$$R = 1\,\text{k}\Omega, \quad L = 10\,\text{mH}, \quad C = 10\,\text{nF}$$

Quickly we see that the selectivity factor is $Q = 1$ using the circuit elements, and $\omega_n = 10^5$ rad/s. The formulae for half power frequencies do not change for the parallel RLC circuit. Hence,

$$\omega_L = \frac{\sqrt{5}-1}{2}10^5 \quad \text{and} \quad \omega_H = \frac{\sqrt{5}+1}{2}10^5 \text{ rad/s}$$

so that the bandwidth is 10^5 rad/s. The maximum stored energy and the energy dissipated per cycle are

$$\tilde{w} = 5\,\text{nJ} \quad \text{and} \quad \tilde{w}_R = 31.42\,\text{nJ}$$

respectively.

Exercise 7.24 Verify that Eqn (7.23) holds good for a series RLC circuit also.

Exercise 7.25 Design a series RLC circuit that has a resonant frequency at 200π krad/s and in response to an applied voltage of $v = 8\cos(2\pi 10^5 t)$ V stores a maximum energy of 10 μJ, and dissipates 2 μJ per cycle.

Example 7.12 Consider the circuit shown in Fig. 7.22.

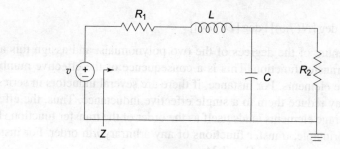

Fig. 7.22

Apparently, this circuit conforms neither to the standard series RLC nor to the standard parallel RLC circuit. But we may bring it to the series form as shown in Fig. 7.23.

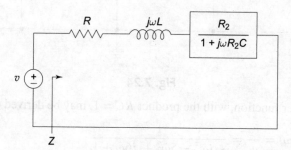

Fig. 7.23

Here, the voltage source 'sees' an impedance

$$Z(j\omega) = R_1 + j\omega L + \frac{R_2}{1 + j\omega R_2 C}$$

We may rewrite this as

$$Z(j\omega) = \frac{R_1 + R_2 + \omega^2 R_1 R_2^2 C^2}{1 + \omega^2 R_2^2 C^2}$$

$$+ j\omega \frac{L + R_2^2 C - \omega^2 L R_2^2 C^2}{1 + \omega^2 R_2^2 C^2}$$

The frequency at which the impedance becomes purely resistive may be obtained by equating the imaginary part to be zero. Accordingly, we get

$$\omega_n = \frac{1}{\sqrt{LC}} \sqrt{1 - \frac{L}{R_2^2 C}}$$

provided, of course, $R_2 \geq \sqrt{L/C}$.

Exercise 7.26 Obtain the selectivity factor for the circuit described in Example 7.12.

So far we have seen first- and second-order circuits. To recapitulate quickly, any arbitrary transfer function is a rational function, i.e., it can be written as a ratio of two polynomials $N(j\omega)$ and $D(j\omega)$. We consider

$$\max \{\deg [N(j\omega)], \deg [D(j\omega)]\}$$

i.e., the maximum of the degrees of the two polynomials, and assign this as the order of the transfer function. This is a consequence of the effective number of energy-storage elements. For instance, if there are several inductors in series in a circuit, we may reduce them to a single effective inductance. Thus, the effective number of storage elements lends itself as the order of the transfer function. There could be, in principle, transfer functions of any arbitrary nth order. For instance, consider the circuit shown in Fig. 7.24.

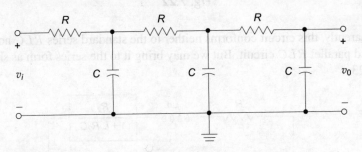

Fig. 7.24

The transfer function, with the product $RC = 1$, may be derived to be

$$H(j\omega) = \frac{v_0}{v_i} = \frac{1}{-j\omega^3 - 5\omega^2 + j6\omega + 1}$$

and the corresponding Bode plot is given in Fig. 7.25.

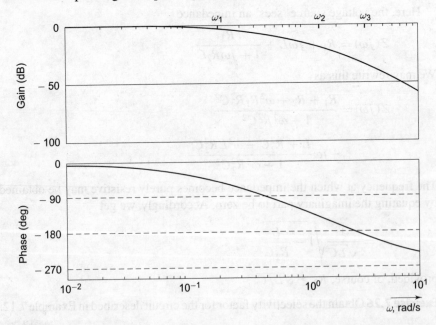

Fig. 7.25

In principle, we may factorize the higher-order transfer functions into first- and second-order functions. And the Bode plots may be obtained by 'algebraically adding' the simpler Bode plots. We shall postpone more details on network functions to Chapter 9. For now, we simply remark that the rate of roll-up or roll-off of an nth-order transfer function would be $\pm 20n$ dB/decade. We will put to use this fact in the next section.

7.4 Frequency Selective Filters

Having seen several interesting transfer functions of circuits and their frequency responses, we are prompted to make a classified study of the circuits that have great practical value. We will accomplish this as follows.

At the beginning of this chapter, while defining the transfer function, we have seen that

$$\text{output} = H(j\omega) \times (\text{input})$$

with both the input and output signals expressed in frequency domain, i.e., in the phasor notation. Suppose we have a $H(j\omega)$ which graphically looks like Fig. 7.26.

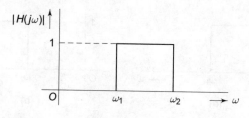

Fig. 7.26

The circuit operation may be summarized as

$$\text{output} = \begin{cases} \text{input} & \text{if } \omega_1 \leq \omega_{\text{in}} \leq \omega_2 \\ 0 & \text{otherwise} \end{cases}$$

In plain words, it says that the output is exactly equal to the input signal if and only if its frequency lies in the interval $[\omega_1, \omega_2]$. We call such a circuit as a *band pass filter*. The term 'band pass' suggests that only a certain *band* or closed interval of the frequency spectrum is found at the output, and the term 'filter' suggests the pruning of signals outside the specified band. The band of frequencies allowed to pass through the filter is called the *pass band* and the other interval is called the *stop band* or *attenuation band*. Figure 7.26 shows a very ideal transfer function—the gain in the pass band is unity (0 dB) and the gain in the stop band is 0 ($-\infty$ dB), and there is a very *sharp* discrimination of the two bands. The frequencies ω_1 and ω_2 are called *cut-off frequencies*. From our discussion towards the end of the previous section, we noted that such a sharp *cut-off* takes place only if the order of the filter is arbitrarily high. Further, comparing with the Bode plots of the RLC circuits of the previous section (cf, Exercise 7.15), one might even guess that the selectivity factor should also be quite high, and indeed this is true. Unfortunately, these ideal characteristics cannot be met with *passive networks* alone. We need to make use of *active devices* such as high-gain operational amplifiers to meet the specifications.

Even then, there are several compromises one has to make in designing a filter with satisfactory performance. We depict some of these active filters later in Chapter 12. However, a detailed treatment on filters is beyond the scope of this text. Interested readers may refer to books on operational amplifiers, communication systems, and signal processing.

7.4.1 Other Ideal Filters

We will show few other ideal filters and their characteristics. If the transfer function $H(j\omega)$ appears as in Fig. 7.27(a), we call the circuit as *low pass filter* since it allows low-frequency input signals to pass through. The RLC circuit with the transfer function given in Example 7.9 approximates the low pass characteristic. We may also have a *high pass filter* if the transfer function appears as in Fig. 7.27(b). The RLC circuit with the transfer function in Exercise 7.17 approximates this characteristic. The transfer function shown in Fig. 7.27(c) corresponds to a *band reject filter*, also called the *Notch filter*. A combination of low pass and high pass filters is required to synthesize a band reject filter.

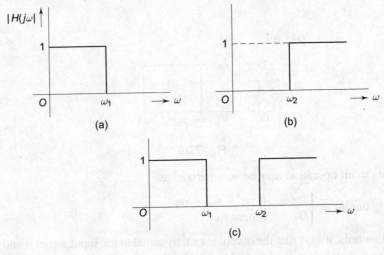

Fig. 7.27

A high Q band pass filter is crucial in as trivial a device as a radio receiver. 'Tuning' to a particular station is choosing the resonant frequency. From this point of view we call the selectivity factor also as *quality factor* since higher the Q, better the reception of the signal by the receiver. It is not an exaggeration to say that present day communication systems and control systems could not have been so rapid had there been no filters.

Summary

In Chapter 6, we have shown how a circuit responds to a simple sinusoidal source of frequency ω and phase angle $0°$. The basic observation was that the response of the circuit will also be a sinusoid, of the same frequency ω. However, the amplitude and phase angle of the response would differ from those of the source. Accordingly, we developed the idea of a phasor, eliminating the trigonometric functions altogether. In this chapter, we have depicted a circuit as an 'operator' that manipulates the input sinusoidal signal and gives an output. This led us to define a transfer function. We have shown the dependency of the amplitude and the phase angle of the output on the input frequency as a Bode plot. The functions are not easily amenable to obtaining the plots, in general. To this end, we have introduced the idea of asymptotes to obtain the Bode plot. We have also shown that the error between the actual plot and the asymptotic plot is negligible. Given any circuit, we would be able to obtain the frequency response by breaking the transfer function into smaller transfer functions, first-order functions and second-order (with $\zeta < 1$) functions in particular. For a first-order transfer function, there is one parameter, the time constant τ, that is of concern to us. Likewise, for a second-order transfer function with complex conjugate roots, we have defined two parameters, ζ and ω_n. We have suggested to take the constant to the end so that we simply shift the magnitude plot. For second-order circuits, we have also introduced the ideas of bandwidth and selectivity factor. In the end, we have briefly introduced frequency-selective circuits, called filters, and their typical magnitude plots. In a later chapter, we will give some more details on these filter circuits.

Problems

7.1 Obtain the transfer function $H(j\omega) = v_R(j\omega)/v_S(j\omega)$ of the circuit in Fig. 7.28.

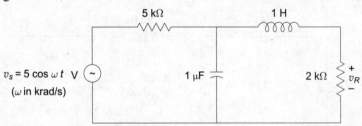

Fig. 7.28

7.2 Obtain the transfer function $H(j\omega) = v_C(j\omega)/i_S(j\omega)$ of the circuit in Fig. 7.29.

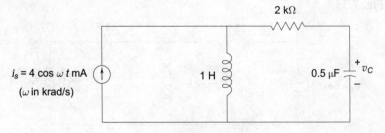

Fig. 7.29

7.3 Obtain the transfer function $H(j\omega) = v_R(j\omega)/v_S(j\omega)$ of the circuit in Fig. 7.30.

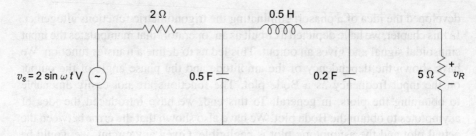

Fig. 7.30

7.4 Obtain the transfer function $H(j\omega) = i_x(j\omega)/v_S(j\omega)$ of the circuit in Fig. 7.31.

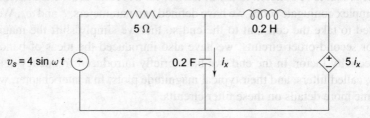

Fig. 7.31

7.5 Obtain the transfer function $H(j\omega) = i_R(j\omega)/v_S(j\omega)$ of the circuit in Fig. 7.32.

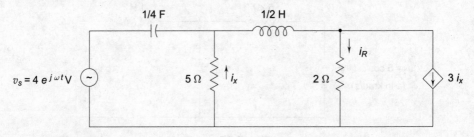

Fig. 7.32

7.6 Obtain the transfer function $H(j\omega) = i_x(j\omega)/v_S(j\omega)$ of the circuit in Fig. 7.33.

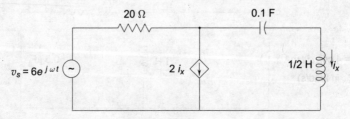

Fig. 7.33

7.7 Obtain the impedance $\mathcal{Z}(j\omega)$ of a series RLC circuit with $R = 1\,\text{k}\Omega$, $L = 10\,\text{mH}$, and $C = 10\,\text{nF}$, and sketch the Bode plot over the frequency range $10^3 \le \omega \le 10^7$ rad/s. Repeat the problem if it were a parallel RLC circuit with the same elements.

7.8 Obtain the transfer function $H(j\omega) = v_R(j\omega)/v_S(j\omega)$ of the circuit in Fig. 7.34, and hence obtain its Bode plot.

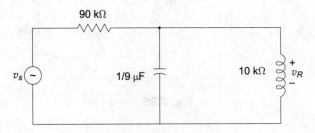

Fig. 7.34

7.9 Repeat Problem 7.8, with the capacitance replaced with a 1.2-mH inductance.

7.10 Given the transfer function

$$H(j\omega) = \frac{1}{1 - \dfrac{\omega^2}{50^2} + j\dfrac{\omega}{125}}$$

compute the frequency at which magnitude $|H(j\omega)| = 1$, and the frequency at which $|H(j\omega)|$ is maximum, and $|H(j\omega)|_{\max}$.

7.11 Obtain the Bode plot of the function

$$H(j\omega) = \frac{10(10 + j\omega)}{100 + j\omega}$$

Show all the details such as corner frequencies, slopes of the asymptotes, phase angles, etc.

7.12 Obtain the Bode plot of the function

$$H(j\omega) = \frac{10^6(10 + j\omega)^2}{(j\omega)^2(10^4 + j\omega)}$$

7.13 Obtain the Bode plot of the function

$$H(j\omega) = \frac{10^3(1 + j\omega)}{(10 + j\omega)(100 + j\omega)}$$

7.14 Obtain the Bode plot of the function

$$H(j\omega) = \frac{25 \times 10^7(90 - \omega^2 + j33\omega)}{j\omega(300 + j\omega)(25 \times 10^6 - \omega^2 + j4 \times 10^4\omega)}$$

7.15 Obtain the Bode plot of the function

$$H(j\omega) = \frac{j10\omega(300 + j\omega)}{(1 + j\omega)(400 - \omega^2 + j6\omega)}$$

7.16 Consider the circuit shown in Fig. 7.35. If $R_1 = 1\,k\Omega$, $R_2 = 1\,\Omega$, $L = 1\,mH$, and $C = 1\,\mu F$, determine the transfer function $H(j\omega) = v_0(j\omega)/i_S(j\omega)$ and hence obtain the Bode plot. What do you observe?

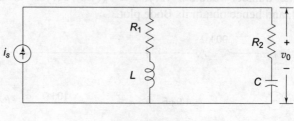

Fig. 7.35

7.17 Consider the circuits shown in Figs 7.36(a) and (b). Obtain the transfer functions $H(j\omega) = i_0/v_S$ in both the circuits and sketch their Bode plots.

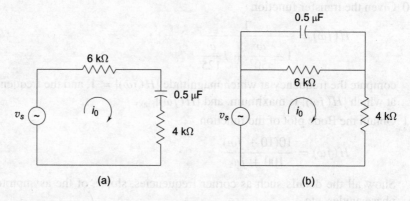

(a)   (b)

Fig. 7.36

7.18 In a series RLC circuit, let $R = 10\,\Omega$, $L = 1\,mH$, and $C = 1\,nF$. If the input is a sinusoidal voltage source with maximum amplitude $V_m = 5\,V$, and the frequency is adjusted until the average power dissipated by R is maximized, determine this power and frequency, as well as the peak voltages across L and C.

7.19 Design a series RLC circuit that satisfies the following specifications: resonant frequency of 100 kHz, bandwidth of 2 kHz, and minimum impedance of $100\,\Omega$. Find the frequencies at which the magnitude of the impedance $|\mathcal{Z}| = 1\,k\Omega$ and the angle $\angle\mathcal{Z} = \pm 90°$.

7.20 A practical source with $v_{oc} = 10\cos(\omega t)$ and internal resistance $R_s = 25\,\Omega$ is connected in series with resistance R, capacitance C, and inductance $L = 100\,\mu H$ with a series resistance of $10\,\Omega$. Compute R and C to make the circuit resonate at 1 MHz, with a bandwidth of 100 kHz.

7.21 Consider the circuit shown in Fig. 7.37.

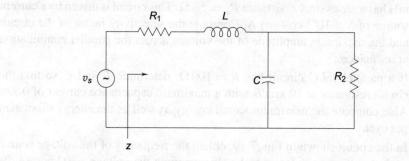

Fig. 7.37

Determine the condition on R_2 under which impedance $\mathcal{Z}(j\omega)$ seen by the voltage source is purely resistive and is equal to

$$\mathcal{Z}(j\omega) = R_1 + \frac{L}{R_2 C}$$

7.22 In a series RLC circuit, if $Q > 0.5$, determine the condition on the frequency under which the peak voltage $|v_{C,\max}|$ across the capacitance is maximum. Also, obtain the expressions for the peak voltage across each element as well as the power factor seen by the source, if $v_S = 10\cos(\omega t)$, $R = 1\,\text{k}\Omega$, $L = 100\,\mu\text{H}$, and $C = 100\,\text{pF}$.

7.23 Design a parallel RLC circuit resonating at a frequency of 10 MHz, a bandwidth of 100 kHz, and a maximum impedance of 100 kΩ.

7.24 For the circuit shown in Fig. 7.38, compute the selectivity factor and bandwidth. If the amplitude of the source is adjusted such that $|v_C| = 1$ V at resonance, determine the current through each of the elements.

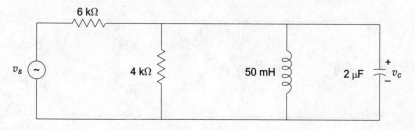

Fig. 7.38

7.25 A current source having $i_{sc} = 10\cos(\omega t)\,\mu\text{A}$ with an internal resistance $R_s = 500\,\text{k}\Omega$ is connected across a parallel RLC circuit with an inductance of 10 mH. Compute R and C so that the circuit resonates at 250 krad/s, with a selectivity factor of 50. What is the bandwidth of the circuit?

7.26 An inductance of 10 mH with a stray series resistance of 50 Ω is connected across a capacitance of $1\,\mu\text{F}$. Obtain the resonant frequency of the circuit and its impedance at resonance.

7.27 In a parallel RLC circuit with $R = 2$ kΩ and $C = 50$ nF, inductance $L = 5$ mH has a series stray resistance $R_s = 50$ Ω. This circuit is driven by a current source of $5 \times 10^{-6} \cos(\omega t)$ A. Compute the selectivity factor of the circuit and the maximum amplitude of the voltage across the parallel combination at resonance.

7.28 If a parallel RLC circuit has $R = 100$ Ω, determine L and C so that the circuit resonates at 10 krad/s with a maximum capacitance current of 0.5A. Also compute the maximum stored energy as well as the energy dissipated per cycle.

7.29 In the circuit shown in Fig. 7.39, obtain the frequency of the voltage source such that the power factor is 1. Also compute the voltage $v_C(t)$ across the capacitance.

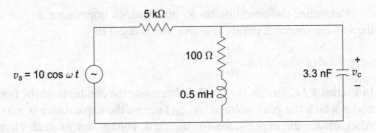

Fig. 7.39

7.30 Consider the circuit shown in Fig. 7.40. Obtain a condition on the resistances R_L and R_C such that the current source sees a power factor of 1.0 for all frequencies. Further, if $L = 12$ mH, $C = 3$ nF, $R = 3$ kΩ, R_L and R_C are according to the above condition, and $i(t) = 5 \times 10^{-3} \cos(\omega t)$, compute the voltage $v_i(t)$ across the current source.

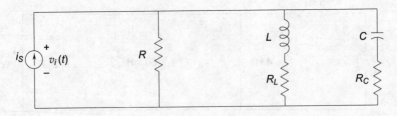

Fig. 7.40

CHAPTER 8

Power Circuits

In this chapter we further discuss the ideas of maximum power transfer and resonance. However, we will not worry about the transfer functions and frequency response. Instead, we characterize the power and energy consumption patterns. The heat generated in a circuit and its dissipation into the surroundings is an important issue, and we ought to minimize this to the possible extent. There is a strong motivation behind this—lowering the electricity bill. This, in turn, motivates us to study how efficiently we can utilize the power. We shall concentrate only on the steady-state response and ignore transients.

8.1 Power and Energy

We will begin with the simple circuit shown in Fig. 8.1.

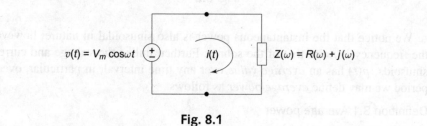

Fig. 8.1

Let us assume that the source

$$v(t) = V_m \cos(\omega t)$$

drives an impedance

$$\mathcal{Z}(\omega) = R(\omega) + jX(\omega)$$

The '+' sign for the second term in $\mathcal{Z}$ has been chosen arbitrarily. In other words, we assume that the current $i(t)$ delivered by the source 'lags' behind the voltage source, i.e.,

$$i(t) = I_m \cos(\omega t - \phi)$$

where

$$\phi = \tan^{-1} \frac{X}{R}$$

We may now compute the power $p(t)$ dissipated by the impedance at any time instant t as

$$p(t) = v(t)i(t)$$
$$= V_m I_m \cos(\omega t) \cos(\omega t - \phi)$$
$$= \frac{1}{2} V_m I_m [\cos \phi + \cos(2\omega t - \phi)] \tag{8.1}$$

This *instantaneous power* is depicted in Fig. 8.2.

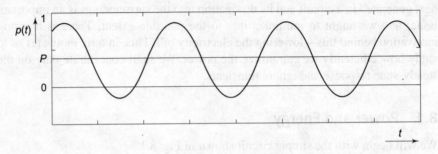

Fig. 8.2

We notice that the instantaneous power is also sinusoidal in nature; however, the frequency is *twice* that of the source. Further, unlike the voltage and current sinusoids, $p(t)$ has an *average value* over any time interval; in particular, over a period we may define *average power* as follows.

Definition 8.1 Average power

$$P = \frac{1}{2} V_m I_m \cos \phi \ \text{W}$$
$$= V_{\text{rms}} I_{\text{rms}} \cos \phi \ \text{W} \tag{8.2}$$

using the relationship between the maximum amplitude and the 'rms' quantity derived in Eqn (6.21) of Chapter 6. We quickly notice that this average power depends upon the amplitudes of the voltage and the current, and their phase difference; it is independent of time. For a given impedance $\mathcal{Z}$, the phase difference ϕ is constant and hence $\cos \phi$ is a constant. The average power is maximum when $\phi = 0°$, i.e., when the source 'sees' a pure resistance connected to it. We now define *apparent power* as follows.

Definition 8.2 Apparent power

$$S = V_{\text{rms}} I_{\text{rms}} \ \text{VA} \tag{8.3}$$

We call this 'apparent' since this is the average power a voltage source can deliver had a pure resistance been connected to it. In other words, this expression is very much similar to that we have when a dc source is connected to a purely resistive circuit. It is simply the product of the readings of a voltmeter and an ammeter connected in the circuit. However, had there been a pure reactance connected to the source, the phase difference is 90° and the average power is zero. Clearly,

$$0 \leq P \leq S \tag{8.4}$$

Notice that we have used two different units—watt and volt-ampere—for the average power and the apparent power, respectively.

Let us now look at the sinusoid in Eqn (8.1). To begin with, if we assume that $\phi = 90°$, we have

$$p(t) = V_{\text{rms}} I_{\text{rms}} \sin(2\omega t)$$

which implies that the reactance *absorbs power* during the first half-cycle and *releases this power back* to the source in the second half-cycle. In other words, 'on an average' no power is dissipated in a reactance. We may now generalize the whole picture and say that

- the resistance part $R(\omega)$ dissipates power on an average, and
- the reactance part $X(\omega)$ does not dissipate any power on an average; it absorbs power during one half-cycle and returns the same back to be source in the next half-cycle.

Let us now obtain some interesting relations. Recall that we are given a source $v(t)$ and an impedance $\mathcal{Z}$ in Fig. 8.1 to start with. Using the generalized Ohm's law, we get the following steps:

$$\mathcal{Z} = R(\omega) + jX(\omega) = |Z|\, e^{j\phi}$$

$$\Rightarrow R = |Z| \cos\phi$$

$$\Rightarrow P = V_{\text{rms}} I_{\text{rms}} \frac{R}{|Z|}$$

$$= I_{\text{rms}}^2 R \ \text{W} \tag{8.5}$$

$$= \frac{(V_{\text{rms}} \cos \phi)^2}{R} \ \text{W} \tag{8.6}$$

We next define another important quantity, the *power factor*.

Definition 8.3 Power factor

$$pf \triangleq \cos \phi \tag{8.7}$$

This power factor plays a very important role as we see ahead. Before that we need to clarify certain ambiguity. Notice that the power factor is the 'cosine' of phase difference. Hence, whether the phase difference is positive or negative, i.e., whether the reactance is inductive or capacitive, the result is same. We shall remove this ambiguity by adopting the following convention.

- If the current 'lags' the voltage in Fig. 8.1, i.e., if $\phi > 0$, we call the corresponding power factor *lagging pf*. The impedance is 'inductive' in this case.
- If the current 'leads' the voltage in Fig. 8.1, i.e., if $\phi < 0$, we call the corresponding power factor *leading pf*. The impedance is 'capacitive' in this case.

Recall that an impedance is a complex number with a 'resistive' real part and a 'reactive' imaginary part. Depending upon the nature of the reactance, it is customary to call the impedance either inductive or capacitive.

Example 8.1 Consider the circuit shown in Fig. 8.3.

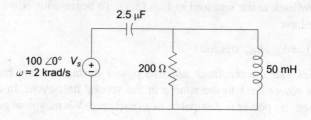

Fig. 8.3

The impedance may be found by inspection as

$$Z = -j200 + (200 \parallel j100) = 40 - j120 = 126.5 \angle -71.56° \ \Omega$$

Clearly, this is a capacitive impedance and

$$pf = \cos(71.56°) = 0.3163 \text{ leading}$$

The current $i(t)$ delivered by the source to the circuit is

$$i(t) = 0.7905 \cos(2000\,t + 71.56°) \ \text{A}$$

Thus, we have

$$P = 12.5 \ \text{W}$$

and

$$S = 39.525 \ \text{V A}$$

Since *energy* is the integral of power, we have

$$w(t) = \int_{-\infty}^{t} p(t)\,dt \tag{8.8}$$

For practical convenience, we assume that the circuit is studied from time $t = 0$ onwards. Accordingly, the lower limit of integration in Eqn (8.8) may be shifted to 0. From the general expression for $p(t)$ in Eqn (8.1), we have

$$w(t) = Pt + \frac{V_{rms}I_{rms}}{2\omega}\left[\sin \phi + \cos(2\,\omega t - \phi - 90°)\right] \ \text{J} \tag{8.9}$$

Exercise 8.1 Repeat Example 8.1 if the source operates at the resonant frequency of the circuit.

Exercise 8.2 Derive Eqn (8.9) by plugging Eqn (8.1) into Eqn (8.8).

A resistance absorbs energy which is dissipated as heat, or converted to some other form of energy. If T is the period of the sinusoids of $v(t)$ and $i(t)$, then

$$\Delta w_R = PT \ \text{J}$$

is the energy absorbed by the resistance $R(\omega)$ in each cycle.

As we mentioned earlier, when $p(t) > 0$ in Eqn (8.1), we interpret it as energy being supplied to the impedance; when $p(t) < 0$, we interpret it as energy being returned to the source. From Eqn (8.9) we see that the energy absorbed by the reactance is a sinusoid and is always positive. However, it is obvious that $w(t)$ returns to zero for every cycle of the voltage and current. This implies that the energy supplied to the reactance is returned completely every cycle. Thus,

$$\Delta w_X \text{ per cycle} = 0$$

Thus, in every cycle, a portion of the energy supplied by the source is dissipated as heat by the resistance and the remaining portion is stored in the reactance. We call this *conservation of energy* and express it as

$$w_{\text{total}} = w_R + w_X$$

Since $p(t) = (d/dt) w(t)$, we also say that there is a *conservation of power*.

In Example 8.1, we notice that only $100 \cos \phi = 31.62\%$ of the apparent power is dissipated as heat in the resistive part of the impedance. What happened to the rest of the power? We shall see this next. As usual, let us first search for some intuitive ideas. We make the following observations from the discussion above.

1. The impedance $R + jX$ can be thought of as a series combination of a resistance and a reactance. Hence, we might assume that the same current $i(t)$ flows through both these elements. But, in the expression for the average power we have accounted only for the resistance $R(\omega)$. Does there exist an expression of the form $I_{\text{rms}}^2 X(\omega)$?

2. Since $P = S \cos \phi$, we cannot but help thinking about a term of the form $Q = S \sin \phi$ so that

$$S = \sqrt{P^2 + Q^2}$$

Let us now investigate in this direction.

8.1.1 Reactive Power and Complex Power

Let us take Eqn (8.1) again and expand the second cosine to get

$$p(t) = V_{\text{rms}} I_{\text{rms}} \left[\cos \phi + \cos(2\omega t) \cos \phi + \sin(2\omega t) \sin \phi \right] \qquad (8.10)$$

Definition 8.4 Reactive Power

$$Q \overset{\Delta}{=} V_{\text{rms}} I_{\text{rms}} \sin \phi \qquad (8.11)$$

Using this definition, we may write

$$p(t) = P[1 + \cos(2\omega t)] + Q \sin(2\omega t)$$

$$\overset{\Delta}{=} p_R(t) + p_X(t) \qquad (8.12)$$

We call $p_R(t)$ the *real instantaneous power*. Notice that it oscillates between 0 and $2P$ and averages to P. From this point of view, we also call P the *real power*. However, $p_X(t)$ averages to 0. The quantity Q is called the *reactive power*, and is measured in volt-ampere-reactive or 'VAR'. Since $\sin \phi$ is involved in the definition of Q, we observe that $Q > 0$ if the impedance is inductive, and $Q < 0$ if the impedance is capacitive. According to the standard convention in power engineering, inductive impedances *consume* reactive power and capacitive impedances *produce* reactive power. Clearly, Q accounts for the energy transfer between the source and the reactance.

Exercise 8.3 What is the relationship between P, Q, and the phase difference ϕ? In other words, given P and ϕ, how do you find Q?

(***Ans.*** $\tan\phi = Q/P$)

Exercise 8.4 Verify that

$$Q = I_{rms}^2 X(\omega)$$

[***Ans.*** Follow Eqns (8.5) and (8.6).]

Example 8.2 Let us consider the circuit in Fig. 8.4. Assuming rms quantities,

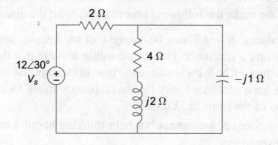

Fig. 8.4

the total impedance seen by the source may be computed to be

$$\mathcal{Z} = 2.474 \angle -25.37° \ \Omega$$

i.e., the sources see a capacitive impedance. The current delivered by the source is

$$\mathcal{I} = 4.85 \angle 55.37° \text{ A (rms)}$$

Thus, the total real power delivered by the source is

$$P_{source} = 12 \times 4.85 \times \cos(-25.37°) = 52.6 \text{ W}$$

We leave it to the reader to verify that this real power delivered by the source is equal to the sum of real powers absorbed by the 2 Ω resistance and the 4 Ω resistance, respectively.

Exercise 8.5 Verify that the network in Example 8.2 'delivers' reactive power to the source.

(***Ans.*** After all $\mathcal{Z}$ is a capacitive impedance, as seen by the source.)

Let us now look at S, P, and Q from a different angle. Observe that all these three have the same dimension, but the units we adopt are different. It is clear that

$$S = \sqrt{P^2 + Q^2}$$

i.e., P, Q, S form a Pythagorean triplet. If we assume that P and Q are the real and imaginary components of a complex variable $\mathcal{S}$, we have

Definition 8.5 Complex power

$$\mathcal{S} \triangleq P + jQ = S \angle \phi \tag{8.13}$$

Notice that the symbol $\mathcal{S}$ is different[1] from that of the apparent power S. We observe that the *magnitude* of complex power equals the apparent power and hence may be measured in V A; otherwise we leave the complex power without assigning any units. Just as we have P and Q in terms of I_{rms}, we may also obtain

$$\mathcal{S} = I_{\text{rms}}^2 \mathcal{Z} \tag{8.14}$$

Further, we observe that

$$I_{\text{rms}}^2 = \mathcal{I}^* \mathcal{I}$$

where

$$\mathcal{I} = I_{\text{rms}} \cos(\omega t - \phi)$$

and $\mathcal{I}^*$ is the complex conjugate of $\mathcal{I}$. Now, using the generalized Ohm's law, we may rewrite Eqn (8.14) as

$$\mathcal{S} = \mathcal{I}^* \mathcal{I} \mathcal{Z}$$

$$= \mathcal{V} \mathcal{I}^* \tag{8.15}$$

Thus, the concept of complex power relates the apparent power S, the real power P, the reactive power Q, and the rms phasors $\mathcal{V}_{\text{rms}}$ and $\mathcal{I}_{\text{rms}}$ in an elegant and succinct mathematical form. Moreover, the complex, real, and reactive powers of the source equal the sum of the complex, real, and reactive powers, respectively, of the individual impedances.

Clearly, it is sufficient to compute the complex powers in each of the elements (including the source) and verify that they are conserved. We emphasize that the apparent power of the source is generally different from the sum of the individual apparent powers.

Thus, with the introduction of Q, everything fits into the bill. We have accounted for the energy consumption. It is clear that the government charges us for the

---

[1] Notationwise, we use $\mathcal{S}$ (read as 'script S') for the complex power and S (read as 'capital S') for apparent power.

complex power, i.e., the apparent power while we consume only the average power dissipated by the resistive components of our household gadgets.

Example 8.3 Consider the circuit given in Fig. 8.5.

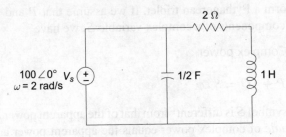

Fig. 8.5

Since we are interested in power calculations, we prefer to work with rms quantities. The source 'sees' an impedance of

$$\mathcal{Z} = -j1 \parallel (2 + j2) = 0.4 - j1.2\,\Omega$$

i.e., a capacitive impedance. The current delivered by the source is

$$\mathcal{I} = \frac{\mathcal{V}}{\mathcal{Z}} = 25\sqrt{10}\,\angle\,72^\circ \; \text{A}$$

Thus, the current 'leads' the voltage. The complex power $\mathcal{S}$ delivered by the source may be computed as

$$\mathcal{S} = \mathcal{V}\mathcal{I}^* = 2500\,(1 - j3)$$

Let us now see the complex power consumed by each of the individual elements. Since the same voltage appears across the parallel combination, we have

$$\mathcal{I}_C = j1\mathcal{V} = 100\,\angle\,90^\circ \; \text{A}$$

$$\mathcal{I}_R = \mathcal{I}_L = \frac{\mathcal{V}}{2 + j2} = 25\sqrt{2}\,\angle\,-45^\circ$$

Consequently, we get

$$\mathcal{S}_C = \mathcal{V}\mathcal{I}_C^* = 10^4\,\angle\,-90^\circ$$

$$\mathcal{S}_R = \mathcal{I}_R^*\mathcal{I}_R 2 = 2500$$

$$\mathcal{S}_L = \mathcal{I}_L^*\mathcal{I}_L j2 = 2500\,\angle\,90^\circ$$

We leave it to the reader to verify that

$$\mathcal{S} = \mathcal{S}_C + \mathcal{S}_R + \mathcal{S}_L$$

Exercise 8.6 Verify the results of Example 8.3. In particular, verify the conservation of the real and the reactive powers. What about the apparent powers?

Example 8.4 Consider the circuit in Fig. 8.6.

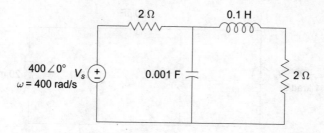

Fig. 8.6

The source 'sees' an impedance of

$$Z = \left[(2 + j4) \parallel \frac{1}{j0.4}\right] + 2 = 4\sqrt{2}\angle - 45°$$

and hence delivers a current of

$$I = 50\sqrt{2}(1 + j1) \text{ A}$$

The complex power delivered by the source may be computed as

$$S = (20 - j20) \times 10^3$$

The average power absorbed by the impedance is

$$P = 400 \times 50\sqrt{2} \times \cos(+45°) = 20 \text{ kW}$$

Since it is a capacitive impedance, the power factor is

$$pf = \frac{1}{\sqrt{2}} \text{ leading}$$

Exercise 8.7 Obtain the average power dissipated by each of the elements and the corresponding power factors in the circuit of Fig. 8.5. Verify that the complex power is conserved.

Exercise 8.8 Verify the conservation of complex power for the circuit shown in Fig. 8.7.

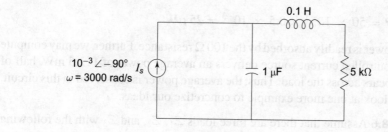

Fig. 8.7

Exercise 8.9 Repeat Exercise 8.7 for the circuit shown in Fig. 8.8.

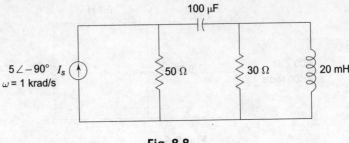

Fig. 8.8

Example 8.5 Consider the circuit shown in Fig. 8.9.

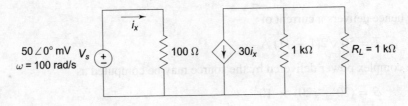

Fig. 8.9

It is trivial to calculate that

$$i_x = 0.5 \, \text{mA}$$

and

$$i_L = -\frac{1}{2} \, 30 \, i_x$$

Clearly, the average power absorbed by the load is

$$P_{\text{load}} = 56.25 \, \text{mW}$$

The voltage source at the far left delivers an average power of

$$P = 50 \times 10^{-3} \times 0.5 \times 10^{-3} = 25 \, \mu\text{W}$$

and this power is readily absorbed by the 100 Ω resistance. Further, we may compute that the controlled current source delivers an average power of 112.5 mW, half of which appears across the load. Thus, the average power is conserved in this circuit.

Let us look at one more example to concretize our ideas.

Example 8.6 Assume that there are three loads Z_1, Z_2, and Z_3 with the following properties:

- load Z_1 absorbs 100 kW and 40 kVAR,
- load Z_2 absorbs 25 kV A at $pf = 0.9$ leading,
- load Z_3 absorbs 80 kW at $pf = 1$.

Let us now determine the equivalent impedance and the power factor if these three loads are connected in parallel across a 240 V (rms) source.

Recall from Exercise 8.3 that

$$\frac{X(\omega)}{R(\omega)} = \tan\phi = \frac{Q}{P}$$

We shall use this relationship. Since the load $\mathcal{Z}_1$ absorbs $P = 100\,\text{kW}$ and $Q = 40$ kVAR, we have

$$\text{phase difference } \phi = \tan^{-1}\frac{40,000}{100,000} = 21.84°$$

From the relationship among V_{rms}, $|Z|$, and the power factor we have

$$P = \frac{V_{\text{rms}}^2}{|Z|}\cos\phi = 100 \times 10^3$$

$$\Rightarrow |Z| = 0.5346$$

and

$$\mathcal{Z} = 0.5346\,\angle\,21.8°\,\Omega$$

Similarly, we may obtain

$$\mathcal{Z}_2 = 1.004\,\angle -25.84°\,\Omega$$

and

$$\mathcal{Z}_3 = 0.72\,\angle\,0°\,\Omega$$

The parallel combination of these three impedances may be found to be

$$\mathcal{Z}_1 \parallel \mathcal{Z}_2 \parallel \mathcal{Z}_3 = 0.247 - j0.016\,\Omega = 0.2475\,\angle -3.7°$$

and the source sees an overall *pf* of 0.997, fairly close to unity.

8.1.2 Optimizing Power Transfer

We shall now discuss an important idea in optmizing power transfer. We shall study the following example first and theoretize the ideas after that.

Example 8.7 Consider the circuit shown in Fig. 8.10.

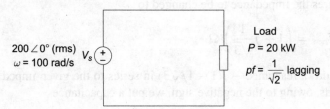

$200\,\angle\,0°$ (rms) V_s
$\omega = 100$ rad/s

Load
$P = 20$ kW
$pf = \dfrac{1}{\sqrt{2}}$ lagging

Fig. 8.10

Given that the power factor is $1/\sqrt{2}$ lagging, i.e., $\phi = +45°$, we may quickly identify that it is an inductive impedance of the form

$$\mathcal{Z} = R(1 + j1)\ \Omega$$

so that

$$\angle\, \mathcal{Z} = \tan^{-1} \frac{+R}{+R} = 45°$$

Given the source and the average power dissipated by the impedance, we have

$$P = \frac{(V \cos \phi)^2}{R}$$

$$\Rightarrow\ R = 1\ \Omega$$

Thus, we have

$$\mathcal{Z} = 1 + j1\ \Omega \quad \text{or} \quad \mathcal{Y} = 0.5 - j0.5\ \Omega^{-1}$$

Suppose we now ask the following question.

- Can we modify the circuit so that the same source, operating at the same frequency, delivers a *higher* average power than before?

From the expression for the average power P in Eqn (8.6), intuitively we find that the average power increases if $\cos \phi / |\mathcal{Z}|$ increases for the same old V_{rms}. First let us look at increasing $\cos \phi$. This happens when the phase difference ϕ decreases. Remembering that

$$\phi = \tan^{-1} \frac{X}{R}$$

we notice that ϕ decreases (i) either by 'increasing' the real part R of the impedance, (ii) or by 'decreasing' the imaginary part X. We rule out the former because it increases $|Z|$ rather sharply, and hence decreases the average power. In the latter case, decreasing X increases the power factor and, at the same time, decreases $|Z|$ and our purpose is served. Let us explore further in this direction.

Suppose, in the above example we wish to raise the power factor from $1/\sqrt{2}$ lagging to $\sqrt{3}/2$ lagging, i.e., we wish to bring down the phase difference to $30°$. This requires the impedance to be changed to

$$\mathcal{Z}_{\text{new}} = \left(1 + j\frac{1}{\sqrt{3}}\right)\ \Omega$$

i.e., we 'add' a reactance $-j(1 - 1/\sqrt{3})$ in series to the given impedance $\mathcal{Z}$. In other words, owing to the negative sign, we put a capacitance

$$C = \frac{\sqrt{3}}{\sqrt{3} - 1} \frac{1}{\omega}\Big|_{\omega=100\ \text{rad/s}} = 23.7\ \text{mF}$$

We find that $|\mathcal{Z}|_{\text{new}} = 2/\sqrt{3}$ and hence,

$$P = \frac{(V_{\text{rms}})^2 \cos\phi}{|Z|} = 30\,\text{kW}$$

An astute reader would have noticed that we are doing something close to the idea of impedance matching for maximum power transfer. For instance, if we have connected a capacitive reactance of $-j1\,\Omega$ in series, the source would have 'seen' a pure resistance. Consequently, the power factor would have become unity and the average power would have been 40 kW, which is greater than 30 kW above. However, there is a subtle difference. Here we are not exactly looking at some load 'as seen by' a Thévenin's equivalent network; rather we are forcing the circuit to 'resonate' at the given frequency by nullifying the reactance part of the impedance.

Proceeding on similar lines as above, we may determine an appropriate element to be connected in series with the given impedance so that

- the power factor is unity where the phase difference is brought down to zero, or
- the power factor is leading where the phase difference is reduced further and made negative.

What exactly do we gain by playing with the power factor? From Eqn (8.14), we have

$$S = P + jQ = I_{\text{rms}}^2 \mathcal{Z}$$

Put in another way, we notice that

$$\tan^{-1}\frac{X}{R} = \phi = \tan^{-1}\frac{Q}{P}$$

i.e., the phase difference ϕ is a 'measure' of the relative size of the reactive power over the average power. Clearly, by making the power factor close to unity we are reducing the size of Q and we pay only for what we actually consume; transfer of energy back and forth, now minimized, is no longer of concern to us. This process is called *power factor correction*. It is easy to see that reducing Q reduces the product $V_{\text{rms}} I_{\text{rms}}$ so that, for a fixed terminal voltage $\mathcal{V}$, the current drawn by the impedance is reduced. In fact, the reader would have noticed that on the electrical gadgets such as music systems and water heaters there would be a 'rated current' indicated. Our domestic ac outlets provide a sinusoidal voltage of 220 V at 100π rad/s (= 50 Hz); and we have choice of plug sockets, either 5 A for low-power appliances such as music systems, televisions, and personal computers, or 15 A for high-power appliances such as refrigerators and air-conditioners.

Here again, we make a comparison with the maximum power transfer theorem. It requires that load impedance $\mathcal{Z}_L$ be the 'complex conjugate' of the source impedance $\mathcal{Z}_s$. In this case, the terminal voltage is halved and the maximum average power is $P = V_{\text{rms}}^2/4R$. The source impedance consumes half of the rated power and we say that the *efficiency of power transfer* is just 50%. Such an efficiency of transmission may be permitted in low-power applications such as communications, but is of little use in power engineering. Here, more or less, we discard the maximum power transfer theorem and settle with power factor correction.

Exercise 8.10 Repeat Example 8.5 for the circuit shown in Fig. 8.11.

$200 \angle 0°$ (rms) V_s
$\omega = 100$ rad/s

Load
$S = 10$ kV A
$pf = 0.8$ lagging

Fig. 8.11

The power factor correction suggested in Example 8.5, although theoretically simple and appealing, has certain interesting issues in terms of practical connections. Adding a new element in series requires breaking the circuit, which is often impracticable. It would be more convenient if we could simply 'plug-in' the new reactance across the existing impedance. The principle is the same, but we need to work with admittance rather than with impedance. We illustrate this in the following example.

Example 8.8 Suppose we apply the household ac voltage to a load of $\mathcal{Z} = 100 + j300 \ \Omega$. We notice that this load has a power factor of 0.316 lagging. In admittance terms, the load may be expressed as

$$\mathcal{Y} = 10^{-3} - j3 \times 10^{-3} \ \Omega^{-1}$$

To achieve unity power factor, we need to nullify the susceptance component. In other words, we need to 'plug-in' a susceptance

$$B_p = +j 3 \times 10^{-3} \ \Omega^{-1}$$

across the load. For the household ac voltage this susceptance would result in a capacitance equal to 9.55 μF.

In terms of this parallel connection of correcting elements across the load, we have yet another interesting aspect. The source now supplies only the real power which is identical to the average power absorbed by the load; the exchange of reactive power takes place between the load and the new element.

Most industrial loads may be classified as being inductive and hence have lagging power factors. These may be 'corrected' by making use of a suitable bank of capacitors. However, homes have power factors approaching unity and hence do not require power factor correction[2].

In practical situations it is enough to make the power factor *close* to unity. Moreover, the load is specified in terms of 'impedance' rather than in terms of 'admittance'. Thus, it is a good idea to find a direct expression for the reactance to be plugged across the load. We leave it to the reader to verify that

$$X_p = \frac{V_{rms}^2/P}{\tan(\pm \cos^{-1} pf_{new}) - \tan(\pm \cos^{-1} pf_{old})} \tag{8.16}$$

where P is the average power, and pf_{old} and pf_{new} are, respectively, the values of the power factor before and after correction.

Exercise 8.11 Derive Eqn (8.16).

Example 8.9 Suppose a load is powered with the domestic ac voltage 220 V (rms) at 50 Hz, and it consumes 50 kW with $pf = 0.75$ lagging. Let us find a suitable correction element to ensure $pf = 0.95$ lagging. From Eqn (8.16), we readily find that

$$X_p = -1.75\,\Omega$$

and hence we need to plug-in a capacitance of 1.82 mF across the load. Since the requirement is in millifarads, a single capacitance may not be available. Hence, we need to make use of a *bank of capacitors*. We also notice that

$$I_{rms} = \frac{220}{1.75} = 125.7\,A$$

is the current through the capacitor bank. This current rating has to be kept in mind while choosing the capacitors for the bank.

Measurement of power is slightly tricky owing to several definitions we had earlier. We might use a multimeter at the terminals of an impedance and measure the rms quantities—V_{rms} and I_{rms}. However, if we simply multiply these two, we get the apparent power measured in volt-amperes. To obtain the real average power consumed by the load, we need to multiply with the power factor also. To this end we have instruments such as the *electrodynamometer movement*, where the mechanical inertia of the system provides the *time-average* and hence the average power. Such an instrument is called a *wattmeter*. We encourage the reader to refer to a book on measurement theory to gain more insight into power measurements.

---

[2] It should be noted that, at present, there is a propensity to use gadgets such as pumps, inverters, etc. in common households, and these gadgets are inductive in nature. It is for this reason, commercial advertisements of these, relatively expensive, gadgets highlight power consumption more than anything else.

8.2 Polyphase Circuits and Power Systems

So far we have seen ac circuits with (effectively) a single source. The sinusoidal nature of the voltages and the currents indicate that they go to zero *twice* in a cycle. Any device, say a motor, powered by such a source would experience an oscillatory torque that disappears twice in a cycle. Fortunately, these electromechanical devices have an inertia and hence the jerky operation is avoided. Nevertheless, it is always desirable to run these devices with *uniform* torque. This is achieved by a clever arrangement of *three* sources as follows. Suppose we have three sources:

$$v_a(t) = V_m \cos(\omega t)$$

$$v_b(t) = V_m \cos(\omega t - 120°)$$

$$v_c(t) = V_m \cos(\omega t + 120°) \tag{8.17}$$

Notice that all the three sources have the same amplitude and frequency, but have different phase angles. Let us now connect these sources across three resistances as shown in Fig. 8.12. The phase angles intuitively suggest the connection, wherein the angle between two adjacent branches is *geometrically* 120°.

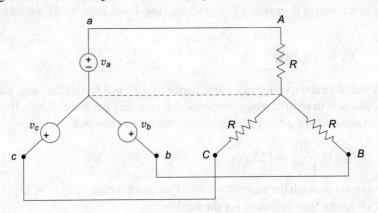

Fig. 8.12

We call the load a *balanced load* since it consists of three *identical* resistances. The individual power components *at any instant of time* may be found to be

$$p_a(t) = v_a^2(t)/R, \quad p_b(t) = v_b^2(t)/R, \quad p_c(t) = v_c^2(t)/R$$

and are, as might be expected, oscillatory with frequency 2ω between 0 and V_{rms}^2/R. Let us add these individual components, *at any instant of time*. We get

$$p(t) = p_a(t) + p_b(t) + p_c(t) = 3\frac{V_{rms}^2}{R} \tag{8.18}$$

This result is rather surprising to us—the total power is a constant *at all times* and is equal to three times the individual averages. As one might easily verify, this result holds for any general *balanced load*, i.e., three identical impedances connected as shown in Fig. 8.12. This arrangement is called a *balanced three-phase power system*. Since the total instantaneous power is a constant, loads driven by such an arrangement of three-phase sources experience a uniform force and hence operate more smoothly and efficiently. This idea is not restricted to electrical systems alone; we prefer automobiles, particularly bikes, with multiple (two or more) cylinders providing a uniform driving force. We may look at the following intuitive example to understand the idea.

Suppose we have a small toy-wheel with 10 blades centered on an axis[3]. Let us number the blades from 1 to 10. Let us now rotate the wheel in the following, say, three ways.

- We may apply enough torque on blade 1 to set the wheel into motion, wait for three complete rotations; then apply the same torque on blade 1 again, wait for three more complete rotations; and so on. We may observe that at the end of the first rotation, the speed of the wheel would have come down; at the end of the second rotation, the speed would have further reduced, and, possibly, at the end of the third rotation the wheel might come to rest. This may be thought of as torque being applied with a delay of $3 \times 2\pi = 6\pi$ rad.

- Let us modify our pattern and apply the same torque to blade 1, but now at the end of one complete rotation. It is easy to see that, in this case, the wheel rotates more uniformly than in the previous case. In this case the torque is being applied with a delay of 2π rad.

- Let us now apply the same torque at the end of one-half rotation, i.e., to blade 1, then to blade 6, to blade 1 again, and so on with a delay of π radians. The rotation would be even more uniform than in the previous case.

Of course, we may asymptotically decrease the delay in the application of torque and observe that the wheel is rotating, virtually, forever at a constant speed, provided the speed does not increase exponentially and the whole system collapses. The polyphase systems work in an analogous way. We encourage the reader to brainstorm on several natural phenomena, for example, the merry-go-round, around us that fit into this scheme.

From an efficiency point of view virtually all bulk electric power is generated, transmitted, and distributed in three-phase form, although domestic consumption is generally one-phase. It is not uncommon these days to have three-phase supply even in homes to handle modern gadgets such as geysers, water pumps, etc. We will spend some time to study these three-phase systems in more detail. We talk about voltage sources, and three-phase current sources are seldom found in practice.

8.2.1 Three-phase Sources

A *three-phase source* consists of a stator with three distinct and symmetrically distributed windings around its periphery, and a rotor electromagnet driven at a synchronous speed by a gas turbine, or a steam turbine, or a hydraulic turbine, or a diesel engine. As it rotates, the electromagnet induces sinusoidal voltage called *phase voltage* in each of the windings. The three voltages are equal in amplitude and frequency, but naturally differ in phase angle by 120° successively.

---

[3] Imagine something like a potter's wheel, or a child's play with a bicycle on stand!

The entire setup is said to form a *balanced voltage set*. There are two possible ways of interconnecting the windings—(i) the Y (wye) or star connection shown in Fig. 8.13(a), and (ii) the Δ (delta) connection shown in Fig. 8.13(b).

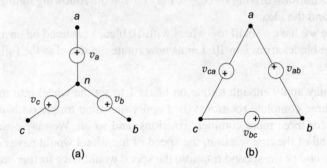

Fig. 8.13

In the Y connection, there is a common node n called the *neutral*. In this connection, we may express the individual voltages as

$$V_{an} = V_{\text{rms}} \angle 0°$$
$$V_{bn} = V_{\text{rms}} \angle -120°$$
$$V_{cn} = V_{\text{rms}} \angle +120°$$

Notice that we use double subscripts to denote the potential difference of a particular winding with respect to the neutral. And, as before, we prefer to work with rms quantities, expressed in the polar form.

Let us quickly do some useful manipulations. The first one is to observe that

$$v_a + v_b + v_c = V_{an} + V_{bn} + V_{cn} = 0 \tag{8.19}$$

i.e., the sum of the instantaneous phase voltages of a balanced source is zero at all times; that is why we use the word *balanced*.

Next, we shall look at the voltage difference between the power lines, i.e.,

$$V_{ab} = V_{an} - V_{bn} = \sqrt{3}\, V_{\text{rms}} \angle 30°$$
$$V_{bc} = V_{bn} - V_{cn} = \sqrt{3}\, V_{\text{rms}} \angle -90°$$
$$V_{ca} = V_{cn} - V_{an} = \sqrt{3}\, V_{\text{rms}} \angle 150° \tag{8.20}$$

We call these *line voltages*. We might as well express these line voltages in terms of the phase voltages as

$$V_{ab} = (\sqrt{3}\angle 30°)V_{an}$$
$$V_{bc} = (\sqrt{3}\angle 30°)V_{bn}$$
$$V_{ca} = (\sqrt{3}\angle 30°)V_{cn} \tag{8.21}$$

i.e., the line voltage set *leads* the phase voltage set by 30°.

Finally, we see that the line voltages also form a balanced set, i.e.,

$$V_{ab} + V_{bc} + V_{ca} = 0 \tag{8.22}$$

Exercise 8.12 Verify Eqns (8.19), (8.20), (8.21), and (8.22).

Notice that the delta connection of Fig. 8.13(b) does not have a neutral node. Owing to this reason the delta connection is generally not used in practice since a slight imbalance in the voltage set would result in an undesirable internal current around the delta loop. However, from a theoretical point of view, a delta connection may be considered as an equivalent of a star connection whose phase voltages are given by the lines from the vertices to the centre of the equilateral triangle having the phase voltages of the delta connection as its sides. In other words, we may transform balanced three-phase sources from Y to Δ, or vice versa, without affecting the load currents or voltages.

8.2.2 Three-phase Loads

In Fig. 8.12 we have shown a *balanced load* consisting of three identical resistances. The reader would have readily identified that it was a star connection. In general, a *three-phase load* consists of three impedances which can be connected either in the star fashion or in the delta fashion. These are shown in Fig. 8.14.

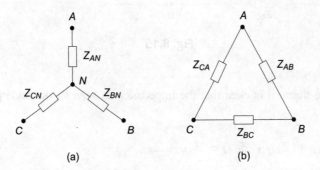

Fig. 8.14

As with the sources we conform to clockwise enumeration of the nodes; however, we use capital letters to avoid ambiguity. The voltages across the impedances are called *phase voltages* and the currents through the impedances are called *phase currents*, regardless of the load type; for instance, the phase voltages of a star connected load are V_{AN}, V_{BN}, and V_{CN} while the phase currents of a delta connected load are I_{AB}, I_{BC}, and I_{CA}. We say that the load is *balanced* if all the three impedances are identical, i.e., if

$$Z_{AN} = Z_{BN} = Z_{CN} = Z_Y \qquad (8.23)$$

for the star connection, and

$$Z_{AB} = Z_{BC} = Z_{CA} = Z_\Delta \qquad (8.24)$$

for the delta connection. Observe that the equality is for the complex quantities, meaning that the magnitude and phase are identical for all the three impedances.

8.2.2.1 *Star–delta conversion*

It is helpful if we perform equivalence transformations between these two types of loads, i.e., given a delta type load we will determine an equivalent star type load that exhibits the same terminal behaviour. We suggest Fig. 8.15 to gain an intuition.

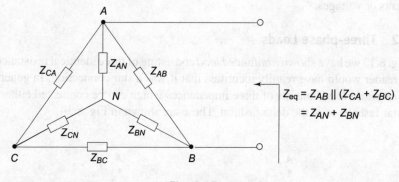

Fig. 8.15

From the figure it is clear that the impedance *as seen by* the node pair AB is given by

$$Z_{AB} \parallel (Z_{BC} + Z_{CA}) = Z_{AN} + Z_{BN} \qquad (8.25)$$

since Z_{CN} is floating and not a part of the circuit. Likewise, we get two more equations as follows *looking into* the node pairs BC and CA, respectively:

$$Z_{BC} \parallel (Z_{CA} + Z_{AB}) = Z_{BN} + Z_{CN} \qquad (8.26)$$

$$Z_{CA} \parallel (Z_{AB} + Z_{BC}) = Z_{CN} + Z_{AN} \qquad (8.27)$$

Solving the system of these three equations we get the *star-to-delta* transformation as follows:

$$Z_{AB} = \frac{Z_{AN}Z_{BN} + Z_{BN}Z_{CN} + Z_{CN}Z_{AN}}{Z_{CN}} \tag{8.28}$$

$$Z_{BC} = \frac{Z_{AN}Z_{BN} + Z_{BN}Z_{CN} + Z_{CN}Z_{AN}}{Z_{AN}} \tag{8.29}$$

$$Z_{CA} = \frac{Z_{AN}Z_{BN} + Z_{BN}Z_{CN} + Z_{CN}Z_{AN}}{Z_{BN}} \tag{8.30}$$

Exercise 8.13 Derive Eqns (8.28), (8.29), and (8.30).

We give the following rule of thumb for the star-to-delta connection in Eqns (8.28), (8.29), and (8.30):

- The numerator is common for all the three equations—sum of the products of the three impedances taken two at a time.
- Looking at Fig. 8.14, the delta connection is enumerated by the sides of an equilateral triangle and the star connection is enumerated by the lines joining the vertices of the triangle to its centre N. The denominator in each of the equations is the impedance connecting the 'opposite' vertex to the neutral at the centre.

8.2.2.2 Delta–star conversion

Exercise 8.14 Obtain the delta-to-star transformation using Eqns (8.25), (8.26), and (8.27). What is the rule of thumb for this transformation?
(*Ans.*

$$Z_{AN} = \frac{Z_{CA}Z_{AB}}{Z_{AB} + Z_{BC} + Z_{CA}}$$

$$Z_{BN} = \frac{Z_{AB}Z_{BC}}{Z_{AB} + Z_{BC} + Z_{CA}}$$

$$Z_{CN} = \frac{Z_{BC}Z_{CA}}{Z_{AB} + Z_{BC} + Z_{CA}}$$

The denominator is common for all the three equations—sum of the three delta impedances. The numerator is the product of impedances that have a common subscript along with the star branch impedance under consideration; for Z_{BN}, for instance, we eschew Z_{CA} in the numerator.)

For balanced loads, these transformations simplify to

$$Z_Y = Z_{AN} = Z_{BN} = Z_{CN} = \frac{1}{3} Z_\Delta \qquad (8.31)$$

$$Z_\Delta = Z_{AB} = Z_{BC} = Z_{CA} = 3 Z_Y \qquad (8.32)$$

Example 8.10 Let us obtain the equivalent resistance of the network as seen by the node pair AB in Fig. 8.16(a).

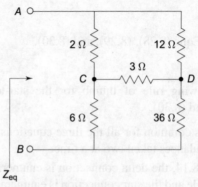

Fig. 8.16(a)

Based on our previous experience with Thévenin's equivalents we might be prompted to apply a test voltage source and measure the current using nodal or mesh analysis. However, an ardent reader would have identified a star as well as a delta connection hidden. Let us then apply delta-to-star transformation to the upper half of the network to obtain the equivalent of Fig. 8.16(b). We shall use the results of Exercise 8.14 to get

$$R_{AN} = \frac{2 \times 12}{2 + 12 + 3} = \frac{24}{17} \, \Omega$$

$$R_{DN} = \frac{12 \times 3}{2 + 12 + 3} = \frac{36}{17} \, \Omega$$

$$R_{CN} = \frac{3 \times 2}{2 + 12 + 3} = \frac{6}{17} \, \Omega$$

Fig. 8.16(b)

We may now apply the simple series/parallel reductions to obtain

$$R_{eq} = R_{AN} + [(R_{BC} + R_{CN}) \| (R_{BD} + R_{DN})] = \frac{48}{7} \, \Omega$$

Exercise 8.15 Verify Example 8.10.

Exercise 8.16 Redo Example 8.10 by (i) transforming the lower half of the network in Fig. 8.16(a) into a star and then applying the series/parallel reductions, and (ii) transforming the star connection A-B-D into a delta connection and then applying the series/parallel reductions.

Exercise 8.17 Repeat Example 8.10 for the network shown in Fig. 8.17.

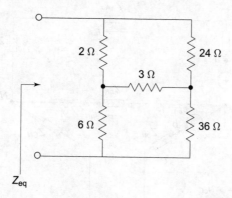

Fig. 8.17

(*Ans.* $R_{eq} = 7\,\Omega$)

Exercise 8.18 Obtain the equivalent impedance of the network shown in Fig. 8.18.

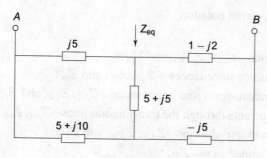

Fig. 8.18

(*Ans.* $j1\,\Omega$)

8.3 Power Systems

When a three-phase source and a (three-phase) load are connected together, we have a *power system*. Clearly, we have four possible power systems—the Y–Y, the Y–Δ, the Δ–Y, and the Δ–Δ. Of these we ignore the last two since we are not interested in Δ connected sources from a practical point of view.

8.3.1 The Y–Y Power System

Let us first study the Y connected source driving a Y connected load. This power system is shown in Fig. 8.19.

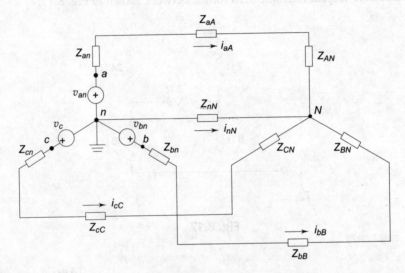

Fig. 8.19

We use the following notation:

- per-phase sources—$\mathcal{V}_{an}$, $\mathcal{V}_{bn}$, and $\mathcal{V}_{cn}$;
- per-phase source impedances—$\mathcal{Z}_{an}$, $\mathcal{Z}_{bn}$, and $\mathcal{Z}_{cn}$;
- per-phase transmission line impedances—$\mathcal{Z}_{aA}$, $\mathcal{Z}_{bB}$, and $\mathcal{Z}_{cC}$;
- per-phase currents through the transmission lines—$\mathcal{I}_{aA}$, $\mathcal{I}_{bB}$, and $\mathcal{I}_{cC}$;
- per-phase load impedances—$\mathcal{Z}_{AN}$, $\mathcal{Z}_{BN}$, and $\mathcal{Z}_{CN}$;
- neutral line impedance—$\mathcal{Z}_{Nn}$

The transmission line impedances account for the actual wires as well as the winding impedances of the step-up and step-down transformers that are necessary at the source and load ends. The currents are enumerated using double subscript notation; for instance, $\mathcal{I}_{cC}$ suggests that the current flows from node 'c' to node 'C'.

We make use of nodal analysis by assuming $V_n = 0$ V. We have

$$\frac{V_{an} - V_N}{Z_{an} + Z_{aA} + Z_{AN}} + \frac{V_{bn} - V_N}{Z_{bn} + Z_{bB} + Z_{BN}}$$

$$+ \frac{V_{cn} - V_N}{Z_{cn} + Z_{cC} + Z_{CN}} + \frac{V_n - V_N}{Z_{Nn}} = 0$$

(8.33)

Let us proceed further assuming that the load is a balanced one and the source and transmission line impedances are all identical. We define per-phase impedance as

$$Z_\phi \overset{\Delta}{=} Z_{\text{source}} + Z_{\text{line}} + Z_Y$$

Recall from Eqn (8.19) that

$$V_{an} + V_{bn} + V_{cn} = 0$$

Thus, Eqn (8.33) simplifies to

$$V_N \left(\frac{3}{Z_\phi} + \frac{1}{Z_{Nn}} \right) = 0$$

which implies that

$$V_N = 0 \text{ and } I_{nN} = 0$$

(8.34)

i.e., in a balanced Y–Y power system the neutral line does not carry any current and two neutral nodes of the source and the load are at zero potential. If the system is perfectly balanced, we may remove the neutral wire completely saving its cost; however, in practice there might be a slight imbalance in which case a very small amount of current flows through the neutral wire, justifying the use of a cheaper neutral wire in practice.

Since $I_{nN} = 0$, we might suspect that

$$I_{aA} + I_{bB} + I_{cC} = 0$$

Indeed this is true since the per-phase currents are

$$I_{aA} = \frac{V_a}{Z_\phi}$$

$$I_{bB} = \frac{V_b}{Z_\phi}$$

$$I_{cC} = \frac{V_c}{Z_\phi}$$

(8.35)

It is now very clear that the analysis of a *balanced* power system can be significantly simplified. For instance, it is sufficient to carry out detailed analysis for only one phase, say, the a-A phase using the equivalent circuit of Fig. 8.20.

Once this a-A phase analysis is over, we may adapt these results for b-B and c-C phases by shifting the phase angle by $-120°$ and $+120°$, respectively.

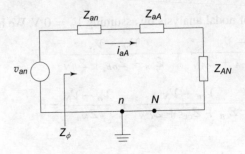

Fig. 8.20

We remark that in the single-phase equivalent of Fig. 8.20 we have shown the connection between the two neutrals n and N with a perfect conductor to close the circuit, i.e., we show that the per-phase current $\mathcal{I}_{aA}$ flows through this. This might cause some concern since we are apparently deviating from the basic idea. However, in a three-phase system the three per-phase currents would be flowing through this conductor simultaneously; since these currents add up to zero, we need not bother about the conductor in the single-phase equivalent circuit.

Example 8.11 In a balanced three-phase Y–Y system we have the following:

- per-phase source—$V_{*n} = 120$ V, rms
- per-phase source impedance—$\mathcal{Z}_{source} = 0.1 + j0.2\ \Omega$,
- per-phase transmission line impedance—$\mathcal{Z}_{line} = 0.5 + j1\ \Omega$,
- per-phase load impedance—$\mathcal{Z}_Y = 15 + j10\ \Omega$

We quickly compute that the phase impedance is

$$\mathcal{Z}_\phi = 15.6 + j11.2\ \Omega$$

Assuming that $v_a = 120\ \angle 0°$, we get

$$\mathcal{I}_{aA} = \frac{120}{\mathcal{Z}_\phi} = 6.25 \angle -35.7°\ \text{A}$$

It is easy to observe that

$$\mathcal{I}_{bB} = \mathcal{I}_{aA} \times 1\angle -120°\ \text{A}$$

and

$$\mathcal{I}_{cC} = \mathcal{I}_{aA} \times 1\angle +120°\ \text{A}$$

Using these currents we may obtain the phase voltages at the load as

$$V_{AN} = \mathcal{I}_{aA} \times \mathcal{Z}_Y = 112.6 \angle -2°\ \text{V}$$

etc.

Similarly, the line voltages at the load are

$$V_{AB} = \sqrt{3}\angle 30° \, V_{AN} = 195 \angle 28° \text{ V}$$

etc.

We suggest the reader to verify the answers by computing

$$\mathcal{I}_{aA} + \mathcal{I}_{bB} + \mathcal{I}_{cC}$$

and

$$V_{AB} + V_{BC} + V_{CA}$$

Exercise 8.19 Verify the results of Example 8.11.

Exercise 8.20 Given the balanced $Y-Y$ power system of Example 8.9, if it is required to have the phase voltage at the load as $V_{AN} = 120 \angle 0°$, what must be the per-phase source voltages v_a, v_b, v_c?

Before proceeding to consider the Δ load power system, we make the following comment with respect to an unbalanced Y load. In practice, the two neutral nodes n and N need not be connected at all since no current flows through them, provided the load is balanced. In Fig. 8.20, we have safely assumed, and justified, that the two nodes n and N can be connected with a perfect conductor to close the circuit. However, if a balanced three-phase source is connected to an unbalanced load, still we may make use of Fig. 8.20 and make the analysis on a per-phase basis, subject to the following two conditions:

- the two nodes n and N must be connected in the power system, and
- the connecting wire is a perfect conductor, without possessing any impedance.

Observe that this is a logical AND condition, meaning that if either of the conditions is not satisfied, we cannot proceed with per-phase analysis using Fig. 8.20 as before. There are alternative schemes developed for the unbalanced power systems, but these are beyond the scope of this textbook. Interested readers may refer to books exclusively devoted to power engineering.

8.3.2 The $Y-\Delta$ Power System

In practice, we generally prefer Δ connected loads to Y connected ones since it is easier to add or subtract impedances from a single phase in the case of slight imbalance. Also, since the load does not contain a neutral node, there is no need for a neutral wire, thus saving the cost to some extent. The $Y-\Delta$ power system is shown in Fig. 8.21.

From the figure we observe that

1. The 'line voltages' of the source are the 'phase voltages' of the load.

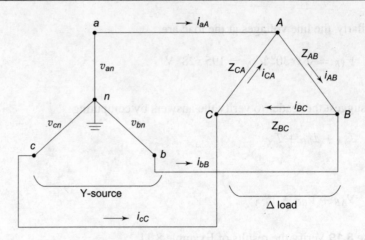

Fig. 8.21

2. The currents bear the following relationship:

$$\mathcal{I}_{aA} = \mathcal{I}_{AB} - \mathcal{I}_{CA}$$
$$\mathcal{I}_{bB} = \mathcal{I}_{BC} - \mathcal{I}_{AB}$$
$$\mathcal{I}_{cC} = \mathcal{I}_{CA} - \mathcal{I}_{BC}$$

i.e., the line currents of the source and the phase currents of the load no longer coincide.

3. In fact, we may derive that

$$\mathcal{I}_{aA} = (\sqrt{3} \angle - 30°)\mathcal{I}_{AB}$$
$$\mathcal{I}_{bB} = (\sqrt{3} \angle - 30°)\mathcal{I}_{BC}$$
$$\mathcal{I}_{cC} = (\sqrt{3} \angle - 30°)\mathcal{I}_{CA} \qquad (8.36)$$

Exercise 8.21 Derive Eqn (8.36).

The analysis of a $Y-\Delta$ system can be made similar to that of the $Y-Y$ system by transforming the delta connected load into a Y connected equivalent and using the one-phase equivalent of Fig. 8.20. The reverse transformation can be done using the above three observations. We illustrate this in the following examples.

Example 8.12 Suppose it is found by measurement that $\mathcal{I}_{bB} = 12 \angle 90°$ in a balanced $Y-\Delta$ power system. Let us compute the load current $\mathcal{I}_{CA}$. Since this is a balanced load, it is easy to observe that $\mathcal{I}_{bB} = 1 \angle 120° \times \mathcal{I}_{cC}$ and $\mathcal{I}_{cC} = (\sqrt{3} \angle - 30°) \mathcal{I}_{CA}$. Accordingly, we get

$$\mathcal{I}_{CA} = 6.93 \angle 0° \text{ A}$$

Exercise 8.22 Verify the results of Example 8.12. Find all other possible quantities of the power system.

Example 8.13 Suppose we have a balanced Y–Δ power system with

- per-phase source voltage—$\mathcal{V}_{*n} = 120$ V, rms;
- per-phase source impedance—$\mathcal{Z}_{\text{source}} = 0.2 + j0.3 \ \Omega$;
- per-phase transmission line impedance—$\mathcal{Z}_{\text{line}} = 1 + j2 \ \Omega$;
- per-phase load impedance—$\mathcal{Z}_{\Delta} = 90 + j60 \ \Omega$

It is easy to compute that the per-phase load impedance of the equivalent Y connection is

$$\mathcal{Z}_Y = \frac{1}{3} \mathcal{Z}_{\Delta} = 30 + j20 \ \Omega$$

and hence

$$\mathcal{Z}_{\phi} = \mathcal{Z}_Y + \mathcal{Z}_{\text{line}} + \mathcal{Z}_{\text{source}} = 31.2 + j22.3 \ \Omega$$

Assuming that $v_{an} = 120 \angle 0°$, we have the line currents

$$\mathcal{I}_{aA} = 3.13 \angle - 35.56°, \quad \mathcal{I}_{bB} = 3.13 \angle - 155.56°,$$

and

$$\mathcal{I}_{cC} = 3.13 \angle 84.44°$$

all in amperes (rms). Using these currents, we may find the phase currents at the load as

$$\mathcal{I}_{AB} = 1.81 \angle - 125.56° \text{ A (rms)}$$

etc. Further, we may compute the phase voltages $\mathcal{V}_{AB}$, etc. using Ohm's law.

Exercise 8.23 Verify the results of Example 8.13.

Example 8.14 Suppose it is found by measurement that at a balanced Δ connected load $|\mathcal{V}_{BC}| = 220$ V (rms) and $|\mathcal{I}_{aA}| = 10$ A (rms), and that $\mathcal{I}_{aA}$ leads $\mathcal{V}_{BC}$ by 60°. Let us compute the load impedance.

Let us assume that $\mathcal{V}_{BC} = 220 \angle \theta°$ V (rms) so that $\mathcal{I}_{aA} = 10 \angle (\theta° + 60°)$ A (rms). From this, we get

$$\mathcal{I}_{AB} = \frac{10}{\sqrt{3}} \angle (\theta° + 90°) \text{ A (rms)} \quad \text{and} \quad \mathcal{I}_{BC} = \frac{10}{\sqrt{3}} \angle (\theta° - 30°) \text{ A (rms)}$$

Accordingly, we get

$$\mathcal{Z}_{\Delta} = \frac{\mathcal{V}_{BC}}{\mathcal{I}_{BC}} = 22\sqrt{3} \angle 30° \ \Omega$$

We end this section reminding the student that we do not deal with unbalanced power systems in this textbook.

8.4 Power Calculations

Based on our observations in the previous sections, we shall simplify our notation as follows:

- For Y connected balanced loads,

$$V_\phi \triangleq |\mathcal{V}_{AN}| = |\mathcal{V}_{BN}| = |\mathcal{V}_{CN}|$$

$$I_\phi \triangleq |\mathcal{I}_{AN}| = |\mathcal{I}_{BN}| = |\mathcal{I}_{CN}| \tag{8.37}$$

$$V_L \triangleq |\mathcal{V}_{AB}| = |\mathcal{V}_{BC}| = |\mathcal{V}_{CA}| = \sqrt{3}\, V_\phi$$

$$I_L \triangleq |\mathcal{I}_{AN}| = |\mathcal{I}_{BN}| = |\mathcal{I}_{CN}| = I_\phi \tag{8.38}$$

- For Δ connected balanced loads,

$$V_\phi \triangleq |\mathcal{V}_{AB}| = |\mathcal{V}_{BC}| = |\mathcal{V}_{CA}|$$

$$I_\phi \triangleq |\mathcal{I}_{AB}| = |\mathcal{I}_{BC}| = |\mathcal{I}_{CA}| \tag{8.39}$$

$$V_L \triangleq |\mathcal{V}_{AB}| = |\mathcal{V}_{BC}| = |\mathcal{V}_{CA}| = V_\phi$$

$$I_L \triangleq \sqrt{3}\,|\mathcal{I}_{AB}| = \sqrt{3}\,|\mathcal{I}_{BC}| = \sqrt{3}\,|\mathcal{I}_{CA}| = \sqrt{3}\, I_\phi \tag{8.40}$$

where the subscripts ϕ and L stand for 'phase' and 'line', respectively. We emphasize that we are working with rms quantities in this section.

Suppose we have a Δ connected load. The *total instantaneous power* may be computed as

$$p_T(t) = v_{AB}i_{AB} + v_{BC}i_{BC} + v_{CA}i_{CA} \tag{8.41}$$

where we are referring to instantaneous voltages and currents. If we assume that

$$i_{AB} = \sqrt{2}I_\phi \cos \omega t$$

$$i_{BC} = \sqrt{2}I_\phi \cos(\omega t - 120°)$$

and

$$i_{CA} = \sqrt{2}I_\phi \cos(\omega t + 120°)$$

and hence

$$v_{AB} = \sqrt{2}V_\phi \cos(\omega t + \theta)$$

$$v_{BC} = \sqrt{2}V_\phi \cos(\omega t + \theta - 120°)$$

and

$$v_{CA} = \sqrt{2}V_\phi \cos(\omega t + \theta + 120°)$$

Notice that we have measured the phase angle of the voltage with reference to that of the current. Also, the factor $\sqrt{2}$ comes into picture while mixing the maximum amplitudes with rms quantities. Plugging these into Eqn (8.41) gives us

$$p_T(t) = 3V_\phi I_\phi \cos\theta \tag{8.42}$$

where we have used the fact that $v_{AB} + v_{BC} + v_{CA} = 0$. Though we have worked out for the Δ connected load, it is easy to verify that the same result holds for the Y connected load and we leave it to the reader. What is interesting to us here is that the total instantaneous power *does not* change with time. In other words, the load experiences a uniform driving force.

Exercise 8.24 Derive Eqn (8.42) and verify that it holds for a Y connected load also.

As with the single-phase circuits, we may express *per-phase real power* P_ϕ and *per-phase reactive power* Q_ϕ and hence the *per-phase complex power* S_ϕ as

$$P_\phi = V_\phi I_\phi \cos\theta_\phi = \frac{1}{\sqrt{3}} V_L I_L \cos\theta_\phi \text{ W}$$

$$Q_\phi = V_\phi I_\phi \sin\theta_\phi = \frac{1}{\sqrt{3}} V_L I_L \sin\theta_\phi \text{ VAR}$$

$$S_\phi = P_\phi + jQ_\phi \text{ V A}$$

where θ_ϕ is the phase difference between the voltage and current of the same phase or of the same line.

Clearly, the per-phase apparent power is $S_\phi = V_L I_L/\sqrt{3}$ and the *total complex power* S_T is given by

$$S_T = 3S_\phi = P_T + jQ_T \text{ V A} \tag{8.43}$$

where P_T is the *total real power* in watts and Q_T is the *total reactive power* in VAR.

An ardent reader should have noticed (in fact, expected) that the total instantaneous power p_T coincides with the total real power P_T.

Example 8.15 Consider the results of the Y–Y power system of Example 8.9 again. At the load we have

$$V_\phi = 112.6 \text{ V (rms)}$$

$$I_\phi = 6.25 \text{ A (rms)}$$

$$\theta_\phi = \tan^{-1}\frac{10}{15} = 33.7°$$

Accordingly, we may compute the total complex power absorbed by the load as

$$S_{T(\text{load})} = 1757 + j1170.6 \text{ V A}$$

Similarly, the total complex power absorbed by the line may be computed to be

$$S_{T(\text{line})} = 58.6 + j117.1 \ \text{VA}$$

and

$$S_{T(\text{source})} = 11.7 + j23.4 \ \text{VA}$$

Thus, the total complex power absorbed is

$$S_{T(\text{absorbed})} = 1827.3 + j1311.1 \ \text{VA}$$

At the source, we have

$$V_\phi = 120 \ \text{V (rms)}$$

$$I_\phi = 6.25 \ \text{A (rms)}$$

$$\theta_\phi = 0 - (-35.7°) = 35.7°$$

and hence, the total complex power delivered is

$$S_{T(\text{source})} = 1827.3 + j1311.1 \ \text{VA}$$

thus validating the conservation of complex power.

Here we define the efficiency of a power system as follows.

Definition 8.6 Efficiency of a power system

$$\eta \triangleq \frac{P_{T(\text{load})}}{P_{T(\text{source})}} \times 100\% \tag{8.44}$$

In this example, we may compute that

$$\eta = \frac{1757}{1827.3} = 96.2\%$$

Exercise 8.25 Verify the results of Example 8.15.

The schematic respresentation discussed for the power measurement in single-phase systems can be extended, in principle, to the power measurement in the balanced three-phase systems; simply obtain the power of one of the phases and multiply by 3. However, in practice, the neutral line may not be accessible in the case of Y loads and is missing in the Δ loads. Hence, we need to perform the measurements on the lines. Some more thinking into the ideas of phase differences would prompt us to conclude that a pair of wattmeters would be sufficient to make the power measurement in three-phase systems. The two wattmeters would measure the power flow on two of the lines with the third line as a reference. The two readings are sufficient to indirectly measure the real power, the reactive power, and the power factor.

Example 8.16 Suppose we have an interconnection of two unknown three-phase systems as shown in Fig. 8.22.

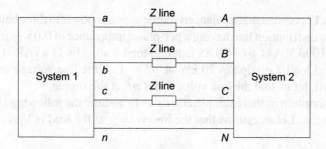

Fig. 8.22

The per-phase line impedance is $1 + j2\Omega$, $V_{ab} = 10 \angle 0°$ V (rms), and $V_{AB} = 10 \angle 6°$ V (rms). Let us now verify the conservation of power. First, we assume that it is a Y–Y system for simplicity. Then we have

$$\mathcal{I}_{aA} = \frac{V_{an} - V_{AN}}{\mathcal{Z}_{\text{line}}} = 270.3 \angle 179.6° \text{ mA (rms)}$$

Looking at the phase angle, we may assume that this current is flowing from system 2 into system 1. Let us see if this assumption is justified.

The (real) power absorbed by system 1 is

$$P_1 = 3 \times |V_{an}||I_{aA}| \cos(\angle V_{an} - \angle I_{aA})$$
$$= 3 \times \frac{10}{\sqrt{3}} \times 270.3 \times 10^{-3} \times \cos[-30° - (179.6° - 180°)]$$
$$= 4.072 \text{ W}$$

We account 180° for the angle of I_{aA} to denote the appropriate direction of flow. Since P_1 turned out to be positive, system 1 indeed absorbs power, justifying our assumption partly. Let us now look at the power absorbed (or, delivered?) by system 2.

$$P_2 = 3 \times |V_{An}||I_{aA}| \cos(\angle V_{An} - \angle I_{aA})$$
$$= 3 \times \frac{10}{\sqrt{3}} \times 270.3 \times 10^{-3} \times \cos(-24° - 179.6°)$$
$$= -4.291 \text{ W}$$

The negative sign confirms our assumption that system 2 is the source and system 1 is the load. The difference in these two powers may be accounted to be the loss in the transmission line, i.e,

$$P_{\text{line}} = 3 \times 1 \times \left(270.3 \times 10^{-3}\right)^2$$
$$= 0.219 \text{ W}$$
$$= |P_2| - |P_1|$$

thus verifying the conservation of power.

Exercise 8.26 Verify the results of Example 8.16.

Example 8.17 Assume that a balanced source supplies power to three balanced loads via a power distribution line having a per-phase impedance of $0.05 + j0.1\ \Omega$. Load 1 absorbs 100 kV A at $pf = 0.85$ lagging, load 2 absorbs 25 kVAR at $pf = 0.75$ leading, and load 3 dissipates 20 kW at $pf = 1$. If the line voltage at the load is 240 V (rms), let us find the line voltage and pf at the source.

The description in this example allows us to assume the following single-phase parallel circuit. Let us assume that the line voltage at the load is $V_L = 240\angle 0°$ V.

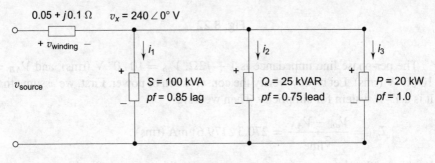

Fig. 8.23

From Fig. 8.23, using the power factors in each of the branches we may deduce that

$$\mathcal{I}_1 = \frac{100 \times 10^3}{240} = 416.67\angle -31.8°\ A$$

$$\mathcal{I}_2 = \frac{25 \times 10^3 / \sin(\cos^{-1} 0.75)}{240} = 157.49\angle 41.4°\ A$$

$$\mathcal{I}_3 = \frac{20 \times 10^3}{240} = 83.33\angle 0°\ A$$

Since $\mathcal{I} = \mathcal{I}_1 + \mathcal{I}_2 + \mathcal{I}_3$ is delivered by the source V_{source}, we have

$$\mathcal{I} = 567.46\angle -11.73°\ A$$

i.e., the source 'sees' an inductive network. Further, since $\mathcal{I}$ flows through the distribution line impedance, we may compute the voltage drop across the winding as

$$V_{winding} = \mathcal{I} \times Z_{winding} = 63.44\angle 51.71°\ V$$

Using KVR, we have

$$V_{source} = 283.72\angle 10.11°\ V$$

i.e., the voltage source 'leads' the net current flowing in the circuit by an angle

$$\psi = 10.11° - (-11.73°) = 21.84°$$

and hence the power factor at the source is $\cos(21.84°) = 0.9283$ *lagging*.

Exercise 8.27 Verify the results of Example 8.17.

Exercise 8.28 Assume that a balanced source supplies power to three balanced loads via a power distribution line having a per-phase impedance of $1 + j2\,\Omega$. Load 1 absorbs 100 kV A at $pf = 0.95$ leading, load 2 absorbs 125 kW at $pf = 0.8$ lagging, and load 3 draws $150 + j50$ kV A. If the line voltage at the load is 300 V (rms), find the line voltage at the load and pf at the source as well as the efficiency of the system.

Example 8.18 Suppose that a balance Y connected load is powered by a balanced source and draws a total of 120 kW at $pf = 0.8$ lagging. The line voltage at the load is 208 V (rms), 50 Hz. If three 5 mF capacitances are connected in parallel with the load in a Y configuration, let us determine the power factor of the combined load.

Once again, the above description suggests us the following single-phase circuit in Fig. 8.24.

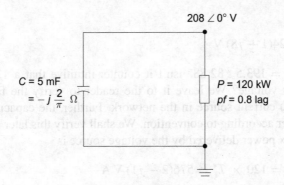

Fig. 8.24

From Fig. 8.24, we have

$$\mathcal{I} = \frac{120 \times 10^3}{208 \times 0.8} = 721.15\,\angle -36.87°\ \text{A}$$

$$\Rightarrow \quad \mathcal{Z}_Y = 0.288\,\angle 36.87°\ \Omega$$

$$= 0.23 + j0.17\ \Omega$$

If we place a capacitance of 5 mF $= j0.5\pi\ (\Omega)^{-1}$ across $\mathcal{Z}_Y$, we get

$$\mathcal{Z}_{Y\text{new}} = 0.35 + j0.06\ \Omega$$

$$= 0.355\,\angle 9.73°\ \Omega$$

and hence the power factor of the combined load is 0.986 *lagging*.

Exercise 8.29 Repeat Example 8.18 with the three capacitances connected in a Δ configuration. Compare the two results.

Example 8.19 Let us consider the circuit shown in Fig. 8.25. This is the single-phase equivalent of a power system.

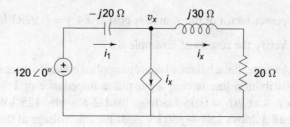

Fig. 8.25

Using KVR, we easily get

$$\mathcal{I}_1 = 4.8(2 + j1) \text{ A}$$

and hence

$$\mathcal{I}_x = 2.4(2 + j1) \text{ A}$$

and

$$\mathcal{V}_x = 24(1 + j8) \text{ V}$$

Looking at $\mathcal{V}_x = 193.5 \angle 82.88°$ isn't it counter-intuitive that a $120 \angle 0°$ source delivers higher voltage? We leave it to the reader to verify the fact looking at the (controlled) current source in the network. Further, the capacitance *delivers* (reactive) power according to convention. We shall verify this later statement.

The complex power delivered by the voltage source is

$$S_{\text{source}} = 120 \times \mathcal{I}_1^* = 576(2 - j1) \text{ V A} \tag{8.45}$$

The capacitance has

$$S_{\text{capacitance}} = \mathcal{I}_1 \times -j20 \times \mathcal{I}_1^* = 460.8(0 - j5) \text{ V A} \tag{8.46}$$

The controlled current source has

$$S_{\text{current source}} = \mathcal{V}_x \times \mathcal{I}_x^* = 57.6(10 + j15) \text{ V A} \tag{8.47}$$

and the inductance-resistance series combination has

$$S_{\text{LR}} = \mathcal{V}_x \times \mathcal{I}_x^* = 57.6(10 + j15) \text{ V A} \tag{8.48}$$

We leave it to the reader to verify the conservation of complex power with a careful attention to the 'signs' of real and imaginary parts in each of Eqns (8.45)–(8.48).

This examples illustrates that

$$\sum_{i=1}^{n} \mathcal{V}_i \mathcal{I}_i^* = 0 \tag{8.49}$$

and this result is popularly known as *Tellegen's theorem*[4]. A proof of this theorem involves writing the summation of powers (instantaneous or complex) for n branches that satisfy both Kirchhoff's rules. Clearly, this demonstrates the conservation of power, a result we are already familiar with. However, the importance of this theorem stems from the fact that it is applicable to a very general class of networks composed of elements that are linear or non-linear, passive or active, time-invariant or time-varying.

Example 8.20 Let us consider the circuit in Fig. 8.26 and verify if Tellegen's theorem holds good.

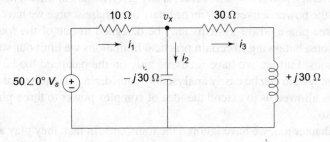

Fig. 8.26

Using nodal analysis we may compute that

$$V_x = 42 - j6 \text{ V}$$

Consequently, we have

$$\mathcal{I}_1 = 0.8 + j0.6 \text{ A}$$
$$\mathcal{I}_2 = 0.2 + j1.4 \text{ A}$$
$$\mathcal{I}_3 = 0.6 - j0.8 \text{ A}$$

satisfying Kirchhoff's rules. Observe that if the currents $\mathcal{I}_i$ obey KCR, so do the currents $\mathcal{I}_i^*$. It is easy to verify now that

$$50 \times (0.8 - j0.6) - (8 + j6) \times (0.8 - j0.6)$$
$$- (42 - j6) \times (0.2 - j1.4) - (42 - j6) \times (0.6 + j0.8) = 0$$

We adhered to the convention that sources deliver power while passive elements absorb power and accordingly placed the +ve and −ve signs in the equation above.

Exercise 8.30 Verify the results of Examples 8.19 and 8.20.

Summary

In this chapter we have introduced one of the most important issues of electrical engineering, namely, power and energy. We have seen that the power delivered/consumed in ac circuits has a different flavour and has to be carefully

---

[4] B.D.H. Tellegen, 'A General Network Theorem with Applications', *Phillips Research Reports*, vol. 7, pp. 259–69, 1952.

dealt with. We have come across average power, apparent power, and complex power. While it is the real power we actually consume, we need to account for the reactive power also. However, to optimize the utility of power, we may resort to power factor correction. We have obtained expressions for computing appropriate elements to be connected in parallel to achieve the desired power factor. And, ac circuits obey the law of conservation of complex power. We have also introduced the idea of poly-phase circuits, three-phase power systems in particular. Apart from the advantages in generation, transmission, and distribution, the three-phase power system has the distinctive advantage that the total power is constant, equal to thrice the average power of each source, at any given instant of time. This is in sharp contrast to the power delivered by an ordinary one-phase source we have dealt with earlier. Three-phase power systems may be designed in one of the four possible configurations, but owing to certain practical limitations we limit our study to two configurations. Further, we have focussed only on the balanced loads. A Y–Y or Y–Δ power system can be easily analysed by considering an equivalent one-phase circuit. This allowed us to extend the idea of complex power to three-phase power systems also.

In this chapter too, we have ignored the transients. In fact, they play a very vital role in actual power systems since the initial power requirement would be generally high. For instance, when we discussed the example of toy-wheel to illustrate the idea of poly-phase circuits, it is quite obvious that the torque required to move the wheel from rest would be higher than the torque required to maintain the speed. Transients in power systems is a subject by itself, and it is beyond the scope of this textbook.

Problems

8.1 A series RLC circuit with $R = 10\ \Omega$, $L = 25$ mH, and $C = 100\ \mu$F is driven by a voltage source $v_S = 10\cos(1000\,t)$ V. Compute the instantaneous power (delivered/absorbed) by each of the four elements at $t = 0$. Verify the conservation of power.

8.2 Consider the circuit in Fig. 8.27. Verify that the average power is conserved if $v_S = 10\,\angle 60°$ V.

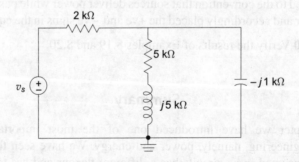

Fig. 8.27

8.3 Consider the circuit in Fig. 8.28. Compute the average power absorbed by each of the branch impedances in the circuit if $v_S = 12 \angle -45°$. Verify that the total power released by the source is absorbed by the passive elements.

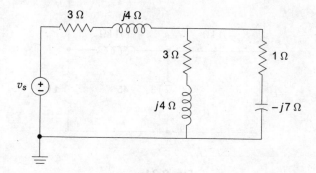

Fig. 8.28

8.4 Consider the circuit in Fig. 8.29. If the independent voltage source $v_S = 24\cos(1000\,t)$ V, determine the average power (absorbed/delivered) by the current controlled current source (CCCS).

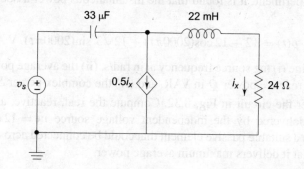

Fig. 8.29

8.5 Consider the circuit in Fig. 8.30. Determine $\mathcal{Z}_L$ that absorbs maximum average power. Verify that the average power is conserved in the circuit.

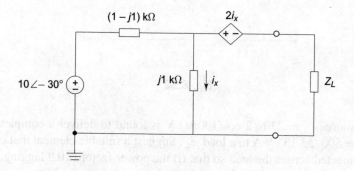

Fig. 8.30

8.6 Repeat Problem 8.5 for the circuit shown in Fig. 8.31.

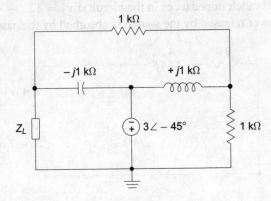

Fig. 8.31

8.7 Assume that a load $\mathcal{Z}_L$ Ω is connected to a Thévenin's equivalent circuit with a $\mathcal{V}_{OC}$ V and a $\mathcal{Z}_{eq}$ Ω. Show that $\mathcal{Z}_L$ draws maximum average power $P_{L,\max}$ if $|\mathcal{Z}_L| = |\mathcal{Z}_{eq}|$, while it is not necessary that $\angle\mathcal{Z}_L = \angle\mathcal{Z}_{eq}$.

8.8 In an experiment, it is found that the instantaneous power associated with a load is

$$p(t) = 12 + 12\cos(2000\pi t) + 12\sqrt{3}\,\sin(2000\pi t)\ \text{V A}$$

Determine (i) the source frequency ω in rad/s, (ii) the average power P in W, (iii) the reactive power Q in VAR, and (iv) the complex power $\mathcal{S}$ in V A.

8.9 Consider the circuit in Fig. 8.32. Compute the real, reactive, and complex power delivered by the independent voltage source $v_S = 12\cos(10t)$ V. Suggest a suitable passive element that could be connected across the source v_S so that it delivers maximum average power.

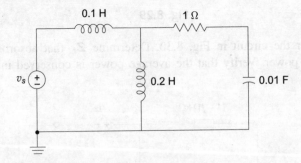

Fig. 8.32

8.10 A source $v_S = 220\sqrt{2}\,\cos(100\pi t)$ V is found to deliver a complex power $\mathcal{S} = 500\angle 53.13°$ V A to a load $\mathcal{Z}_L$. Suggest a suitable element that could be connected across the load so that (i) the power factor is 0.9 lagging, (ii) the power factor is 0.75 leading, and (iii) the power factor is 1.0.

8.11 Two loads are connected in parallel and are found to absorb a total average power of 5 kW at 100 V (rms) and 100π rad/s with a power factor $pf = 0.8$ leading. If load 1 itself is found to absorb 2 kW with a power factor 0.7 leading, compute the power factor of load 2. Suggest a suitable element that could be connected across the two loads so that the combined power factor of the parallel combination is 1.0.

8.12 An ac source v_S drives a parallel combination of two loads via a transmission line having an impedance of $\mathcal{Z}_{line} = 0.25 - j0.25\ \Omega$. The common voltage across the parallel combination of the two loads is measured to be $v_L = 100\angle 45°$. If load 1 absorbs an average power of 20 kW at $pf = 0.6$ lagging, and load 2 absorbs an average power of 40 kW at $pf = 0.8$ lagging, determine the source voltage $v_S(t)$.

8.13 An ac source drives an inductive load via a transmission line having an impedance of $\mathcal{Z}_{line} = 0.3 + j0.4\ \Omega$. If the power absorbed by the transmission line is 1 kW and the load demands an average power of 10 kW at a voltage of $V_L = 100\angle -30°$, compute the complex power delivered by the source.

8.14 Across a port whose potential difference is $100\angle 0°$ (rms), three loads are connected and the following information is obtained in an experiment:
- load 1 absorbs $S = 24 + j10$ V A,
- load 2 absorbs $S = 24 - j10$ V A,
- load 3 absorbs $S = 6 + j8$ V A.

Compute the current delivered by the source, and the combined power factor of the load as seen by the source.

8.15 A three-phase source is constructed with the following: $v_{an} = 100\cos(\omega t + 75°)$, $v_{bn} = 100\cos(\omega t + 45°)$, and $v_{cn} = 100\cos(\omega t - 165°)$. Obtain an expression for the total instantaneous power delivered by this source if it is connected to a balanced Y connection of three identical resistances R. What is the average power? Is the total instantaneous voltage zero?

8.16 Consider a Y connected balanced load. (i) If the line voltage $V_{AN} = 100\angle 45°$, compute the phase voltage V_{BC}. (ii) If the line voltage $V_{CN} = 100\angle -45°$, compute the phase voltage V_{AB}.

8.17 Three balanced loads, two Y connections, and one Δ connection: $\mathcal{Z}_{Y_1} = 25\ \Omega$, $\mathcal{Z}_{Y_2} = j25\ \Omega$, and $\mathcal{Z}_\Delta = -j25\ \Omega$ are connected in parallel with $\mathcal{Z}_\Delta$ in the middle flanked on either sides by $\mathcal{Z}_{Y_1}$ and $\mathcal{Z}_{Y_2}$. Compute the Y equivalent, as well as the Δ equivalent, of the composite load.

8.18 In a Y–Y balanced power system, $V_{an} = 100\angle 0°$ and $\mathcal{Z}_Y = 10 + j5\ \Omega$. If the transmission line has a per-phase impedance of $0.1 - j0.1\ \Omega$, compute the line currents and the phase voltages at the load.

8.19 In a Y–Y power system, the transmission line has an impedance of $0.5 + j0.25\ \Omega$ and absorbs an average power of 50 W. If the per-phase impedance of the balanced load is $30 + j40\ k\Omega$, compute the total complex power delivered by the source.

8.20 A balanced Y connected three-phase source with $V_{an} = 100\angle 0°$ is connected to an unbalanced Y connected load with $\mathcal{Z}_{AN} = 3 + j4\ \Omega$, $\mathcal{Z}_{BN} = 5\ \Omega$, and $\mathcal{Z}_{CN} = 3 - j4\ \Omega$. If the source-load connection is made with four perfect conductors having zero impedance, what is the current flowing through the connecting wire between the nodes n and N?

8.21 If in Problem 8.20 the neutral wire is removed, compute the voltage $\mathcal{V}_{Nn}$, i.e., the potential of the load neutral node with reference to that of the source neutral node.

8.22 A Y connected balanced three-phase source with $\mathcal{V}_{an} = 100\angle 0°$ is connected to a balanced Δ connected load with $\mathcal{Z}_\Delta = 30 + j40\,\Omega$ via transmission lines having an impedance of $0.5 - j0.25\,\Omega$. Compute the line currents and the phase voltages at the load.

8.23 A balanced Y–Δ power system has

$$\mathcal{Z}_{\text{winding}} = 2 + j\sqrt{3}\,\Omega, \quad \mathcal{Z}_{\text{line}} = 0.5 - j0.5\,\Omega, \quad \text{and} \quad \mathcal{Z}_\Delta = 6 + j8\,\text{k}\Omega$$

If one of the line currents $\mathcal{I}_{AB}$ at the load is $4\angle -30°$ (rms), compute the source voltage.

8.24 In a balanced Y–Δ power system, $\angle \mathcal{V}_{an} = 30°$ and $\mathcal{I}_{aA} = 2\angle -30°$. If the load absorbs 1 kW at a power factor of 0.8 lagging, and the transmission line absorbs 25 W per phase, compute (i) $\mathcal{V}_{an}$, (ii) $\mathcal{Z}_{\text{line}}$, (iii) $\mathcal{Z}_\Delta$, and (iv) $\mathcal{V}_{AB}$.

8.25 In a balanced Y–Y system, $\mathcal{V}_{an} = 100\angle 0°$, $\mathcal{Z}_Y = 30\angle 60°$, and $\mathcal{Z}_{\text{line}} = 0.1\angle 45°$. Compute the total complex power delivered by the source, and the ratio of the average power absorbed by the load to that delivered by the source.

8.26 A balanced load consumes $S = 1.2 + j0.9$ kV A at a line voltage of 3 kV (rms), and 100π rad/s. Assuming that the load is a (i) Δ connection, find a simple parallel equivalent for its phase impedance, (ii) Y connection, find a simple series equivalent for its phase impedance.

8.27 For the load in Problem 8.26, suggest (i) a suitable Δ connection, and (ii) a suitable Y connection of elements to be connected in parallel with the load to achieve unity power factor. Assume that the line voltage remains unchanged.

8.28 A power system has a per-phase transmission line impedance of $\mathcal{Z}_{\text{line}} = 0.01 + 0.01\,\Omega$, and a balanced load absorbs 360 kW at $pf = 0.6$ leading with $\mathcal{V}_{\text{load}} = 0.5$ kV (rms). What is the power factor at the source?

8.29 Three balanced loads are found to absorb the following power when connected in a three-phase power system:
- load 1 absorbs 100 kV A at a power factor of 0.9 leading;
- load 2 absorbs 125 kW at a power factor of 0.6 leading; and
- load 3 absorbs $150 + j100$ kV A

If the line voltage at the load is 4 kV (rms), and the transmission line has a per-phase impedance of $\mathcal{Z}_{\text{line}} = 2 + j\sqrt{3}\,\Omega$, compute the power factor at the source. Also, compute the efficiency of the power system.

8.30 An unbalanced Δ connected load with

$$\mathcal{Z}_{AB} = 4 + j3\,\Omega, \quad \mathcal{Z}_{BC} = 4 - j3\,\Omega, \quad \text{and} \quad \mathcal{Z}_{CA} = 5 + j5\,\Omega$$

is driven by a balanced set of line voltages of 500 V (rms) each. Compute the complex power of the entire load.

CHAPTER 9

Generalized Frequency Response

In Chapter 3, we have introduced the notion of a circuit, and then went on to develop, in Chapter 4, some techniques[1], for circuits that exclusively have resistors and dc sources; we were doing real algebra. Next, we went on to study the transients caused by energy-storage elements in the circuits; we were solving *homogeneous* and *forced* ordinary differential equations and found that the transients are of the form e^{st}. In Chapter 6 we dealt with sinusoidal sources and found that both the sources as well as the responses could be *conveniently* expressed in terms of complex exponentials; we were doing complex algebra. Let us remember that we have 'ignored' the transients while studying the sinusoidal response.

In this chapter, we shall further generalize our ideas and develop an approach wherein we would obtain both the transient response and the steady-state response for *any* source. It is useful to do so because, ultimately, we are interested in *designing* circuits, and general 'systems'. We begin with a quick recapitulation, of course, from a different perspective.

9.1 Two Questions

Let us consider a series RLC circuit once again. This is shown again in Fig. 9.1 for convenience.

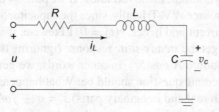

Fig. 9.1

As it is, if we are interested in obtaining the voltage v_C across the capacitor as our output $y(t)$, then we might get the following ODE:

$$\frac{d^2}{dt^2} y(t) + 2\zeta\omega_n \frac{d}{dt} y(t) + \omega_n^2 y(t) = \omega_n^2 u(t) \tag{9.1}$$

---

[1] Tricks, if you prefer to call them so!

Alternatively, if we are interested in obtaining the voltage v_R across the resistor as our output $y(t)$, then the ODE would be

$$\frac{d^2}{dt^2} y(t) + 2\zeta\omega_n \frac{d}{dt} y(t) + \omega_n^2 y(t) = 2\zeta\omega_n \frac{d}{dt} u(t) \tag{9.2}$$

We observe that there is no change in the left-hand side of the ODE, but the right-hand side changes if the variables of our interest in the circuit change. Our earlier technique of 'guessing and justifying' still works here. We will go a little deeper in this direction and make some interesting observations.

We choose the input $u(t)$ to be

$$u(t) = A\, e^{st} \tag{9.3}$$

This choice should not surprise the reader since (i) putting $s = 0$ would take us to dc steady-state response (cf. Chapter 5), and (ii) putting $s = j\omega$ would take us to sinusoidal steady-state response (cf. Chapters 6 and 7). Let us now guess the response to be

$$y(t) = B\, e^{st} \tag{9.4}$$

We will use these guesses in Eqn (9.2) and see what is in store for us. Plugging Eqn (9.4) in the lhs of the ODE, and Eqn (9.3) in the rhs, and cancelling out the common e^{st}, we get

$$\frac{B}{A} = \frac{\text{output magnitude}}{\text{input magnitude}}$$

$$= \frac{2\zeta\omega_n s}{s^2 + 2\zeta\omega_n s + \omega_n^2} \tag{9.5}$$

We observe that the *ratio* of the magnitudes turns out to be a *rational function*, i.e., a ratio of two polynomials, in the variable s. If we put $s = 0$, we get $y(t) = 0$, the response due to a dc source A V. Trivially, since the capacitor acts as an open circuit, there would be no current and hence $y(t) = 0$. Likewise, if we put an 'imaginary number' $s = j5$, we get the *steady-state* response (ignoring the transient response) due to a sinusoidal source $A\cos(5t)$. In other words, we get the forced response quite easily. An immediate question should be: What happens if we put a complex number (that has both real and imaginary parts) $s = \sigma + j\omega$?

In these two dc and ac steady-state cases, the value of s is given to us in terms of the input; we need not determine it. We need to determine the magnitude and phase angle of the response. In sharp contrast, to get the natural response, we cannot directly meddle with the rational function we have obtained; we need to go back to our ODE and put $u(t) = 0$ so that after plugging in $y(t) = B\, e^{st}$ we are left with

$$(s^2 + 2\zeta\omega_n s + \omega_n^2)\, y(t) = 0$$

wherein we define the *characteristic polynomial* and determine what values would the complex variable s take. Now, another question is: Is there any way where we can extract the complete response, i.e., both the transient response and the steady-state response, at once? We shall answer these two questions in Sec. 9.3.

We might extend the ideas in the above series RLC circuit example to any general nth order circuit. Consequently, we shall get a general nth-order ODE of the form:

$$y^{(n)} + a_{n-1}\, y^{(n-1)} + a_{n-2}\, y^{(n-2)} + \cdots + a_0\, y$$
$$= \quad b_m\, u^{(m)} + b_{m-1}\, u^{(m-1)} + b_{m-2}\, u^{(m-2)} + \cdots + b_0\, u \qquad (9.6)$$

where $y^{(i)}$ denotes $(d^i/dt^i)\, y$, the ith derivative of y, $i = 1, 2, \ldots, n$. The same holds for $u(i)$.

9.2 State Variables

Let us resort to the series RLC circuit again. The two energy-storage elements can be *independently* charged in the past $(t < 0)$ and placed in the circuit at $t = 0$. Recall that we have defined a *state vector* $x(t)$, in Eqn (5.21), as

$$\overline{x}(t) \triangleq \begin{bmatrix} i(t) \\ v_C(t) \end{bmatrix} \qquad (9.7)$$

Let us denote the time differentiation d/dt using a 'dot' above the variable, i.e.,

$$\dot{\overline{x}}(t) \triangleq \frac{d}{dt}\, \overline{x}(t), \quad \ddot{\overline{x}}(t) \triangleq \frac{d^2}{dt^2}\, \overline{x}(t),$$

and so on

In terms of this vector, we may obtain the *matrix-vector form* as

$$\dot{\overline{x}}(t) = \begin{bmatrix} -R/L & -1/L \\ 1/C & 0 \end{bmatrix} \overline{x}(t) + \begin{bmatrix} 1 \\ 0 \end{bmatrix} u$$
$$= A\, \overline{x}(t) + B\, \overline{u}(t) \qquad (9.8)$$

where $\overline{u}$ generally denotes the input *vector* of dimension $r \times 1$; in this particular case the dimension is 1×1. Any m signals in the circuit may be obtained as a linear combination of the state variables and the input:

$$\overline{y} = E\, \overline{x} + D\, \overline{u} \qquad (9.9)$$

The matrix E is called the *output matrix*[2]. The matrix D of dimension $m \times r$ is called the *transmission matrix* and appears only under certain special circumstances, as we see in a while.

The ODE in Eqn (9.8) is called the *state equation* that describes the *evolution* of the state variables with respect to time. The *algebraic equation*, i.e., the linear combination of the state variables and the input variables, in Eqn (9.9) is called the *output equation*. We remind the reader that we call $\overline{x}$ the 'state vector' because

- these are the variables that give us the 'information' or 'status' of the circuit *at any instant of time*, and
- this is the *minimum set* of *independent* variables that are necessary and sufficient to tell us the status of the circuit.

---

[2] In general systems theory, this output matrix, of dimension $m \times n$, is designated as C. To avoid any confusion between this designation and the capacitance, we prefer E in this textbook.

We further note that the sequence in which the state variables are stacked in $\bar{x}$ is immaterial. All we have is the set of four matrices:

$$\Sigma = \{A, B, E, D\}$$

We further note that the elements of all these four matrices are *real numbers*. To put it more formally, we have

$$\Sigma = \left\{A \in \Re^{n \times n}, B \in \Re^{n \times r}, E \in \Re^{m \times n}, D \in \Re^{m \times r}\right\} \qquad (9.10)$$

Let us look at the following examples to gain more experience.

Example 9.1 Consider the circuit in Fig. 9.2.

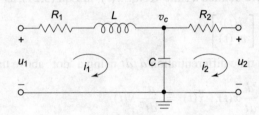

Fig. 9.2

We have the following immediate observations:

1. There are two independent voltage sources providing input to the circuit, and
2. There are only two energy-storage elements.

Applying KVR in the two meshes gives us

$$v_{R1} + v_{L1} + v_C = u_1$$

$$v_C - v_{R2} = u_2$$

The relationship among the mesh currents i_1 and i_2 and the capacitor voltage v_C is

$$i_1 - i_2 = C \dot{v}_C$$

Taking the inductor current i_1 and the capacitor voltage v_C as the state variables, we have

$$\dot{i}_1 = -\frac{R_1}{L} i_1 - \frac{1}{L} v_C + \frac{1}{L} u_1$$

$$\dot{v}_C = \frac{1}{C} i_1 - \frac{1}{R_2 C} v_C + \frac{1}{R_2 C} u_2$$

or equivalently, we have

$$\dot{\bar{x}} = A \bar{x} + B \bar{u}$$

where

$$A = \begin{bmatrix} -\dfrac{R_1}{L} & -\dfrac{1}{L} \\ \dfrac{1}{C} & -\dfrac{1}{R_2 C} \end{bmatrix}, \quad B = \begin{bmatrix} \dfrac{1}{L} & 0 \\ 0 & \dfrac{1}{R_2 C} \end{bmatrix}$$

with an appropriately defined state vector $\overline{x}$ and input vector $\overline{u}$:

$$\overline{x} \triangleq \begin{bmatrix} i_1 \\ v_C \end{bmatrix}, \quad \overline{u} \triangleq \begin{bmatrix} u_1 \\ u_2 \end{bmatrix}$$

respectively.

Consistent with our observations earlier, we have *two* first-order differential equations (= no. of energy storage elements), and *two* columns in the B matrix (= no. of independent input sources).

We may use the state vector $\overline{x}$ and the input vector $\overline{u}$ to *compute* any quantity in the circuit as follows:

$$v_{R1} = \begin{bmatrix} R_1 & 0 \end{bmatrix} \overline{x} + \begin{bmatrix} 0 & 0 \end{bmatrix} \overline{u}$$

$$v_L = \begin{bmatrix} -R_1 & -1 \end{bmatrix} \overline{x} + \begin{bmatrix} 1 & 0 \end{bmatrix} \overline{u}$$

$$v_{R2} = \begin{bmatrix} 0 & 1 \end{bmatrix} \overline{x} + \begin{bmatrix} 0 & -1 \end{bmatrix} \overline{u}$$

$$i_2 = \begin{bmatrix} 0 & \dfrac{1}{R_2} \end{bmatrix} \overline{x} + \begin{bmatrix} 0 & -\dfrac{1}{R_2} \end{bmatrix} \overline{u}$$

Observe that the D matrix is zero only for v_{R1}. We may *pack m* number of quantities that we prefer to call outputs as follows. For instance, if we are interested in v_L, v_C, and i_2, in that order, we may obtain them as

$$\begin{bmatrix} v_L \\ v_C \\ i_2 \end{bmatrix} = \begin{bmatrix} -R_1 & -1 \\ 0 & 1 \\ 0 & \dfrac{1}{R_2} \end{bmatrix} \overline{x} + \begin{bmatrix} 1 & 0 \\ 0 & 0 \\ 0 & -\dfrac{1}{R_2} \end{bmatrix} \overline{u}$$

where E has the dimension 3×2 and D has the dimension 3×2.

We just make a subtle observation that the ODE is always a first-order one; we need to be alert on the dimensions of the matrices.

Exercise 9.1 Verify the results of Example 9.1.

An ardent reader might ask whether the principle of superposition would be a better choice since we have more than one input source. We simply answer that we have actually made use of the principle of superposition, albeit in disguise, as follows. Consider the following pair of linear equations:

$$\begin{bmatrix} z_1 \\ z_2 \end{bmatrix} = \begin{bmatrix} b_{11} & b_{12} \\ b_{21} & b_{22} \end{bmatrix} \begin{bmatrix} u_1 \\ u_2 \end{bmatrix}$$

We may resort to a clever, but legal, manipulation and look at this pair of linear equations as

$$\begin{bmatrix} z_1 \\ z_2 \end{bmatrix} = \begin{bmatrix} b_{11} & b_{12} \\ b_{21} & b_{22} \end{bmatrix} \begin{bmatrix} u_1 \\ 0 \end{bmatrix} + \begin{bmatrix} b_{11} & b_{12} \\ b_{21} & b_{22} \end{bmatrix} \begin{bmatrix} 0 \\ u_2 \end{bmatrix}$$

which might be interpreted as the sum of *two* partial responses—due to u_1 acting alone with $u_2 = 0$ and due to u_2 acting alone with u_1 suppressed.

Example 9.2 Consider the circuit in Fig. 9.3.

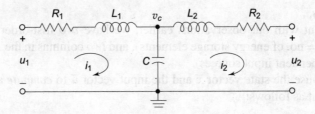

Fig. 9.3

Applying KVR in the two meshes gives us

$$v_{R1} + v_{L1} + v_C = u_1$$

$$v_{L2} + v_{R2} = v_C - u_2$$

The relationship among the mesh currents i_1 and i_2 and the capacitor voltage v_C is

$$i_1 - i_2 = C \dot{v}_C$$

Taking the inductor currents i_1 and i_2 as x_1 and x_2, respectively, and the capacitor voltage v_C as x_3, we have

$$\dot{\bar{x}} = A\bar{x} + B\bar{u}$$

where

$$A = \begin{bmatrix} -\dfrac{R_1}{L_1} & 0 & -\dfrac{1}{L_1} \\ 0 & -\dfrac{R_2}{L_2} & \dfrac{1}{L_2} \\ \dfrac{1}{C} & -\dfrac{1}{C} & 0, \end{bmatrix}, \quad B = \begin{bmatrix} \dfrac{1}{L_1} & 0 \\ 0 & -\dfrac{1}{L_2} \\ 0 & 0 \end{bmatrix}$$

Observe the *coupling* among the state variables—the currents x_1 and x_2 through the inductors depend on the voltage x_3 across the capacitor; in turn, x_3 depends on x_1 and x_2. We may use the state vector $\bar{x}$ and the input vector $\bar{u}$ to *compute* any quantity in the circuit.

Exercise 9.2 Verify the results of Example 9.2. Obtain the matrices E and D for all the possible signals in the circuit.

Example 9.3 Consider the circuit in Fig. 9.4.

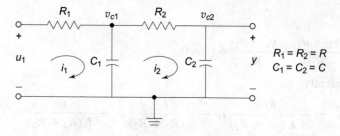

Fig. 9.4

We immediately observe that this circuit has two independent, identical capacitors. Enumerating the voltages across the capacitors as x_1 and x_2 as shown in the figure, we have

$$\dot{x}(t) = \begin{bmatrix} -\dfrac{1}{RC} & \dfrac{1}{RC} \\ \dfrac{1}{RC} & -\dfrac{2}{RC} \end{bmatrix} x + \begin{bmatrix} 0 \\ 1 \end{bmatrix} u$$

Exercise 9.3 Verify the results of Example 9.3. Obtain the E and D matrices for different signals in the circuit.

Exercise 9.4 Repeat Example 9.3 and Exercise 9.3 with two different capacitors C_1 and C_2.

Example 9.4 Consider the circuit in Fig. 9.5.

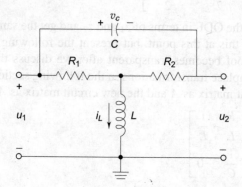

Fig. 9.5

Looking at the circuit, we immediately observe that

$$v_C = u_1 - u_2$$

We might expect that if either u_1 or u_2, or both, are time varying then the capacitor has a current through it and it influences the voltage drops across the resistors, and hence the current i_L. Let us examine this, by applying nodal analysis: Assuming that the voltage across the inductor with respect to the ground as $v_L = L \, di_L/dt$, we have

$$\frac{v_L - u_1}{R_1} + \frac{v_L - u_2}{R_2} + i_L = 0$$

or, equivalently,

$$\dot{i}_L = -\frac{R_1 R_2}{L(R_1 + R_2)} i_L + \frac{R_2}{L(R_1 + R_2)} u_1 + \frac{R_1}{L(R_1 + R_2)} u_2$$

Quite surprisingly, the current i_L does not depend on v_C at all.

Exercise 9.5 Verify the results of Example 9.4.

Let us conclude this section with a quick look at the *linear algebra*. For the sake of an example, let us resort to the same old series RLC circuit with the physical state variables $x_1 = i_L$ and $x_2 = v_C$. Suppose now we define two variables $z_1 \triangleq x_1 + x_2$ and $z_2 \triangleq x_1 - x_2$, so that $x_1 = (z_1 + z_2/2)$ and $x_2 = (z_1 - z_2/2)$ We have the following state equation:

$$\dot{z} = \begin{bmatrix} \left(-\dfrac{R+1}{2L} + \dfrac{1}{2C} \right) & \left(-\dfrac{R-1}{2L} + \dfrac{1}{2C} \right) \\[2ex] \left(-\dfrac{R+1}{2L} - \dfrac{1}{2C} \right) & \left(-\dfrac{R-1}{2L} - \dfrac{1}{2C} \right) \end{bmatrix} z + \begin{bmatrix} \dfrac{1}{L} \\[2ex] \dfrac{1}{L} \end{bmatrix} u$$

and, the voltage across the inductor may be expressed as

$$v_L = \begin{bmatrix} -\dfrac{R+1}{2} & \dfrac{R-1}{2} \end{bmatrix} z + [1]\, u$$

We may solve the ODE in terms of z_1 and z_2 and get the same results as before. We do not prove this at this point, but present the following simple convincing argument; the proof becomes transparent after we discuss the solution of state equation using Laplace transformations in the following section. Let us designate the original circuit matrix as A and the new circuit matrix as A', i.e.,

$$A = \begin{bmatrix} -\dfrac{R}{L} & -\dfrac{1}{L} \\[2ex] \dfrac{1}{C} & 0 \end{bmatrix}$$

and

$$A' = \begin{bmatrix} \left(-\dfrac{R+1}{2L} + \dfrac{1}{2C} \right) & \left(\dfrac{1}{2L} + \dfrac{1}{2C} \right) \\[2ex] \left(-\dfrac{R+1}{2L} - \dfrac{1}{2C} \right) & \left(\dfrac{1}{2L} - \dfrac{1}{2C} \right) \end{bmatrix}$$

We say that A and A' are similar since

1. the *trace* of $A \equiv$ the trace of A',
2. the *rank* of $A \equiv$ the rank of A', and
3. the *eigenvalues* of $A \equiv$ the eigenvalues of A'

Exercise 9.6 Verify that A and A' are similar using the three properties listed above.

Let us explain the algebra. The new variables z_1 and z_2 are related to the old variables x_1 and x_2 as

$$\overline{z} = \begin{bmatrix} 1 & 1 \\ 1 & -1 \end{bmatrix} \overline{x}$$

$$= T \overline{x}$$

Since T is a non-singular matrix, we may write

$$\overline{x} = T^{-1} \overline{z}$$

Given,

$$\dot{\overline{x}} = A \overline{x} + B \overline{u}$$

We may obtain the following sequence:

$$T \dot{\overline{x}} = T A \overline{x} + T B \overline{u}$$

$$\Rightarrow \dot{\overline{z}} = T A T^{-1} \overline{z} + T B \overline{u}$$

$$= A' \overline{z} + B' \overline{u}$$

Any transformation such that

$$T A = A' T \tag{9.11}$$

with a non-singular matrix T is called a *similarity transformation*. We will not explore on these issues any further in this textbook; we suggest that students in early semesters ought to appreciate the power of mathematical techniques, despite their abstract nature, and use mathematics in engineering rather habitually. Courses such as control systems and digital signal processing in later semesters rely heavily on this type of algebra.

We conclude this section with a note that, in fact, *any* linear combination of the physical variables x_1 and x_2, i.e.,

$$\overline{z} = T \overline{x}$$

would do the same job, provided, of course, T is non-singular. What is most important for us is that we *generate* two 'independent' variables z_1 and z_2 using both x_1 and x_2, and any two independent variables should suffice since there are only two energy-storage elements in the circuit and the order of the ODE describing the circuit is 2.

In other words, variables such as $z_1 = x_1$, and $z_2 = -x_1$ are not valid since, in this case, z_1 is *not* independent of z_2. In general, any linear combination of x_1 and x_2 such that the transformation matrix T is non-singular would generate the necessary and sufficient pair of independent variables z_1 and z_2 from x_1 and x_2. Of course, z_1 and z_2 are also called state variables.

9.3 The Laplace Transformation

We have come across ordinary differential equations in Chapters 5 and 6. We solved them by guessing a possible general solution and determining the parameters for the complete solution. In Chapter 6, Sec. 6.1.1, we have also noted that our guesses might be incomplete. In this section, we shall present a more systematic way of solving ODE's using a linear transformation called *Laplace transformation*. We just remark that this transformation is much like the algebraic transformation discussed in the preceding section; the theory of complex variables is needed in addition to the matrix methods. Nevertheless, we assure the reader that this is quite an enjoyable section. In fact, the reader should have been familiar with this in an early or concurrent course in mathematics. Accordingly, the treatment is geared to motivate the reader.

First, let us understand what a transformation is, and why is it required. Suppose we need to multiply two large, say hundred digit, numbers N_1 and N_2. If we go about multiplying them the way we are generally trained, it requires 100×100 multiplications and one hundred additions. Even if we are doing this on a computer that would perform a multiplication in just a nanosecond, $10\,\mu s$ for multiplication alone is an enormous time and, even worse, this time 'blows up' exponentially with every additional digit in the multiplier and the multiplicand. Isn't it a more *efficient* way of computing the product taking the logarithms of these numbers, adding them, and obtaining the antilogarithm of the sum? This would simplify our problem by 'transforming (= converting)' the multiplication into an addition. All we need is a table of logarithms. However, the most important issue is the one-to-one mapping, i.e., for a given number the logorithm is uniquely defined, and for a given number its anti-logarithm is also uniquely defined; this would ensure that there is no ambiguity in the results obtained. Thus, we are already familiar with one variety of 'transformation'. We should consider transformations as tools that simplify complicated operations into simpler ones[3]. Of course, given a problem, we should *cleverly* devise a transformation, that has one-to-one mapping. For instance, the matrix T in the preceding section is non-singular, i.e., its inverse exists and is unique; moreover, it is an orthogonal matrix, i.e., its inverse is a scalar multiple of its transpose.

One such clever transformation useful for us in solving differential equations, and further looking at the circuits in a better perspective, is the Laplace[4] transformation:

$$\mathcal{L} : f(t) \longleftrightarrow F(s)$$

Let us now define this formally:

Definition 9.1: Laplace Transformation

$$F(s) \overset{\Delta}{=} \mathcal{L}\{f(t)\}$$
$$= \int_0^\infty f(t)\, e^{-st}\, dt \tag{9.12}$$

i.e., to construct the Laplace transform of a given time function $f(t)$, we first

---

[3] Can you paint a big aeroplane with just a litre of colour? Yes sir. First, I shall take it into the sky and make it small. Isn't painting easy?!

[4] Pierre Simon Marquis De Laplace (1749–1827) was a great French mathematician who made important contributions to celestial mechanics, probability theory, etc. It is interesting to note that Napolean Bonaparte was his student for a year!

multiply it by e^{-st} and then integrate the product with respect to time over the entire time interval 0 to $+\infty$. Here

$$s = \sigma + j\omega$$

is a complex variable with units nepers/second, abbreviated as Np/s hereafter[5].

Let us do a few examples so that we are familiar with the integral.

Example 9.5 To start with let us look at the unit step function $q(t)$ defined in Eqn (5.58) in Chapter 5. We repeat it here for convenience:

$$f(t) = \begin{cases} 1 & \text{for } t \geq 0 \\ 0 & \text{otherwise} \end{cases}$$

Plugging this into the integral in Eqn (9.12), we get

$$F(s) = \int_0^\infty e^{-st}\, dt$$

i.e.,

$$\mathcal{L}\{q(t)\} = \frac{1}{s}$$

Example 9.6 Let us compute the Laplace transform of e^{at} for $t \geq 0$. Plugging this in Eqn (9.12), we get

$$\mathcal{L}\{e^{at}, t \geq 0\} = \frac{1}{s-a}$$

We may extend this example to compute

$$\mathcal{L}\{e^{jat}, t \geq 0\} = \frac{1}{s-ja}$$

so that

$$\mathcal{L}\{\sin at, t \geq 0\} = \frac{a}{s^2 + a^2}$$

and

$$\mathcal{L}\{\cos at, t \geq 0\} = \frac{s}{s^2 + a^2}$$

At the end of Example 9.6 above, the reader would have become skeptical as to what happens if we try to compute the Laplace transform of e^{at} for $t < 0$; more generally, is any arbitrary function $f(t)$ transformable? Yet another question would be in terms of computing $f(t)$ *uniquely* given $F(s)$. We will address these questions now.

Any function $f(t)$ is transformable to a corresponding $F(s)$ if

$$\int_0^\infty |f(t)|e^{-\sigma t}\, dt < \infty \tag{9.13}$$

where σ is a real, positive number.

It is easy to see that this sufficient (but not necessary) condition is violated if we consider e^{at} for $t < 0$. Any function which is bounded in absolute value for all $t \geq 0$, such as the sine and cosine functions, satisfies the sufficient condition above.

[5] The word Neper honours a 16th century mathematician Napier.

A counter-example is e^{t^2}. Seeing from a different angle, since we give signals to the circuits from $t = 0$ onwards, we may safely assume that this condition is satisfied by most of the signals we encounter in engineering. In practice, we may forget about the condition. However, it is recommended that the reader be well informed that such a sufficient condition does exist.

Some Laplace transform pairs

Function	Symbol	Laplace transform
Unit impulse	$\delta(t)$	1
Unit step	$q(t)$	$\dfrac{1}{s}$
Unit ramp	$r(t)$	$\dfrac{1}{s^2}$
Exponential	e^{at}	$\dfrac{1}{s-a}$
Decaying ramp	te^{-at}	$\dfrac{1}{(s+a)^2}$
Cosine	$\cos \omega t$	$\dfrac{s}{s^2+\omega^2}$
Sine	$\sin \omega t$	$\dfrac{\omega}{s^2+\omega^2}$
Decaying cosine	$e^{-at}\cos \omega t$	$\dfrac{s+a}{(s+a)^2+\omega^2}$
	$\dfrac{t^{n-1}}{(n-1)!}$	$\dfrac{1}{s^n}$
	$\dfrac{d^n}{dt^n}\delta(t)$	s^n

Coming to the second question, there is a rich theory of complex variables behind this transformation. The *inverse Laplace transformation* of $F(s)$ is given by

$$f(t) = \mathcal{L}^{-1}\{F(s)\} = \frac{1}{j\,2\pi} \int_{\sigma-j\infty}^{\sigma+j\infty} F(s)\,e^{st}\,ds \qquad (9.14)$$

which a contour integral. This integral relationship assures us that there cannot be two different functions having the same Laplace transformation $F(s)$. In other words, the mapping

$$\mathcal{L} : f(t) \longleftrightarrow F(s)$$

is indeed a one-to-one mapping as desired. We call $f(t)$ and $F(s)$ as a Laplace transform pair. Just as we have Clark's tables for logarithms, there are available several compilations of extensive tables of Laplace transform pairs and the two integral equations, Eqn (9.12) and Eqn (9.14), are seldom used in practice. In this section we have given a smaller table of Laplace transforms of function we come across more frequently in this chapter. We shall uncover the significance of this Laplace transformation in the following section.

9.3.1 Properties of Laplace Transformation

We shall tap on a few very important properties of this transformation and see how we simplify our computations.

The most important property of the Laplace transformation is *linearity*. Looking at the definition in Eqn (9.12), we may observe that

$$\mathcal{L}\{\alpha\, f(t) + \beta\, g(t)\} = \alpha\, \mathcal{L}\{f(t)\} + \beta\, \mathcal{L}\{g(t)\} \tag{9.15}$$

The reader may recollect that differentiation and integration also satisfy this property of linearity. We have already made use of this property in Example 9.6 while trying to compute the transforms of $\cos \omega t$ and $\sin \omega t$, i.e., we have written

$$\mathcal{L}\{e^{j\omega t}\} = \mathcal{L}\{\cos \omega t + j \sin \omega t\}$$
$$= \mathcal{L}\{\cos \omega t\} + j\, \mathcal{L}\{\sin \omega t\}$$
$$= \frac{s}{s^2 + \omega^2} + j\, \frac{\omega}{s^2 + \omega^2}$$

so that we can segregate the real and imaginary parts.

Next in the order of importance come the Laplace transformation of (i) the differentiation and (ii) the integration of a function $f(t)$ with respect to time.

Suppose that a given arbitrary function $f(t)$ is continuous for all $t \geq 0$ and satisfies the sufficient condition in Eqn (9.13). Then

$$\mathcal{L}\left\{\frac{d}{dt}\, f(t)\right\} = \int_0^\infty e^{-st}\left(\frac{d}{dt}\, f(t)\right) dt$$
$$= s \int_0^\infty e^{-st}\, f(t)\, dt + e^{-st}\, f(t)\big|_0^\infty$$
$$= s\, F(s) - f(0) \tag{9.16}$$

Example 9.7 Let $f(t) = \sin^2 t$. Since

$$\frac{d}{dt}\, f(t) = 2 \sin t \, \cos t = \sin 2t \quad \text{and} \quad f(0) = 0, \text{ we have}$$

$$\mathcal{L}\left\{\frac{d}{dt}\,f(t)\right\} = \frac{2}{s^2 + 4}$$

$$\Rightarrow \quad s\,F(s) - f(0) = \frac{2}{s^2 + 4}$$

$$\text{or} \quad \mathcal{L}\left\{\sin^2 t\right\} = \frac{2}{s(s^2 + 4)}$$

Notice that the initial condition $f(0)$ is important. At this juncture, we shall adopt the following notation. We are interested in circuits, energized by external sources, in the present and in the future, i.e., for all $t \geq 0$. Accordingly, very much in agreement with our notion in Chapter 5, we shall make use of $f(0^-)$ as the initial condition, instead of $f(0)$.

Let us now extend to higher derivatives. To begin with, if we consider

$$\mathcal{L}\left\{\frac{d^2}{dt^2}\,f(t)\right\}$$

we may use the previous result in Eqn (9.16) and write

$$\mathcal{L}\left\{\frac{d^2}{dt^2}\,f(t)\right\} = s\left[\mathcal{L}\left\{\frac{d}{dt}\,f(t)\right\}\right] - \frac{d}{dt}\,f(t)|_{t=0}$$

$$= s^2\,F(s) - s\,f(0^-) - f^1(0^-) \qquad (9.17)$$

where we define

$$f^k(0^-) = \frac{d^k}{dt^k}\,f(t)|_{t=0^-}, \quad k = 0,\,1,\,2,\,\ldots$$

By induction, we might obtain the following general extension to the nth derivative of a function:

$$\mathcal{L}\left\{\frac{d^n}{dt^n}\,f(t)\right\} = s^n\,F(s) - s^{n-1}\,f(0^-)$$

$$- s^{n-2}\,f^1(0^-) - \cdots - f^{n-1}(0^-) \qquad (9.18)$$

Example 9.8 Let $f(t) = t^2$. For this, we have

$$f(0^-) = f^1(0^-) = 0, \quad \text{and} \quad \frac{d^2}{dt^2}\,f(t) = 2$$

Hence,

$$s^2\,F(s) = \mathcal{L}\{2\}$$

$$= \frac{2}{s}$$

$$\Rightarrow \quad \mathcal{L}\{f(t)\} = \frac{2}{s^3}$$

Exercise 9.7 Verify the results of Examples 9.7 and 9.8 by directly using the definition of Laplace transformation in Eqn (9.12).

Coming to the integration, we may derive easily that

$$\mathcal{L}\left\{\int_0^\infty f(\lambda)\,d\lambda\right\} = \frac{1}{s} F(s) \tag{9.19}$$

Exercise 9.8 Derive Eqn (9.19) using Eqn (9.12).

We might induce this result to the transformation of multiple integrals. We illustrate this in the following example.

Example 9.9 Suppose it is given that

$$\mathcal{L}\{f(t)\} = \frac{1}{s^2\,(s^2+9)}$$

and we need to determine $f(t)$, $t \geq 0$. Using linearity and Example 9.6, we know that

$$\mathcal{L}\left\{\frac{1}{3}\sin 3t\right\} = \frac{1}{s^2+9}$$

Hence,

$$\mathcal{L}^{-1}\left\{\frac{1}{s^2\,(s^2+3^2)}\right\} = \frac{1}{3}\int_0^t\int_0^t \sin(3\lambda)\,d\lambda$$

$$= \frac{t}{9} - \frac{\sin 3t}{27}$$

In this example, we have also introduced the idea of inverse Laplace transformation. We present more details after a while.

So far, we have observed that differentiation in time $\equiv$ multiplication by 's' and integration in time $\equiv$ division by 's', i.e., we were able to *transform* the operations of differentiation and integration into simple algebraic operations of multiplication and division.

It is interesting to note that the above observations work in the reverse direction also. In other words,

- differentiation with respect to 's' $\equiv$ multiplication by 't',

$$\mathcal{L}\{t\,f(t)\} = -\frac{d}{ds} F(s) \tag{9.20}$$

- integration with respect to 's' $\equiv$ division by 't',

$$\mathcal{L}\left\{\frac{f(t)}{t}\right\} = \int_0^\infty F(\rho)\,d\rho \tag{9.21}$$

Exercise 9.9 Derive Eqns (9.20) and (9.21).

We were looking at shifting on the time axis earlier. Let us then see the transformation of a *time-shifted* function $f(t - T)$. Once again, from the basic definition in Eqn (9.12), we have

$$\mathcal{L}\{f(t - T)\} = e^{-Ts} F(s) \tag{9.22}$$

Example 9.10 Let us compute the transform of a unit step delayed by 3 s:

$$\mathcal{L}\{q(t - 3)\} = \int_3^\infty e^{-st} \, dt$$

$$= \frac{e^{-3s}}{s}$$

Conversely, we have the 'shifting on the s-axis' as follows:

$$\mathcal{L}\{e^{s_o t} f(t)\} = F(s - s_o) \tag{9.23}$$

Exercise 9.10 Derive Eqn (9.23).

Example 9.11 Earlier we had the underdamped sinusoidal response given by

$$y(t) = e^{-\zeta \omega_n t} \sin(\omega_d t + \theta)$$

The transformation for this may be obtained by first expanding $\sin(A + B)$ and then applying the linearity property of the transformation as follows:

$$\mathcal{L}\{y(t)\} = \frac{(s + \zeta \omega_n)}{(s + \zeta \omega_n)^2 + \omega_d^2} \sin \theta + \frac{\omega_d}{(s + \zeta \omega_n)^2 + \omega_d^2} \cos \theta$$

Exercise 9.11 Verify the results of Example 9.11.

Next comes another irresistible property. Given two time functions $f_1(t)$ and $f_2(t)$, we often come across an integral of the form

$$g(t) = \int_0^\infty f_1(\lambda) f_2(t - \lambda) \, d\lambda \tag{9.24}$$

$$= \int_0^\infty f_2(\lambda) f_1(t - \lambda) \, d\lambda$$

$$\triangleq f_1(t) * f_2(t)$$

We use the symbol $*$ to denote this integration. As it is, this is quite a messy integral involving the following four steps:

- flip one of the functions, say $f_2(\lambda)$, to get $f_2(-\lambda)$
- shift it by t to get $f_2(t - \lambda)$
- multiply with $f_1(\lambda)$
- find the area under the curve $f_1(\lambda) f_2(t - \lambda)$

We call this integral *convolution integral*.

Example 9.12 Suppose we have that

$$f_1(\lambda) = f_2(\lambda) = q(\lambda),$$

the unit step function. Flipping the unit step function and shifting by t gives us

$$q(t - \lambda)$$

Clearly, $q(t - \lambda) = 0$ if $\lambda > t$. Accordingly,

$$f_1(t) * f_2(t)$$

$$= \int_0^\infty 1 \, q(t - \lambda) \, d\lambda$$

$$= \int_0^t d\lambda$$

$$= t$$

We illustsrate the four steps graphically in Figs 9.6(a)–9.6(d). The area of the rectangular box in Fig. 9.6(d) is $1 \times t = t$.

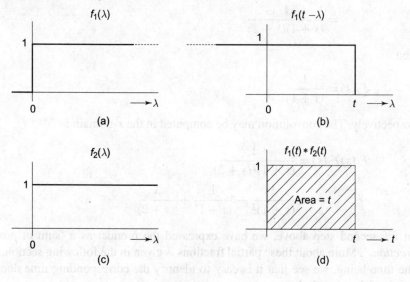

Fig. 9.6

Noting that t and λ are two independent variables, we can tame this integral using Laplace transformation as follows:

$$\mathcal{L}\{g(t)\} = \int_0^\infty \left(\int_0^\infty f_1(\lambda) f_2(t - \lambda) \, d\lambda \right) e^{-st} \, dt$$

$$= \int_0^\infty f_1(\lambda) \left(\int_0^\infty f_2(t - \lambda) e^{-s(t-\lambda)} \, dt \right) e^{-s\lambda} \, d\lambda$$

$$= \left(\int_0^\infty f_1(\lambda) e^{-s\lambda} \, d\lambda \right) \left(\int_0^\infty f_2(t - \lambda) e^{-s(t-\lambda)} \, dt \right)$$

$$= F_1(s) \, F_2(s)$$

$$= F_2(s) \, F_1(s)$$

(9.25)

In other words, the messy convolution integral has become a multiplication. We examine Example 9.12 in this light.

$$\mathcal{L}\{q(t)\} = \frac{1}{s}$$

$$\Rightarrow \quad \mathcal{L}\{q(t) * q(t)\} = \frac{1}{s}\frac{1}{s} = \frac{1}{s^2}$$

From the table we have given earlier, we recognize that

$$q(t) * q(t) = \mathcal{L}^{-1}\left\{\frac{1}{s^2}\right\} = t$$

Example 9.13 Consider the two functions $f_1(t) = e^{-t} - e^{-2t}$ and $f_2(t) = e^{-t}$. The corresponding Laplace transforms are

$$F_1(s) = \frac{1}{s+1} - \frac{1}{s+2}$$

$$= \frac{1}{(s+1)(s+2)}$$

and

$$F_2(s) = \frac{1}{(s+1)}$$

respectively. The convolution may be computed in the s-domain as

$$F_1(s)F_2(s) = \frac{1}{(s+1)^2(s+2)}$$

$$= \frac{1}{(s+1)^2} - \frac{1}{(s+1)} + \frac{1}{(s+2)}$$

In the second step above, we have expressed the product as a 'sum of *partial fractions*'. More about these partial fractions is given in the following section. For the time being, we see that it is easy to identify the corresponding time domain result of the convolution as

$$f_1(t) * f_2(t) = te^{-t} - e^{-t} + e^{-2t}$$

using the examples and properties of Laplace transformation discussed earlier.

Exercise 9.12 Verify the results of Example 9.13 using the four steps to evaluate the convolution integral.

It is evident that the transformation of a function *in the time domain* into a function *in the s-domain* simplifies our computations significantly. We will see two more properties of the transformation a little while later. At the moment, we shall concentrate on the application of this transformation to solve our circuit problems.

9.3.2 Solving ODE

A first-order ODE has the following general form:

$$\dot{y}(t) + a_0 y(t) = b_1 \dot{u}(t) + b_0 u(t) \tag{9.26}$$

where, for convenience, we call $y(t)$ the output and $u(t)$ the input. Applying Laplace transformation on both sides of the equation, and remembering that the transformation is linear, we get

$$s\,Y(s) - y(0^-) + a_0\,Y(s) = b_1\left[s\,U(s) - u(0^-)\right] + b_0 U(s)$$

or

$$Y(s) = \frac{y(0^-) - b_1 u(0^-) + (b_1 s + b_0)\,U(s)}{s + a_0} \tag{9.27}$$

Observe that we need the Laplace transformation of the input signal $u(t)$, the initial conditions $y(0^-)$ and $u(0^-)$, and the coefficients a_0, b_0, b_1 of the ODE to solve for $Y(s)$. Later we may obtain the inverse Laplace transformation to determine the output signal $y(t)$ in time domain.

Earlier we have agreed upon that we apply signals to the circuit only for the time $t \geq 0$, and hence we may assume that $u(0^-) = 0$. We will rewrite the above equation in the following more readable way:

$$Y(s) = \underbrace{\frac{y(0^-)}{s + a_0}}_{\text{natural response}} + \underbrace{\frac{b_1 s + b_0}{s + a_0} \cdot U(s)}_{\text{forced response}} \tag{9.28}$$

If we examine each of the terms—the natural response and the forced response—we understand that the natural response due to the initial condition $y(0^-)$ is a decaying exponent; after all, this is what we have learnt in Chapter 5. Further, if we define

$$H(s) \triangleq \frac{b_1 s + b_0}{s + a_0} \tag{9.29}$$

then we see that the forced response is a *product* of two functions $H(s)$ and $U(s)$ which means that in the time domain there is a convolution integral to be evaluated in order to solve the ODE completely. Indeed, using the method of integrating factors, as was done in Eqn (5.66) and Eqn (5.67) of Chapter 5, we can show that the complete solution in the time domain has a convolution integral. We will illustrate this point in the following section while discussing the inverse transformation. Until then, however, the reader might get convinced that the ODE we have seen in Chapter 5 is a special case of Eqn (9.26) with $a_0 = b_0 = \frac{1}{\tau}$ and $b_1 = 0$ and, with $u(t) = q(t)$, the unit step function, i.e., $U(s) = 1/s$, we have

$$Y(s) = \frac{y(0^-)}{s + \frac{1}{\tau}} + \frac{\frac{1}{\tau}}{s\left(s + \frac{1}{\tau}\right)}$$

and

$$y(t) = y(0^-)e^{-t/\tau} + \left(1 - e^{-t/\tau}\right)$$
$$= \left[y(0^-) - y(\infty)\right]e^{-t/\tau} + y(\infty)$$

Well, if a transformation could make things simpler for us for a first-order ODE, why not for the higher-order ones? Let us look at the general form of a 2 second-order ODE as an extension of Eqn (9.26) as follows:

$$y^{(2)} + a_1 y^{(1)} + a_0 y = b_2 u^{(2)} + b_1 u^{(1)} + b_0 u \tag{9.30}$$

Applying Laplace transformation on both sides and rearranging the terms, we get

$$Y(s) = \overbrace{\frac{(s + a_1)y(0^-) + y^{(1)}(0^-)}{s^2 + a_1 s + a_0}}^{\text{natural response}}$$
$$+ \underbrace{\frac{b_2 s^2 + b_1 s + b_0}{s^2 + a_1 s + a_0} U(s)}_{\text{forced response}} \tag{9.31}$$

Once again, we might extend the definition of $H(s)$ in Eqn (9.29) as follows:

$$H(s) \triangleq \frac{b_2 s^2 + b_1 s + b_0}{s^2 + a_1 s + a_0} \tag{9.32}$$

and end up evaluating a convolution integral in the s domain, this time with some more complicated functions. We leave it to the reader to study these equations in more detail vis á vis the results we have got for RLC circuits in Chapters 5 and 6, and in this chapter earlier.

Exercise 9.13 Compare Eqn (9.30) to Eqn (9.32) with those we obtained for the RLC circuits and comment on the complete response $y(t)$.

We shall now present the general form of an nth order ODE as an extension of Eqn (9.26) as follows:

$$y^{(n)} + a_{n-1} y^{(n-1)} + a_{n-2} y^{(n-2)} + \cdots + a_0$$
$$= b_m u^{(m)} + b_{m-1} u^{(m-1)} + b_{m-2} u^{(m-2)} + \cdots + b_0 \tag{9.33}$$

We leave it to the reader to expand this using Eqn (9.18). We simply show that the transfer function $H(s)$ looks like

$$H(s) = \frac{b_m s^m + b_{m-1} s^{m-1} + \cdots + b_1 s + b_0}{s^n + a_{n-1} s^{n-1} + \cdots + a_1 s + a_0} \tag{9.34}$$

We remark here that, as far as the circuit theory is concerned, we need not really take the trouble of deriving a higher-order ODE for the given circuit. Making use of state variables turns out to be immensely advantageous as we see in the following third-order ODE. We solve it completely to gain an insight into what's going on.

Example 9.14 Consider the circuit in Fig. 9.7.

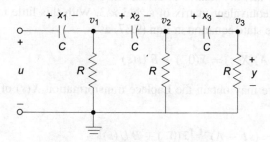

Fig. 9.7

This is a popular circuit used for phase shifting in oscillators. Defining the input, the state variables x_1, x_2, x_3; and the output y as shown in the figure, we have

$$u = x_1 + x_2 + x_3 + y$$

$$C\dot{x}_1 = \frac{u - x_1}{R} + C\dot{x}_2$$

$$C\dot{x}_2 = \frac{u - x_1 - x_2}{R} + C\dot{x}_3$$

$$C\dot{x}_3 = \frac{u - x_1 - x_2 - x_3}{R}$$

We may eliminate the derivatives in the right-hand side by back substitution and get the following state and output equations:

$$\dot{\overline{x}} = \frac{1}{RC} \begin{bmatrix} -3 & -2 & -1 \\ -2 & -2 & -1 \\ -1 & -1 & -1 \end{bmatrix} \overline{x} + \frac{1}{RC} \begin{bmatrix} 3 \\ 2 \\ 1 \end{bmatrix} \tag{9.35}$$

$$y = \begin{bmatrix} -1 & -1 & -1 \end{bmatrix} \overline{x} + [1]\, u \tag{9.36}$$

respectively. Of course, we can manipulate, juggle around, and obtain a third-order ODE. But, we profess that the above methodology using state variables offers much more. Applying Laplace transformation to each of the three first-order differential equations, we get

$$s\overline{X}(s) - \overline{x}(0^-) = A\,\overline{X}(s) + B\,U(s)$$

$$Y(s) = E\,\overline{X}(s) + D\,U(s) \tag{9.37}$$

where $\overline{x}(0^-)$ is the vector of initial capacitor voltages. The first term $s\,\overline{X}(s)$ on the left-hand side may be cleverly written as

$$s\overline{X}(s) \equiv \begin{bmatrix} s & 0 & 0 \\ 0 & s & 0 \\ 0 & 0 & s \end{bmatrix} \overline{X}(s) = sI\overline{X}(s) \tag{9.38}$$

where I is the identity matrix of size 3. In other words, we have converted[6] a vector into an equivalent matrix of size 3×3. With this little manipulation, we may rewrite the state equation in Eqn (9.37) as

$$(sI - A)\,\overline{X}(s) = \overline{x}(0^-) + B\,U(s) \tag{9.39}$$

from which, we may obtain the Laplace transformation $\overline{X}(s)$ of the state vector $\overline{x}(t)$ as

$$\overline{X}(s) = (sI - A)^{-1}\left[\overline{x}(0^-) + B\,U(s)\right] \tag{9.40}$$

and hence,

$$Y(s) = \overbrace{E\,(sI - A)^{-1}\,x(0^-)}^{\text{natural response}}$$
$$+ \underbrace{\left[E\,(sI - A)^{-1}\,B + D\right]U(s)}_{\text{forced response}} \tag{9.41}$$

We may identify the transfer function $H(s)$ as

$$H(s) = \left[E\,(sI - A)^{-1}\,B + D\right]$$
$$= \frac{E\,\mathrm{adj}\,(sI - A)\,B}{\mathrm{determinant}\,(sI - A)} + D \tag{9.42}$$

Let us plug in the numericals to get a better picture. We assume, for the sake of simplicity, that the initial voltages of all the capacitors are zero, that the product $RC = 1$, and that the input is a unit step signal applied at $t = 0$. With the matrices A, B, E, D given in Eqns (9.35) and (9.36), we get

$$\overline{X}(s) = \begin{bmatrix} (s+3) & 2 & 1 \\ 2 & (s+2) & 1 \\ 1 & 1 & (s+1) \end{bmatrix}^{-1} \begin{bmatrix} 3 \\ 2 \\ 1 \end{bmatrix} U(s)$$

$$= \frac{1}{s^3 + 6s^2 + 5s + 1} \begin{bmatrix} 3s^2 + 4s + 1 \\ 2s^2 + s \\ s^2 \end{bmatrix} U(s)$$

---

[6] transformed!

Apparently, there are three convolutions as might be expected out of the solution of a third-order ODE. For a unit step input $U(s) = 1/s$, each of the state variables may be written as

$$X_1(s) = \frac{3s^2 + 4s + 1}{s(s^3 + 6s^2 + 5s + 1)}$$

$$X_2(s) = \frac{2s^2 + s}{s(s^3 + 6s^2 + 5s + 1)}$$

and

$$X_3(s) = \frac{s}{(s^3 + 6s^2 + 5s + 1)}$$

With these state variables, the output may be written as

$$Y(s) = - [X_1(s) + X_2(s) + X_3(s)] + U(s)$$

We leave it to the reader to obtain $Y(s)$.

Exercise 9.14 Obtain $Y(s)$ in Example 9.14.

Exercise 9.15 Obtain $Y(s)$ in Example 9.14 with the initial conditions $x_1(0^-) = -1$, $x_2(0^-) = 3$, $x_3(0^-) = -2$.

9.3.3 The Matrix Exponential

The matrix $(sI - A)^{-1}$ plays a very crucial role in the computation of the state variables and output. First, we need to make sure that this matrix is non-singular. It is clear that the matrix becomes singular if we substitute the eigenvalues of A for the complex variable s. We shall explore this a little later. Second, if it were a simple first-order circuit, then $(sI - A)^{-1}$ would look like $1/(s - a)$, the Laplace transform of e^{at}. Following this convention, we may write

$$\mathcal{L}^{-1}\{(sI - A)^{-1}\} \triangleq e^{At} \qquad (9.43)$$

and call this *matrix exponential*. Observe that it has the same dimension as that of the matrix A. There are certain interesting properties of this matrix exponential; for instance, it is not obtained just by raising each of the elements of the matrix A. These properties are more useful in studying general system theory and hence we avoid them here. We conclude this section giving the following algorithm to compute the state vector. In general, we may consider the input as the vector $\bar{u}$ and the output as the vector $\bar{y}$, indicating that there could be several inputs and several outputs.

Step 1	Enumerate the state variables systematically.
Step 2	Apply nodal or mesh analysis to the circuit and obtain the state and output equations: $\dot{\overline{x}} = A\overline{x} + B\overline{u}; \quad \overline{y} = E\overline{x} + D\overline{u}$
Step 3	Obtain the matrix $sI - A$ and its inverse, in terms of the complex variable s.
Step 4	The solution for the state vector $\overline{X}(s)$ is $\overline{X}(s) = (sI - A)^{-1}\left[\overline{x}(0^-) + B\,\overline{U}(s)\right]$
Step 5	The required solution is $\overline{Y}(s) = E\overline{X}(s) + D\overline{U}(s)$

9.3.4   The Linear Algebra

Let us do Example 9.14 again, this time from a different perspective. Towards the end, we shall point out some important issues.

Example 9.14(a) Applying KCR to node v_1 gives us

$$C\,(\dot{v}_1 - \dot{u}) + C\,(\dot{v}_1 - \dot{v}_2) + \frac{v_1}{R} = 0$$

Applying Laplace transformation to this equation, assuming zero initial conditions, gives us

$$\frac{v_1 - u}{1/sC} + \frac{v_1 - v_2}{1/sC} + \frac{v_1}{R} = 0 \tag{9.44}$$

Exercise 9.16 Obtain the state equation and the output equation of the form

$$\dot{\overline{v}}(t) = A'\overline{v}(t) + B'u(t), \quad y(t) = E'\overline{v}(t) + D'u(t)$$

in terms of the new set of variables v_1, v_2, v_3. How are A and A' related?

A closer look at Eqn (9.44) suggests us the following: The term $1/sC$ is similar to resistance since it divides the potential difference to give us current. This is due to the fact that

$$i_C = C\,\dot{v}_C$$
$$\Rightarrow \quad I_C(s) = sC\,V_C(s)$$

and hence

$$Z_C(s) \triangleq \frac{V_C(s)}{I_C(s)} = \frac{1}{sC}$$

We may proceed ahead applying nodal analysis to the circuit, just as we did in Chapter 6 using $j\omega$. We get the following linear system of equations:

$$\begin{bmatrix} 2sC + \frac{1}{R} & -sC & 0 \\ -sC & 2sC + \frac{1}{R} & -sC \\ 0 & -sC & sC + \frac{1}{R} \end{bmatrix} \begin{bmatrix} v_1 \\ v_2 \\ v_3 \end{bmatrix} = \begin{bmatrix} sC \\ 0 \\ 0 \end{bmatrix} u \tag{9.45}$$

from which we may solve for the node voltages v_1, v_2, and v_3 as

$$\bar{V}(s) = \frac{1}{s^3 + 6s^2 + 5s + 1} \begin{bmatrix} s^3 + 3s^2 + s \\ s^3 + s^2 \\ s^3 \end{bmatrix} U(s) \tag{9.46}$$

using $RC = 1$.

Exercise 9.17 Verify Eqn (9.45) and Eqn (9.46).

Let us conclude this example with the following algebra. Consider the coefficient matrix in Eqn (9.45); hereafter, let us call it $sI - A'$:

$$sI - A' = \begin{bmatrix} 2sC + \frac{1}{R} & -sC & 0 \\ -sC & 2sC + \frac{1}{R} & -sC \\ 0 & -sC & sC + \frac{1}{R} \end{bmatrix}$$

We perform the following operations[7] on the columns of this matrix in the order given:

(1) $C_1 \longleftarrow C_1 + C_2 + C_3$

(2) $C_2 \longleftarrow C_2 + C_3 + C_1$

(3) $C_3 \longleftarrow C_3 + C_2$

i.e., we replace each of the columns with a linear combination of all the three columns. The result would be

$$sI - A' = \begin{bmatrix} s+1 & 1 & 1 \\ 1 & s+2 & 2 \\ 1 & 2 & s+3 \end{bmatrix}$$

with $RC = 1$. Without altering the determinant we may *permute* this matrix and obtain

$$sI - A' = \begin{bmatrix} (s+3) & 2 & 1 \\ 2 & (s+2) & 1 \\ 1 & 1 & (s+1) \end{bmatrix}$$

---

[7] It is expected that the reader is familiar with this in an elementary course on matrices. These are called column reductions. These operations are extremely useful, for instance, in computing the determinant, rank, and inverse of a given matrix. Of course, similar operations on rows (called row reductions) are also allowed.

Let us subtract this from sI and extract A' as

$$A' = \begin{bmatrix} -3 & -2 & -1 \\ -2 & -2 & -1 \\ -1 & -1 & -1 \end{bmatrix}$$

which we discover is the same as the A matrix in Eqn (9.35). What did we accomplish? We simply emphasize that the two sets of variables $\{x_1, x_2, x_3\}$ and $\{v_1, v_2, v_3\}$ are *two* different ways of looking at the same circuit. We often call this *change of basis*[8]. As we have shown in Chapter 4, nodal analysis and mesh analysis are two different perspectives of the same circuit; so are Thèvenin's and Norton's theorems. Depending on the context, each of these perspectives offers tremendous computational advantage.

Let us make some important observations.

- The state variables x_1, x_2, x_3 and the node voltages v_1, v_2, v_3 are linearly related:

$$x_1 = u - v_1$$

$$x_2 = v_1 - v_2$$

$$x_3 = v_2 - v_3$$

or equivalently, in a matrix form,

$$\bar{x} = \begin{bmatrix} -1 & 0 & 0 \\ 1 & -1 & 0 \\ 0 & 1 & -1 \end{bmatrix} \bar{v} + \begin{bmatrix} u \\ 0 \\ 0 \end{bmatrix}$$

Hence, there should be same (linear) relationship between the vectors $\bar{X}(s)$ and $\bar{V}(s)$. We leave it to the reader to verify this.

- Earlier, we found that

$$\det(sI - A) = s^3 + 6s^2 + 5s + 1$$

which appears as the (common) denominator in $\bar{X}(s)$. The same denominator reappears in Eqn (9.44). This suggests that the eigenvalues of A are not altered and we are talking about similarity transformation between the coefficient matrices A of $\bar{x}(t)$ and A' of $\bar{v}(t)$. This was demonstrated just now. A *similarity transformation* is just performing a sequence of operations such as column (and/or row) reductions and permutations, without altering the trace, the rank, and hence the eigenvalues of a given matrix. If these quantities are invariant in a transformation, the characteristic polynomial (recall the Cayley–Hamilton theorem) remains invariant.

- Looking back at the way we have defined (earlier in Chapters 5 and 6) the natural response and the forced response, we may easily see that this common denominator is none else than the *characteristic polynomial*. We discuss more about this in Sec. 9.4.

---

[8] On the lighter side, one may wonder why so many mirrors are put in a barber's shop; the barber may choose his vantage point and do the hair well!

- In other words, as was already mentioned earlier in this chapter, any minimal (three in this example) independent variables suffice to describe a circuit, be it $\overline{x}$ or $\overline{v}$, since they would be *linearly related*. However, it is noteworthy that the initial conditions are given in terms of $\overline{x}(0)$ and we need to transform them into $\overline{v}(0)$, perhaps an additional computation if $\overline{x}(0) \neq 0$. We may still define the transfer function $H(s)$ in the same old fashion and obtain the results. One major advantage of this latter approach is to directly apply any analysis technique we have learnt in Chapter 4 by replacing the components with their *s-domain equivalents*. We will come back to this point more formally after learning about inverse Laplace transformation.

- If the circuit has initial conditions, we should redraw the circuit to include them. We will come back to this point in Example 9.25 after developing some more ideas.

The idea behind introducing linear algebra in this chapter is not to bemuse the reader. Rather, we strongly feel that the reader should be *prepared* thoroughly for more abstract courses in later semesters. We suggest the reader to keep handy the material presented in this section for easy reference while doing, say, digital signal processing or control systems during later semesters.

Its time for us to look at the inverse Laplace transformation, a little formally, to obtain the time functions. We focus on the solutions of ODE's we have been encountering in this text.

9.3.5 Inverse Laplace Transformation

We begin this section with the following simple observations we have made so far.

- We make use of Laplace transformation to compute *signals*, i.e., time-dependent functions such as voltages and currents.

- We learnt in Chapters 5 and 6 that arbitrary waveforms can be synthesized as linear combinations of unit impulse functions (or, equivalently, unit step functions, and unit ramp functions), exponential functions of the form e^{pt}, and sinusoids represented as complex exponentials of the form $e^{j\omega t}$.

- The solution $Y(s)$ of any ODE consists of two parts—the natural response and the forced response. Either of these parts appear as *rational functions* of the form

$$F(s) = \frac{N(s)}{D(s)}$$

where $N(s)$ is the 'numerator' polynomial in the complex variable s, and $D(s)$ is the 'denominator' polynomial in s.

- From the second observation above, we understand that $F(s)$ must be expressed in the following form

$$F(s) = k_{11} + \frac{k_{21}}{s} + \frac{k_{22}}{s^2} + \cdots + \frac{k_{31}}{s - p_1} + \frac{k_{32}}{s - p_2} + \cdots + \frac{k_{41}}{s^2 + \omega_1^2} + \cdots$$

so that we can identify the unit step functions, unit impulse functions, and the sinusoids.

- Notice that, in the previous observation, we have written a linear combination of the Laplace transforms of *known* signals. These individual terms are once again *rational functions*, and we prefer to call them *partial fractions*. The reader would have been familiar with these fractions, albeit in a different name, such as *indeterminate coefficients*.

- Thus, the whole idea of obtaining inverse Laplace transformation boils down to resolving a given function $F(s)$ into its constituent partial fractions, and this is the theme of this section.

Let us continue Example 9.14. For simplicity, we take up $V_1(s)$ and leave the others to the reader.

Example 9.15 In Example 9.14, we found that

$$V_1(s) = \frac{s^3 + 3s^2 + s}{s(s^3 + 6s^2 + 5s + 1)}$$

if the input to the circuit is a unit step. Suppose that, somehow, we were able to factorize the denominator polynomial as

$$D(s) = s(s + 5.05)(s + 0.65)(s + 0.30)$$

It is easy to see that we might rewrite $V_1(s)$ as

$$V_1(s) = \frac{s^2 + 3s + 1}{s^3 + 6s^2 + 5s + 1}$$

$$= \frac{s^2 + 3s + 1}{(s + 5.05)(s + 0.65)(s + 0.30)}$$

$$= \frac{k_1}{s + 5.05} + \frac{k_2}{s + 0.65} + \frac{k_3}{s + 0.3}$$

so that the LCM is the product of the three factors and the numerator is a sum of three *quadratic polynomials* as in the original rational function (after cancelling the common s). The numerators k_1, k_2, and k_3 are called *residues*.

Now the question is how to extract the residues. Before we present a standard result on computing these residues, we apply a little logic and see how things turn in favour. First, we notice that

$$V_1(s) = \frac{s^2 + 3s + 1}{s^3 + 6s^2 + 5s + 1} = \frac{k_1(s + 0.65)(s + 0.3)}{\cdots} + \frac{k_2(\cdots)}{\cdots} + \frac{k_3(\cdots)}{\cdots}$$

Then, let us multiply both sides by $s + 5.05$ to have

$$\frac{s^2 + 3s + 1}{(s + 0.65)(s + 0.30)} = k_1 + (s + 5.05)\left[\frac{k_2(\cdots)}{\cdots} + \frac{k_3(\cdots)}{\cdots}\right]$$

If we substitute $s = -5.05$, we are left with

$$k_1 = 0.5432$$

The same procedure may be repeated to obtain k_2 and k_3 as

$$k_2 = [(s + 0.65)V_1(s)]\,|_{s=-0.65} = 0.3425$$

$$k_3 = [(s + 0.30)V_1(s)]\,|_{s=-0.30} = 0.1143$$

Thus,

$$V_1(s) = \frac{0.5432}{s + 5.05} + \frac{0.3425}{s + 0.65} + \frac{0.1143}{s + 0.30}$$

and hence, we may quickly identify that

$$v_1(t) = 0.5432\,e^{-5.05t} + 0.3425\,e^{-0.65t} + 0.1143\,e^{-0.30t}$$

Exercise 9.18 Verify the results of Example 9.15 by back substitution.

One of the most important questions here is—*how did we factorize a cubic polynomial*? While we have a standard formula for the roots of a quadratic polynomial, no such formulae are available for higher-order polynomials. One has to necessarily resort to *numerical methods* such as the bisection method or Newton–Raphson method. Today, we have several high-level software packages such as MATLAB, Mathematica, MATRIX-x, etc., which work as extended calculators with built-in subroutines for almost all the numerical methods. The reader may get himself/herself familiar with one such package, which should be available in his/her institution. In the rest of the textbook, we focus only on the method with simpler examples without going deep into numerical methods.

Another question is in terms of multiple roots and complex conjugate roots of the denominator polynomial, such as those we have encountered for the critically damped and underdamped RLC circuit, respectively. We now present a standard result followed by several examples.

9.3.5.1 *Heaviside's expansion theorem*

The idea we have *discovered* in Example 9.15 was originally given by Oliver Heaviside[9]. Let

$$F(s) = \frac{N(s)}{D(s)} = \frac{k_1}{s - p_1} + \frac{k_2}{s - p_2} + \cdots \qquad (9.47)$$

where the denominator polynomial $D(s)$ can be split into several *distinct* factors. Then, any of the residues k_i may be computed as

$$k_i = (s - p_i) \times F(s)|_{s=p_i}, \quad i = 1, 2, \ldots \qquad (9.48)$$

---

[9] Heaviside (1850–1925) was an English engineer gifted with an insight into physical problems. His point of view drew severe criticism from the mathematicians of his time. However, his publication based on heuristics was followed up later and the required mathematical rigour was furnished by Bromwich and others. No important errors have been found in Heaviside's methods.

Here, p_i could be a real number or a complex number; in the latter case, the corresponding residue k_i also turns out to be a complex number.

There are a few important variants of the general case presented in Eqn (9.47).

Case 1

If $N(s)$ and $D(s)$ have the same degree, we may rewrite $F(s)$ as

$$F(s) = k_\infty + \frac{N'(s)}{D(s)} \tag{9.49}$$

with the degree of $N'(s)$ lesser than that of $D(s)$. In this case, we may compute k_∞ by simply taking the limit $s \to \infty$ on either side.

Example 9.16 Let

$$F(s) = \frac{s+1}{s+2}$$

We quickly observe that

$$F(s) = \frac{s+2-1}{s+2} = 1 - \frac{1}{s+2}$$

Thus,

$$k_\infty = \lim_{s \to \infty} F(s) = 1$$

Case 2

If $D(s)$ has multiple roots at $s = p_j$, i.e., $D(s)$ has a factor of the form $(s - p_j)^r$ with the root p_j repeating r number of times,

$$F(s) = \frac{k_1}{s - p_j} + \frac{k_2}{(s - p_j)^2} + \frac{k_3}{(s - p_j)^3} + \cdots + \frac{k_r}{(s - p_j)^r} \tag{9.50}$$

Let us begin with a simple example wherein $D(s)$ is a quadratic polynomial with repeated roots. We assume that the degree of $N(s)$ is less than that of $D(s)$.

Example 9.17 Let

$$F(s) = \frac{N(s)}{(s - p)^2}$$

Here, we may write $F(s)$ as

$$F(s) = \frac{k_1}{s - p} + \frac{k_2}{(s - p)^2}$$

so that the numerator may be written as

$$N(s) = k_1 s + k_2 - k_1 p$$

It is easy to see that k_2 may be quickly computed by multiplying either side of the equation by $(s - p)^2$ and then substituting $s = p$. More generally,

$$k_r = (s - p_j)^r \times F(s)|_{s=p_j} \qquad (9.51)$$

Looking at the numerator, we identify that k_1 may be computed by differentiating $N(s)$ with respect to s since the remaining terms on the rhs are constants. In other words,

$$k_{r-i} = \frac{1}{i!} \frac{d^i}{ds^i} \left[(s - p_j)^r \times F(s) \right], \quad i = 1, 2, \ldots, (r - 1) \qquad (9.52)$$

We observe that we get Eqn (9.51) if we put $i = 0$ in Eqn (9.52).

Exercise 9.19 Derive Eqn (9.52).

Let us look at the following examples.

Example 9.18 Let

$$F(s) = \frac{s^2 + 3s + 2}{s^2 + 2s + 2} = \frac{(s + 1)(s + 2)}{(s + 1 + j1)(s + 1 - j1)}$$

We may write this as

$$F(s) = k_\infty + \frac{k_1}{s + 1 + j1} + \frac{k_2}{s + 1 - j1}$$

where

$$k_\infty = \lim_{s \to \infty} F(s) = 1$$

$$k_1 = (s + 1 + j1) \times F(s)|_{s=-1-j1} = \frac{1 - j1}{2}$$

$$k_2 = (s + 1 - j1) \times F(s)|_{s=-1+j1} = \frac{1 + j1}{2}$$

We remark the following:

- According to the fundamental theorem of algebra, if all the $n + 1$ coefficients of a nth degree polynomial are *non-zero*, and have the same sign, then the roots of the polynomial are either all real or complex conjugates. We find this theorem extensively useful in Chapter 11 on network synthesis.
- If there is a pair of complex-conjugate roots for the denominator polynomial, the corresponding residues also turn out to be complex conjugates as was demonstrated in Example 9.18.

From the partial fractions, we get

$$f(t) = \delta(t) + \frac{1 - j1}{2} e^{(-1-j1)t} + \frac{1 + j1}{2} e^{(-1+j1)t}$$

$$= \delta(t) + e^{-t} \left[\frac{e^{jt} + e^{-jt}}{2} - \frac{e^{jt} - e^{-jt}}{j2} \right]$$

$$= \delta(t) + e^{-t} [\cos t - \sin t]$$

Let us make a careful note that the complex-conjugate roots and residues resulted in sinusoids.

Example 9.19 Let

$$F(s) = \frac{2s^2 + 3s + 2}{(s + 1)^3}$$

We write

$$F(s) = \frac{k_1}{s + 1} + \frac{k_2}{(s + 1)^2} + \frac{k_3}{(s + 1)^3}$$

It would be convenient for us to compute the residues in the following order:

$$k_3 = (2s^2 + 3s + 2)|_{s=-1} = 1$$

$$k_2 = \frac{1}{1!} \frac{d}{ds} (2s^2 + 3s + 2)|_{s=-1} = -1$$

$$k_1 = \frac{1}{2!} \frac{d^2}{ds^2} (2s^2 + 3s + 2)|_{s=-1} = 2$$

Thus,

$$F(s) = \frac{2}{s + 1} - \frac{1}{(s + 1)^2} + \frac{1}{(s + 1)^3}$$

and

$$f(t) = 2 e^{-t} - t e^{-t} + \frac{t^2}{2} e^{-t}$$

Example 9.20 Let

$$F(s) = \frac{s^3 - 5s^2 + 9s + 9}{s^2(s^2 + 9)}$$

$$= \frac{k_1}{s} + \frac{k_2}{s^2} + \frac{k_3}{s - j3} + \frac{k_4}{s + j3}$$

Here,

$$k_2 = \frac{s^3 - 5s^2 + 9s + 9}{s^2 + 9} \bigg|_{s=0} = 1$$

$$k_1 = \frac{1}{1!} \frac{d}{ds} \frac{s^3 - 5s^2 + 9s + 9}{s^2 + 9} \bigg|_{s=0} = 1$$

$$k_3 = \left. \frac{s^3 - 5s^2 + 9s + 9}{s^2(s + j3)} \right|_{s=j3} = j1$$

$$k_4 = k_3^* = -j1$$

Exercise 9.20 Obtain the time function $f(t)$ for $F(s)$ in Example 9.20.

Example 9.21 For $F(s)$ given below,

$$F(s) = \frac{1}{s^3(s^2 - 1)} = \frac{k_1}{s} + \frac{k_2}{s^2} + \frac{k_3}{s^3} + \frac{k_4}{s + 1} + \frac{k_5}{s - 1}$$

Here,

$$k_3 = \left. \frac{1}{s^2 - 1} \right|_{s=0} = -1$$

$$k_2 = \left. \frac{1}{1!} \frac{d^2}{ds^2} \frac{1}{s^2 - 1} \right|_{s=0} = -0$$

$$k_1 = \left. \frac{1}{2!} \frac{d^2}{ds^2} \frac{1}{s^2 - 1} \right|_{s=0} = -1$$

$$k_4 = \left. \frac{1}{s^3(s - 1)} \right|_{s=1} = \frac{1}{2}$$

$$k_5 = \left. \frac{1}{s^3(s + 1)} \right|_{s=1} = \frac{1}{2}$$

Exercise 9.21 Verify that

$$f(t) = -1 - \frac{t^2}{2} + \cosh t$$

in Example 9.21.

Example 9.22 Let

$$F(s) = \frac{2s^3 + 11s^2 + 14s - 5}{s^2 + 5s + 6}$$

Here we notice that the degree of the numerator polynomial $N(s)$ is greater than that of the denominator by 1. We need not worry. We just proceed as follows. We may write $F(s)$ as

$$F(s) = \frac{k_1}{s + 2} + \frac{k_2}{s + 3} + k_3 s + k_4$$

The residues k_1 and k_2 may be easily obtained as

$$k_1 = (s + 2)F(s)|_{s=-2} = -5$$
$$k_2 = (s + 3)F(s)|_{s=-3} = 2$$

The residue k_3 may also be found using the same formula, with a slight modification, using our intuition:

$$k_3 = \lim_{s \to \infty} \frac{F(s)}{s} = 2$$

Why do we take the limit as $s \to \infty$? Observe that, for a factor like $s - p$ in the denominator, $F(s)$ at $s = p$ evaluates to ∞; for a factor like s in the 'numerator', $F(s)$ evaluates to ∞ in the limit $s \to \infty$.

For k_4, we use the blind rule—if everything else is known, why not this one? We simply put $s = 0$ in the equation to get

$$\frac{k_1}{2} + \frac{k_2}{3} + k_4 = -\frac{5}{6}$$

from which we deduce that

$$k_4 = 1$$

Thus,

$$f(t) = -5\,e^{-2t} + 2\,e^{-3t} + 2\,\frac{d}{dt}\,\delta(t) + 1$$

Observe that

$$\mathcal{L}^{-1}\{s\} = \frac{d}{dt}\,\delta(t)$$

Exercise 9.22 Verify the results of Example 9.22.

How do we know that these results are correct? We answer this question in a different manner towards the end of this section.

In the following examples, let us see how an ODE is solved completely using Laplace transformation.

Example 9.23 We shall solve the state equation and the output equation in Eqn (9.35) and Eqn (9.36).

$$\overline{X}(s) = \begin{bmatrix} (s+3) & 2 & 1 \\ 2 & (s+2) & 1 \\ 1 & 1 & (s+1) \end{bmatrix}^{-1} \begin{bmatrix} 3 \\ 2 \\ 1 \end{bmatrix} U(s)$$

$$= \frac{1}{s^3 + 6s^2 + 5s + 1} \begin{bmatrix} 3s^2 + 4s + 1 \\ 2s^2 + s \\ s^2 \end{bmatrix} U(s)$$

$$Y(s) = \begin{bmatrix} -1 & -1 & -1 \end{bmatrix} X(s) + [1]\, U(s)$$

Assuming that the circuit is given a unit step input, we have

$$X_1(s) = \frac{3s^2 + 4s + 1}{s(s^3 + 6s^2 + 5s + 1)}$$

$$= \frac{k_1}{s} + \frac{k_2}{s + 5.05} + \frac{k_3}{s + 0.65} + \frac{k_4}{s + 0.3}$$

$$= \frac{1}{s} + \frac{-0.5431}{s + 5.05} + \frac{-0.3493}{s + 0.65} + \frac{-0.1076}{s + 0.3}$$

so that

$$x_1(t) = 1 - 0.5431\, e^{-5.05t} - 0.3493\, e^{-0.65t} - 0.1076\, e^{-0.3t}$$

We leave to the reader to compute $x_2(t)$ and $x_3(t)$, and hence the output $y(t)$.

Suppose we have the following initial conditions for the state equation (from Exercise 9.15):

$$x_1(0) = -1, \quad x_2(0) = 3, \quad x_3(0) = -2$$

Then, the natural response is

$$\overline{X}_{\text{natural}}(s) = \begin{bmatrix} (s+3) & 2 & 1 \\ 2 & (s+2) & 1 \\ 1 & 1 & (s+1) \end{bmatrix}^{-1} \begin{bmatrix} -1 \\ 3 \\ -2 \end{bmatrix}$$

$$= \frac{1}{s^3 + 6s^2 + 5s + 1} \begin{bmatrix} -s^2 & -7s & -4 \\ 3s^2 & +16s & +9 \\ -2s^2 & -12s & -7 \end{bmatrix}$$

We may obtain $\overline{x}_{\text{natural}}(t)$ by solving three sets of partial fractions. Thus, we need to add

$$\overline{X}_{\text{natural}}(s) = (sI - A)^{-1} [\overline{x}(0)]$$

to $\overline{X}_{\text{forced}}(s)$ to obtain the complete state response $\overline{X}(s)$ and hence $\overline{x}(t)$, and hence $y(t)$.

Exercise 9.23 Do Example 9.23 completely and verify your results.

Example 9.24 Let us solve the following ODE subject to the given initial conditions:

$$\ddot{y}(t) + 3\dot{y}(t) + 2y(t) = 2\dot{u}(t) + u(t)$$

with $y(0) = 1$ and $y^1(0) = 2$.

We get the following natural response:

$$Y_{\text{natural}}(s) = \frac{(s+3)y(0) + y^1(0)}{s^2 + 3s + 2} = \frac{s+5}{(s+1)(s+2)}$$

We leave it to the reader to obtain $y_{\text{natural}}(t)$. The forced response would look like

$$Y_{\text{forced}} = \frac{2s + 1}{(s+1)(s+2)} U(s)$$

Notice that we need to evaluate a convolution integral had we chosen to solve the ODE using time-domain techniques. Given the input (forcing function) $u(t)$, we may obtain its Laplace transform $U(s)$ and then evaluate the forced response by simple multiplication in the s-domain. For instance, if $u(t)$ is a unit ramp function,

$$Y_{\text{forced}}(s) = \frac{k_1}{s} + \frac{k_2}{s^2} + \frac{k_3}{s+1} + \frac{k_4}{s+2}$$

Exercise 9.24 Do Example 9.24 completely. Repeat it for $u(t) = 2e^{-3t}$.

Let us now look at the following circuits and learn how to make use of Laplace transformation, rather directly.

9.3.5.2   Solving circuit problems directly

Example 9.25 Consider the simple first-order circuit in Fig. 9.8.

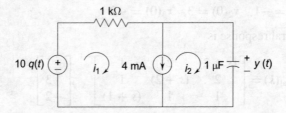

Fig. 9.8

Let us apply mesh analysis. For $t \geq 0$, we get

$$10^3 \times i_1(t) + y(t) = 10\,q(t)$$
$$i_1(t) - i_2(t) = 4 \times 10^{-3}$$
$$10^{-6} \times \dot{y}(t) = i_2(t)$$

Let us now apply Laplace transformation to each of the equations above to get

$$1000\,I_1(s) + Y(s) = \frac{10}{s}$$

$$I_1(s) - I_2(s) = \frac{4 \times 10^{-3}}{s}$$

$$10^{-6} \times [s\,Y(s) - y(0)] = I_2(s)$$

Looking at the circuit, for $t < 0$, we observe that the voltage source is a short circuit and the current source delivers 4 mA to the resistance from left to right; hence, $y(0^-) = y(0^+) = -4$ V. Thus, we get

$$s\,Y(s) - y(0) = 10^6 \times I_2(s)$$
$$= 10^6 \times \left(I_1(s) - \frac{4 \times 10^{-3}}{s} \right)$$
$$= \frac{-1000\,s\,Y(s) + 6000}{s}$$

or,

$$s^2 Y(s) + 1000\, s\, Y(s) = s\, y(0) + 6000$$

$$= \frac{y(0)}{s + 1000} + \frac{6000}{s(s + 1000)}$$

$$\Rightarrow \qquad y(t) = \underbrace{-4\, e^{-1000t}}_{\text{natural response}} + \underbrace{6 \times \left(1 - e^{-1000t}\right)}_{\text{forced response}} \text{ V}$$

$$= 6 - 10\, e^{-1000t} \text{ V}$$

Let us now look at this example from a state variable perspective. In fact, this is as good as applying nodal analysis. For $t \geq 0$, KCR gives us

$$10^{-6} \times \dot{x}(t) + 4 \times 10^{-3} + \frac{x - 10}{1000\,\Omega} = 0$$

i.e., the state equation is

$$\dot{x}(t) = -1000\, x(t) + 6000$$

with $x(0) = -4$ V, and the output equation is

$$y(t) = x(t)$$

In this *first-order* case, we have

$$(s + 1000)X(s) = x(0) + \frac{6000}{s}$$

and we get the same answer $y(t)[= x(t)]$ as before.

Having convinced that we are using Laplace transformation to solve the ODE of a given circuit, we now look at an even more simplified technique. Since the capacitor $1\,\mu$F is the cynosure in the circuit, let us spend some time on that. We have the following sequence of information:

$$i(t) = C\, \dot{v}(t)$$

$$\Rightarrow \quad I(s) = C\,[s V(s) - v(0)]$$

$$= sC\, V(s) - C\, v(0)$$

$$\Rightarrow \quad I(s) + \underbrace{C\, v(0)}_{\triangleq\, I(0)} = \frac{V(s)}{Z(s)} \qquad\qquad (9.53)$$

where

$$Z(s) \triangleq \frac{1}{sC}$$

We may interpret Eqn (9.53) as KCR at a node and obtain the *Laplace equivalent* of capacitance as shown in Fig. 9.9.

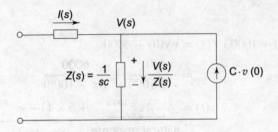

Fig. 9.9

Observe that the initial condition $v(0)$ on the capacitor is *transformed*[10] into a current source $I(0)$ 'in parallel with' the capacitor; if the initial condition $v(0) = 0$, then there is no such parallel current source. Here, in fact, we slightly abuse the notation. Strictly speaking, the product Cv should have the units of charge q. However, since we are interested in this product only at the time instant $t = 0$, we might as well consider it as initial current in lieu of initial charge. Similar remarks apply to the case of an inductor with initial voltage source $v(0)$ as given in the next example.

With this Laplace equivalent of capacitance plugged in the circuit of Fig. 9.8, we have the following nodal equation 'in the s-domain':

$$\frac{Y(s)}{1/10^{-6}s} - I(0) + \frac{4 \times 10^{-3}}{s} + \frac{Y(s) - U(s)}{1\,k\Omega} = 0$$

which reduces to

$$Y(s) = \underbrace{\frac{-4}{s + 1000}}_{\text{natural response}} + \underbrace{\frac{6000}{s(s + 1000)}}_{\text{forced response}}$$

with $U(s) = 10/s$ and $I(0) = -4\,\mu A$.

We shall next look at a higher-order circuit.

Example 9.26 Consider the circuit in Fig. 9.10.

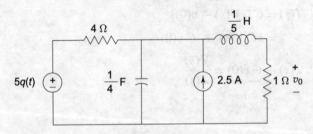

Fig. 9.10

Let us first discover that the 'Laplace equivalent' of inductance has a voltage source in series as shown in Fig. 9.11.

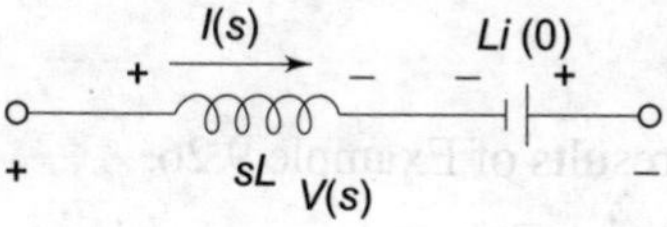

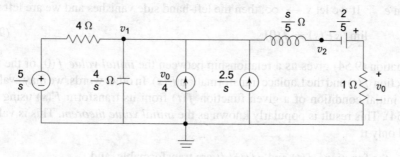

Fig. 9.11

Exercise 9.25 Obtain the 'Laplace equivalent' of resistance.

Next, let us determine the initial conditions. By simple current division rule, we get $v_C(0) = 2\,\text{V}$ and $i_L(0) = 2\,\text{A}$. However, we recommend that these initial conditions be plugged into our solution only towards the end. The rationale for this will become clear as we go ahead.

With the Laplace equivalents plugged in, we get the circuit shown in Fig. 9.12.

Fig. 9.12

Let us apply nodal analysis to get the following equations:

$$\frac{V_1(s) - \dfrac{5}{s}}{4} + \frac{V_1(s)}{\dfrac{4}{s}} - \frac{v_C(0)}{4} - \frac{2.5}{s} + \frac{V_1(s) - V_2(s)}{0.2\,s} = 0$$

$$\frac{V_2(s) - V_1(s)}{0.2} + \frac{V_2 + 0.2\,i_L(0)}{1} = 0$$

We might simplify these equations and obtain the matrix-vector equation for the two node voltages v_1 and v_2 as

$$\begin{bmatrix} s^2 + s + 20 & -20 \\ -20 & 4s + 20 \end{bmatrix} \begin{bmatrix} V_1(s) \\ V_2(s) \end{bmatrix} = \begin{bmatrix} s & 0 \\ 0 & -0.8\,s \end{bmatrix} \begin{bmatrix} v_C(0) \\ i_L(0) \end{bmatrix} + \begin{bmatrix} 15 \\ 0 \end{bmatrix}$$

In particular, if we solve for $V_0(s) = V_2(s) + 0.4$, we get

$$V_0(s) = \frac{2s^2 + 12s + 75}{s(s^2 + 6s + 25)}$$

with the intitial conditions substituted. Obtaining the partial fractions and hence the inverse transformation, we get

$$v_0(t) = 3 + \frac{5}{4}e^{-3t}\cos(4t + 143.13°)$$

Exercise 9.26 Verify the results of Example 9.26.

In the following section we will come back to these circuits more formally.

9.3.5.3 Initial and final values

It is interesting to note that *two* more properties of the Laplace transformation help us verify the results of inverse transformation.

Earlier, we have derived that

$$\mathcal{L}\left\{\frac{d}{dt}f(t)\right\} = sF(s) - f(0)$$

On the left-hand side, we need to evaluate an integral (with respect to t) that has a term e^{-st}. If we let $s \to \infty$, then the left-hand side vanishes and we are left with

$$\lim_{s \to \infty} sF(s) = f(0) \tag{9.54}$$

Equation (9.54) gives us a relationship between the *initial value* $f(0)$ of the time function $f(t)$ and the Laplace transformation $F(s)$. In other words, we may evaluate the initial condition of a given function $f(t)$ from its transform $F(s)$ using Eqn (9.54). This result is popularly known as the *initial value theorem*. This is valid if and only if

- the functions $f(t)$ and $df(t)/dt$ are transformable, and
- the degree of the numerator polynomial is less than that of the denominator polynomial if $F(s)$ is a rational function, as is the case in most of the circuit problems. This condition is to ensure that the limit exists.

In a similar way, we may show that

$$\lim_{s \to 0} \mathcal{L}\left\{\frac{d}{dt}f(t)\right\} = \lim_{s \to 0}[sF(s) - f(0)]$$

$$= \int_0^\infty \frac{d}{dt}f(t)\,dt$$

$$= f(\infty) - f(0)$$

and hence

$$\lim_{s \to 0} sF(s) = f(\infty) \tag{9.55}$$

This result is popularly called the *final value theorem*. Once again this theorem holds if and only if the limit exists.

It is interesting to derive a condition under which the final value theorem holds. Earlier, we have learnt that any function $F(s)$ can be split into several partial fractions; each of the partial fractions can be inverted into exponential functions of the form e^{pt} in the time domain. Clearly, if $p > 0$, then these exponentials 'blow up' to infinity as $t \to \infty$ and hence the limit in the s-domain does not exist. Thus, if $F(s)$ is a rational function, the necessary condition for the limit in Eqn (9.55) to exist is that all the roots of the denominator polynomial must have *negative real parts*. We elaborate on these ideas in Chapter 11 again.

Example 9.27 Consider the function

$$f(t) = 5 + 6t\,e^{-5t} + 7e^{-2t}\sin 4t$$

Clearly, the initial and final values are

$$f(0) = 5 = f(\infty)$$

Now let us look at the transform $F(s)$,

$$F(s) = \frac{5}{s} + \frac{6}{(s+5)^2} + \frac{28}{(s+2)^2 + 16}$$

From Eqn (9.54), we have

$$\lim_{s \to \infty} sF(s) = 5 + 0 + 0 = 5$$

Likewise, from Eqn (9.55), we have

$$\lim_{s \to 0} sF(s) = 5 + 0 + 0 = 5$$

Exercise 9.27 Find the inverse Laplace transformation of

$$\frac{-s^4 + 5s^3 + 20s^2 + 19s + 8}{s(s+1)^3(s+4)}$$

and check your results by using the initial value and final value theorems.

9.4   The Transfer Function

All of our effort in this text is to emphasize the *input–output* paradigm for any general network. Unless an input signal is given to the circuit, it would not respond. Of course, we may get a *transient* response due to the initial conditions, but the initial conditions might have been caused by some inputs in the past. In other words, if we take *fresh* components straight from the factory and build a circuit, there will not be any response if no input signal is provided. We also call this the *cause–effect* paradigm, and the circuit is said to be *causal*. On the contrary, if at all, we build a circuit that responds *in anticipation*, we call it *anticipatory*. An example could be a telephone that decides not to ring if the caller is an annoying salesman, since we do not like such calls. Observe that we are not *programming* the telephone, by creating a database of important numbers, to behave that way; rather, we build a telephone that exhibits this property naturally, as if it has *understood* our likes and dislikes. Clearly, at least so far, there is no physical system that is anticipatory[11].

---

[11] Perhaps politicians belong to this category since we often hear about their taking 'anticipatory' bails!

So far in this text, we were pronouncing two important properties—(i) linearity and (ii) time-invariance. We need to add the property of causality to the list. Of course, it is understood otherwise. When we build circuits for a specific purpose, the *input–output* paradigm is inevitable. For instance, we wish to build a radio. Broadly speaking, the circuit receives some arbitrary electrical signal that represents the audio. And, there are many radio stations broadcasting different signals at the same time. Now the question is how to select a particular station. Further, once a station is selected, we need to convert the electrical signal into audio form because our ears cannot make out music from electrical signals. Thus, we have an electrical signal as an input and an audio signal as an output. Several such examples may be generated. For instance, in a public-address system, we speak into a microphone; this is an input. The microphone converts it into an electrical signal and sends this to an amplifier, which in turn gives a louder replica of our voice through a speaker.

Exercise 9.28 Make a list of several practical circuits and systems that fit into the input–output paradigm.

For (causal), linear time-invariant circuits, the initial conditions may be ignored hereafter for the following reasons:

- the natural response is a part of transient response that does not last longer;
- once we have the matrix A, we may always add $(sI - A)^{-1} x(0)$ to the forced response and compute the complete response; moreover,
- it does not represent the 'effect' of any external input.

We focus on the transfer function:

$$H(s) \triangleq \frac{\mathcal{L}\{\text{output}\}}{\mathcal{L}\{\text{input}\}}$$

It is customary to call $H(s)$ the *driving-point function*, if the output is measured at the same place where the input has been applied. We discuss more about this in the following chapter.

9.4.1 Generalized Impedance

Having noted that we can *conveniently* ignore the initial conditions, we may go ahead defining a *generalized impedance* $Z(s)$ of an element as

$$Z(s) \triangleq \frac{V(s)}{I(s)} \tag{9.56}$$

where we understand that the input is the Laplace transform of the current function $i(t)$ passed through an element and the output is the Laplace transform of the voltage function $v(t)$ that is measured across the element. Obviously, we are referring to the driving-point impedance here. We may further extend this and define the generalized impedance of a circuit itself. For instance, the generalized impedance of our building blocks—R, L, and C—may be defined as

$$Z_R(s) = R, \quad Z_L(s) = sL, \quad Z_C(s) = \frac{1}{sC}$$

respectively. If there is a series combination of all these three elements, we may define

$$Z_{RLC}(s) = R + sL + \frac{1}{sC} = \frac{LCs^2 + RCs + 1}{sC}$$

Observe that all these impedance functions are *rational functions* in the complex variable s.

Example 9.28 Let us quickly obtain the impedance as seen by a source across the port 1-1' of Fig. 9.13.

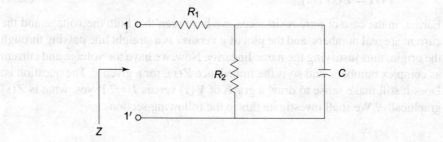

Fig. 9.13

The parallel combination of R_2 and C gives us an equivalent impedance of

$$R_2 \parallel \frac{1}{sC} = \frac{R_2}{sR_2C + 1}$$

and hence the impedance as seen by the port 1-1' is

$$Z(s) = R_1 + \frac{R_2}{sR_2C + 1} = \frac{sR_1R_2C + R_1 + R_2}{sR_2C + 1}$$

Exercise 9.29 Obtain the driving point impedance of the circuit in Fig. 9.14.

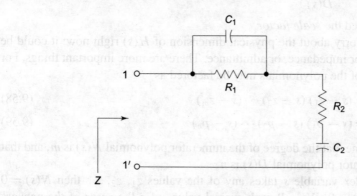

Fig. 9.14

We may define, yet another rational function, the *generalized admittance* function, as

$$Y(s) \overset{\Delta}{=} \frac{1}{Z(s)} = \frac{I(s)}{V(s)}$$

9.4.2 Generalized Ohm's Law

With the generalized impedances and admittances defined above, we may look at the voltage–current relationship in the *s*-domain as *generalized Ohm's law* given by

$$V(s) = Z(s)\,I(s) \tag{9.57}$$

Earlier, in the case of pure resistances, we have seen that both the voltage and the current are real numbers, and the plot of v versus i is a straight line passing through the origin, thus justifying the name linearity. Now, we have the voltage and current as complex numbers, and so is the impedance $Z(s)$, for a given s. The question is: Does it still make sense to draw a graph of $V(s)$ versus $I(s)$? If yes, what is $Z(s)$ graphically? We shall investigate this in the following section.

9.5 Network Functions

The rational functions, transfer functions, and driving-point functions are generally called network functions. We elaborate to a large extent on the *analysis*[12] of these in the rest of this chapter. Subsequently, we will carry forward these ideas to Chapter 11 to *synthesize* networks.

Let us first make a few observations to consolidate the ideas. The network function $H(s)$ by definition has two *co-prime* polynomials, i.e., polynomials that do not have common factors. We designate these two polynomials as $N(s)$ and $D(s)$ so that

$$H(s) = K\,\frac{N(s)}{D(s)}$$

where K is called the *scale factor*.

Let us not worry about the physical dimension of $H(s)$ right now; it could be simply a gain, or impedance, or admittance. There are more important things. For instance, each of the polynomials may be factored as

$$N(s) = (s - z_1)\,(s - z_2)\cdots(s - z_m) \tag{9.58}$$

$$D(s) = (s - p_1)\,(s - p_2)\cdots(s - p_n) \tag{9.59}$$

where we assume that the degree of the numerator polynomial $N(s)$ is m, and that of the denominator polynomial $D(s)$ is n.

If the complex variable s takes any of the values $z_1, z_2, \ldots$ then $N(s) = 0$ and hence $H(s) = 0$. We call such special values of s the *zeros of the network function*. Likewise, if the complex variable s takes any of the values $p_1, p_2, \ldots,$

---

[12] by this we mean dissecting the functions mathematically to probe deeper.

then $D(s) = 0$ and hence $H(s) = \infty$. We call such special values of s the *poles of the network function*. For a given network function, there could be r poles or zeros having the same value. If that were the case, the pole or zero is said to be *of multiplicity r*. If the pole or zero is not repeated, it is said to be *simple*. Let us look at a plot of $H(s)$ versus s in Fig. 9.15 below.

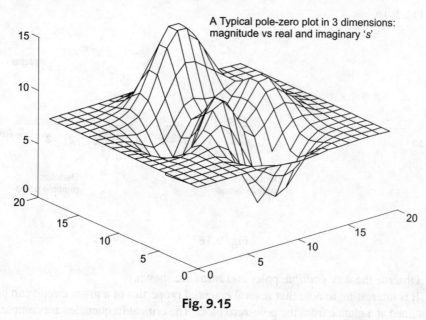

Fig. 9.15

We notice that at the location of a pole in the three-dimensional plot, the magnitude of $H(s)$ 'blows up' to infinity. In fact, we have a stunning landscape with spectacular peaks[13] (at the location of the poles) and beautiful springs (at the location of zeros). The zeros and the poles are collectively called *critical frequencies*; recall that s has the physical dimension of frequency. We might be tempted to think that these critical frequencies have to be avoided altogether. Fortunately, it is not so. It is noteworthy that these critical frequencies depend solely on the circuit, independent of the external signals applied, or the initial conditions of the elements. Looking backwards, we also observe that each of the polynomials, in general, is a result of applying KVR or KCR and hence involves the product of the elemental value (R, L, or C) and the conventional current, or voltage, or their derivatives. As a natural consequence, we see that the coefficients of the polynomials are real and positive[14]. By the fundamental theorem of algebra, the roots of such polynomials are either real or complex conjugates.

It is customary to depict the critical frequencies of the network function $H(s)$ in a two-dimensional plane called the *complex s plane* with the horizontal axis as the real axis and the vertical axis as the imaginary axis. In these so-called *pole-zero*

---

[13] A pole is literally a 'pole' standing tall in the complex plane!

[14] Strictly speaking, all the coefficients have the *same* sign. However, we may prefer positive sign for these coefficients and choose the scale factor K accordingly.

plots, a zero is denoted by 'o' (small oh), and a pole is denoted by '×' (cross). As an example, we show the pole-zero plot of

$$H(s) = 10 \frac{(s+5)^2(s^2+4s+13)}{s^2(s+13)(s^2+2s+2)}$$

in Fig. 9.16.

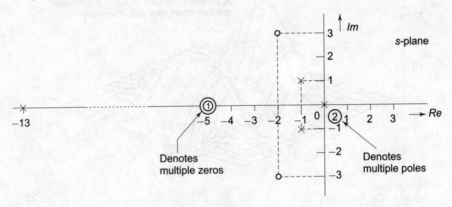

Fig. 9.16

Observe the way *multiple* poles and zeros are shown.

It is interesting to note that several essential properties of a given circuit can be obtained at a glance from the pole-zero plots. The critical frequencies are complex in general, though they turn out to be real in many cases. They all appear to be finite on the face of the network function. However, if the degree m of the numerator polynomial $N(s)$ is greater than that of the denominator $D(s)$, then we have that

$$\lim_{s \to \infty} H(s) = \infty$$

and hence we conclude that there are $m - n$ poles at infinity in addition to the n poles [the n roots of the nth degree polynomial $D(s)$] that are finite. Likewise, if $m < n$, then we say that there are $n - m$ zeros at infinity in addition to the m zeros that are finite. Counting this way, the total number of zeros is equal to the total number of poles. For instance, in Fig. 9.16, there are five zeros including the one at infinity.

The significance of the poles is straightforward. We have observed that, given any rational function, each of the partial fractions contains a pole, i.e., each partial fraction is of the form

$$\frac{k_i}{s - p_i}$$

The zeros come into the picture while evaluating the residues. Further, we also notice that each partial fraction gives us an exponential $k_i\, e^{p_i t}$ in the time domain. Thus, we may infer that the poles directly contribute to the time response of the circuit. In other words, we simply look at the pole-zero plot and tell precisely how the time response is going to look like. As an illustration, let us try the pole-zero plot in Fig. 9.16. The poles are located at $s = 0, 0, -1 + j1, -1 - j1$, and -13. Accordingly, the response would be of the form

$$y(t) = \frac{k_1}{s} + \frac{k_2}{s^2} + \frac{k_3}{s+1+j1} + \frac{k_4}{s+1-j1} + \frac{k_5}{s+13}$$

The residues may be evaluated using Heaviside's theorem.

It's not just the *shape*, i.e., the waveform, of the response that can be obtained from the pole-zero plot; we may also obtain a more useful information—whether there exists a network with so and so transfer function. For instance, in the above example, we notice that the time response has a term of the form $k_2 t$, which implies that the *impulse response* of the network approaches a very large magnitude as time passes, i.e.,

$$\lim_{t \to \infty} h(t) \to \infty$$

We will show later in Chapter 11 that it is impossible to synthesize passive networks with this behaviour. It is necessary and sufficient that the time response be *bounded* [i.e., the magnitude $|H(s)| < \infty$] for all time $t \geq 0$.

Broadly speaking, the *s-plane* in which we show the poles and zeros can be divided into two halves— the left half of the plane (lhp) which contains poles and zeros with strictly negative real parts, and the right half of the plane (rhp) which contains poles and zeros with non-negative real parts. It is also customary to call the lhp the 'open-lhp' since there is no rigid boundary containing this half. Likewise, the rhp is called the 'closed-rhp' since it is rigidly bounded by the imaginary axis to its left. Putting it simply, in the example above, the transfer function has three poles in the lhp and two poles, the poles on the imaginary axis, in the rhp.

In what follows, we describe the natural response of a network given its pole locations.

Case 1

If the pole is real and simple, i.e., it is the only pole at a given location on the real axis, and it is in the lhp, the response is an exponential decay as shown in Fig. 9.17(a). Our first-order circuits exhibit this kind of a response. On the other hand, if it is in the rhp, the response is either a constant as in Fig. 9.17(b), or an exponential rise as in Fig. 9.17(c).

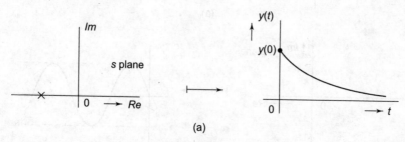

(a)

Fig. 9.17

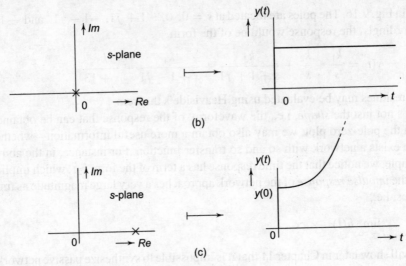

(b)

(c)

Fig. 9.17

Case 2

If there is a pair of complex-conjugate simple poles in the lhp, the natural response is an exponentially decaying sinusoid. Our second-order *underdamped* circuits exhibited this response; poles with greater displacement from the imaginary axis may be associated with more rapidly decaying exponentials and poles with greater displacement from the real axis may be associated with higher frequencies of oscillation. If the poles are simple-conjugate pairs on the imaginary axis, the response is oscillatory; the *LC* circuit exhibited this response. And, if the pair of complex-conjugate poles is in the rhp, we have an exponentially growing sinusoid. We show this case in Fig. 9.18(a, b, c).

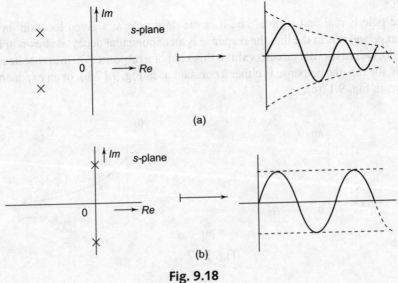

(a)

(b)

Fig. 9.18

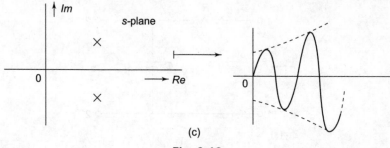

(c)

Fig. 9.18

Case 3

If the transfer function has poles of multiplicity r, we leave it to the reader to verify that the response grows exponentially with time if the poles are in the rhp.

Exercise 9.30 Verify Case 3 above. Draw relevant waveforms.

Case 4

For any arbitrary pattern of poles, we may classify them according to the three cases above. The total response corresponding to these poles may be simply found by *adding* each of the individual factors expressed as (real or complex) exponentials weighted by the residues.

Looking at the four cases above, it is clear that for the impulse response $h(t)$ to be bounded for all $t \geq 0$, the poles must lie in the lhp. As a consequence, if we apply any arbitrary bounded input signal, the output remains bounded. This is an important property which deserves a detailed treatment in Chapter 11.

We have some more information about the network from the poles and zeros of the network function. Consider the impedance of an inductor for instance. Its network function is sL. Hence, it has got a zero at origin and a pole at infinity. Likewise, the impedance of a capacitor has a pole at origin and a zero at infinity. Turning around this information, we may characterize a given network function as either an impedance function or an admittance function. For example, if a given network function has a pole at origin, i.e., one of the partial fractions has the term k/s, we may treat this network function as an impedance function to synthesize a network with a capacitor $C = 1/k$ as a series element, or we may treat the network function as an admittance function and synthesize a network with an inductor $L = 1/k$ as a shunt element. We will make use of observations of this nature later in Chapter 11.

We look at a few examples to summarize the ideas of this chapter.

Example 9.29 Consider the simple RC network in Fig. 9.19.

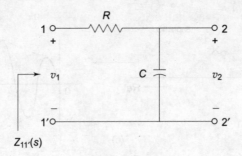

Fig. 9.19

Seen from the port 1-1', we may obtain the following driving-point impedance and admittance functions:

$$Z_{11'}(s) = R + \frac{1}{sC} = \frac{sRC + 1}{sC}$$

$$Y_{11'}(s) = \frac{1}{Z_{11'}(s)} = \frac{sC}{sRC + 1}$$

We may also obtain the following transfer functions from the input voltage to the voltage across the resistor or the voltage across the capacitor, using the voltage division rule:

$$T_1(s) = \frac{R}{R + \dfrac{1}{sC}} = \frac{sRC}{sRC + 1}$$

$$T_2(s) = \frac{\dfrac{1}{sC}}{R + \dfrac{1}{sC}} = \frac{1}{sRC + 1}$$

Example 9.30 Consider the *RC* network in Fig. 9.20.

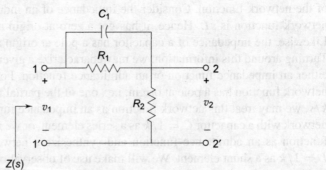

Fig. 9.20

We have a parallel combination of a resistance R_1 and a capacitance C_1 in the series arm. Accordingly, the driving-point impedance is

$$Z(s) = \frac{R_1}{sR_1C_1 + 1} + R_2 = \frac{sR_1R_2C_1 + R_1 + R_2}{sR_1C_1 + 1}$$

Exercise 9.31 Obtain the admittance function of the network in Fig. 9.20.

The voltage across R_2 in terms of the input voltage may be obtained as

$$T(s) = \frac{R_2}{Z(s)} = \frac{sR_1R_2C_1 + R_2}{sR_1R_2C_1 + R_1 + R_2}$$

Exercise 9.32 Obtain all possible network functions of the network given in Fig. 9.21.

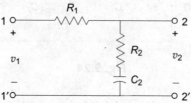

Fig. 9.21

Example 9.31 Let us consider the RL network of Fig. 9.22.

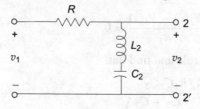

Fig. 9.22

If we wish to obtain the voltage across the shunt arm in terms of the input voltage, we get

$$T(s) = \frac{sL_2 + R_2}{sL_2 + R_2 + R_1}$$

We simply make an observation that this transfer function is *similar* to that in Example 9.30. It is not only the transfer functions but also the impedance and/or admittance functions that would be similar among several networks.

Exercise 9.33 Obtain the transfer function of the network shown in Fig. 9.23. Comment on your discovery.

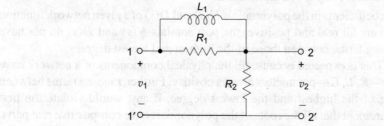

Fig. 9.23

The above examples are simple only to the point of illustrating the ideas. For more complex networks, for instance in the following example, we may resort to nodal/mesh analysis to obtain the network functions.

Example 9.32 Consider the network shown in Fig. 9.24.

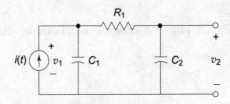

Fig. 9.24

Let us determine the ratio $V_2(s)/I_1(s)$. Using nodal analysis, we get

$$sC_1\, V_1(s) + \frac{V_1(s) - V_2(s)}{R_1} = I_1(s)$$

$$sC_2\, V_2(s) + \frac{V_2(s) - V_1(s)}{R_1} = 0$$

We may use Cramer's rule and find out

$$\frac{V_2(s)}{I_1(s)} = \frac{1/R_1C_1C_2}{s\left(s + \dfrac{C_1 + C_2}{R_1C_1C_2}\right)}$$

Owing to its units, we may call this transfer function trans-impedance function.

Let us summarize our observations regarding the driving-point functions. At this point, we remark that these observations tell us only the necessary conditions but not sufficient. We elaborate these issues in Chapter 11.

1. The real part of all poles and zeros is non-positive. Further, if the real part is zero, that pole or zero is simple. This is to ensure that the response is always bounded, for any given input signal.

2. The coefficients in the polynomials $N(s)$ and $D(s)$ of a given network function $H(s)$ are all real and positive; the polynomials $N(s)$ and $D(s)$ do not have missing terms between those of the highest and lowest degrees.

   This is expected because all the physical components of a network have their—R, L, C—parameters always positive. Further, missing terms between those of the highest and the lowest degree, if any, would violate the first requirement that all the roots of the polynomials have non-positive real parts. There is one exception, however. We shall explore this in Chapter 11.

3. The degree of $N(s)$ and $D(s)$ may differ by either 0 or 1 only.

   This is obvious looking at any driving-point function. If a driving-point function is to be an impedance function, we may, at the most, pull out an inductor if the numerator polynomial's degree is greater by 1; if the denominator has a higher degree, we may pull out a capacitor. If the difference in degrees is greater than 1, we cannot identify any element that could be pulled out. If the degrees are equal, we may pull out a resistance.

4. There may be a simple zero or a simple pole at origin, and accordingly, the terms of the lowest degree in $N(s)$ and $D(s)$ may differ in degree by 1 at most.

Let us look at the following examples to learn further.

Example 9.33 Examine the following network function, assumed to be an impedance function:

$$Z(s) = \frac{4s^4 + s^3 - 3s + 1}{s^3 + 2s^2 + 2s + 40}$$

Let us see more details. The given function can be written as

$$Z(s) = \frac{k_1}{s - 1 + j3} + \frac{k_2}{s - 1 - j3} + \frac{k_3}{s + 4}$$

Clearly, there is a pair of complex-conjugate poles in the rhp. Further, we also notice that the numerator polynomial has a term with a negative sign, indicating the presence of a zero in the rhp. Hence this function does not qualify as an impedance function.

Exercise 9.34 Verify that the impedance function of Example 9.33 has a zero of multiplicity 2 in the rhp.

Example 9.34 Let us now consider the function

$$F(s) = 15 \frac{s^3 + 2s^2 + 3s + 2}{s^4 + 6s^3 + 8s^2}$$

Clearly, this function has a pole of multiplicity 2 in the rhp (at the origin). Hence this does not qualify as an impedance function. Also, we observe that

$$\lim_{s \to 0} F(s) \to \frac{2}{8s^2}$$

implying that a resistor, a capacitor, or an inductor cannot be pulled out. To explain this little more clearly, we obtain the following partial fractions:

$$F(s) = \frac{30/8}{s^2} + \frac{45/16}{s} + \frac{-15/2}{s+2} + \frac{315/16}{s+4}$$

If this were to be an impedance function, each of the partial fractions should represent elements (or parallel combination of elements) put in series. Clearly $30/8s^2$ does not qualify as a series branch. Further, we have a 'negative impedance' term which is physically not possible with passive elements.

Example 9.35 Let us now look at the following function:

$$F(s) = \frac{s^2 + 4s + 3}{s^2 + 2s}$$

This function matches all of our observations. However, we cannot guarantee that there exists a network that has $F(s)$ as an impedance or an admittance function. We reiterate that our observations are only necessary conditions. We will present the sufficient conditions in Chapter 11. However, we complete this example in the following way:

$$F(s) = \frac{3/2}{s} + \frac{1/2}{s + 2}$$

Assuming that $F(s)$ is an impedance function, we identify each of the partial fractions also to be impedance functions. Hence we have a capacitance $C_1 = 2/3$ F in series with a parallel combination of a resistance $R = 1/4\ \Omega$ and a capacitance $C_2 = 1/2$F. This parallel combination can be synthesized by noting that the second term looks like

$$\frac{1/2}{s + 2} = \frac{1}{2s + 4} = \frac{1}{Y(s)} = \frac{1}{Y_1(s) + Y_2(s)}$$

Now it is easy to identify that $Y_1(s) = 2s$ is a capacitance of 2 F, and that $Y_2(s) = 4$ is a resistance of $1/4\ \Omega$. The network is shown in Fig. 9.25.

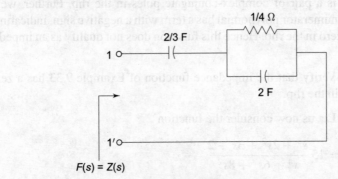

Fig. 9.25

Fortunately, we could realize a network. We just make a passing note that this function has its poles and zeros alternating.

Exercise 9.35 Obtain a network using $F(s)$ in Example 9.35 as an admittance function.

If the network function is a transfer function, in lieu of an impedance/admittance function, we have the following. In general, $H(s)$ may have some meaning with respect to the network, but its reciprocal $1/H(s)$ may not have any sense. Accordingly, the restrictions on the pole-zero locations may be slightly relaxed as follows.

1. While the poles are restricted to lhp and simple ones on the imaginary axis, there is no such restriction on the zeros; zeros can be anywhere in the s-plane.
2. The degree of the numerator polynomial $N(s)$ may be as small as zero, that is, the numerator is a constant, independent of the degree of the denominator polynomial $D(s)$. However, the maximum degree of $N(s)$ cannot exceed the degree of $D(s)$ unless $H(s)$ is either a trans-impedance function or a trans-admittance function.

We will meet these functions again in Chapter 11.

Example 9.36 Consider the network in Fig. 9.26.

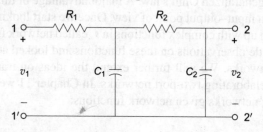

Fig. 9.26

Applying nodal analysis, we may obtain the transfer function as

$$H(s) = \frac{V_2(s)}{V_1(s)} = \frac{1/R_1R_2C_1C_2}{s^2 + s\dfrac{R_1C_1 + R_1C_2 + R_2C_2}{R_1R_2C_1C_2} + \dfrac{1}{R_1R_2C_1C_2}}$$

It is easy to see that all of our observations match here.

Exercise 9.36 Obtain the transfer function of the network shown in Fig. 9.27. Verify whether it matches with the observations.

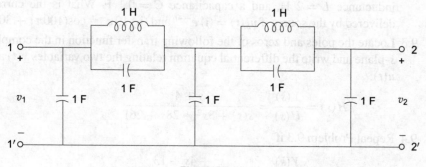

Fig. 9.27

We are not delving into further examples synthesizing networks given transfer functions. We will do this in the following chapter.

Summary

In this chapter, we have attempted to *generalize* networks mathematically. By this we mean looking at arbitrary networks (containing any source, any element, or combinations/networks thereof) in a unified framework. To this end, we have systematically introduced the ideas of state variables, the actual variables of any network that evolve with time. Further, we have introduced a very important *technique* called the Laplace transformation that would help us solve the problems in a very elegant way, furthering our generalization. In terms of the complex Laplace variable s, which happens to have the units of frequency, we have defined the impedance/admittance of individual elements as well as networks, thereby postulating the generalized Ohm's law. A major advantage of this is looking at any network from an input–output point of view. Once we start looking at the networks this way, we end up with complex functions in s, called network functions. We have made a few subtle observations on these functions and looked at the possibility of synthesis of networks. We will further extend the ideas on transfer functions in Chapter 10 by elaborating two-port networks. In Chapter 11 we shall concentrate on synthesizing networks given network functions.

Problems

9.1 Given
- $v(t) = 5$ V and (ii) $v(t) = 2\,e^{-0.1t}$ V, determine the complex frequency s.
- $i(t) = 2\,e^{-5t}\cos(100t - 45°)$ A and (iv) $i(t) = 2\sin(100\pi t + 75°)$ A, determine the complex frequency s.
- $\mathcal{V} = 10\angle 60°$ V and (vi) $\mathcal{V} = 10\angle -45°$ V, determine $v(t)$ if the complex frequency is $s = 10\angle(180° - \tan^{-1}\frac{5}{3})$ Np/s.
- $\mathcal{I} = 10\angle 120°$ A and $s = 0$, and (viii) $\mathcal{I} = 10\angle 0°$ V, with $s = j1$ Mrad/s, determine $i(t)$.

9.2 A series RLC circuit consists of a source v_S, a resistance $R = 5$ Ω, an inductance $L = 2$ H, and a capacitance $C = 0.1$ F. What is the current delivered by the source if $v_S(t) = $ (i) e^{-4t} and (ii) $10\,e^{-2t}\cos(100\pi t + 30°)$.

9.3 Locate the poles and zeros of the following transfer function in the complex s-plane and write the differential equation relating the two variables $y(t)$ and $u(t)$:

$$H(s) = \frac{Y(s)}{U(s)} = \frac{s+4}{s(s^3 + 3s^2 + 28s + 26)}$$

9.4 Repeat Problem 9.3 if

$$H(s) = \frac{Y(s)}{U(s)} = \frac{s^2 + 4s + 13}{s(s^3 + 3s^2 + 28s + 26)}$$

Also, determine $\angle H(s)$ for $s = -1 + j1$ Np/s.

9.5 A network function has poles at $s = 0, -5, -5, -7$ and zeros at $s = -1, -6$ with $H_0 = 10$. Express the network function as a rational function $N(s)/D(s)$. How about a network function with poles at $s = \pm j2, -3 \pm j1$ and zeros $s = 0, -5 \pm j5$?

9.6 Find the transfer function $V_0(s)/I_S(s)$ for the circuit shown in Fig. 9.28. Where are the poles and zeros located?

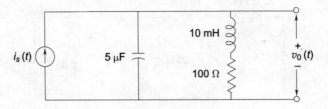

Fig. 9.28

9.7 Find the transfer function $V_0(s)/V_S(s)$ for the circuit of Fig. 9.29. Sketch the pole-zero plot in the complex s-plane.

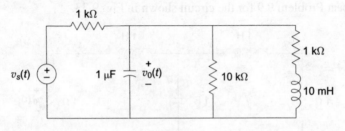

Fig. 9.29

9.8 Find the voltage $v(t)$ across the resistance if the initial current $i_L(0)$ through the inductor is 10 mA in the direction shown in the circuit in Fig. 9.30.

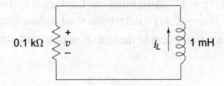

Fig. 9.30

9.9 Consider the circuit in Fig. 9.31. Obtain a state-space description of the circuit.

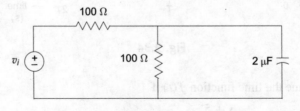

Fig. 9.31

9.10 Repeat Problem 9.9 for the circuit shown in Fig. 9.32.

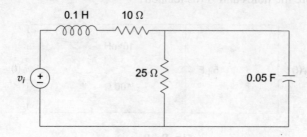

Fig. 9.32

9.11 Repeat Problem 9.10 with the capacitor and the inductor interchanged from their respective locations.

9.12 Repeat Problem 9.9 for the circuit shown in Fig. 9.33.

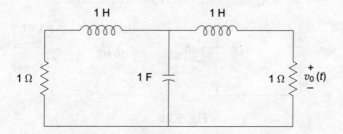

Fig. 9.33

9.13 (i) If $F(s) = 1/(s^4 + 1)$, determine the Laplace transform $\mathcal{L}\{f(2t)\}$.
(ii) If $F(s) = 1/(s + 1)^3$, determine the Laplace transform $\mathcal{L}\{t^2 f(2t)\}$.
(iii) If $F(s) = 1/(s^2 + s + 1)$, determine the Laplace transform $\mathcal{L}\{f(100t)\}$.

9.14 Obtain the Laplace transform of the time function, called the *triangular wave*, shown in Fig. 9.34.

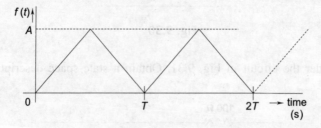

Fig. 9.34

9.15 Determine the time function $f(t)$ if

$$F(s) = \frac{s+5}{s^2 + 4s + 3}$$

Check your results using the initial-value and final-value theorems.

9.16 Determine the time function $f(t)$ if

$$F(s) = 6 \frac{s^2 + 5}{(s^2 + 1)(s^2 + 4)}$$

Check your results using the initial-value and final-value theorems.

9.17 Given the ODE

$$\frac{d^2}{dt^2} y(t) + 3 \frac{d}{dt} y(t) + 2 y(t) = 2\cos(2t) \quad \forall t \geq 0$$

Use Laplace transformation and obtain the natural and forced components of the solution $y(t)$ if

$$y(0^-) = -1$$

and

$$\frac{d}{dt} y(t)|_{t=0^-} = 1$$

9.18 The network function of a circuit is given by

$$H(s) = 7 \frac{s + 1}{s^2 + 2s + 2}$$

Obtain the network's step response if the initial conditions are

$$y(0^-) = -1$$

and

$$\frac{d}{dt} y(t)|_{t=0^-} = 1$$

9.19 The network function of a circuit is given by

$$H(s) = 5 \frac{s + 2}{(s + 1)(s + 3)}$$

Obtain the impulse response and the step response. How are these two responses related?

9.20 Use Laplace transformation to determine $v(t)$ for $t \geq 0$ in the circuit shown in Fig. 9.35.

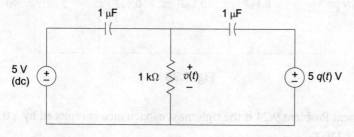

Fig. 9.35

9.21 In the circuit shown in Fig. 9.35, if the resistance is replaced by an inductance of 1 mH, determine $v(t \geq 0)$.

9.22 Repeat Problem 9.20 for the circuit shown in Fig. 9.36, if $v_S = 10\cos(10^6 t)$ V.

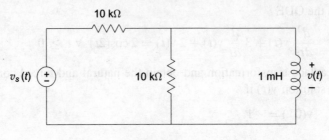

Fig. 9.36

9.23 Determine the current $i(t \geq 0)$ in the circuit shown in Fig. 9.37, if $v_S = 10t\, q(t)$ V.

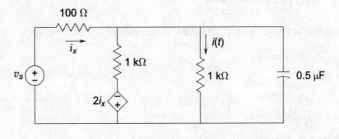

Fig. 9.37

9.24 Determine $v(t \geq 0)$ in the circuit shown in Fig. 9.38.

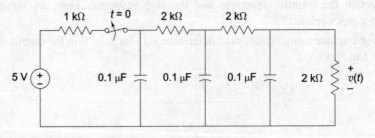

Fig. 9.38

9.25 Repeat Problem 9.24 if the rightmost capacitance is replaced by a 0.01 mH inductance.

9.26 If the source v_S jumps from $+10$ V (dc) to $+5$ V (dc) at $t = 0$, determine $v(t \geq 0)$ in the circuit shown in Fig. 9.39.

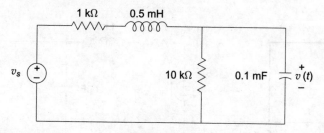

Fig. 9.39

9.27 In the circuit shown in Fig. 9.40, determine the voltage $v(t \geq 0)$ across the parallel RLC network.

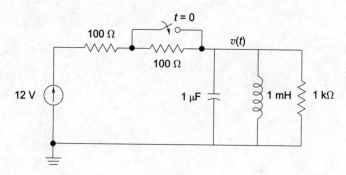

Fig. 9.40

9.28 In the circuit in Fig. 9.41, determine the current $i(t \geq 0)$ if the current source is $i_S = 2e^{-t}$, $t \geq 0$ mA.

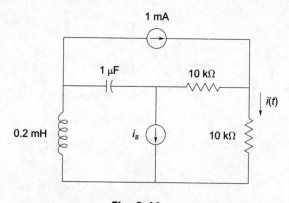

Fig. 9.41

9.29 Repeat Problem 9.28, with the 1 mA source at the top replaced with a 3 V dc source, with the positive terminal at the right.

9.30 For the circuit in Fig. 9.42(b), use convolution to determine the output $v_0(t)$ of the circuit if the input $v_S(t)$ is as shown in Fig. 9.42(a).

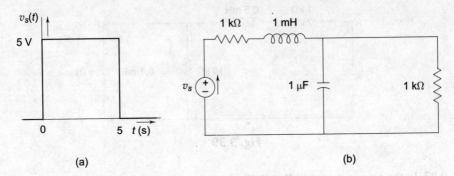

(a)

(b)

Fig. 9.42

CHAPTER 10

Two-port Parameters

In the preceding chapter we have discussed the input–output paradigm of an arbitrary network. We have defined driving point functions, wherein the input and the output are *measured* at the same port. We have also seen that the measurements may be made anywhere and defined transfer functions accordingly. In this chapter we will delve into transfer functions little more formally. We will see that the ideas presented in this chapter help us 'identify' any arbitrary network (or system in general) by making just the required number of measurements systematically.

10.1 Two-port Networks

Recall from Chapter 3 that a network is an interconnection of several elements, passive and active. We referred to a pair of nodes as a *port* in the sense that any element can be 'docked' across this pair of nodes. Obviously, a network may have umpteen ports. In this chapter we will consider just two-ports, the input port and the output port. The first is designated as 1-1' and the second is designated as 2-2'. We assume that a network can be drawn as in Fig. 10.1 with these two-ports protruding outside, thereby allowing us to access them for applying signals or making measurements. We may even use these ports to interconnect another two-port network. Notice that we are not much bothered about the network *per se*; it is a linear network that could have any topology (series, parallel, *T*-network, or, ...) with arbitrarily several elements. Accordingly, we just show a rectangular box in Fig. 10.1 with two pairs of nodes hanging outside. Such networks are called two-port networks. Clearly, a two-port network is a generalization of any passive element from the point of view of linearity and generalized impedance we have discussed in the preceding chapter.

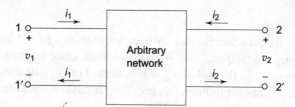

Fig. 10.1

Let us further fix the polarities of voltages and currents, either applied or measured, at these ports. As seen in the figure, the potentials of nodes 1 and 2 are considered to be higher than those at 1' and 2', respectively, and the currents *entering into* the network are considered positive. This is strictly according to the sign convention we have adopted in this textbook. While at port 1-1' it appears as if we have defined a mesh, we do not intend to do so; hence the choice of the direction of current at port 2-2'.

Since we project a two-port network as a building block for complex networks, we need to put forth certain assumptions. First, we would like to consider the box as an operator of some sort, i.e., a signal at one of the ports is transformed (or converted) into a different signal at the other port by virtue of its transfer function. Second, we assume that the current entering the upper node is the same as the current leaving the lower node at either port, i.e., current entering node 1 leaves node 1', and current entering node 2 leaves node 2'. We quickly notice that, for this second assumption to hold, there cannot be external connections between nodes of different ports, though such connections are allowable between the nodes of the same port.

A note on the notation employed In this chapter we *depict* the voltages and currents in the networks with lowercase symbols, i.e., $v(t)$ and $i(t)$, respectively, with appropriate subscripts and/or superscripts wherever necessary. However, we *compute* using either real algebra for the dc signals or complex algebra for the Laplace transformed ac signals; the symbols would be either uppercase or script ones, conforming to our notation in the earlier chapters.

10.2 Two-port Descriptions

With the definition and assumptions of a two-port network, we observe that of the four quantities—v_1, i_1, v_2, i_2, only two of them are independent. For instance, we may choose v_1 and i_2 and compute v_2 and i_1 in terms of v_1 and i_2. Proceeding this way, we find that there are $^4C_2 = 6$ different choices for the independent quantities. We will enumerate them as follows:

1. i_1 and i_2
2. v_1 and v_2
3. v_1 and i_1
4. v_2 and i_2
5. i_1 and v_2
6. v_1 and i_2

Once again, we quickly observe that we need to generate a pair of linear equations to express the dependent quantities. We will now enumerate the *expressions* for the dependent quantities in the following subsections. For each of these expressions we give specific names for future reference. The rationale behind the names should be clear from the expressions themselves.

10.2.1 The Z Parameters

Let us first take up the currents i_1 and i_2 as the independent quantities and express the voltages v_1 and v_2. Since the network is considered to be linear, we get each of the voltages v_1 and v_2 as a *linear combination* of i_1 and i_2 as follows:

$$v_1 = z_{11}i_1 + z_{12}i_2 \tag{10.1}$$

$$v_2 = z_{21}i_1 + z_{22}i_2 \tag{10.2}$$

It is customary to pack this pair of linear equations into a matrix-vector equation as

$$\begin{bmatrix} v_1 \\ v_2 \end{bmatrix} = \begin{bmatrix} z_{11} & z_{12} \\ z_{21} & z_{22} \end{bmatrix} \begin{bmatrix} i_1 \\ i_2 \end{bmatrix} \tag{10.3}$$

It is now easy to see generalized Ohm's law as

$$\overline{v} = Z\,\overline{i}$$

where $\overline{v}$, Z, and $\overline{i}$ are shown in Eqn (10.3). Owing to the physical dimensions, the four unknown parameters, z_{ij} of matrix Z, have the units of resistance/impedance (Ω) and hence are called Z *parameters* or *impedance parameters*. These parameters may be computed, using the linearity, from Eqns (10.1) and (10.2) as follows.

Step 1	Connect an independent current source i_1 at port 1-1' and leave port 2-2' open-circuited.
Step 2	Measure v_1 and v_2.
Step 3	Since $i_2 = 0$, we have from Eqn (10.1) $z_{11} = \frac{v_1}{i_1}$ and $z_{21} = \frac{v_2}{i_1}$
Step 4	Connect an independent current source i_2 at port 2-2' and leave the port 1-1' open-circuited.
Step 5	Measure v_1 and v_2.
Step 6	Since $i_1 = 0$, we have from Eqn (10.2) $z_{12} = \frac{v_1}{i_2}$ and $z_{22} = \frac{v_2}{i_2}$

Observe that one of the ports is always open-circuited while making the measurements at the other port. Accordingly, we also call these parameters more formally as *open-circuit impedance parameters*. Let us illustrate this with the following example.

Example 10.1 Consider the following network in Fig. 10.2.

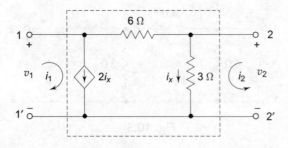

Fig. 10.2

The network is a delta connection and has a controlled current source. However, we need not bother about these details. Let us connect an independent current source i_1 across port 1-1′ and leave port 2-2′ open-circuited. Applying KCR we get

$$i_1 = 2i_x + i_x = 3i_x$$
$$v_2 = 3i_x$$
$$v_1 - v_2 = 6i_x$$

These three equations give us

$$v_1 = 3i_1, \quad v_2 = 1i_1$$

and hence

$$z_{11} = 3\,\Omega, \quad z_{21} = 1\,\Omega$$

Let us next connect a current source i_2 across port 2-2′ and leave port 1-1′ open-circuited. Applying KCR we get

$$i_2 = i_x + 2i_x = 3i_x$$
$$v_2 = 3i_x$$
$$v_2 - v_1 = 12i_x$$

These three equations give us

$$v_1 = -3i_2, \quad v_2 = 1i_2$$

and hence

$$z_{12} = -3\,\Omega, \quad z_{22} = 1\,\Omega$$

Thus, we have

$$Z = \begin{bmatrix} 3 & -3 \\ 1 & 1 \end{bmatrix} \Omega$$

We should not be surprised at this juncture to have a parameter with *negative* resistance; the controlled current source makes this possible.

Exercise 10.1 Obtain the Z parameters of the network shown in Fig. 10.3.

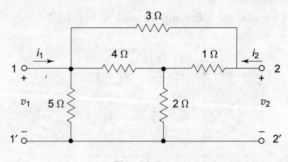

Fig. 10.3

The description of a network in terms of the Z parameters and the algorithm we have formulated above to compute the parameters suggests us the inverse problem, i.e., synthesizing a network given the Z parameters. We will get into this in the next chapter.

10.2.2 The Y Parameters

Let us now take up the voltages v_1 and v_2 as the independent quantities and express the currents i_1 and i_2 as a *linear combination* of v_1 and v_2 as follows:

$$i_1 = y_{11}v_1 + y_{12}v_2 \qquad (10.4)$$
$$i_2 = y_{21}v_1 + y_{22}v_2 \qquad (10.5)$$

Once again, we may pack this pair of linear equations into a matrix-vector equation as

$$\begin{bmatrix} i_1 \\ i_2 \end{bmatrix} = \begin{bmatrix} y_{11} & y_{12} \\ y_{21} & y_{22} \end{bmatrix} \begin{bmatrix} v_1 \\ v_2 \end{bmatrix} \qquad (10.6)$$

Here we have the *inverse* of generalized Ohm's law as

$$\overline{i} = Y\,\overline{v}$$

Owing to the physical dimensions, the four unknown parameters—y_{ij}—have the units of conductance/admittance $[(\Omega)^{-1}]$ and hence are called the Y *parameters* or *admittance parameters*. These parameters may be computed, much like the Z parameters, from Eqs (10.4) and (10.5) as follows.

Step 1	Connect an independent voltage source v_1 at port 1-1$'$ and short-circuit port 2-2$'$.
Step 2	Measure i_1 and i_2.
Step 3	Since $v_2 = 0$, we have from Eqn (10.4) $y_{11} = \frac{i_1}{v_1}$ and $y_{21} = \frac{i_2}{v_1}$
Step 4	Connect an independent voltage source v_2 at port 2-2$'$ and short-circuit port 1-1$'$.
Step 5	Measure i_1 and i_2.
Step 6	Since $v_1 = 0$, we have from Eqn (10.5) $y_{12} = \frac{i_1}{v_2}$ and $y_{22} = \frac{i_2}{v_2}$

A hawk-eyed reader would observe the duality between the Z parameters and the Y parameters—$Z = Y^{-1}$—provided, of course, the inverse exists. Since one of the ports is always short-circuited while making the measurements, it is customary to call these parameters as *short-circuit admittance parameters*.

Let us see the following example.

Example 10.2 Consider the network shown in Fig. 10.4.

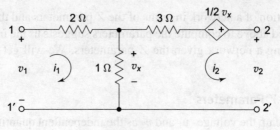

Fig. 10.4

Let us connect a voltage source v_1 across port 1-1' and short-circuit port 2-2'. Applying KVR we get

$$v_1 = 2i_1 + 1(i_1 + i_2)$$

$$0 = 1(i_1 + i_2) + 3i_2 + 0.5v_x$$

$$v_x = 1(i_1 + i_2)$$

These three equations give us

$$i_1 = \frac{3}{8}v_1, \quad i_2 = -\frac{1}{8}v_1$$

and hence

$$y_{11} = \frac{3}{8} \, (\Omega)^{-1}, \quad y_{21} = -\frac{1}{8} \, (\Omega)^{-1}$$

Let us now connect a voltage source v_2 across port 2-2' and short-circuit port 1-1'. Applying KCR we get

$$0 = 2i_1 + 1(i_1 + i_2)$$

$$v_2 = 0.5v_x + 1(i_1 + i_2) + 3i_2$$

$$v_x = 1(i_1 + i_2)$$

These three equations give us

$$i_1 = -\frac{1}{12}v_2, \quad i_2 = \frac{1}{4}v_2$$

and hence

$$y_{12} = -\frac{1}{12} \, (\Omega)^{-1}, \quad y_{22} = \frac{1}{4} \, (\Omega)^{-1}$$

Thus, we have

$$Y = \begin{bmatrix} \dfrac{3}{8} & -\dfrac{1}{12} \\[2ex] -\dfrac{1}{8} & \dfrac{1}{4} \end{bmatrix} (\Omega)^{-1}$$

Exercise 10.2 Obtain the Y parameters of the network shown in Fig. 10.5.

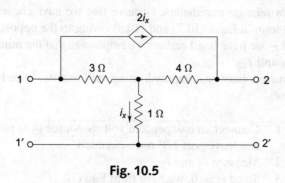

Fig. 10.5

Exercise 10.3 Obtain the Y parameters of the network shown in Fig. 10.2 (Example 10.1) and verify the inverse relationship between Y and Z.

Exercise 10.4 Obtain the Z parameters of the network shown in Fig. 10.4 (Example 10.2) and verify the inverse relationship between Z and Y.

Exercise 10.5 Obtain the Z parameters of the network shown in Fig. 10.6 and verify that the Y parameters do not exist for this network. Synthesize a network for which Y parameters exist but Z parameters do not exist.

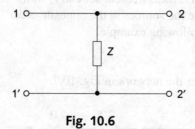

Fig. 10.6

10.2.3 The *T* Parameters

Let us take up the currents v_1 and i_1 of port 1-1' as independent quantities and express the voltage v_2 and current i_2 of port 2-2' as follows:

$$v_2 = T_{11}v_1 - T_{12}i_1 \tag{10.7}$$

$$i_2 = T_{21}v_1 - T_{22}i_1 \tag{10.8}$$

or, equivalently,

$$\begin{bmatrix} v_2 \\ i_2 \end{bmatrix} = \begin{bmatrix} T_{11} & T_{12} \\ T_{21} & T_{22} \end{bmatrix} \begin{bmatrix} v_1 \\ -i_1 \end{bmatrix} \tag{10.9}$$

Owing to the physical dimensions, T_{11} and T_{22} have no units, T_{12} has the units of Ω, and T_{21} has the units of $(\Omega)^{-1}$. These parameters are called the *T parameters* or *forward transmission* parameters. Observe that we have chosen a minus sign for the second term in Eqns (10.7) and (10.8) owing to the opposite directions of currents i_1 and i_2 we have fixed earlier; we emphasize that the minus sign is for i_2 and not for T_{21} and T_{22}.

These parameters may be computed, using the linearity, from Eqns (10.7) and (10.8) as follows.

Step 1	Connect an independent voltage source v_2 at port 2-2' and leave port 1-1' open circuited.
Step 2	Measure v_1 and i_2.
Step 3	Since $i_1 = 0$, we have from Eqn (10.7) $T_{11} = \frac{v_2}{v_1}$ and $T_{21} = \frac{i_2}{v_1}$
Step 4	Connect an independent current source i_2 at port 2-2' and short-circuit port 1-1'.
Step 5	Measure v_2 and i_1.
Step 6	Since $v_1 = 0$, we have from Eqn (10.8) $T_{12} = -\frac{v_2}{i_1}$ and $T_{22} = -\frac{i_2}{i_1}$

We remark that we connect sources only at port 2-2' and we have arbitrarily chosen them to be v_2 in the first measurement set and i_2 in the second measurement set. We may as well choose the sources in the opposite order.

Let us look at the following example.

Example 10.3 Consider the network in Fig. 10.7.

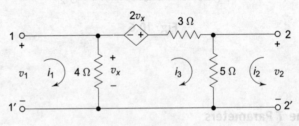

Fig. 10.7

Let us first leave port 1-1' open circuited and apply a voltage source v_2 to port 2-2'. Mesh analysis gives us the following equations:

$$-3v_x + 8i_3 + 5i_2 = 0$$
$$-5(i_2 + i_3) + v_2 = 0$$

With $v_x = -4i_3$, we have

$$i_2 = -4i_3 = v_x \quad i_3 = -\frac{1}{15}v_2$$

from which we may compute

$$T_{11} = \frac{v_2}{v_1} = \frac{15}{4}, \quad T_{21} = \frac{i_2}{v_1} = 1$$

Let us now connect a current source i_2 to port 2-2′ and short-circuit port 1-1′. This gives us the following mesh equations:

$$4(i_1 - i_3) = 0$$

$$-4i_1 + 12i_3 + 5i_2 = 0$$

Solving this pair of equations, we get

$$i_1 = i_3 = -\frac{5}{8}i_2, \quad v_2 = \frac{15}{8}i_2$$

and hence,

$$T_{12} = 3, \quad T_{22} = \frac{8}{5}$$

Thus, we get

$$\begin{bmatrix} v_2 \\ i_2 \end{bmatrix} = \begin{bmatrix} \dfrac{15}{4} & 3\,\Omega \\ 1\,\Omega^{-1} & \dfrac{8}{5} \end{bmatrix} \begin{bmatrix} v_1 \\ -i_1 \end{bmatrix}$$

Exercise 10.6 Obtain the T parameters of the network shown in Fig. 10.2.

Exercise 10.7 Obtain the T parameters of the network shown in Fig. 10.3.

Exercise 10.8 Obtain the T parameters of the network shown in Fig. 10.4.

Exercise 10.9 Obtain the T parameters of the network shown in Fig. 10.5.

Exercise 10.10 Obtain the T parameters of the network shown in Fig. 10.6.

Just as we have duals, the Z and Y parameters, we also have the dual of forward transmission parameters, namely, the *reverse transmission parameters*, or T' parameters in short. They may be written as the following equations:

$$v_1 = T'_{11}v_2 - T'_{12}i_2 \tag{10.10}$$

$$i_1 = T'_{21}v_2 - T'_{22}i_2 \tag{10.11}$$

or, equivalently,

$$\begin{bmatrix} v_1 \\ i_1 \end{bmatrix} = \begin{bmatrix} T'_{11} & T'_{12} \\ T'_{21} & T'_{22} \end{bmatrix} \begin{bmatrix} v_2 \\ -i_2 \end{bmatrix} \tag{10.12}$$

Owing to the physical dimensions, T'_{11} and T'_{22} have no units, T'_{12} has the units of Ω, and T'_{21} has the units of $(\Omega)^{-1}$. These parameters may be computed, using the linearity, from Eqns (10.10) and (10.11). We leave it to the reader as an exercise to write the algorithm for computing these parameters.

Exercise 10.11 Write the algorithm for computing the T' parameters from Eqns (10.10) and (10.11).

We shall look at the inverse relationship between T and T'. Owing to the negative sign in the second terms, we need to be a little careful. Let us follow the steps below in the sequence:

$$\begin{bmatrix} v_2 \\ i_2 \end{bmatrix} = \begin{bmatrix} T_{11} & T_{12} \\ T_{21} & T_{22} \end{bmatrix} \begin{bmatrix} v_1 \\ -i_1 \end{bmatrix}$$

$$\Rightarrow \begin{bmatrix} v_1 \\ -i_1 \end{bmatrix} = \begin{bmatrix} T_{11} & T_{12} \\ T_{21} & T_{22} \end{bmatrix}^{-1} \begin{bmatrix} v_2 \\ i_2 \end{bmatrix}$$

$$= \frac{1}{\Delta T} \begin{bmatrix} T_{22} & -T_{12} \\ -T_{21} & T_{11} \end{bmatrix} \begin{bmatrix} v_2 \\ i_2 \end{bmatrix}$$

$$\Rightarrow \begin{bmatrix} v_1 \\ i_1 \end{bmatrix} = \frac{1}{\Delta T} \begin{bmatrix} T_{22} & T_{12} \\ T_{21} & T_{11} \end{bmatrix} \begin{bmatrix} v_2 \\ -i_2 \end{bmatrix}$$

In other words,

$$\begin{bmatrix} T'_{11} & T'_{12} \\ T'_{21} & T'_{22} \end{bmatrix} = \frac{1}{\Delta T} \begin{bmatrix} T_{22} & T_{12} \\ T_{21} & T_{11} \end{bmatrix}$$

where ΔT is the determinant of matrix T. Thus, unlike Z and Y parameters, the T and T' parameters do not bear an exact inverse relationship. It would be convenient to formulate the following three steps to compute T' from T:

- Given T, negate the sign in the second column to get T_1.
- Invert T_1.
- Negate the sign in the second column of T_1^{-1} to get T'.

Example 10.4 Let us consider the network in Fig. 10.7 again to compute the T' parameters. The algorithm to compute the T' parameters, as was asked in Exercise 10.11, is given below. One can easily observe the duality.

Step 1	Connect an independent voltage source v_1 at port 1-1' and leave port 2-2' open circuited.
Step 2	Measure i_1 and v_2.
Step 3	Since $i_2 = 0$, we have $T'_{11} = \frac{v_1}{v_2}$ and $T'_{21} = \frac{i_1}{v_2}$
Step 4	Connect an independent current source i_1 at port 1-1' and short-circuit port 2-2'.
Step 5	Measure v_1 and i_2.
Step 6	Since $v_2 = 0$, we have $T'_{12} = -\frac{v_1}{i_2}$ and $T'_{22} = -\frac{i_1}{i_2}$

We will redraw the network as Fig. 10.8 here for convenience.

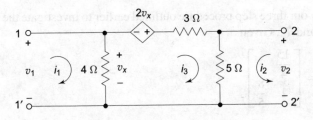

Fig. 10.8

Let us first connect an independent voltage source v_1 at port 1-1' and leave port 2-2' open circuited. Applying mesh analysis we get

$$4(i_1 - i_3) = v_1$$

$$-4i_1 + 12i_3 = 2v_1$$

$$5i_3 = v_2$$

Eliminating i_3, we get

$$i_1 = \frac{5}{8}v_1, \quad v_2 = \frac{15}{8}v_1$$

and hence

$$T'_{11} = \frac{8}{15}, \quad T'_{21} = \frac{1}{3}\,\Omega^{-1}$$

Connecting an independent current source i_1 at port 1-1' and short-circuiting port 2-2' and applying nodal analysis (considering the supernode containing the dependent source) we get

$$\frac{v_1}{4} + \frac{3v_1 - v_2}{3} = i_1$$

$$\frac{v_2 - 3v_1}{3} + \frac{v_3}{5} = i_2$$

Noting that $v_2 = 0$, we get

$$v_1 = \frac{4}{5} i_1, \quad i_2 = -\frac{4}{5} v_1$$

and hence

$$T'_{12} = 1 \, \Omega, \quad T'_{22} = \frac{5}{4} \, \Omega^{-1}$$

Thus, we have

$$T' = \begin{bmatrix} \dfrac{8}{15} & 1 \\ \dfrac{1}{3} & \dfrac{5}{4} \end{bmatrix}$$

Let us verify our three step procedure outlined earlier to investigate the relationship between T and T'. Given

$$T = \begin{bmatrix} \dfrac{15}{4} & 3 \\ 1 & \dfrac{8}{5} \end{bmatrix}$$

Negating the elements of the second column we get

$$T_1 = \begin{bmatrix} \dfrac{15}{4} & -3 \\ 1 & -\dfrac{8}{5} \end{bmatrix}$$

Inverting T_1, we get

$$T_1^{-1} = -\frac{1}{3} \begin{bmatrix} -\dfrac{8}{5} & 3 \\ -1 & \dfrac{15}{4} \end{bmatrix}$$

Notice that $\Delta T_1 = -\Delta T$. It is readily seen that negating the elements of column 2 of T_1^{-1} we get T'.

Exercise 10.12: Obtain the T' parameters of the networks shown in Figs 10.2–10.6.

We will go ahead with the last set of parameters called the hybrid parameters:

10.2.4 The *h* Parameters

We will take up current i_1 and voltage v_2 as the independent quantities and express v_1 and i_2 as a *linear combination* as follows.

$$v_1 = h_{11}i_1 + h_{12}v_2 \tag{10.13}$$

$$i_2 = h_{21}i_1 + h_{22}v_2 \tag{10.14}$$

or, equivalently,

$$\begin{bmatrix} v_1 \\ i_2 \end{bmatrix} = \begin{bmatrix} h_{11} & h_{12} \\ h_{21} & h_{22} \end{bmatrix} \begin{bmatrix} i_1 \\ v_2 \end{bmatrix} \tag{10.15}$$

Since we have chosen the independent quantities from both the ports, we call these parameters *hybrid parameters*, or *h* parameters in short. Owing to the physical dimensions, h_{11} has the units of Ω, h_{12} and h_{21} are dimensionless, and h_{22} has the units of Ω^{-1}. These parameters may be computed as follows.

Step 1	Connect an independent current source i_1 at port 1-1′ and short-circuit port 2-2′.
Step 2	Measure v_1 and i_2.
Step 3	Since $v_2 = 0$, we have from Eqn (10.13) $h_{11} = \frac{v_1}{i_1}$ and $h_{21} = \frac{i_2}{i_1}$
Step 4	Connect an independent voltage source v_2 at port 2-2′ and leave port 1-1′ open-circuited.
Step 5	Measure v_1 and i_2.
Step 6	Since $i_1 = 0$, we have from Eqn (10.14) $h_{12} = \frac{v_1}{v_2}$ and $h_{22} = \frac{i_2}{v_2}$

We may also choose v_1 and i_2 as the independent parameters and get the *inverse hybrid parameters*. These may be called as *g* parameters. However, the *h* parameters defined as in Eqns (10.13) and (10.14) are more conventional in the literature. Hence we concentrate on the hybrid parameters alone and leave the inverse hybrid parameters as an exercise to the reader.

Let us obtain the *h* parameters of the following network.

Example 10.5 Consider the following network in Fig. 10.9.

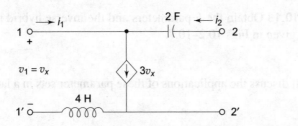

Fig. 10.9

Owing to the presence of the energy-storage elements, we will work in the *s*-domain.

Connecting an independent current source $I_1(s)$ to port 1-1' and short-circuiting port 2-2' gives us the following mesh equations (considering the super mesh):

$$I_1 + I_2 = 3V_1$$
$$- V_1 - \frac{1}{2s}I_2 + 4sI_1 = 0$$

and we get

$$V_1 = \frac{8s^2 + 1}{2s + 3} I_1$$

and

$$I_2 = \frac{2s(12s - 1)}{2s + 3} I_1$$

Likewise, connecting an independent voltage source v_2 to port 2-2' and leaving port 1-1' open-circuited gives us

$$V_2 - V_1 = \frac{1}{2s}I_2$$
$$I_2 = 3V_1$$

from which we get

$$V_1 = \frac{2s}{2s + 3} V_2$$

and

$$I_2 = \frac{6s}{2s + 3} V_2$$

We cannot help but observe that the denominator $2s + 3$ is common for all the four parameters. Thus, we have

$$H = \frac{1}{2s + 3} \begin{bmatrix} 8s^2 + 1 & 2s \\ 2s(12s - 1) & 6s \end{bmatrix}$$

Exercise 10.13 Obtain the h parameters and the inverse hybrid parameters of the networks given in Figs 10.2–10.7.

We will discuss the applications of these parameter sets in a later section of this chapter.

10.2.5 Relationship Among Parameters

The six sets of parameters we have discussed so far are not independent; they all describe the same two-port network. We may find it necessary to convert from one set of parameters to another to get a better insight into a problem. Accordingly, if one set of parameters is known, it should be possible to compute any other set via some algebraic manipulations involving the linearity of the network. For instance, let us consider the relationship between the Z parameters and T parameters. Assume that we already know the open-circuit Z parameters, i.e.,

$$z_{11} = \frac{v_1}{i_1}\Big|_{i_2=0}$$

$$z_{12} = \frac{v_1}{i_2}\Big|_{i_1=0}$$

$$z_{21} = \frac{v_2}{i_1}\Big|_{i_2=0}$$

$$z_{22} = \frac{v_2}{i_2}\Big|_{i_1=0}$$

By definition, we readily get

$$T_{11} = \frac{v_2}{v_1}\Big|_{i_1=0} = \frac{z_{22}}{z_{12}}$$

and

$$T_{21} = \frac{i_2}{v_1}\Big|_{i_1=0} = \frac{1}{z_{12}}$$

However, the other two T parameters demand the short-circuit parameters, i.e., with $v_1 = 0$. These may be better expressed in terms of Y parameters first as

$$T_{12} = \frac{v_2}{i_1}\Big|_{v_1=0} = \frac{1}{y_{12}}$$

and

$$T_{22} = \frac{i_2}{i_1}\Big|_{v_1=0} = \frac{y_{22}}{y_{12}}$$

Since there is an inverse relationship between the Z parameters and the Y parameters, we get

$$T_{12} = \frac{v_2}{i_1}\Big|_{v_1=0} = \frac{\Delta Z}{z_{12}}$$

and

$$T_{22} = \frac{i_2}{i_1}\Big|_{v_1=0} = \frac{z_{11}}{z_{12}}$$

where $\Delta Z = z_{11}z_{22} - z_{12}z_{21}$ is the determinant of the Z matrix.

Applying similar techniques, we can express any parameter set in terms of any other set. We leave it as an exercise to the reader.

Exercise 10.14 Develop a 6×6 table illustrating the relationship among all the six parameters sets.

10.3 Reciprocal and Symmetric Networks

In this section we will look at two special cases of networks. A two-port network is said to be *reciprocal* if interchanging an ideal independent voltage (current) source at one port with an ideal ammeter (voltmeter) at the other port does not change the ammeter reading. We depict this in Figs 10.10(a) and (b).

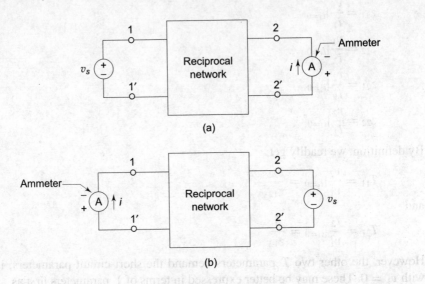

(a)

(b)

Fig. 10.10

Since we do not care as to what the box contains, we may determine the conditions on the parameters for a two-port network to be qualified as a reciprocal network. We observe from Figs 10.10(a) and (b) that, for the same current i to pass through the ammeter given the voltage source v, $y_{12} = y_{21}$ since the current to voltage ratio is invariant. Equivalently, we may also observe that $z_{12} = z_{21}$. Using the relationship among the six sets of parameters we may derive the following conditions equivalent to the two observed above:

$$h_{12} = -h_{21} \tag{10.16}$$

$$\Delta T = 1 \tag{10.17}$$

$$\Delta T' = 1 \tag{10.18}$$

In general, it is easy to know by inspection whether a network is symmetrical.

Exercise 10.15 Verify Eqns (10.16)–(10.18).

From these conditions it is evident that we need only *three* parameters to characterize a reciprocal network; the fourth parameter is either already known in the case of Z, Y, and h parameters, or may be computed by using Eqns (10.17) and (10.18) for T and T' parameters. We also say that if a network satisfies one (and hence all) of the above conditions it is said to obey the principle of *reciprocity*.

Example 10.6 Consider the following network in Fig. 10.11.

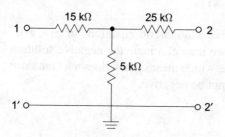

Fig. 10.11

Proceeding to compute the Z parameters, we get

$$z_{11} = 15 + 5 = 20 \, \text{k}\Omega, \quad z_{22} = 25 + 5 = 30 \, \text{k}\Omega, \quad z_{12} = z_{21} = 5 \, \text{k}\Omega$$

Clearly, this is a reciprocal network. One may easily verify that the conditions on the other parameter sets are also met. We emphasize that reciprocity is the property of the network itself rather than that of the concerned parameter set.

Exercise 10.16 Verify the conditions on the Y, T, T', and the h parameters for the network given in Fig. 10.11.

A reciprocal network is said to be *symmetrical* if its ports can be interchanged without altering its terminal currents and voltages. If we consider the Z parameters, the condition for a reciprocal network to be symmetrical would be simply $z_{11} = z_{22}$ since the voltage-to-current ratio should be independent of a particular port. We leave it to the reader to derive conditions on the other parameter sets for a reciprocal network to be symmetrical.

Exercise 10.17 Derive the conditions on the parameter sets for a reciprocal network to be symmetrical.

Example 10.7 Suppose it is given that a certain reciprocal symmetrical passive network has the following measurements: $v_1 = 8$ V and $i_1 = 2$ mA with port 2-2' open, and $v_1 = 7$ V and $i_2 = -3$ mA with port 2-2' short-circuited.

From the first set of measurements, we readily get $z_{11} = 8/2 = 4 \, \text{k}\Omega$. Since it is given that the network is symmetrical, $z_{22} = 4 \, \text{k}\Omega$. From the second set of measurements, we readily get $y_{21} = -3/7 \, \text{m}\Omega^{-1} = y_{12}$ since the network is reciprocal. Further, using the inverse relationship, we get

$$y_{12} = -\frac{z_{12}}{\Delta Z} = -\frac{z_{12}}{z_{11}^2 - z_{12}^2}$$

Solving for z_{12}, we get the Z parameters as

$$Z = \begin{bmatrix} 4 & 3 \\ 3 & 4 \end{bmatrix} k\Omega$$

Here we need to solve a quadratic equation for z_{12}, and we get two possible solutions. However, we may eliminate the negative solution since it is given that the network is passive, which means that it does not contain any source whatsoever. Accordingly, z_{12} cannot be negative.

10.4 Applications of Two-port Parameters

Let us now look at larger systems as interconnections of several two-port networks. These interconnections may be made in different ways, as described in the following subsections. First we will look at the amplifiers, the fundamental building blocks of several electronic gadgets.

10.4.1 Amplifiers

An amplifier accepts a signal, typically a voltage, and delivers a *magnified replica*. The magnification is only in terms of the amplitude; frequency and phase shift are maintained intact between the input and output. Further, this amplification is provided by the device *independent* of the particular source from which it accepts the signal and of the particular load to which it delivers the output. Amplifiers, particularly those built with bipolar junction transistors (BJTs), field effect transistors (FETs), etc., are better described using the h parameters, because of the properties of the transistor. The two-port network description using h parameters in Eqns (10.13) and (10.14) suggests us the following *model* shown in Fig. 10.12.

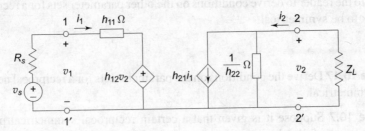

Fig. 10.12

Notice that the input port 1-1' looks like Thevenin's equivalent and the output port 2-2' looks like Norton's equivalent. The usefulness of this set of parameters in representing amplifiers comes from the ease with which the measurements can be made. Typically, a voltage source v_s, in series with an internal resistance R_s, is connected to port 1-1' and a load Z_L is connected across port 2-2' so that $v_2 = -Z_L i_2$. The following quantities are important in the case of amplifiers:

1. Current gain A_I:

$$A_I \triangleq \frac{i_L}{i_1} = -\frac{i_2}{i_1} \tag{10.19}$$

From Norton's equivalent of the output port, we get

$$i_2 = h_{21}i_1 + h_{22}v_2$$
$$= h_{21}i_1 - h_{22}Z_Li_2$$

Rearranging the terms, we get

$$A_I = -\frac{h_{21}}{1 + h_{22}Z_L} \tag{10.20}$$

2. Input impedance Z_i:

$$Z_i \triangleq \frac{v_1}{i_1} \tag{10.21}$$

This is the input impedance 'as seen by' the source v_s. From Thevenin's equivalent of the input port, we get

$$v_1 = h_{11}i_1 + h_{12}v_2$$
$$= h_{11}i_1 + h_{12}Z_Li_L$$

Dividing both sides by i_1 and noting that $A_I = i_L/i_1$, we have

$$Z_i = h_{11} + h_{12}A_IZ_L \tag{10.22}$$

3. Voltage gain A_V:

$$A_V = \triangleq \frac{v_2}{v_1} \tag{10.23}$$

Since $v_2 = -i_2/i_1$ from the output port, it is easy to write A_V as

$$A_V = \frac{v_2}{v_1}$$
$$= \frac{v_2}{i_L}\frac{i_L}{i_1}\frac{i_1}{v_1}$$

Identifying each of the fractions we have

$$A_V = A_I\frac{Z_L}{Z_i} \tag{10.24}$$

4. Output admittance Y_0:

$$Y_0 = \frac{i_2}{v_2} \tag{10.25}$$

with $v_s = 0$ and $Z_L = \infty$. Replacing the voltage source v_s with a short circuit and leaving port 2-2' open-circuited, we have

$$(R_s + h_{11})i_1 + h_{12}v_2 = 0$$

or

$$i_1 = -\frac{h_{12}}{R_s + h_{11}}v_2$$

so that

$$i_2 = h_{22}v_2 + h_{21}i_1$$

$$= \left(h_{22} - \frac{h_{12}h_{21}}{R_s + h_{11}} \right) v_2$$

Thus, we have

$$Y_0 = h_{22} - \frac{h_{12}h_{21}}{R_s + h_{11}} \qquad\qquad (10.26)$$

We may also compute the voltage gain and current gain including the source resistance.

Amplifiers could be single stage or multistage; for instance the popular μ A741 operational amplifier is a complex circuit consisting of 24 transistors, each working as an amplifier on its own. The key to any amplifier operation is the so-called operating point (also called the quiescent point) obtained by driving the transistor(s) into the active region using dc sources. In this region the transistor behaves like a linear element, and the hybrid parameter model discussed above comes handy.

The four quantities discussed above suffice to fully characterize the terminal behaviour of any linear amplifier, independent of its internal circuitry. We may also deduce that, for the ideal operation of amplifying a signal independent of the source and the load, $Z_i = Y_0 = \infty$. We caution the reader not to mistake the h parameter *model* as shown in Fig. 10.12 for an actual amplifier. It is only a model, which *mimics* the behaviour, that we are studying here; the presence of controlled sources should testify this. The actual device that *achieves* this behaviour has a more complicated circuit, which is beyond the scope of this book.

10.4.2 Cascade Connected Two-ports

In Chapter 9 we considered the *input–output* paradigm and discussed an example of a communication system, namely, the radio and the public address system. Evidently, we considered an *interconnection* of a few circuits. Let us look at the public address system example once again. We speak into a microphone. A microphone is a *transducer* that converts our voice into an electrical signal. Clearly, this may be *modelled* using a two-port network. Further, the electrical signal is fed to an amplifier which increases the amplitude of the signal. This amplified electrical signal is fed to another transducer, namely loaudspeaker, that converts the electrical signal into the voice signal. These three two-port networks are connected in *cascade* to make a more complex two-port network. We will depict this in Fig. 10.13.

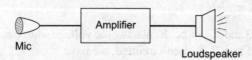

Fig. 10.13

Clearly, the input–output paradigm is a very convenient description to conceive several general systems. The choice of a particular parameter set depends upon the application at hand. It is easy to see that a cascade connection such as the one shown in Fig. 10.13, where the output of a network is fed as an input to the following network, could be easily analysed using the transmission parameters. Indeed they were first used in the analysis of power transmission lines. These are also called *general circuit parameters*.

Given two 2-port networks with the T parameters T^a and T^b, respectively, we will obtain the 'overall' description of the cascade connection as shown in Fig. 10.14.

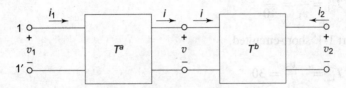

Fig. 10.14

With reference to the directions of currents defined in Fig. 10.14, we have

$$\begin{bmatrix} v_2 \\ i_2 \end{bmatrix} = T^b \begin{bmatrix} v \\ -i \end{bmatrix}$$

$$= T^b T^a \begin{bmatrix} v_1 \\ -i_1 \end{bmatrix} \tag{10.27}$$

or, equivalently,

$$T = T^b T^a \tag{10.28}$$

We remark that the order in which the matrix multiplication is performed is important—T^a is *pre-multiplied* by T^b.

Example 10.8 Let us consider the cascade of the two networks shown in Fig. 10.15.

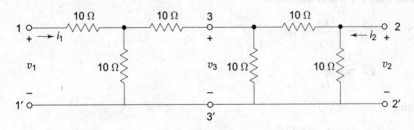

Fig. 10.15

The first network is a star connection (also called a T network[1], and is reciprocal and symmetrical by inspection). The T parameters may be computed by connecting a

---

[1] just like a T-shirt

voltage source v at *its* output port and measuring the required voltages and currents by leaving port 1-1' open-circuited and short-circuiting port 1-1' as follows:

With port 1-1' open-circuited,

$$T_{11}^a = \frac{v_3}{v_1} = 2$$

$$T_{21}^a = \frac{i_3}{v_1} = \frac{1}{10}$$

With port 1-1' short-circuited,

$$T_{12}^a = -\frac{v_3}{i_1} = 30$$

$$T_{22}^a = -\frac{i_3}{i_1} = 2$$

Thus,

$$T^a = \begin{bmatrix} 2 & 30 \\ \frac{1}{10} & 2 \end{bmatrix}$$

confirming that this network is indeed reciprocal since $\Delta T^a = 1$.

The second network is a reciprocal and symmetrical delta network for which we might compute the T parameters as follows:

With port 1-1' open-circuited,

$$T_{11}^b = \frac{v_2}{v_3} = 2$$

$$T_{21}^b = \frac{i_2}{v_3} = \frac{3}{10}$$

With port 1-1' short-circuited,

$$T_{12}^b = -\frac{v_2}{i_3} = 10$$

$$T_{22}^b = -\frac{i_2}{i_3} = 2$$

Thus,

$$T^b = \begin{bmatrix} 2 & 10 \\ \frac{3}{10} & 2 \end{bmatrix}$$

confirming that this network is indeed reciprocal since $\Delta T^a = 1$. The overall T parameters may be computed as

$$T = T^b T^a = \begin{bmatrix} 5 & 80 \\ \frac{8}{10} & 13 \end{bmatrix}$$

We observe that the composite network in Fig. 10.15 is reciprocal but not symmetrical, which confirms our intuition.

Exercise 10.18 Look at the network shown in Fig. 10.15 as a cascade of three 2-port networks and obtain the overall T parameters. Verify your answer using the results of Example 10.8. The network is a ladder network.

10.4.3   Series Connected Two-ports

The connection of two 2-port networks shown in Fig. 10.16 is called a *series connection* because the two input ports carry the same current i_1 and the two output ports carry the same current i_2.

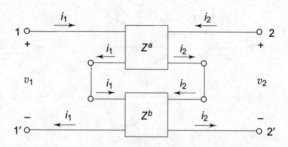

Fig. 10.16

Since $v_1 = v_{a1} + v_{b1}$ and $v_2 = v_{a2} + v_{b2}$, we may immediately conclude that the overall Z parameters are the sum of the individual Z parameters, i.e.,

$$Z = Z^a + Z^b \qquad\qquad (10.29)$$

Example 10.9 Consider the two-port network shown in Fig. 10.17(a) as a cascade connection of two 2-ports shown in Fig. 10.17(b).

The overall Z parameters may be computed by adding the Z parameters of each of the two-port networks as

$$
\begin{aligned}
Z &= Z^a + Z^b \\
&= \begin{bmatrix} 10 & 0 \\ -300 & 100 \end{bmatrix} + \begin{bmatrix} 50 & 50 \\ 50 & 50 \end{bmatrix} \\
&= \begin{bmatrix} 60 & 50 \\ -250 & 150 \end{bmatrix}
\end{aligned}
$$

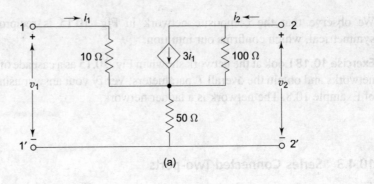

(a)

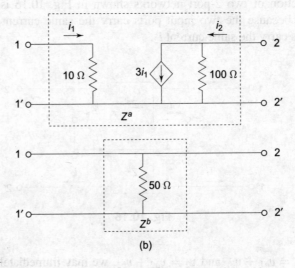

(b)

Fig. 10.17

10.4.4 Parallel Connected Two-ports

The reader would have guessed by now that a parallel connection of two 2-ports results in the addition of the individual Y parameters. We formally show the connection in Fig. 10.18 and leave it to the reader as an exercise to verify.

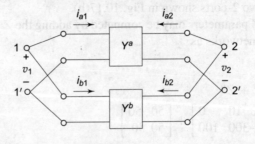

Fig. 10.18

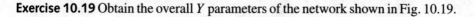

Exercise 10.19 Obtain the overall Y parameters of the network shown in Fig. 10.19.

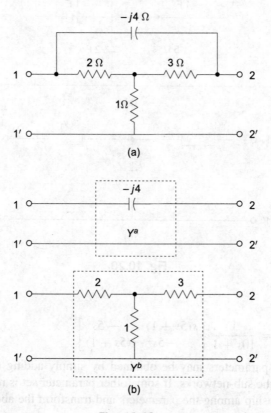

Fig. 10.19

Example 10.10 Consider the network shown in Fig. 10.20.
We see that two symmetrical T networks, such as the one shown in Fig. 10.20(b), are connected in parallel. For each of the networks, the Y parameters may be computed by applying a voltage source at one port and short-circuiting the other port. For the general network in Fig. 10.20(b) with y_1 and y_2 as the series and shunt admittances, respectively, we have the following. Owing to the presence of capacitances, we work in the s-domain.

$$y_{11}(s) = \frac{I_1}{V_1} = \frac{y_1(y_1 + y_2)}{2y_1 + y_2} = y_{22}(s)$$

$$y_{21}(s) = \frac{I_2}{V_1} = \frac{y_1^2}{2y_1 + y_2} = y_{22}(s)$$

Substituting appropriate values for y_1 and y_2 for each of the networks, we get

$$Y^a(s) = \frac{1}{20(10s + 1)} \begin{bmatrix} 20s + 1 & -1 \\ -1 & 20s + 1 \end{bmatrix}$$

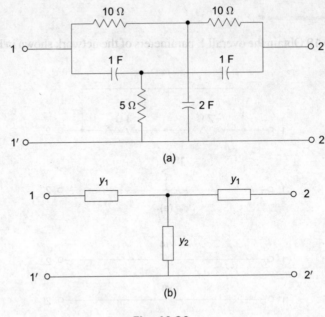

Fig. 10.20

and

$$Y^b(s) = \frac{1}{10s + 1}\begin{bmatrix} s(5s + 1) & -5s^2 \\ -5s^2 & s(5s + 1) \end{bmatrix}$$

The overall Y parameters may be obtained by simply adding the individual Y parameters of the sub-networks. If some other parameter set is required, we may use the relationship among the parameters and transform the above Y parameter set into the required parameter set.

Exercise 10.20 Verify the results of Example 10.10 by obtaining the Z parameters of the overall network and inverting the Z matrix to get the Y parameters.

10.4.5 Magnetically Coupled Coils

We end this chapter with another interesting example of a two-port network, the magnetically coupled coil. In Chapter 2 we have introduced an element, called inductor. We have seen that if a current i is forced through the coil wound around a core, a magnetic flux ϕ results. The rate at which the flux linkage $N\phi$ varies with the current is the inductance L. We will now call this parameter as *self-inductance* of the coil.

If two coils C_1 and C_2 (with inductances L_1 and L_2, respectively) are placed closer to each other, as shown in Fig. 10.21, they share the magnetic flux.

Looking at the figure carefully, we may observe the following.

1. The current i_1 through coil C_1 produces magnetic flux

$$\phi_1 = \phi_{11} + \phi_{21}$$

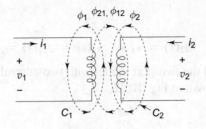

Fig. 10.21

where ϕ_{11} is called the *leakage flux* inside coil C_1 alone, and ϕ_{21} is called the *mutual flux* coupling both C_1 and C_2 coils. Using Faraday's law we may express the voltages v_{11} and v_{21} as

$$v_{11} = L_1 \frac{di_1}{dt} \qquad (10.30)$$

$$v_{21} = M_{21} \frac{di_1}{dt} \qquad (10.31)$$

where M_{21} is called the *mutual inductance* representing the ability of coil C_2 to develop voltage v_{21} in response to the changing current i_1 in the C_1 coil. Observe that the unit of mutual inductance is also Henry.

2. In a similar fashion, we can see that the current i_2 through the C_2 coil produces magnetic flux

$$\phi_2 = \phi_{12} + \phi_{22}$$

where ϕ_{22} is called the *leakage flux* inside the C_2 coil alone, and ϕ_{12} is the *mutual flux* coupling both C_1 and C_2 coils. Once again, using Faraday's law we may express the voltages v_{22} and v_{12} as

$$v_{22} = L_2 \frac{di_2}{dt} \qquad (10.32)$$

$$v_{12} = M_{12} \frac{di_2}{dt} \qquad (10.33)$$

where M_{12} may also be called *mutual inductance*.

3. If an inductor is considered to be a bilateral element, why not a coupled coil? It is indeed so and, using the law of conservation of power, it can be shown that a pair of magnetically coupled coils form a *reciprocal* two-port with

$$M_{12} = M_{21}$$

Thus, we can *model* a pair of coupled coils in terms of the Z parameters as follows:

$$v_1 = L_1 \frac{di_1}{dt} + M \frac{di_2}{dt}$$

$$v_2 = M \frac{di_1}{dt} + L_2 \frac{di_2}{dt} \qquad (10.34)$$

so that

$$z_{11}(s) = sL_1, \quad z_{22}(s) = sL_2, \quad z_{12}(s) = z_{21}(s) = sM$$

The circuit symbol and the two-port model with two controlled sources of a pair of coupled coils are shown in Fig. 10.22.

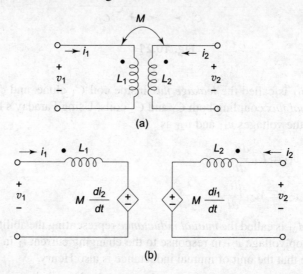

(a)

(b)

Fig. 10.22

The reader would have quickly observed a *dot* at each of the inductance symbols in Fig. 10.22(a). Along with our sign convention, we need to say something about the 'dot convention' too. We use these dots to spell out what Fig. 10.21 shows: *when the reference current enters the dotted terminal of one coil, the reference polarity of the voltage induced in the other coil is positive at its dotted terminal.*

Let us now look at an example.

Example 10.11 Consider the circuit shown in Fig. 10.23 with $v_S = 10$ V, $R = 10\,\Omega$, $L_1 = 2$ H, $L_2 = 4$ H, and $M = 2$ H. Let us assume that the initial current $i_1(0^-) = 0$. Compute $v_2(t)$.

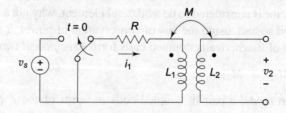

Fig. 10.23

If we apply KVR around mesh m1 at $t = 0^+$, we get

$$10i_1 + 2\frac{di_1}{dt} + 2\frac{di_2}{dt} = 10$$

from which we may solve for $i_1(t)$ as

$$i_1(t) = [i_1(0^+) - i_1(\infty)] e^{-t/\tau} + i_1(\infty)$$

$$= 1 - e^{-5t} \text{ A}$$

where we considered the fact that $i_2 = 0$ since the port 2–2′ is open-circuited, and the time-constant is simply the ratio of L_1 to R.

Applying KVR around mesh m2, we get

$$v_2(0^+) = L_2 \frac{di_2}{dt} + M \frac{di_1}{dt}$$

$$= 2 \frac{di_1}{dt}$$

$$= 10 e^{-5t}$$

Notice how we have made use of the dot convention in this example. In practice, we do not have an easy access to the terminals of a coupled coil since they are packed commercially. This demands us to determine the 'dots' experimentally, and the above example suggests the way. We can make use of our three step strategy—guess, work out, justify—to determine the dots.

Example 10.12 Consider the circuit shown in Fig. 10.24. Here we show a different way of connecting a two-port with an external network.

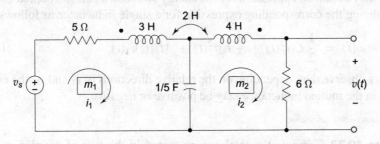

Fig. 10.24

What is most important to us in a network containing a pair of coupled coils is the dot markings. We may readily identify two meshes and apply KVR to get the following pair of linear equations.

$$5i_1 + 3 \frac{di_1}{dt} - 2 \frac{di_2}{dt} + \frac{1}{0.2} \int (i_1 - i_2) \, dt = v_S$$

$$\frac{1}{0.2} \int (i_2 - i_1) \, dt + 4 \frac{di_2}{dt} - 2 \frac{di_1}{dt} + 6i_2 = 0$$

Notice the way we have used the dot convention. For the first equation, we are bothered about the dot in the second mesh—since i_2 is assumed to *leave* the dotted terminal, the polarity at the dot of 3 H inductance in mesh m1 is negative. Similarly, for the second equation, the polarity at the dot of 4 H inductance in mesh m2 is positive; compare this with the Z parameter model of Eqn (10.34).

Rest everything is routine, and we may use Laplace transformation to determine the unknown voltage $v(t)$ as follows. First, we eliminate i_1 and compute

$$I_2(s) = \frac{2s^2 + 5}{8s^3 + 38s^2 + 45s + 55} V_S(s)$$

and hence

$$V(s) = \frac{12s^2 + 30}{8s^3 + 38s^2 + 45s + 55} V_S(s)$$

Observe that this is a third-order circuit since there are three independent energy-storage elements. Further, if $v_S(t) = 10\cos(t)$, i.e.,

$$V_S(s) = \frac{10s}{s^2 + 1}$$

then

$$v(t) = 4.42\cos(t - 77.55°)$$

Exercise 10.21 Verify the results of Example 10.12. Check your results using initial value and final value theorems.

Exercise 10.22 Repeat Example 10.12 with the terminals of the 4 H inductance in the second mesh interchanged so that the dot marking appears at the left.

We may obtain an expression for the energy stored in a pair of coupled coils by generalizing the corresponding expression for a single inductance as follows:

$$w(t) = \frac{1}{2} L_1 i_1^2(t) + \frac{1}{2} L_2 i_2^2(t) + M i_1(t) i_2(t) \tag{10.35}$$

We may observe that, depending on the relative directions of i_1 and i_2, the energy stored in the mutual inductance may be positive or negative.

Exercise 10.23 Compute the total energy stored in the pair of coupled coils of Example 10.12 at $t = 1$ s. How do you check your result?

The equation for the total energy stored may be written in the following manner to obtain additional insight:

$$w(t) = \frac{1}{2}\left[\left(L_1 - \frac{M^2}{L_2} \right) i_1^2(t) + L_2 \left(i_2 + \frac{M}{L_2} i_1 \right)^2 \right] \tag{10.36}$$

Since the energy can never be negative, it is clear that the following condition has to be satisfied:

$$M \leq \sqrt{L_1 L_2} \tag{10.37}$$

We use this inequality and define the *coefficient of coupling* between the two inductances as

$$k \triangleq \frac{M}{\sqrt{L_1 L_2}} \qquad (10.38)$$

Once again, it is easy to see that

$$0 \leq k \leq 1$$

i.e., the pair of coils is either completely uncoupled or perfectly coupled at their extremes. In other words, we say that the two coils behave as two independent inductances if $k = M = 0$. Similarly, if the mutual inductance is the geometric mean of the self-inductances, i.e., if $k = 1$, the coils are said to be perfectly coupled and form a *perfect transformer*. In practice, to achieve perfect coupling, the coils are usually wound around a common core of a highly permeable material, such as iron. If $0 < k < 1$, we say that the coils are partially coupled. In practice, this is achieved by using a nonmagnetic material as a common core. Here the linear relationship between the current and the flux is retained. A pair of partially coupled coils is also called as a *linear transformer*.

Transformers are electrical devices that we come across around us everyday in different sizes and shapes. For instance, our domestic electrical power of 220 V magnitude is the output of a *step-down* transformer at the street corner; our walkmans work with, typically, 3 V dc, which can be obtained by passing our domestic 220 V sinusoid through another step-down transformer and then rectifying it. We will not discuss further these devices.

Exercise 10.24 For the circuit shown in Fig. 10.25, containing a pair of coupled coils, obtain the transfer function $V_0(s)/V_S(s)$.

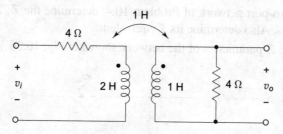

Fig. 10.25

Exercise 10.25 Repeat Exercise 10.24 if the both the dots are at the bottom.

Summary

In this chapter we have introduced a linear two-port network and established a set of linear relationships among its terminal voltages and currents, independent of the network topology. We have found that there are six different ways of choosing the parameter sets and have investigated five of them. We left the inverse hybrid parameters as an exercise to the reader. Each parameter set may be computed via suitable open-circuit and/or short-circuit tests. And, once a parameter set is known, it is possible to find any other parameter set by making use of appropriate relationships among the parameters. We have also observed that, in some cases, some parameter sets may not exist. Further, we have defined reciprocal networks and symmetrical networks and have explored the interconnections of two or more 2-port networks. These interconnections allow us to analyse/synthesize several general systems. In this chapter we have given the examples of amplifiers and public address systems. Other examples, the reader would come across in different courses, include magnetically coupled coil pairs and linear transformers. We will use some of the ideas of two-port networks in the next chapter, particularly in the synthesis of networks given the transfer functions.

Problems

10.1 In an experiment it was found that a two-port T network has the following Z parameters at a frequency of $\omega = 10^6$ rad/s.

$$Z = \begin{bmatrix} 2 - j1 & 1 \\ 1 & 2 + j1 \end{bmatrix}$$

Determine the elements of the network.

10.2 For the two-port network of Problem 10.1, determine the Y, T, and hybrid parameters.

10.3 Determine the π equivalent of the T network of Problem 10.1.

10.4 Given the following Y parameters, determine a π network:

$$Z = \begin{bmatrix} 2s + 3 & -1 \\ -1 & 3s + 2 \end{bmatrix}$$

10.5 For the two-port network of Problem 10.4, determine the Z, T, and hybrid parameters. Also determine its T equivalent.

10.6 Obtain the Z parameters of the network shown in Fig. 10.26.

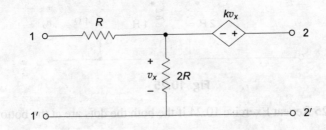

Fig. 10.26

10.7 Obtain a T equivalent of the π network shown in Fig. 10.27.

10.8 On a two-port network, the following measurements are obtained in an experiment: (i) With port 1 open-circuited and a source $v_2 = 10$ V across port 2, $v_1 = 5$ mV and $i_2 = 0.5$ mA, and (ii) with port 1 short-circuited and with the same source $v_2 = 10$ V across port 2, $i_1 = -\mu$A and $i_2 = 1$ mA. Obtain the Z parameters of the two-port network, and hence its Y parameters.

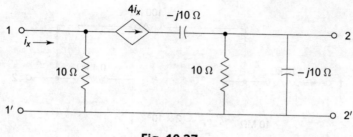

Fig. 10.27

10.9 Given the two-port network in Fig. 10.28, determine its hybrid parameters.

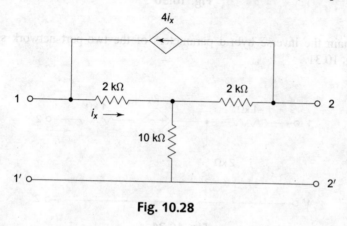

Fig. 10.28

10.10 In an experiment, a 10-V dc source is connected across port 1, and the following measurements are recorded: (i) $v_2 = 20$ V and $i_1 = 5$ mA with port 2 open-circuited, and (ii) $i_1 = 3$ mA and $i_2 = -2$ mA with port 2 short-circuited. Obtain the Y parameters of the networks.

10.11 Obtain the transmission parameters of the network shown in Fig. 10.29.

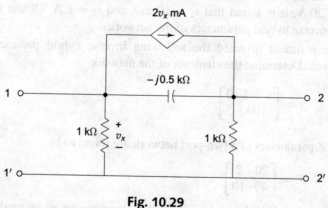

Fig. 10.29

10.12 Obtain the hybrid parameters of the two-port network shown in Fig. 10.30.

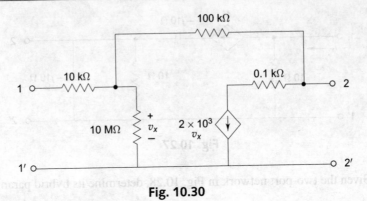

Fig. 10.30

10.13 Obtain the inverse hybrid parameters of the two-port network shown in Fig. 10.31.

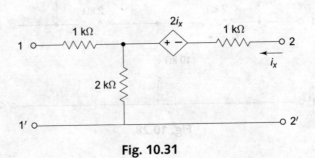

Fig. 10.31

10.14 Obtain the transmission parameters of the two-port network of Problem 10.13.

10.15 In an experiment on a two-port network, the following observations are recorded: (i) with port 1 open-circuited and $v_2 = 10$ V, it is found that $v_1 = 25$ mV and $i_2 = 5$ mA, and (ii) with port 2 short-circuited, and $v_1 = 20$ V, it is found that $i_1 = 10$ mA and $i_2 = 1$ A. Obtain the hybrid and inverse hybrid parameters of the network.

10.16 An experiment revealed the following inverse hybrid parameters for a network. Determine the elements of the network.

$$g = \begin{bmatrix} s+1 & 0 \\ 100 & 5 \end{bmatrix}$$

10.17 The Z parameters of a two-port network are found to be

$$Z = \begin{bmatrix} 20 & 2 \\ 40 & 10 \end{bmatrix}$$

If this network is driven by a 10 V dc source having an internal resistance of 1 Ω and is terminated by a 100 Ω load resistance, obtain Thévenin's equivalent as seen by (i) the source and (ii) the load.

10.18 The circuit shown in Fig. 10.32 is a *model* of the common-base BJT amplifier. Obtain an expression for the source-to-voltage gain.

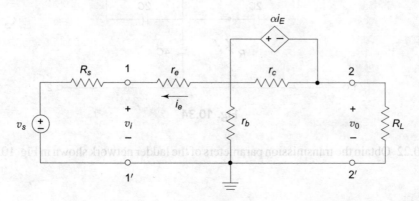

Fig. 10.32

Determine the gain if $R_S = 100\,\Omega$, $r_e = 20\,\Omega$, $r_b = 0.25\,\text{k}\Omega$, $r_0 = 1\,\text{M}\Omega$, $R_L = 1\,\text{k}\Omega$, and $\alpha = 0.995$.

10.19 A two-port network has the following Y parameters:

$$Y = \begin{bmatrix} \dfrac{1}{100} & -\dfrac{1}{2000} \\ -\dfrac{1}{10} & \dfrac{1}{50} \end{bmatrix}$$

If this port is driven by a 5-V dc source whose internal resistance is $0.5\,\Omega$, determine the load R_L for maximum power transfer. Find the maximum power delivered to the load.

10.20 For the two-port network shown in Fig. 10.33, obtain the Y parameters.

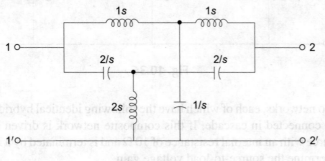

Fig. 10.33

10.21 For the two-port network shown in Fig. 10.34, obtain the transmission parameters.

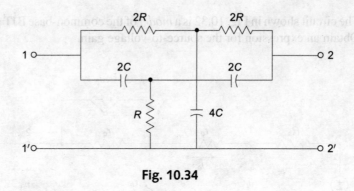

Fig. 10.34

10.22 Obtain the transmission parameters of the ladder network shown in Fig. 10.35.

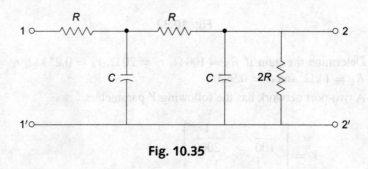

Fig. 10.35

10.23 What is the transfer function $V_2(s)/V_1(s)$ of the circuit shown in Fig. 10.36?

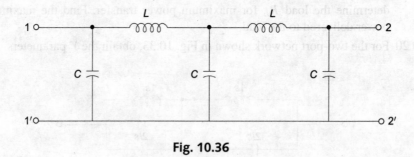

Fig. 10.36

10.24 Two networks, each of which have the following identical hybrid parameters, are connected in cascade. If this composite network is driven by a voltage source with an internal resistance of $10\,\Omega$ and is terminated by a load of $2\,\mathrm{k}\Omega$, determine the source-to-load voltage gain.

$$ h = \begin{bmatrix} 1000 & 0.005 \\ 50 & 10^{-5} \end{bmatrix} $$

10.25 Determine the driving point impedance $z_{11}(s)$ of the two-port network shown in Fig. 10.37.

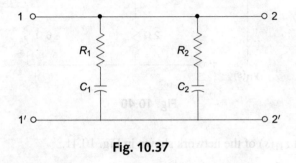

Fig. 10.37

10.26 Determine the driving point impedance $z_{11}(s)$ and the driving point admittance $y_{11}(s)$ of the network shown in Fig. 10.38.

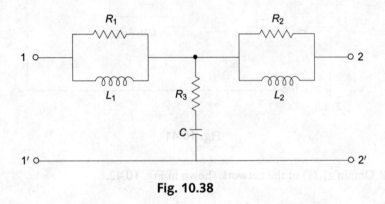

Fig. 10.38

10.27 Obtain $y_{11}(s)$ of the network shown in Fig. 10.39.

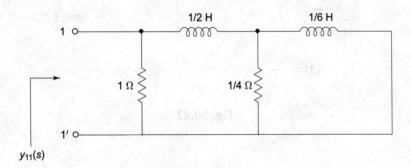

Fig. 10.39

10.28 Obtain $y_{11}(s)$ of the network shown in Fig. 10.40.

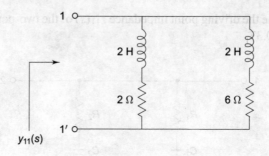

Fig. 10.40

10.29 Obtain $z_{11}(s)$ of the network shown in Fig. 10.41.

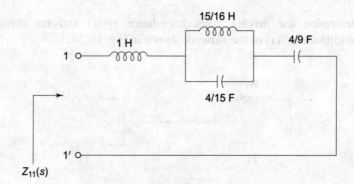

Fig. 10.41

10.30 Obtain $z_{11}(s)$ of the network shown in Fig. 10.42.

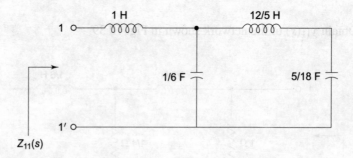

Fig. 10.42

CHAPTER 11

Synthesis of Networks

Chapters 1–10 of this book have been devoted to a fairly exhaustive treatment of what may be called 'analysis' of networks. This chapter presents certain ideas that help the reader 'synthesize' networks given the specifications and/or constraints. From the input–output paradigm we are familiar with by now, we have three objects—input, network, and output. What we have been doing until the preceding chapter was to obtain the output, given the network and its input. This is called the *analysis problem*. The *synthesis problem* is to determine a network, if there exists one, given the input and output. We are often told that *design* is primarily the concern of engineers. However, there are two different interpretations that may be given to the word design. The first one is the traditional way of relying upon handbooks (networks or systems created by an earlier generation of engineers out of some intuition and experience), and/or the sheer excitement in trial and error. A modern interpretation is to look at the prescribed input–output characteristics from a more scientific (rather than mere intuitive) perspective and 'build up' complex networks from parts or elements.

Synthesis is a subject by itself and cannot be dealt with exhaustively here in one chapter. The subject encompasses a range of subjects, from abstract mathematics to day-to-day practical problems. Synthesis has become a scientific basis for innovations in design. It was first applied to passive networks and later to control systems and active networks. It is not any exaggeration to say that familiarity with network synthesis is a prerequisite to understand modern electrical engineering. In this chapter, we will focus on looking back into the previous chapters on analysis, Chapter 9 in particular, and formalize some fundamental ideas.

11.1 An Intuitive Beginning

Example 11.1 The starting point of any synthesis problem is a set of specifications. Typically, it may be spelt as follows: for a sinusoidal input signal of unity magnitude, the output magnitude is required to be unity until $\omega = \omega_c$ and thereafter decrease linearly with frequency, as shown in Fig. 11.1.

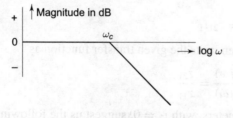

Fig. 11.1

Notice that we have made use of *Bode's magnitude plot*; magnitude of unity is 0 dB. The first step in synthesis is to obtain a suitable, albeit approximate, 'functional model' of the specifications. The magnitude plot of Fig. 11.1 tells us that the gain is unity for lower frequencies and is extremely small for high frequencies. Accordingly, from what we have learnt in Chapter 7, we may think of

$$\frac{\text{output}(j\omega)}{\text{input}(j\omega)} \approx \frac{k}{j\dfrac{\omega}{\omega_c} + k} \tag{11.1}$$

so that

$$\lim_{\omega \to 0} \frac{\text{output}(j\omega)}{\text{input}(j\omega)} = 1$$

and

$$\lim_{\omega \to \infty} \frac{\text{output}(j\omega)}{\text{input}(j\omega)} \to 0$$

and the output magnitude decreases linearly at the rate of 20 dB/decade. Further, from Eqn (11.1) we may observe that it is only an *approximate* function since at $\omega = \omega_c$ the magnitude is -3 dB and not equal to 0 dB as was *specified* in Fig. 11.1. In other words, we attempt to fit a continuous curve that is *as close as possible* to the piece-wise linear specification. This sort of a *tolerance* has to be given, and the approximate function has to be tested, at least, in the limits.

As a second step, we look at the network realization. This appears to be tricky at this moment, but we have a plethora of techniques to do this as we explore in the subsequent sections. Still, it is rewarding to solve the problem here based on our earlier experience. Let us think in the following direction.

- If we assume a two-port network with v_1 as the input and v_2 as the output, it would be convenient for us to keep port 2-2' open-circuited so that $i_2 = 0$. (Obviously, we cannot assume port 2-2' to be short-circuited. Why?)

- With the port 2-2' open-circuited, we have the following parameters:

$$z_{11} = \frac{v_1}{i_1}, \quad z_{21} = \frac{v_2}{i_1}, \quad \frac{1}{t'_{11}} = \frac{v_2}{v_1}$$

  i.e.,

$$v_1 = z_{11} i_1$$

  and

$$v_2 = z_{21} i_1$$

- We might then write the given transfer function as

$$\frac{v_2(j\omega)}{v_1(j\omega)} = \frac{z_{21}}{z_{11}}$$

- The Z parameters, with $i_2 = 0$, suggest us the following network (shown in Fig. 11.2). It looks likes a voltage division circuit.

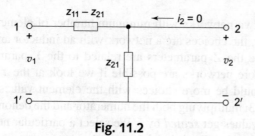

Fig. 11.2

We could have thought of a reciprocal and symmetric T network, but we do not have any information on z_{12} and z_{22}. Further, this network has the minimum number of elements and hence is more economical.

- It follows that

$$z_{11} = k + j\frac{\omega}{\omega_c}$$

and

$$z_{21} = k$$

and the network is an RL network as shown in Fig. 11.3(a), with $R = k\,\Omega$ and $L = 1/\omega_c$ H.

- Alternatively, we also have, by dividing the numerator and denominator by $j\omega$,

$$z_{11} = \frac{1}{\omega_c} + \frac{1}{j\omega/k}$$

and

$$z_{21} = \frac{1}{j\omega/k}$$

and the network is a RC network as shown in Fig. 11.3(b) with $R = 1/\omega_c\,\Omega$ and $C = 1/k$ F.

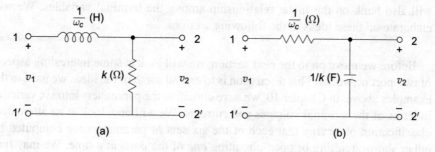

Fig. 11.3

The last step in synthesis is a final choice from a number of alternatives. In the previous step itself (of course with reference to this particular problem alone) we have shortlisted two networks with minimum number of elements with cost as a criterion. Further, the choices are a network with an inductor and a network with a capacitor. Since the Z parameters are related to the Y parameters, at least in principle, two more networks are possible if we look at the Y parameters also. Perhaps, there could be more choices with the element values other than those shown in Fig. 11.3; multiplying both the numerator and the denominator by a factor l, the elemental values get *scaled* by l. We select a particular network among the alternatives based on the criteria involving convenience and engineering judgement, in addition to the cost. For instance, if we are talking about an integrated circuit (IC), the network with capacitance is obvious; if we are talking about a high-voltage electrical circuit, the network with an inductor is obvious. We emphasize that a synthesis problem, in general, has several solutions in contrast to the unique solution in an analysis problem; sometimes there may not exist any solution at all to the synthesis problem.

Example 11.2 Suppose that we are given the following transfer function:

$$\frac{v_2(j\omega)}{v_1(j\omega)} = \frac{j\dfrac{\omega}{\omega_c}}{j\dfrac{\omega}{\omega_c} + k}$$

It is easy to see that the driving point impedance z_{11} is identical to that of the network in Example 11.1. Also, if we stress on the network rather than on a particular output signal, we readily observe that it is the same network that satisfies the transfer function of this example; for instance, the output is the either voltage across the inductance in Fig. 11.3(a) or the voltage across the resistance in Fig. 11.3(b). From this, we note that once there is a network, we may arbitrarily define the two-ports 1-1' and 2-2' anywhere we wish; a transfer function needs at least three parameters (assuming a reciprocal network) to be specified to realize a network. Nevertheless, the driving point impedance plays a key role. By duality, the driving point admittance also plays a key role. In other words, a network with a driving point impedance or admittance function is *sufficient* to describe the transfer function. We will also bank on the linear relationship among the terminal variables. We will elaborate on these ideas in the following sections.

Before we move on to the next section, we will look at some interesting aspects of two-port networks. This discussion is to extend some of the ideas we used in the examples above. In Chapter 10, we have classified the parameters into six varieties in terms of the terminal voltages and currents. We will now look at an alternative classification observing that each of the six sets of parameters are computed by either short-circuiting or open-circuiting *one* of the ports at a time. We may turn around and classify the two-port networks into four sets:

1. Input port open-circuited, i.e., with $i_1 = 0$
2. Input port short-circuited, i.e., with $v_1 = 0$
3. Output port open-circuited, i.e., with $i_2 = 0$
4. Output port short-circuited, i.e., with $v_2 = 0$

Classifying this way, we have a specific advantage. For instance, with $i_2 = 0$, the T' parameters give us

$$v_1 = t'_{11} v_2$$

and the Y parameters give us

$$0 = y_{21} v_1 + y_{22} v_2$$

Together, we have

$$y_{21} = -y_{22} \frac{v_2}{v_1} = -\frac{y_{22}}{t'_{11}}$$

Observe that the driving point and transfer parameters are *linearly* related owing to the fundamental hypothesis that a two-port network is described as a linear combination of the terminal voltages and currents. We leave it as an exercise to the reader to express all the transfer parameters of the form $\alpha_{ij}\, i \neq j$ in terms of the driving point parameters of the form β_{ii}.

Exercise 11.1 Express all the transfer parameters of the form α_{ij} with $i \neq j$ in terms of the driving point parameters of the form β_{ii}, with appropriate conditions such as $i_1 = 0$, etc.

Although the specifications are often given in terms of transfer functions, it is far more convenient to synthesize networks in terms of the driving point Z and Y parameters. As we have seen in our earlier examples, parameters such as z_{12}, y_{21}, etc. can be handled easily by appropriately choosing the second port. In this chapter we focus on the synthesis using the driving point functions. Towards the end, we will discuss briefly on the synthesis given a transfer function. It is time for us to look at a rather fundamental issue—whether there exists[1] a network given some rational function?

11.2 Existence of Networks

As a first step in the synthesis of networks, we have obtained an approximate function that represents the given specifications as closely as possible. This is not a trivial problem, though we did it with some intuition in the previous section. However, we will not focus on this aspect anymore; interested readers are recommended to look at 'curve fitting' in mathematics. If, somehow, we have got a function representing the specifications, there is no guarantee that there would exist a corresponding network built solely with passive elements. In other words, there is a subset called the *set of admissible functions* that we should look at. In this section we will delve into these issues and formulate a list of necessary and sufficient conditions. We will assume that an approximating function is given to us. A comparison of this function with the list of conditions is known as *testing*.

---

[1] Mathematicians are sometimes notorious for their prayer: 'Oh God! If you *exist*, save me, if you can.'

We further assume that the approximating function is a function of the Laplace variable s instead of $j\omega$ so that we can talk about generalized frequency response and network functions.

11.2.1 Mathematical Characterization of Existence

Let us assume that the specifications have been transformed into a driving point impedance function, say, $z_{11}(s)$. As we have observed in Chapter 9, a complex nth-order network will have a rational function

$$z_{11}(s) = H_0 \frac{N(s)}{D(s)} = H_0 \frac{s^m + b_{m-1}s^{m-1} + \cdots + b_0}{s^n + a_{n-1}s^{m-1} + \cdots + a_0}$$

as a network function. Here the denominator $D(s)$ would be an nth-order polynomial. We have also deduced that the difference in the highest and lowest degrees of $N(s)$ and $D(s)$ is not more than 1.

Let us now assume that we are able to factorize $D(s)$ and express the driving point function as

$$z_{11} = k_\infty s + z_0 + \frac{k_1}{s + p_1} + \cdots + \frac{k_n}{s + p_n} \tag{11.2}$$

where we have considered that the numerator $N(s)$ has degree $n + 1$. Clearly, this could represent a series network with z_0 as a resistance, k_∞ as an inductance, and so on. At this moment let us not worry what elements would give us an impedance of the form $k_i/(s + p_i)$, but let us observe that each such term must represent a passive impedance with Ω as its physical unit. Clearly, the scale factor H_0 should be positive and real; none of the residues k_i is negative and, none of p_i is negative. We may also carefully observe that one (and only one) of the roots of $D(s)$ can be zero, for such a term would represent a capacitance. These observations lead us to the fact that the denominator

$$D(s) = (s + p_1)(s + p_2) \cdots (s + p_n) = s^n + a_{n-1}s^{n-1} + \cdots + a_0$$

is a polynomial with all of its coefficients a_i non-negative and that all the roots have negative real parts, with possibly one (and only one) pole at the origin. Analysing further, according to the fundamental theorem of algebra, we may observe that a root may be complex, of course with a negative real part, provided its conjugate is also a root. Such a polynomial with none of its roots lying in the right half of the s-plane is called a *Hurwitz polynomial*.

If we are able to build a network given $z_{11}(s)$, the driving point admittance $y_{11}(s)$ of the network should also be an admissible admittance function in the sense that we should be able to build an equivalent network. Accordingly, we may extend the properties enumerated above to $y_{11}(s)$ and say that the polynomial $N(s)$ should also be a Hurwitz polynomial.

There is another facet of rational functions having poles strictly in the left half of the s-plane. For instance, if we consider the same $z_{11}(s)$, it helps us compute the current i_1, given the voltage v_1. If the applied input voltage v_1 is of a finite magnitude, we say that it is bounded, i.e.,

$$|v_1| \leq M < \infty$$

If the current i_1 also turns out to be a bounded quantity, we say the network is *stable*, more precisely *bounded-input-bounded-output stable*, or BIBO stable in short. Naturally, looking at the partial fractions of $z_{11}(s)$, we notice that the response would be a linear combination of several exponential signals and hence the current would be bounded *if and only if* the poles of $z_{11}(s)$ are restricted to the left half of the s-plane. Intuitively, we feel that if at all a driving point function $F(s)$ and its inverse $1/F(s)$ yield networks, they are necessarily stable in the BIBO sense. To determine if a given polynomial is Hurwitz, we have an elaborate criterion called *Routh–Hurwitz criterion*. We will not discuss it here. Interested readers may see the appendix. This criterion would be taught in the first course on control systems in a later semester.

11.2.2 Positive Real Functions

The requirements imposed on driving point functions were first given in complete form by Otto Brune in his doctoral thesis at MIT Masachussets in 1930[2]. Brune's work had extended some of the earlier work by Foster[3], Cauer[4], and others and laid the scientific foundations of network synthesis. The driving point functions $F(s)$ with the properties

$$\text{Re } F(s) \geq 0 \quad \text{for Re } s \geq 0 \tag{11.3}$$

$$F(s) \text{ is real when } s \text{ is real} \tag{11.4}$$

were given the name *positive real* functions, or *pr* in short, by Brune. He showed that the requirement that a driving point function be positive real was sufficient by demonstrating that a network could always be found to correspond to the positive real driving point function. His sufficiency condition proof is a general method for network synthesis and is beyond the scope of this chapter. In what follows we will first present the synthesis ideas from an analysis point of view.

Before we proceed to the synthesis of networks, let us eliminate certain trivial features. First, if we imagine a *long chain* of passive elements, R_i, L_i, and C_i, connected end to end in series, it is easy to observe that this series network boils down to a network of just three elements—all the resistors collapse to an equivalent

---

[2] 'Synthesis of a finite two terminal network whose driving point impedance is a prescribed function of frequency,' MIT ScD thesis in 1930. This work has also appeared as a paper in *Journal of Mathematics and Physics*, vol. 10, 1931.

[3] The paper 'A Reactance Theorem' in Bell Systems Technical Journal, vol. 3, pp. 259–67, by Ronald M. Foster in 1924 is considered to be the first publication to throw light on the synthesis of networks in the modern sense. Foster was initially with Bell System and later with Polytechnic Institute of Brooklyn.

[4] Wilhelm Cauer received his doctoral degree from Technischen Hochschule, Berlin with the dissertation, 'The realization of impedances with prescribed frequency dependence' in 1924. He served as a university professor in Berlin and also as an engineer in aircraft and telephone industries. Thirty-eight papers and a textbook published in 1940 stand to testify his productivity. His work on ladder networks is influential in the synthesis of networks given a transfer function.

resistance R_{eq}, all the capacitors collapse to an equivalent capacitance C_{eq}, and all the inductors collapse to L_{eq}. And, the driving point functions look like

$$z_{11}(s) = \frac{L_{eq}C_{eq}s^2 + R_{eq}C_{eq}s + 1}{C_{eq}s}$$

and

$$y_{11}(s) = \frac{C_{eq}s}{L_{eq}C_{eq}s^2 + R_{eq}C_{eq}s + 1}$$

Similarly, if we imagine a long parallel combination of passive elements, R_i, L_i, and C_i, connected across a pair of nodes, it is easy to observe that this parallel network, once again, boils down to a network of just three elements—R_{eq}, C_{eq}, and L_{eq}. And, by duality, the driving point functions look like

$$y_{11}(s) = \frac{L_{eq}C_{eq}s^2 + \dfrac{L_{eq}}{R_{eq}}s + 1}{L_{eq}s}$$

and

$$z_{11}(s) = \frac{L_{eq}s}{L_{eq}C_{eq}s^2 + \dfrac{L_{eq}}{R_{eq}}s + 1}$$

We avoid these two trivial cases and go ahead in a different way. We choose two elements at a time, say R and C, and have either a series connection of parallel-connected elements or a parallel connection of series-connected elements. We will investigate these possibilities in the next few sections.

11.3 Synthesis of *RC* Networks

Let us first consider the parallel connection of a resistance R_i and a capacitance C_i as shown in Fig. 11.4.

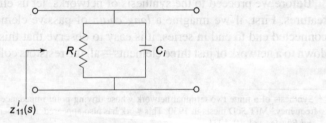

Fig. 11.4

The driving point impedance may be quickly computed to be

$$z_{11}^i(s) = \frac{R_i}{R_i C_i s + 1} \tag{11.5}$$

11.3.1 Foster Forms

Let us now connect $n + 2$ of these parallel sections in series as shown in Fig. 11.5.

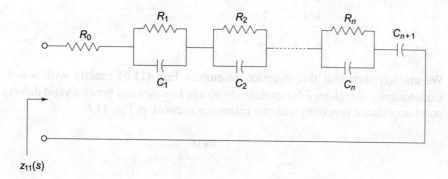

$z_{11}(s)$

Fig. 11.5

This network is called *Foster-1 network*. Notice that we have connected the two-port networks of Fig. 11.4 *in series* and hence we can algebraically add the Z parameters. Including the limiting cases of $C_i = 0$ and $R_i = \infty$ for $i = 0$ and $i = n + 1$, respectively, the driving point impedance $z_{11}(s)$ of the series connection may be computed as

$$z_{11}(s) = \sum_i z_{11}^i$$

$$= R_0 + \sum_{i=1}^{n} \frac{R_i}{R_i C_i s + 1} + \frac{1}{C_{n+1}s} \tag{11.6}$$

It is clear that to realize a Foster-1 network, the given driving point impedance function must satisfy certain conditions; it has to be a pr function. We will not derive any conditions for testing whether a driving point function is pr or not. Instead, we will look at a few examples and then enumerate the conditions based on certain observations in these examples.

Example 11.3 Consider the following function:

$$F(s) = \frac{(s + 1)(s + 3)}{s(s + 2)}$$

To begin with, let us assume[5] that this is a driving point impedance function $z_{11}(s)$ of a possible *RC* network. As it is, we do not know whether the function is pr or

---

[5] Remember our strategy—guess, work out, justify.

not. Let us examine the partial fraction expansion of $F(s)$:

$$F(s) = 1 + \frac{1/2}{s+2} + \frac{3/2}{s}$$

$$= 1 + \frac{1/4}{\frac{1}{2}s + 1} + \frac{1}{\frac{2}{3}s}$$

We are fortunate that this equation resembles Eqn (11.6) exactly with $n = 1$. Consequently, the given $F(s)$ qualifies to be a pr function and hence a valid driving point impedance function, with the following network in Fig. 11.6.

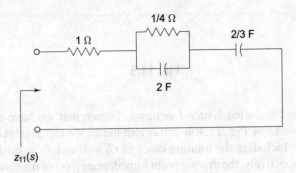

Fig. 11.6

Exercise 11.2 Verify the values of the elements in the network above.

At this moment, let us not be taken aback by capacitances of the order of few farads. Towards the end of this chapter, we will justify everything.

We suggest the reader to observe the following points in the driving point function of the previous example.

1. The poles and zeros are simple (i.e., multiplicity is 1), and are located on the negative real axis of the s-plane. In particular, the pole at the origin (on the imaginary axis) is simple.

2. The poles and zeros interlace, i.e., a pair of poles has a zero in between and a pair of zeros has a pole in between.

3. The first corner frequency we come across as we traverse the negative real axis from the origin is a pole; in this example, it is the pole at origin. The last corner frequency along the negative real axis is a zero; in this example, it is the zero at $s = -3$. In general it could be at $-\infty$.

4. The above two observations suggest us that $z_{11}(\infty) < z_{11}(0)$.

5. The residues are, obviously, real and positive.

All the above observations are, in fact, the properties of positive real driving point *impedance* functions, which have to be satisfied for a network to exist. Let us look at another example.

Example 11.4 Consider the following function:

$$F(s) = \frac{(s+2)}{(s+1)(s+3)}$$

Let us assume that this is a driving point impedance function $z_{11}(s)$ of a possible RC network and get the partial fraction expansion of $F(s)$ as

$$z_{11}(s) = \frac{1/2}{s+1} + \frac{1/2}{s+3}$$

$$= \frac{1/2}{s+1} + \frac{1/6}{\frac{1}{3}(s+1)}$$

This equation resembles Eqn (11.6) exactly with $n = 2$ and no limiting cases. Thus, the given $F(s)$ qualifies to be a valid driving point impedance function with the following network in Fig. 11.7.

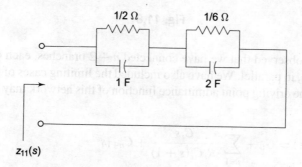

Fig. 11.7

Exercise 11.3 Verify the values of the elements in the network shown in Fig. 11.7.

Exercise 11.4 Verify that the driving point function of Example 11.4 has all the properties of a pr function.

Example 11.5 Let us now consider

$$F(s) = \frac{s+1}{(s+2)(s+3)}$$

Expanding this into its partial fractions, we see that

$$F(s) = \frac{-1}{s+2} + \frac{2}{s+3}$$

Clearly, there is no network possible since one of the residues is negative. Of course, looking at $F(s)$ itself we might disqualify it since the poles and zeros are not interlacing, as was suggested by property no. 2 listed earlier.

Exercise 11.5 Synthesize a Foster-1 network for the impedance function

$$z_{11}(s) = \frac{(s+2)(s+5)}{(s+1)(s+3)}$$

Having seen the driving point impedance functions, let us see the driving point admittance functions of RC networks. As a dual of Foster-1 network, we have the following *Foster-2 network* as shown in Fig. 11.8.

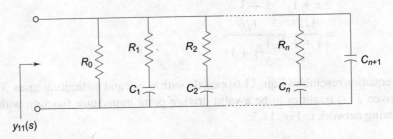

Fig. 11.8

It may be observed that we have connected $n+2$ branches, each of which is a series section, in parallel. We have also included the limiting cases of $C_0 = \infty$ and $R_{n+1} = 0$. The driving point admittance function of this network may be computed as

$$y_{11}(s) = G_0 + \sum_{i=1}^{n} \frac{C_i s}{R_i C_i (s+1)} + C_{n+1}s \qquad (11.7)$$

where $G_0 = 1/R_0$ is the conductance of the leftmost branch. In comparison with Eqn (11.6) one conspicuous difference that may be noted in Eqn (11.7) is that the admittance function of each of the n branches has an s in the numerator.

We will take up the reciprocal of the function $F(s)$ of Example 11.3 and see if we can synthesize a network.

Example 11.6 Consider the following function:

$$F(s) = \frac{s(s+2)}{(s+1)(s+3)}$$

To begin with, let us assume that this is a driving point admittance function $y_{11}(s)$ of a possible RC network. Let us get the partial fraction expansion of $F(s)$ *routinely* as

$$y_{11}(s) = 1 - \frac{1/2}{s+1} - \frac{3/2}{s+3}$$

We end up with negative residues. But, more seriously, we did not get any term with an s in the numerator. To this end, it would be convenient to expand $F(s)/s$ into partial fractions as

$$\frac{F(s)}{s} = \frac{1/2}{s+1} + \frac{1/2}{s+3}$$

so that we get

$$F(s) = \frac{\frac{1}{2}s}{s+1} + \frac{\frac{1}{2}s}{s+3}$$

$$= \frac{1}{2 + \dfrac{1}{(1/2)s}} + \frac{1}{2 + \dfrac{1}{(1/6)s}}$$

Observe that here the residues are positive and the equation resembles Eqn (11.7) exactly. The corresponding network is given in Fig. 11.9.

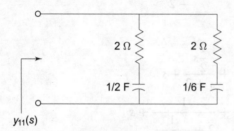

$y_{11}(s)$

Fig. 11.9

Exercise 11.6 Verify the values of the elements in the network shown in Fig. 11.9.

We avouch that the properties of a driving point admittance function would be *dual* to those of a driving point impedance function with poles substituted for zeros and vice versa. We leave it as an exercise to enumerate the properties of driving point admittance functions and tabulate them vis-à-vis those of impedance functions. In addition, an admittance function $y_{11}(s)$ has to possess the property that the residues need to be negative[6], and the residues of $y_{11}(s)/s$ need to be positive.

Exercise 11.7 Enumerate the properties of driving point admittance functions and tabulate them vis-à-vis those of impedance functions.

There are two things the reader has to bear in mind. First, given a pr driving point function $F(s)$ we may consider it either as an impedance function and synthesize a

---

[6] This is a consequence of duality!

Foster-1 network or as an admittance function and synthesize a Foster-2 network.
These two networks are independent of each other; in other words, there are *at least*
two different networks possible out of a given pr driving point function. Second,
while dealing with the given pr driving point admittance function, it is necessary
to expand $y_{11}(s)/s$ into partial fractions, rather than $y_{11}(s)$.

Let us look at another example.

Example 11.7 Consider the following function

$$F(s) = \frac{(s+1)(s+3)}{s+2}$$

Let us assume that this is a driving point admittance function $y_{11}(s)$ of a possible
RC network and get the partial fraction expansion of $F(s)/s$ as

$$\frac{y_{11}(s)}{s} = \frac{3/2}{s} + \frac{1/2}{s+2} + 1$$

or

$$y_{11}(s) = \frac{3}{2} + \frac{\frac{1}{2}s}{s+2} + s$$

$$= \frac{3}{2} + \frac{1}{2 + \frac{1}{(1/4)s}} + s$$

The network is as shown in Fig. 11.10.

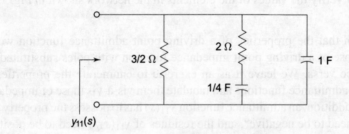

Fig. 11.10

Exercise 11.8 Verify the values of the elements in the network shown in Fig. 11.10.

Exercise 11.9 Verify that the driving point function of Example 11.7 has all the
properties of a pr function.

Example 11.8 Let us now consider

$$F(s) = \frac{(s+2)(s+3)}{s+1}$$

Expanding this into its partial fractions, we see that

$$\frac{F(s)}{s} = 1 + \frac{6}{s} - \frac{2}{s+1}$$

Clearly, there is no network possible since one of the residues is negative. Of course, looking at $F(s)$ itself we might disqualify it since the poles and zeros are not interlacing.

Exercise 11.10 Verify whether the following function is a valid driving point pr impedance/admittance function:

$$F(s) = \frac{(s+3)(s+5)}{(s+1)(s+4)}$$

If it is a pr function, synthesize Foster-1 and Foster-2 networks.

Exercise 11.11 Synthesize Foster networks for the function

$$F(s) = \frac{(s+1)(s+3)}{(s+2)(s+5)}$$

We have suggested duality between Foster-1 and Foster-2 networks *with resistances and capacitances*. The reader might wonder how networks with resistances and inductances would look like, and how do we go about synthesizing such networks. We encourage the reader to speculate on this idea for a while. In the mean time, we present two more networks that could be synthesized using a given pr driving point function.

11.3.2 Cauer Forms

Given a rational function, perhaps it appears natural to look at its partial fractions since we are familiar with them for quite some time. However, there is another category of fractions called *continued fractions*. Continued fractions is, once again, a subject in itself, primarily in number theory. In this section we will look into these very briefly and arrive quickly at some interesting networks.

To begin with some intuition, let us consider a fraction

$$\rho = \frac{43}{30}$$

With a little bit of effort everyone gets the answer $\rho = 1.43333\ldots$. We say that the decimal is recurring. Some of us might not like this and prefer to write

$$\rho = 1\frac{13}{30}$$

Let us now learn to write this fraction in the following way:

$$\frac{43}{30} = 1 + \frac{13}{30} = 1 + \frac{1}{30/13}$$

$$\frac{30}{13} = 2 + \frac{4}{13} = 2 + \frac{1}{13/4}$$

$$\frac{13}{4} = 3 + \frac{1}{4}$$

Combining these individual equations, we obtain an expression of the form

$$\rho = 1 + \cfrac{1}{2 + \cfrac{1}{3 + 1/4}}$$

An expression of the form

$$\rho = \rho_1 + \cfrac{1}{\rho_2 + \cfrac{1}{\rho_3 + \cfrac{1}{\ddots + \cfrac{1}{\rho_n}}}} \tag{11.8}$$

where each of the coefficients ρ_i is a positive integer, is called a *continued fraction*. It is interesting to note that, unlike the recurrence in the decimal representation, the expansion terminates after a few terms. Further, recall that we have developed this for resistive ladder circuits in Chapter 3.

We will make use of continued fractions of this sort in our synthesis. Consider a driving point impedance function:

$$z_{11}(s) = H_0 \frac{N(s)}{D(s)} = H_0 \frac{s^m + b_{m-1}s^{m-1} + \cdots + b_0}{s^n + a_{n-1}s^{m-1} + \cdots + a_0}$$

We will express this as the following continued fraction:

$$z_{11}(s) = \rho_1(s) + \cfrac{1}{\rho_2(s) + \cfrac{1}{\rho_3(s) + \cfrac{1}{\ddots + \cfrac{1}{\rho_n(s)}}}} \tag{11.9}$$

Here, we assume that there exists such an expansion with the full set of n coefficients. In fact, the given function is pr, if there exists a continued fraction expansion. Observe that this is only a sufficient condition. In other words, there may exist pr functions that do not lend themselves to nice continued fractions as above.

Next, we will identify the units and dimensions of each of the coefficients. Since we have considered an *impedance* function, $\rho_1(s)$ must have the units of impedance. This $\rho_1(s)$ is added to the *reciprocal* of $\rho_2(s)$ + something. Hence $\rho_2(s)$ must have the units of admittance. Proceeding this way, we see that $\rho_{odd}(s)$ must have the units of impedance and $\rho_{even}(s)$ must have the units of admittance. Assuming that n is odd, say 7, we may build a network by starting at the rightmost end of the network as follows.

- The reciprocal of $\rho_7(s)$ is added to $\rho_6(s)$, i.e., two branches at the rightmost end are connected in parallel. We account this parallel combination as an admittance.
- The reciprocal of the admittance in the previous step is now added to $\rho_5(s)$, i.e., a branch with $\rho_5(s)$ as an impedance is connected in series with the parallel connection of the previous step.
- This procedure is repeated over and over until we have the driving point impedance.

We illustrate this in Fig. 11.11 below.

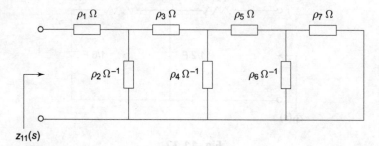

Fig. 11.11

Such a network is called a *ladder network*. We have seen purely resistive ladder networks in Chapter 3. The ladder networks developed using pr driving point functions are called *Cauer networks*. Let us take up an example to see things more transparently.

Example 11.9 Let us consider the driving point impedance of the RC network we have considered earlier:

$$z_{11}(s) = \frac{(s+1)(s+3)}{s(s+2)} = \frac{s^2 + 4s + 3}{s^2 + 2s}$$

The formation of the continued fraction for $z_{11}(s)$ is easily done by the 'divide-invert-divide' procedure. We will make use of the steps carried out for 43/13.

$$z_{11}(s) = 1 + \cfrac{1}{\cfrac{s^2 + 2s}{2s + 3}}$$

$$\frac{s^2 + 2s}{2s + 3} = \frac{1}{2}s + \cfrac{1}{\cfrac{2s + 3}{(1/2)s}}$$

$$\frac{2s + 3}{\cfrac{1}{2}s} = 4 + \cfrac{1}{\cfrac{(1/2)s}{3}}$$

$$\frac{\cfrac{1}{2}s}{3} = \frac{1}{6}s$$

These steps may be neatly packaged into

$$z_{11}(s) = 1 + \cfrac{1}{\frac{1}{2}s + \cfrac{1}{4 + \cfrac{1}{(1/6)\,s}}}$$

The continued fraction we have obtained is often called *expansion of $F(s)$ about pole at infinity*. The resulting ladder network is shown in Fig. 11.12, and is called the *Cauer-1 network*.

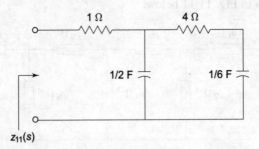

Fig. 11.12

Exercise 11.12 Verify the values of the elements in the network above.

Exercise 11.13 Look at the Foster-1 and Cauer-1 networks of the same driving point function and comment on the element values.

Thus far, we have three different networks for the same driving point impedance function. A quick reader would have guessed by now that a Cauer-2 network may be synthesized using the reciprocal of $F(s)$ in the continued fraction expansion. However, unfortunately, things turn out to be *counter-intuitive*[7] at times, and here is one such instance. The reasons are beyond the scope of this chapter.

Exercise 11.14 Obtain the continued fraction expansion of

$$y_{11}(s) = \frac{s(s+2)}{(s+1)(s+3)}$$

and synthesize a passive network[8] by appropriately identifying the coefficients.

---

[7] Speed thrills, but kills at times!
[8] You are entitled to a patent to call the network by your own name!

We shall take up the same driving point impedance function, but rewrite it as

$$z_{11}(s) = \frac{b_0' + b_1's + b_2's^2 + \cdots + b_{m-1}'s^{m-1} + b_m's^m}{a_0 + a_1s + \cdots + a_{n-1}s^{n-1} + s^n} \qquad (11.10)$$

taking the scale factor inside $N(s)$. We now expand this *rewritten function* as a continued fraction. Folk wisdom suggests that there is no change in the end result. But, let us verify it ourselves with an example.

Example 11.10 Considering the same driving point impedance function,

$$z_{11}(s) = \frac{(s+1)(s+3)}{s(s+2)} = \frac{3 + 4s + s^2}{2s + s^2}$$

we now have

$$z_{11}(s) = \frac{3}{2}\frac{1}{s} + \frac{1}{\dfrac{2s + s^2}{(5/2)s + s^2}}$$

$$\frac{2s + s^2}{\dfrac{5}{2}s + s^2} = \frac{4}{5} + \frac{1}{\dfrac{(5/2)s + s^2}{(1/2)s^2}}$$

$$\frac{(5/2)s + s^2}{(1/2)s^2} = \frac{25}{2}\frac{1}{s} + \frac{1}{\dfrac{(1/5)s^2}{s^2}}$$

$$\frac{\dfrac{1}{5}s^2}{s^2} = \frac{1}{5}$$

This expansion, in contrast to the earlier one, is called *expansion about pole at origin*. These steps may be neatly packaged into

$$z_{11}(s) = \frac{3}{2}\frac{1}{s} + \cfrac{1}{\dfrac{4}{5} + \cfrac{1}{\dfrac{25}{2}\dfrac{1}{s} + \cfrac{1}{\dfrac{1}{5}}}}$$

The resulting ladder network is shown in Fig. 11.13, and is called *Cauer-2 network*. Observe the duality between Cauer-1 and -2 networks.

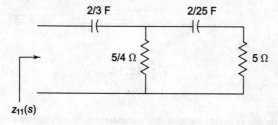

Fig. 11.13

Exercise 11.15 Verify the values of the elements in the network above.

Thus, we have four standard networks with resistances and capacitances for the same driving point function. We have a very important observation here:

Three of them (Foster 1, Cauer 1 and Cauer 2) are obtained directly from the impedance function, and Foster 2 is obtained from the admittance function.

Let us summarize the synthesis of RC networks.

1. Given a driving point function, we will obtain the partial fractions and verify if the properties match with those of impedance or admittance functions enumerated earlier. If there is a match, we call it a pr function, else we say there exists no RC network.

2. If the given function is a driving point impedance function, we may readily obtain a Foster-1 network using the partial fractions of $F(s)$, a Cauer-1 network using the continued fraction expression of $F(s)$ with the polynomials written in the descending order of power, and a Cauer-2 network using the continued fraction expression of $F(s)$ with the polynomials written in the ascending order of power.

3. A driving point impedance function may be inverted to get the corresponding driving point admittance function $y_{11}(s)$ from which we may readily obtain a Foster-2 network using the partial fractions of $y_{11}(s)/s$.

4. Since RC driving point impedance functions have a pole closer to the origin, the continued fraction expansion, in either of the ways, is possible and we get Cauer networks. However, the RC driving point admittance functions do not[9] allow us to synthesize Cauer networks since the zero is closer to the origin rather than the pole.

Let us now look at the synthesis of RL networks.

11.4 Synthesis of *RL* Networks

We will do this section rather quickly with the duality principle in front of us. Let us first consider the parallel connection of a resistance R_i and an inductance L_i as shown in Fig. 11.14.

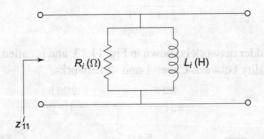

Fig. 11.14

---
[9] In a way the principle of duality suggests that a counterpart should exist; in a *counter-intuitive* way it suggests that there does not exist a counterpart.

The driving point impedance may be quickly computed to be

$$z_{11}^{i}(s) = \frac{R_i L_i s}{L_i s + R_i} \tag{11.11}$$

11.4.1 Foster Forms

Let us now connect $n + 2$ of these parallel sections in series as shown in Fig. 11.15.

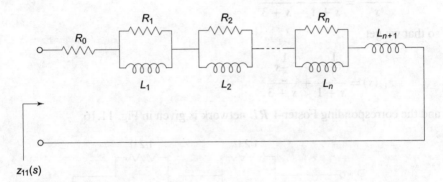

Fig. 11.15

This network is also called *Foster-1 network*. Including the limiting cases of $L_i = \infty$ and $R_i = \infty$ for $i = 0$ and $i = n + 1$, respectively, the driving point impedance of the series connection may be computed as

$$z_{11} = \sum_i z_{11}^{i}$$

$$= R_0 + \sum_{i=1}^{n} \frac{R_i L_i s}{L_i s + R_i} + L_{n+1} s \tag{11.12}$$

Once again, to realize a Foster-1 network, the given driving point impedance function has to be a pr function. If we observe little more closely, we notice that the driving point impedance function of a Foster-1 RL network is identical to the driving point admittance function of a Foster-2 RC network. Thus, the conditions to be satisfied in the present case are as follows:

1. The poles and zeros are simple (i.e., multiplicity is 1) and are located on the negative real axis of the s-plane. The zero at the origin is simple in particular.
2. The poles and zeros interlace, i.e., a pair of poles has a zero in between and a pair of zeros has a pole in between. The first corner frequency we come across traversing from the origin on the negative real axis is a zero. In general this could be located at the origin.
3. The last corner frequency along the negative real axis is a pole. In general it could be at $-\infty$.
4. The above two observations suggest us that $z_{11}(\infty) > z_{11}(0)$.
5. The residues are, obviously, real and positive.

Example 11.11 Consider the following function:

$$F(s) = \frac{s(s+2)}{(s+1)(s+3)}$$

The partial fraction expansion of $F(s)/s$ is

$$\frac{z_{11}(s)}{s} = \frac{1/2}{s+1} + \frac{1/2}{s+3}$$

so that we get

$$z_{11}(s) = \frac{\frac{1}{2}s}{s+1} + \frac{\frac{1}{2}s}{s+3}$$

and the corresponding Foster-1 RL network is given in Fig. 11.16.

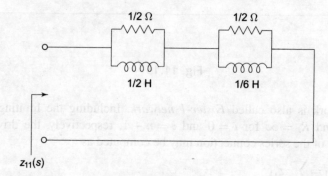

Fig. 11.16

Exercise 11.16 Verify the values of the elements in the network shown in Fig. 11.16.

We observe here that, given an impedance function, we get either an RC network or an RL network, but not both. Let us look at another example.

Example 11.12 Consider the following function:

$$F(s) = \frac{(s+1)(s+3)}{(s+2)}$$

Let us assume that this is a driving point impedance function $z_{11}(s)$ of an RL network and get the partial fraction expansion of $F(s)/s$ as

$$\frac{z_{11}(s)}{s} = \frac{3/2}{s} + \frac{1/2}{s+2} + 1$$

or,

$$z_{11}(s) = \frac{3}{2} + \frac{\frac{1}{2}s}{s+2} + s$$

and hence the network is as shown in Fig. 11.17.

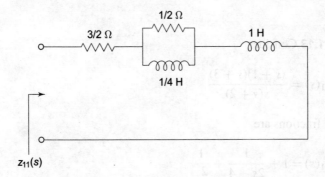

Fig. 11.17

Exercise 11.17 Verify the values of the elements in the network shown in Fig. 11.17.

Exercise 11.18 Verify that the driving point function of Example 11.12 has all the properties of a pr function.

Exercise 11.19 Synthesize a Foster-1 RL network for the impedance function

$$z_{11}(s) = \frac{(s+1)(s+3)}{(s+2)(s+5)}$$

Having seen the driving point impedance functions for RL, let us see the following *Foster-2 network* as shown in Fig. 11.18.

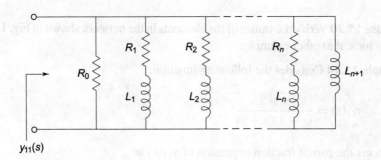

Fig. 11.18

The driving point admittance function of this network may be computed as

$$y_{11}(s) = G_0 + \sum_{i=1}^{n} \frac{1}{L_i s + R_i} + \frac{1}{L_{n+1}s} \tag{11.13}$$

where $G_0 = 1/R_0$ is the conductance of the leftmost branch.

We will take up the reciprocal of the function $F(s)$ of Example 11.11 and see if we can synthesize a network.

Example 11.13 Consider

$$y_{11}(s) = \frac{(s+1)(s+3)}{s(s+2)}$$

The partial fractions are

$$y_{11}(s) = 1 + \frac{1}{2s+4} + \frac{1}{\frac{2}{3}s}$$

We have the following network in Fig. 11.19.

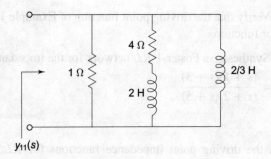

4 Ω

1 Ω

2 H

2/3 H

$y_{11}(s)$

Fig. 11.19

Exercise 11.20 Verify the values of the elements in the network shown in Fig. 11.19. Let us look at another example.

Example 11.14 Consider the following function:

$$y_{11}(s) = \frac{s+2}{(s+1)(s+3)}$$

Let us get the partial fraction expansion of $y_{11}(s)$ as

$$y_{11}(s) = \frac{1/2}{s+1} + \frac{1/2}{s+3}$$

$$= \frac{1}{2s+2} + \frac{1}{2s+6}$$

and we have the following network in Fig. 11.20.

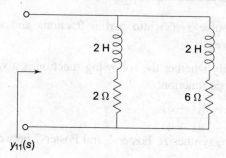

Fig. 11.20

Exercise 11.21 Verify the values of the elements in the network shown in Fig. 11.20.

Exercise 11.22 Verify that the driving point function of Example 11.14 has all the properties of a pr function.

We emphasize that the properties of a driving point admittance function would be *similar* to those of a driving point impedance function with poles substituted for zeros and vice versa. We leave it as an exercise to enumerate the properties of driving point admittance functions and tabulate them vis-à-vis those of impedance functions.

Exercise 11.23 Enumerate the properties of driving point admittance functions of RL networks and tabulate them vis-à-vis those of impedance functions. Compare with the corresponding properties of RC networks.

Once again, with reference to the Foster networks with resistances and inductances, there are two things the reader has to bear in mind. First, given a pr driving point function $F(s)$ we may consider it either as an impedance function and synthesize a Foster-1 network, or as an admittance function and synthesize a Foster-2 network. These two networks are independent of each other. Second, while dealing with the given pr driving point impedance function of a possible RL network, it is necessary to expand $z_{11}(s)/s$ into partial fractions, rather than $z_{11}(s)$.

Example 11.15 Let us now consider

$$z_{11}(s) = \frac{(s+2)(s+3)}{s+1}$$

Clearly, there is no network possible since one of the residues is negative. Of course, looking at $F(s)$ itself we might disqualify it since the poles and zeros are not interlacing.

Exercise 11.24 Resolve $z_{11}(s)$ into partial fractions and verify the results of Example 11.15 above.

Exercise 11.25 Verify whether the following function is a valid driving point pr impedance/admittance function:

$$F(s) = \frac{(s+3)(s+5)}{(s+1)(s+4)}$$

If it is a pr function, synthesize Foster-1 and Foster-2 networks with resistances and inductances.

Exercise 11.26 Synthesize all possible (using RC and RL elements) Foster networks for the function

$$F(s) = \frac{(s+1)(s+3)}{(s+2)(s+5)}$$

How about the reciprocal of this function?

11.4.2   Cauer Forms

Let us take up a few examples to see the Cauer forms of RL networks.

Example 11.16 Let

$$F(s) = \frac{(s+1)(s+3)}{s(s+2)}.$$

We immediately observe that this is an RL admittance function since the lowest corner frequency is a pole and the highest is a zero.

The continued fraction for $y_{11}(s)$, from Example 11.9, is

$$y_{11}(s) = 1 + \cfrac{1}{\frac{1}{2}s + \cfrac{1}{4 + \cfrac{1}{\frac{1}{6}s}}}$$

The resulting ladder network is shown in Fig. 11.21 and is called the *Cauer-1 network*.

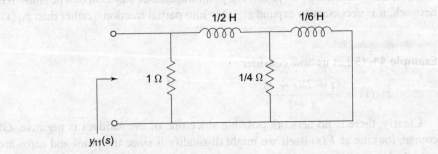

Fig. 11.21

Exercise 11.27 Verify the values of the elements in the network given in Fig. 11.21.

Exercise 11.28 Obtain the driving point impedance of the network in Fig. 11.21 and verify the results of Example 11.21.

Example 11.17 Considering the same driving point admittance function,

$$y_{11}(s) = \frac{(s+1)(s+3)}{s(s+2)} = \frac{3+4s+s^2}{2s+s^2}$$

Expanding this into a continued fraction about $s = 0$, we have

$$y_{11}(s) = \frac{3}{2}\frac{1}{s} + \cfrac{1}{\frac{4}{5} + \cfrac{1}{\frac{25}{2}\frac{1}{s} + \cfrac{1}{\frac{1}{5}}}}$$

The resulting *Cauer-2 network* is shown in Fig. 11.22.

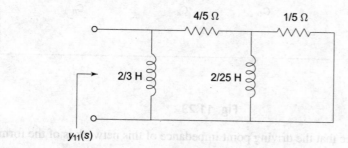

Fig. 11.22

Exercise 11.29 Verify the values of the elements in the network shown in Fig. 11.22.

Let us summarize the synthesis of *RL* networks.

1. Given a driving point function, we will obtain the partial fractions and verify if the properties match with those of impedance or admittance functions enumerated earlier.

2. If the given function is a driving point admittance function, we may readily obtain a Foster-1 network using the partial fractions of $F(s)/s$, a Cauer-1 network using the continued fraction expression of $F(s)$ with the polynomials written in the descending order of power, and a Cauer-2 network using the continued fraction expression of $F(s)$ with the polynomials written in the ascending order of power.

3. A driving point admittance function may be inverted to get the corresponding driving point impedance function $z_{11}(s)$, from which we may readily obtain the Foster-2 network using the partial fractions of $z_{11}(s)$.

4. Cauer networks cannot be obtained from a driving point impedance function.

Let us now look at the synthesis of *LC* networks.

11.5 Synthesis of *LC* Networks

In this section, we shall look at the synthesis of *LC* networks. We shall adopt the Foster and Cauer forms for this synthesis once again. However, the results turn out to be totally independent. For instance, while the poles and zeros of *RC* and *RL* networks are all real, the poles and zeros of *LC* networks turn out to be purely imaginary and appear as complex conjugates.

11.5.1 Foster Forms

The Foster-1 network of an *LC* network, without any resistances, is shown in Fig. 11.23.

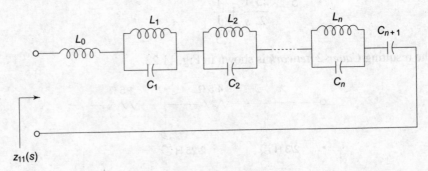

Fig. 11.23

It is easy to see that the driving point impedance of this network is of the form

$$z_{11}(s) = L_0 s + \sum_{i=1}^{n} \frac{L_i s}{L_i C_i s^2 + 1} + \frac{1}{C_{n+1} s} \qquad (11.14)$$

It is easy to observe that this driving point impedance function has

- a pole at infinity (which requires the numerator polynomial to have a degree higher than that of the denominator by 1),
- a pole at origin, and
- pairs of simple and conjugate poles on the imaginary axis of the *s*-plane.

Owing to the third observation, the polynomials in the numerator and denominator assume the following form:

$$P(s) = s^{2r} + \alpha_{2r-2} s^{2r-2} + \cdots + \alpha_2 s^2 + \alpha_0 \qquad (11.15)$$

i.e., all the odd powers are missing. However, the coefficients α_i should all be real and positive. In fact, such polynomials also qualify as Hurwitz polynomials. We state without proof that the driving point impedance function qualifies to be a pr function with, once again, the zeros and the poles interlacing on the imaginary axis. In addition, we may also observe that

$$\text{Real part of } z_{11}(j\omega) = 0 \qquad (11.16)$$

since the network consists of several _reactance_ elements but no resistance. We leave it to the reader to look at a Foster-2 network.

Exercise 11.30 Draw a Foster-2 network with inductances and capacitances. Obtain the driving point admittance function. Compare this with the driving point impedance function of Eqn (11.14).

Let us quickly try some examples.

Example 11.18 Consider the following driving point impedance function of a possible _LC_ network:

$$z_{11}(s) = \frac{(s^2+1)(s^2+9)}{s(s^2+4)}$$

Although we have quadratic terms, we can straight away use Heaviside's expansion theorem to compute the residues:

$$k_\infty = \lim_{s \to \infty} z_{11}(s) = 1$$

$$k_1 = \frac{(s^2+4)}{s} z_{11}(s)|_{s^2=-4} = \frac{15}{4}$$

$$k_0 = s \cdot z_{11}(s)|_{s=0} = \frac{9}{4}$$

The impedance function may be written as

$$s + \frac{1}{\frac{4}{15}s + \frac{16}{15}\frac{1}{s}} + \frac{1}{\frac{4}{9}s}$$

and the Foster-1 network is shown in Fig. 11.24.

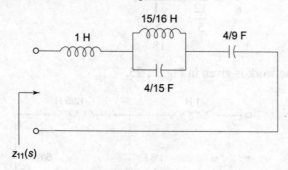

15/16 H
1 H
4/9 F
4/15 F
$z_{11}(s)$

Fig. 11.24

Exercise 11.31 Verify the values of the elements in the network shown in Fig.11.24.

Exercise 11.32 Obtain the Foster-2 network by inverting the driving point impedance function of Example 11.18.

11.5.2 Cauer Forms

The Cauer networks may also be readily synthesized. We will look at the following examples.

Example 11.19 Consider the same impedance function

$$z_{11}(s) = \frac{(s^2+1)(s^2+9)}{s(s^2+4)}$$

Since this has a pole at the origin, we have the following continued fraction:

$$z_{11}(s) = s + \cfrac{1}{\cfrac{s^3+4s}{6s^2+9}}$$

$$\frac{s^3+4s}{6s^2+9} = \frac{1}{6}s + \cfrac{1}{\cfrac{6s^2+9}{\dfrac{5s}{2}}}$$

$$\frac{6s^2+9}{\dfrac{5s}{2}} = \frac{12}{5}s + \cfrac{1}{\cfrac{5s}{\dfrac{2}{9}}}$$

$$\frac{5s/2}{9} = \frac{5s}{18}$$

i.e.,

$$z_{11}(s) = s + \cfrac{1}{\cfrac{1}{6}s + \cfrac{1}{\cfrac{12}{5}s + \cfrac{1}{\cfrac{5}{18}s}}}$$

The Cauer-1 network is given in Fig. 11.25.

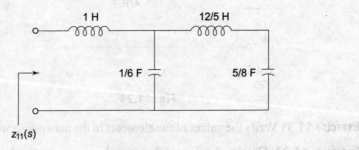

Fig. 11.25

Exercise 11.33 Verify the values of the elements in the network shown in Fig. 11.25.

Example 11.20 For the same impedance function let us obtain a Cauer-2 network also. We leave it to the reader to verify that the continued fraction in this is

$$z_{11}(s) = \frac{9}{4} \times \frac{1}{s} + \cfrac{1}{\frac{16}{31} \times \frac{1}{s} + \cfrac{1}{\frac{961}{60} \times \frac{1}{s} + \cfrac{1}{\frac{15}{31} \times \frac{1}{s}}}}$$

Accordingly, the Cauer-2 network is given in Fig. 11.26.

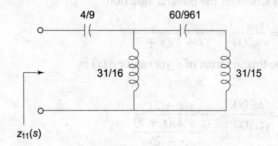

Fig. 11.26

Exercise 11.34 Verify the results of Example 11.20. When do driving point functions have LC Cauer networks?

Exercise 11.35 Obtain Cauer LC networks for the following driving point pr function:

$$F(s) = \frac{s(s^2 + 4)(s^2 + 36)}{(s^2 + 1)(s^2 + 25)(s^2 + 81)}$$

The Foster and Cauer networks we were synthesizing so far contain minimum number of elements and hence are also called as *canonical forms*. For a given driving point function these networks are completely equivalent; if we enclose all of these networks in black boxes, one could not be distinguished from another. Often, we prefer one form to another form looking at the element size (for instance, inductors are bulky), parasitic effects (for instance, stray charges in the capacitors), etc. It is recommended that all the four forms are synthesized and compared before making a choice.

Synthesis of networks containing all the three passive elements—R, L, and C—is little more complicated. Intuitively one might look at the Foster and Cauer LC networks with some (or all) of the branches of the network sharing resistance or conductance along with capacitances and inductances. There are primarily two synthesis procedures developed by Brune and by Bott and Duffin. We will not go into further details here; instead we suggest the reader to look at the references cited in the bibliography (Cauer 1958, Guillemin 1957, Truxal 1955, Tuttle 1958, Van Valkenburg 1960).

11.6 Synthesis Given Transfer Functions

Earlier in this chapter, we have seen certain interesting features of two-port parameters in connection with Examples 11.1 and 11.2. From the given transfer function we may select a driving point parameter such as z_{11} and a transfer parameter such as z_{21}. As shown in Example 11.1, we may assume a *ladder* network to satisfy the requirements. In fact, this method is due to Cauer and is called *Cauer ladder development*. We will present an example here and close the section. An exhaustive treatment is beyond the scope of this book.

Example 11.21 Consider the transfer function

$$T(s) = \frac{v_{\text{out}}(s)}{v_{\text{in}}(s)} = \frac{3}{(s+1)(s+3)}$$

We may rewrite this in terms of $z_{21}(s)$ and $z_{11}(s)$ as

$$T(s) = \frac{z_{21}(s)}{z_{11}(s)} = \frac{\dfrac{3}{s(s+2)}}{\dfrac{(s+1)(s+3)}{s(s+2)}}$$

Observe that we have *chosen* to make $z_{11}(s)$ a pr function by locating the poles at $s = 0$ and at $s = -2$; other choices do exist, as long as the properties of RC driving point impedance functions are satisfied. Further, we may also observe that $z_{21}(s)$ need not be pr in general. Now we may proceed to obtain an RC Cauer network using the driving point function $z_{11}(s)$ as shown in Fig. 11.27.

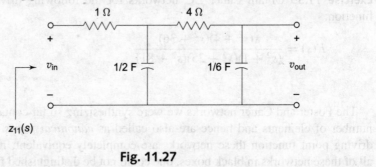

Fig. 11.27

Exercise 11.36 Verify that the network shown in Fig. 11.27 has the required transfer parameter $z_{21}(s)$.

Exercise 11.37 Verify the values of the elements in the network shown in Fig. 11.27.

Exercise 11.38 Synthesize another network for the transfer function of Example 11.21 by choosing a different set of poles for $z_{11}(s)$ and $z_{21}(s)$.

Exercise 11.39 For the transfer function in Example 11.21 synthesize a network using Y parameters rather than the Z parameters.

In the next section, we will come to the practical synthesis. By this we mean to synthesize networks with *real values* of the components, rather than elements such as capacitors with a few farads we have been looking all through the examples of this chapter.

11.7 Scaling

In the examples of this chapter we have come across *impractical* elements such as a 1-F capacitor or a 10-H inductor. What does this mean? At the outset we will convince ourselves that a city in a country is no bigger than a bright dot in the world atlas. If we intend to *see* several things in a small place we need to *scale* the picture, such as 1 cm represents 10 km; of course, we may choose different scales on different axes. Perhaps, the idea of scaling becomes more intuitive if we read the following from Jonathan Swift's *Gulliver's Travels*[10]: 'His Majesty's mathematicians, having taken the height of my body by the help of a *quadrant*, and finding it to exceed theirs in the proportion of twelve to one, concluded from the similarity of their bodies that mine must contain at least 1728 of theirs, and consequently would require as much food as was necessary to support that number of Lilliputians.'

Much the same way, we may scale our network functions, along the magnitude axis as well as along the frequency axis. For instance, scaling of frequency such that 1 MHz becomes just 1 rad/s allows us to avoid numbers such as 2π and powers of 10 in our computations. That we are able to do so is an important consequence of linearity.

Let us take up the simple capacitance. The reactance offered by this element is

$$X_c = \frac{1}{j\omega C} \text{ i.e., } X_c \propto \frac{1}{\omega C} \tag{11.17}$$

This relationship suggests that, for instance, a larger capacitance $C' = m \times C$ offers the same reactance at a lower frequency ω/m. In other words, two different capacitances C and C' exhibit the same behaviour at two different frequencies ω and ω'.

Example 11.22 Consider a simple network function:

$$z_{11}(s) = 10^3 \frac{s + 10^3}{s}$$

We may identify this as a Foster-1 RC network with $R = 1\,k\Omega$ and $C = 1\,\mu F$. However, the powers of 10 may cause some difficulty in the arithmetic. If we choose $\omega' = 10^{-6}\omega$, i.e., $s' = 10^{-6}s$, the network function becomes

$$z_{11}(s') = \frac{10^3 s' + 1}{s'}$$

---

[10] Source: chapter 3, *Gulliver's travels* by Jonathan Swift.

which is somewhat simpler than the original one. Here we identify that the capacitance is 1 F. Observe that there is no change in the resistance; resistance is independent of frequency. This example suggests us that we synthesize *practical* networks as follows:

- Given a driving point function $F(s)$, *scale* this to $F(s')$ with $s' = f \times s$ so as to avoid irrational numbers such as 2π and large powers of 10.
- Synthesize the network with $F(s')$ and hence exaggerated element values.
- The synthesized element values are related to the exact values as follows:
  - the resistances remain unchanged,
  - the capacitances are divided by the factor f, and
  - the inductances are divided by the factor f.

In the algorithm above, we have used scaling on the frequency axis alone. Consequently, there will be shift in the poles and zeros of the network function, i.e., the Bode plot is either stretched or compressed on the ω axis.

We will look at Example 11.22 to examine the scaling on the magnitude axis.

Example 11.23 Given

$$z_{11}(s) = 10^3 \, \frac{s + 10^3}{s}$$

Let us obtain

$$z'_{11}(s) = m \times z_{11}(s) = 10^{-3} \times z_{11}(s) = \frac{s + 10^3}{s}$$

The corresponding elements are $R = 1\,\Omega$ and $C = 1$ mF. These elements can be directly obtained by multiplying the resistance by the scale factor m and dividing (why?) the capacitance by the scale factor m.

More generally, we may have scaling both on the frequency axis (i.e., $s' = f \times s$) and on the magnitude axis [i.e., $F'(s) = m \times F(s)$]. Let us look at the network in Example 11.24.

Example 11.24 Given

$$z_{11}(s) = 2 \, \frac{2s + 1}{2s^2 + s + 1}$$

corresponding to the network shown in Fig. 11.28.

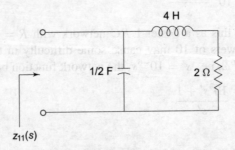

Fig. 11.28

Suppose we choose

$$m = \frac{1}{4} \quad \text{and} \quad f = \frac{1}{2}$$

The scaled driving point function is

$$z'_{11}(s') = \frac{1}{2} \frac{4s' + 1}{8s'^2 + 2s' + 1}$$

It may be verified that this driving point impedance corresponds to the network shown in Fig. 11.29.

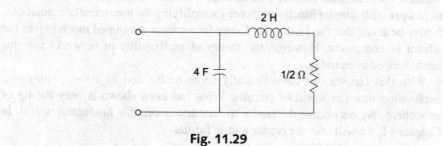

Fig. 11.29

Exercise 11.40 Verify the results of Example 11.24.

Exercise 11.41 Repeat Example 11.24 with $m = 4$ and $f = 2$.

Exercise 11.42 Verify that for transfer functions such as the voltage ratio or the current ratio functions, the frequency scaling may be readily applied, but there is no effect of magnitude scaling.

We summarize the effect of scaling as follows.

Given a network function $F(s)$, we obtain the following scaled function:

$$F'(s') = m \times F(s'/f)$$

The new elements R', C', and L' are related to the older elements R, C, and L as

$$R' = mR \qquad\qquad\qquad\qquad\qquad (11.18)$$

$$C' = \frac{1}{m \times f} C \qquad\qquad\qquad\qquad (11.19)$$

$$L' = \frac{m}{f} L \qquad\qquad\qquad\qquad\qquad (11.20)$$

Summary

In this chapter we have looked at the problem of synthesizing a network, if there exists one, given a network function. Broadly speaking, there are two categories of network functions—driving point functions and transfer functions. We have observed that, while the analysis problem has a unique solution, the synthesis problem need not have a unique solution. We have enumerated, albeit using intuition and prior experience in analysis, certain conditions on the driving point network functions—only the *positive real* functions qualify as the network functions of realizable networks. Given a pr driving point function, these conditions may be utilized to synthesize *four* different networks—the Foster forms and the Cauer forms. The Cauer forms of ladder networks provide a basis for the synthesis of networks given the transfer functions. We have discussed this very briefly. We have not touched upon the synthesis of RLC networks owing to the limited space in this text; as has been mentioned earlier, network synthesis is a course by itself. We have concluded the chapter with an important idea called scaling, which allows us to work with simpler functions, thereby simplifying the numerical computations. It may be noted that the Foster and Cauer forms were developed much before the advent of computers. However, the theory of realizability of networks remains unchallenged even today.

With this chapter, we have virtually come to the end of a long, enjoyable, exploration into the world of circuits. What has been shown is only the tip of an iceberg. We encourage the reader to further explore this fascinating world. In Chapter 14, we will take the reader a little further.

Problems

11.1 The function

$$\tilde{F}(\omega) = \frac{1}{a_4\omega^4 + a_2\omega^2 + a_0}$$

is chosen to approximate the specification: $F(\omega) = 1$ for $0 \le \omega \le 1$ and $F(\omega) = 0$ for $\omega > 1$. The chosen approximation $\tilde{F}$ is required to have the following values: $\tilde{F}(0) = 1$, $\tilde{F}(1/\sqrt{2}) = 4/3$, and $\tilde{F}(1) = 1$. Compute the values of the coefficients a_0, a_2, and a_4. Plot $\tilde{F}(\omega)$ for $0 \le \omega \le 2$ and compare with the actual function $F(\omega)$.

11.2 Compute the magnitude and the phase angle of the following complex functions, using a ruler and a protractor:

$$F_1(s) = \frac{1}{(s+1)(s+2)} \quad \text{and} \quad F_2(s) = 5\frac{s^2 - s + 1}{s^2 + s + 1}$$

11.3 Compare the magnitude and the phase angle of the following complex functions. Comment on the effect of the pole at $s = -30$.

$$G_1(s) = \frac{s + 0.5}{s + 30} \quad \text{and} \quad G_2(s) = \frac{s + 0.5}{30}$$

11.4 In Fig. 11.30, three RC networks are shown. Verify that the driving point impedance functions are identical.

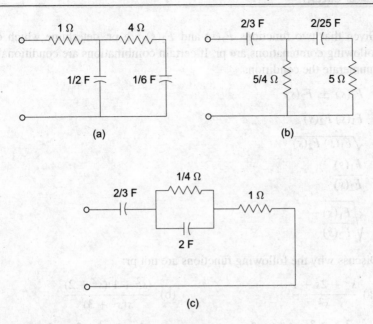

(a)

(b)

(c)

Fig. 11.30

11.5 For the two cases—switch open and switch closed—of the circuit in Fig. 11.31, determine the driving point impedance functions. Compare the orders of numerator and denominator polynomials in these cases.

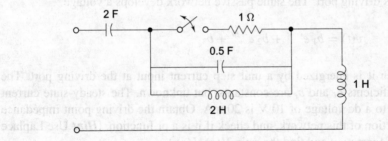

Fig. 11.31

11.6 A network has five nodes and 1 Ω resistances are connected between each of the $^5P_2 = 10$ node pairs. If a dc voltage v_S is applied across nodes 1 and 2, and the voltage v_0 across the nodes 3 and 4 is measured, determine the gain v_0/v_S.

11.7 How does a network whose short circuit transfer admittance parameter y_{12} has a pole at infinity look like?

11.8 An admittance function has just one zero at $z = -1$ and a complex conjugate pair of poles at $s = \alpha \pm j\beta$. Determine the range of allowable values for α and β so that the admittance function is pr.

11.9 Given that two functions $F_1(s)$ and $F_2(s)$ are pr, determine which of the following combinations are pr. If certain combinations are conditionally pr, enumerate the conditions.

- $F_1(s) \pm F_2(s)$

- $F_1(s) \, F_2(s)$

- $\sqrt{F_1(s) \, F_2(s)}$

- $\dfrac{F_1(s)}{F_2(s)}$

- $\sqrt{\dfrac{F_1(s)}{F_2(s)}}$

11.10 Discuss why the following functions are not pr:

(a) $\dfrac{s^2 + 2s + 1}{s^2}$

(b) $\dfrac{(s^2 + 1)(s^2 + 2)}{s(s^2 + 3)}$

(c) $\dfrac{s^3 + 7s^2 + 15s + 9}{s^4 + 6s^2 + 9}$

(d) $\dfrac{s^3 + 6ss^2 + 2s + 1}{s + 3}$

11.11 A unit step voltage input to a passive network results in a current

$$i(t) = a_1 e^{-t} + a_2 e^{-2t} + a_3$$

at its driving port. The same passive network develops a voltage

$$v(t) = b_1 e^{-4t} + b_2 e^{-6t} + b_3$$

when it is energized by a unit step current input at the driving port. The coefficients a_i and b_i are constants, but unknown. The steady-state current due to a dc voltage of 10 V is 20 mA. Obtain the driving point impedance function of this network, and check if it is a pr function. [*Hint* Use Laplace transformation and find the ratio $V(s)/I(s)$.]

11.12 Four pole-zero plots are given in Fig. 11.32. Determine which of them correspond to pr functions. Why?

11.13 Given the admittance function

$$Y(s) = H_0 \frac{s^2 + b_1 s + b_0}{s^2 + a_1 s + a_0}$$

obtain a condition on the coefficients a_0, b_0, a_1, and b_1 so that $Y(s)$ is pr.

11.14 Consider the network in Fig. 11.33 where the first element value is shown. If the driving point impedance function of this network is required to have zeros at $s = -2$ and $s = -4$, and poles at $s = -1$ and $s = -3$, determine the element values of the other elements of the network.

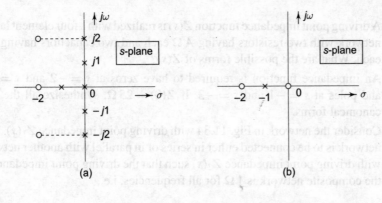

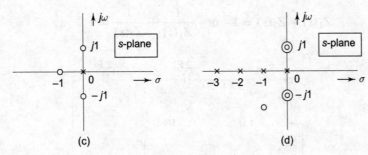

Fig. 11.32

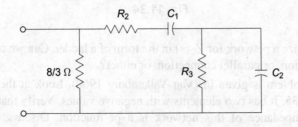

Fig. 11.33

11.15 Given the impedance function

$$Z(s) = \frac{(s+2)(s+6)}{(s+1)(s+5)}$$

synthesize the Foster networks.

11.16 Repeat Problem 11.15 for the following impedance function:

$$Z(s) = \frac{(s+2.5)(s+7.5)}{s(s+5)}$$

11.17 Synthesize the Cauer networks given the following impedance function:

$$Z(s) = \frac{(s+1)(s+5)}{(s+0.5)(s+4)}$$

11.18 A driving point impedance function $Z(s)$ is realized with a four element ladder network with two resistors having A Ω each and two capacitors having A F each. What are the possible forms of $Z(s)$?

11.19 An impedance function is required to have zeros at $s = -2$ and $s = -4$, and poles at $s = -1$ and $s = -3$. If $Z(0) = 25\,\Omega$, synthesize all the four canonical forms.

11.20 Consider the network in Fig. 11.34 with driving point impedance $Z_1(s)$. This network is to be connected either in series or in parallel with another network with driving point impedance $Z_2(s)$ such that the driving point impedance of the composite network is $1\,\Omega$ for all frequencies, i.e.,

$$Z_1(s) + Z_2(s) = 1 \quad \text{or} \quad \frac{1}{Z_1(s)} + \frac{1}{Z_2(s)} = 1$$

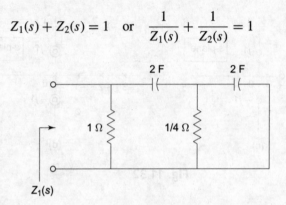

Fig. 11.34

Synthesize a network for $Z_2(s)$ in the form of a ladder. Can we use the series connection, or parallel connection, or either?

11.21 This problem is given in (Van Valkenburg 1960). Look at the network in Fig. 11.35. It has two elements with negative values. Verify that the driving point impedance of this network is a pr function. Discuss, under what conditions do pr functions end up in non-physical networks.

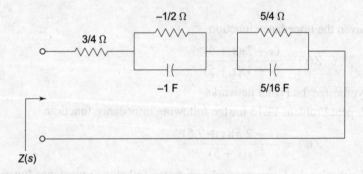

Fig. 11.35

11.22 Given the driving point impedance function

$$Z(s) = \frac{s(s^2 + 2)}{(s^2 + 1)(s^2 + 3)}$$

synthesize the Foster forms of LC networks.

11.23 Repeat Problem 11.22 for

$$Z(s) = 80 \frac{s(s^2 + 2)(s^2 + 5)}{(s^2 + 1)(s^2 + 3)}$$

11.24 For the impedance functions in Problems 11.22 and 11.23, synthesize the Cauer forms.

11.25 Obtain the driving point impedance of the network shown in Fig. 11.36, and synthesize an equivalent network that contains fewer elements.

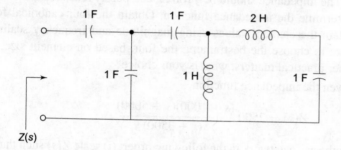

Fig. 11.36

11.26 It is proposed to build a canonical LC network using four inductors and five capacitors. Show how the poles and zeros would be located in the complex s-plane. How about a canonical network using five capacitors and six inductors?

11.27 Consider the driving point impedance function

$$Z(s) = \frac{s(s^2 + 4)}{(s^2 + 1)(s^2 + 9)}$$

- Synthesize a Cauer-1 network.
- Synthesize another network by first removing a network corresponding to one-half of the residue of the poles at $s = \pm j3$ and then synthesize the remaining impedance as a ladder network.

Compare the two realizations.

11.28 Synthesize a network with 3 elements, having the same driving point impedance as the network shown in Fig. 11.37. Generalize your result and show that in a canonical LC network, the number of L and C elements cannot differ by more than 1.

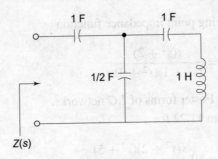

Fig. 11.37

11.29 An *LC* network is to be synthesized with the following specifications:
- The driving point impedance should be infinite at $\omega = 2.4\pi$ krad/s and at $\omega = 10\pi$ krad/s.
- The impedance should be zero at $\omega = 4.8\pi$ krad/s.
- The impedance should be $j100\ \Omega$, i.e., purely reactive, at $\omega = 6\pi$ krad/s.

Determine the impedance function. Obtain the four canonical forms, and obtain the element values using magnitude and frequency scaling. If you were to choose the best among the four, based on element size, cost, and other practical matters, what is your choice?

11.30 Given the impedance function

$$Z(s) = 2500\,\frac{(s + 1000)(s + 5000)}{s(s + 3500)}$$

synthesize a network in the following order: (i) scale $Z(s)$ such that the poles and zeros are reduced to, possibly, single digits, and the scale factor is unity, (ii) synthesize the down-scaled network, and (iii) determine the actual values of the elements by up-scaling.

CHAPTER 12

Operational Amplifiers

In this chapter we introduce an integrated circuit, called *operational amplifier*, popularly known as *Op-Amp*. It is not a circuit element per se, but it has acquired a similar status owing to its size, cost, and, more important, its linear characteristics. An Op-Amp is made up of several (typically 24) transistors and an associated *RC* network. We eschew the details here, suggesting the reader to get them in an electronics text and/or course. In this textbook we would like to see it from an input–output point of view.

As the name indicates, an Op-Amp, more specifically called a *high gain voltage amplifier*, amplifies the voltage applied across its input port. We *model* this as a two-port network as shown in Fig. 12.1.

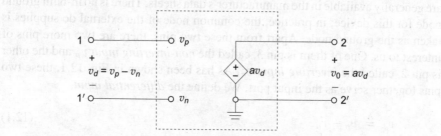

Fig. 12.1

The amplification is only in the magnitude part of the signal, and the frequency and phase angle of an ac signal are left untouched. Owing to this proportionality between the input and the output, the device behaves fairly linearly under given operating conditions. This device has become a workhorse of modern electronics, and hence it is mass produced by integrated-circuit manufacturers. Consequently, the size and the cost of the device, despite the presence of several elements within, are comparable to that of discrete circuit elements such as resistors. There are numerous interesting circuits that are of practical interest, and we will study a few of them in this chapter.

12.1 The Op-Amp as a Circuit Element

An Op-Amp is generally available as a *dual-in-line package* (DIP) with eight nodes[1], called *pins*. A schematic representation is shown in Fig. 12.2(a), and a typical circuit symbol is shown in Fig. 12.2(b). The enumeration of the nodes is standard.

---

[1] like an octopus!

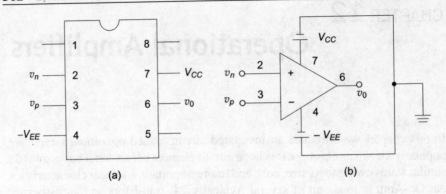

(a)   (b)

Fig. 12.2

Of these eight nodes, two are permanently reserved for *dc bias*. An Op-Amp requires a steady (dc) voltage to work as an amplifier. Accordingly, we must apply a dc voltage of $+ V_{CC}$ at node 7 and a dc voltage of $- V_{EE}$ at node 4. Typically, $V_{CC} = V_{EE} = 15$ V, though other values are also permissible, the details of which are generally available in the manufacturer's data sheets. There is no in-built ground node for this device; in practice, the common node of the external dc supplies is taken as the ground node. Apart from these two pins, there are two more pins of interest to us. One of them is pin 3, called the *non-inverting input* v_p and the other is pin 2, called the *inverting input* v_n. As has been shown in Fig. 12.1, these two pins together serve as the input port. We define the *differential input*,

$$v_d \overset{\Delta}{=} v_p - v_n \tag{12.1}$$

and the output of the device is

$$v_0 = a \times v_d \tag{12.2}$$

where the parameter a is the *high gain* of the Op-Amp. The output is available to us at pin 6. For this device to be used as a circuit element, the five pins described above are sufficient. Since the element does not work without the dc bias, we consider them as omnipresent and avoid showing them explicitly. With the remaining three nodes, we use the symbol shown in Fig. 12.3 for the Op-Amp as a circuit element.

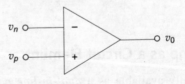

Fig. 12.3

The gain a of the device is extremely high. Typically it is in the range of $10^5 \le a \le 10^7$. Practically, this is infinite, and an extremely small differential voltage v_d across the input port is enough to produce a substantial output v_0;

nevertheless, notice that if $v_p = v_n$, i.e., $v_d = 0$, then $v_0 = 0$ V. Of course, owing to the law of conservation of power, the output cannot exceed the external dc voltages, i.e., $-V_{EE} \leq v_0 \leq +V_{CC}$. In other words, we need to limit the differential input v_d such that the output would remain in this limit. Otherwise, the output remains *saturated* at either of these extremes. Also, we may notice that it is only the differential voltage that matters, and not the individual magnitudes of v_p and v_n.

Further, the impedance as seen by the input port is also infinite, typically few megaohms. And, the impedance as seen by the output is extremely small, typically few ohms. High input impedance and low output impedance, as might be expected, eliminate the loading problem. In other words, the amplifier simply multiplies the magnitude of the input signal without any loss. The characteristics of an Op-Amp, which we will use in this chapter, may be summarized below:

1. High input impedance.
2. Low output impedance.
3. High voltage gain, $a \rightarrow \infty$.
4. Large bandwidth; typically any signal is uniformly amplified without any distortion.
5. Perfect balance, $v_0 = 0 \Leftrightarrow v_d = 0$ V.
6. Operation is fairly independent of the ambient temperature.

All in all, we may consider an Op-Amp as an active, linear, and lumped circuit element. It is interesting to note that this device conforms to our ideal expectations very closely, and this is the secret behind its widespread acceptance in circuit design.

12.2 Basic Configurations

There are a few more important characteristics of the Op-Amp. We will uncover them gradually in this section. The circuits we see here can be generalized to develop more interesting circuits.

12.2.1 The Non-inverting Amplifier

First, let us consider the circuit in Fig. 12.4.

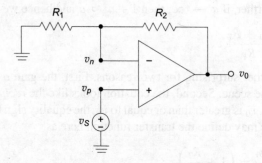

Fig. 12.4

This circuit is called *the non-inverting amplifier*. An input signal v_S is applied to the node v_p. We see a voltage division circuit between the node v_0 and the ground via the two resistances R_1 and R_2, with the node v_n in between. According to the voltage division rule,

$$v_n = \frac{R_1}{R_1 + R_2} v_0$$

and hence the differential input is

$$v_d = v_S - \frac{R_1}{R_1 + R_2} v_0 \tag{12.3}$$

Notice that a fraction of the output signal is present in the differential input signal, with a negative sign. This is an extremely important observation. If a is the gain of the Op-Amp, then the output is

$$v_0 = a v_d$$

$$= a \left(v_S - \frac{R_1}{R_1 + R_2} v_0 \right) \tag{12.4}$$

Equation (12.4) is a simple algebraic equation wherein we can collect terms together so that

$$\left(1 + a \frac{R_1}{R_1 + R_2} \right) v_0 = a v_S$$

and hence

$$v_0 = \frac{a}{1 + a \dfrac{R_1}{R_1 + R_2}} v_S$$

$$= \frac{a(R_1 + R_2)}{(1 + a)R_1 + R_2}$$

$$\approx \frac{a}{1 + a} \frac{(R_1 + R_2)}{R_1} \tag{12.5}$$

The approximation holds because a is extremely large and the resistances may be chosen so that R_2 may be neglected in comparison to $(1 + a) \times R_1$ in the denominator. Further, if $a \to \infty$, then $1 + a \approx a$ and hence we may write

$$v_0 = \frac{R_1 + R_2}{R_1} v_S \tag{12.6}$$

This result is rather surprising for two reasons. First, the gain a of the Op-Amp vanishes from the scene. Second, the equation looks like the *reciprocal* of voltage division. Clearly, v_0 is greater than or equal to v_S, the equality sign being applicable if $R_2 = 0 \ \Omega$. We may define the transfer function here as

$$A \triangleq \frac{v_0}{v_S} = \left(1 + \frac{R_2}{R_1} \right) \tag{12.7}$$

where A is called the *closed-loop gain*, for reasons we explain in a moment. The events happen this way owing to the connections we have made in the circuit, together with the fact that $a \to \infty$. As has been observed in the beginning, a portion of the output is made available back in the differential input, with a negative sign. This is called *negative feedback* for obvious reasons. The R_1-R_2 network is called the *feedback network*. Recalling the definition of closed path (or loop) from Chapter 3, it is clear that we are able to achieve a gain A because of the feedback network from the output to the input. Hence the name closed-loop gain for A. In contrast, the gain a is called *open-loop gain*.

Example 12.1 To amplify a signal v_S by a factor of 3, we may use the circuit with an Op-Amp as shown in Fig. 12.4 and choose the resistances to be in the ratio 1 : 2.

Example 12.2 Suppose that we have an Op-Amp with $a = 10^6$, and a non-inverting amplifier circuit of Fig. 12.4 with $R_1 = 10\,\mathrm{k\Omega}$, and $R_2 = 20\,\mathrm{k\Omega}$. Let us now examine how good were our approximations in computing the closed-loop gain. First, strictly according to the relationship, we have

$$\frac{v_0}{v_S} = \frac{a(R_1 + R_2)}{(1 + a)R_1 + R_2}$$

$$= \frac{10^6 \times 3 \times 10^4}{(1 + 10^6) \times 10^4 + 2 \times 10^4}$$

$$= 2.999991$$

Second, if we consider Eqn (12.5) and ignore R_2 in the denominator, we get

$$\frac{v_0}{v_S} = \frac{a}{(1 + a)} \frac{R_1 + R_2}{R_1}$$

$$= \frac{10^6}{1 + 10^6} \frac{3 \times 10^4}{10^4}$$

$$= 2.999997$$

Finally, in the limit as $a \to \infty$, we have

$$\frac{v_0}{v_S} = A = 1 + \frac{20\,\mathrm{k\Omega}}{10\,\mathrm{k\Omega}} = 3$$

Clearly, the error arising out of our approximations is negligible in practice. Despite the apparent mathematical approximations, the device behaves very closely to our expectation.

We shall now give a more quantitative explanation. Let us consider an amplifier with gain a and a passive network with a transfer function β, connected schematically as shown in Fig. 12.5.

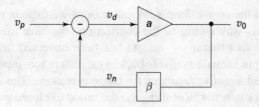

Fig. 12.5

In Fig. 12.5, we have also shown a small circle with a 'minus' sign inside. This element is assumed to simply perform the subtraction operation on the two signals v_p and v_n. To begin with, suppose that a signal v_p is applied to the subtracting element. Since this signal has not yet passed through the amplifier, $v_n = 0$, and hence $v_d = v_p$. Accordingly,

$$v_0 = a v_p$$

In the next instant of time,

$$v_n = \beta v_0 = a \beta v_p$$

and hence

$$v_0 = a(v_p - a\beta v_p) = (a - a^2 \beta) v_p$$

Subsequently, v_n becomes $\beta(a - a^2\beta)v_p$ and hence

$$v_0 = a\left[v_p - (a\beta - a^2\beta^2)v_p\right] = (a - a^2\beta + a^3\beta^2) v_p$$

Looking at this pattern, it is easy to see that

$$v_0 = (a - a^2\beta + a^3\beta^2 - a^4\beta^3 + \cdots) v_p \tag{12.8}$$

asymptotically. In other words, the output starts with v_p but quickly grows up and *converges* to v_0 given by Eqn (12.8). We observe an interesting geometric progression in Eqn (12.8), which can be written in the closed form as

$$v_0 = \frac{a}{1 + a\beta} v_p$$

If we now apply the limit, we get

$$\frac{v_0}{v_p} = \lim_{a \to \infty} \frac{a}{1 + a\beta} = \frac{1}{\beta} \tag{12.9}$$

In the non-inverting amplifier, the β network is simply a voltage divider and hence we get

$$v_0 = \frac{1}{\beta} v_p = \left(1 + \frac{R_2}{R_1}\right) v_p = A v_p$$

with $v_p = v_S$. Thus, the output builds up and converges to a *steady-state* value, even though we may not be able to visualize this. The high open-loop gain a is a catalyst to this build-up process. There is another subtle observation here. Had there been a '+' sign in the small circle of Fig. 12.5, the output would simply soar up unbounded, and the Op-Amp may be dubbed as an *unstable* device, according to our earlier discussion on BIBO stability in Chapter 11. It is the negative feedback that stabilizes (i.e., it does not allow the output signal to grow unbounded) the device and forces the output to converge to a steady-state value.

In the light of this observation, let us examine the node potential v_n once again. Earlier, we have noted that

$$v_n = \frac{R_1}{R_1 + R_2} v_0 = \beta v_0$$

Thus, as v_0 builds up from v_p towards $Av_p = v_p/\beta$, the node potential v_n builds up from βv_p to v_p. In other words, the node potential v_n *tracks* the node potential v_p eventually. Equivalently, we may say that an Op-Amp with a negative feedback network will put out v_0 such that its differential input v_d is driven to zero asymptotically. This is another important observation in the study of Op-Amps. In fact, we take it as yet another rule, in addition to the element and circuit rules of the previous chapters. If the two node potentials coalesce, then there is *virtually* a short circuit at the input port of the Op-Amp. Notice that, owing to the infinite input impedance, there is no flow of current between v_p and v_n, and hence it is like an open circuit. Conventionally, this phenomenon, where there is short circuit in one sense and an open-circuit in another sense, is called a *virtual short*.

Example 12.3 Suppose that resistance $R_2 = 0 \ \Omega$ in the non-inverting amplifier circuit of Fig. 12.4. It is easy to see that the closed-loop gain is

$$A = 1$$

i.e., the output v_0 *follows* the input v_S for all times. Accordingly, the circuit is called *voltage follower*. It is not out of fancy that we choose $R_2 = 0 \ \Omega$, but there is an important reason behind this. Suppose we have a 5 V dc source with an internal resistance of $1 \, k\Omega$ and we need to drive a $4 \, k\Omega$ resistive load. Clearly, owing to the internal resistance, the source can deliver only 4 V and hence 4 mW of power to the load, losing 1 mW to the internal resistance. Now suppose we have put a voltage follower between the source and the load as shown in Fig. 12.6.

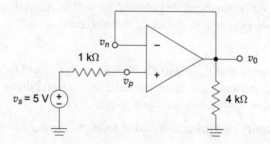

Fig. 12.6

Owing to the high input impedance, no current is drawn by the Op-Amp from the source, and hence the source virtually delivers no power. Therefore, there is no loss across the internal resistance. Consequently, despite the internal resistance, $v_p = v_S = 5$ V, and $v_0 = 5$ V. In other words, the load now appears to draw all the 5 V from the source, and hence dissipates 6.25 mW. Here is a tricky question: When the 5 V dc source is not delivering any power, how come the load absorbs as much as 6.25 mW? In other words, aren't we violating the 'law' of conservation of power? In the actual circuit, there is a voltage follower sitting between the source and the load. Let us remind ourselves that the Op-Amp needs external dc supply. Hence, power is delivered to the load by the voltage follower, relieving the dc source from delivering any power. Owing to such a behaviour of the voltage follower, it is also called a *voltage buffer circuit*.

Summing up the discussion on this circuit, we are interested in the following two issues.

1. The closed-loop gain A is totally under our control, since we are free to choose the resistances R_1 and R_2. Further, A depends only on the ratio of the resistances, and hence can be easily customized to suit a particular application with a desired accuracy. It is sufficient to choose the resistances that track each other with time and/or temperature so that the ratio is always constant.

2. The polarities of the output and the input are one and the same since A, as defined in Eqn (12.7), is always positive.

12.2.2 The Inverting Amplifier

Consider the circuit shown in Fig. 12.7.

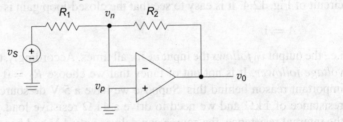

Fig. 12.7

Compared to the non-inverting amplifier of Fig. 12.4, the node v_p is connected to the ground, and hence the node potential $v_p = 0$ V. According to the Op-Amp rule, that the node potential v_n tracks v_p, we also have $v_n = 0$ V.

Exercise 12.1 Verify that v_n is initially $R_2/(R_1 + R_2) v_S$ and then asymptotically reduces to 0 V, tracking the node potential v_p. (**Hint** Follow the quantitative argument provided for the non-inverting amplifier.)

Exercise 12.2 Verify that the actual closed-loop gain without any approximations is

$$\frac{v_0}{v_S} = -\frac{R_2}{R_1} \frac{a}{1 + a + R_2/R_1}$$

Applying nodal analysis at node v_n, we have

$$\frac{v_n - v_S}{R_1} + \frac{v_n - v_0}{R_2} = 0$$

Using the fact that $v_n = 0$ V, we get

$$A \triangleq \frac{v_0}{v_S} = -\frac{R_2}{R_1} \tag{12.10}$$

Exercise 12.3 Verify Eqn (12.10).

Equation (12.10) tells us why we call this circuit an inverting amplifier—if the input is positive, the output is negative, and vice versa. We may also notice that the gain can go down to 0, if we choose $R_2 = 0$. If $R_2 < R_1$, we have attenuation; if $R_2 > R_1$, we have amplification. However, for the case $R_2 = R_1$, we cannot have a voltage follower like the non-inverting case owing to two reasons. First, trivially, the sign is always opposite. Second, a careful inspection of the circuit shown in Fig. 12.7 suggests that the impedance, as seen by the input port, is R_1, and not infinity. And, the output impedance R_0, which is typically a few ohms, will come into the picture, unlike the non-inverting case. If we look at the two-port model of Fig. 12.8, we may account for each of these things.

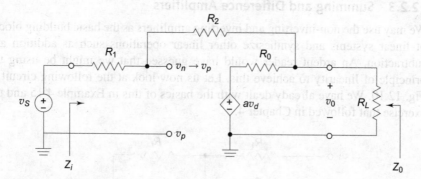

Fig. 12.8

From the two-port model, it is quite clear that

$$\text{input impedance} = R_1 + \text{internal resistance of the source, if any} \tag{12.11}$$

$$\text{output impedance} = R_0 \parallel R_L \tag{12.12}$$

for the inverting amplifier. Thus, loading cannot be eliminated here; there is always a loss of power at the source. Let us look at the following numerical example.

Example 12.4 Let us consider the same old 5 V source with an internal resistance of $1\,\text{k}\Omega$ driving a load of $4\,\text{k}\Omega$. Let us put an inverting amplifier in between, as shown in Fig. 12.9, with $R_1 = R_2 = 1\,\text{k}\Omega$.

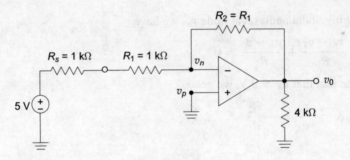

Fig. 12.9

Applying nodal analysis, we may find that $v_0 = -2.5$ V, i.e., one-half of the source, and the power dissipated by the load is $25/16$ mW. There is a heavy loss. The reason may be investigated as follows. First, the input impedance is $R_s + R_1 = 2\,\text{k}\Omega$, and clearly there is a drop of 5 V across this series combination, since $v_n = 0$ V. Second, the output impedance is $R_0 \parallel R_L = 1\,\text{k}\Omega$ in order to satisfy KCR at the node v_0. With $R_L = 4\,\text{k}\Omega$, we see that $R_0 = 4/3\,\text{k}\Omega$. Thus, the inverting amplifier cannot be used as a voltage buffer circuit.

12.2.3 Summing and Difference Amplifiers

We may use the non-inverting and inverting amplifiers as the basic building blocks of linear systems and synthesize other linear operations such as addition and subtraction. An ardent reader would have guessed that we might be using the principle of linearity to achieve this. Let us now look at the following circuit in Fig. 12.10. We have already dealt with the basics of this in Example 4.15 and the exercise that followed in Chapter 4.

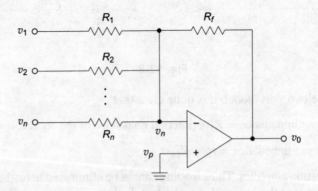

Fig. 12.10

Since $v_p = 0$ V, so is v_n. Applying KCR at node v_n, it is easy to see that the total current i_f flowing through the feedback resistance R_f from left to right is

$$i_f = \frac{v_1}{R_1} + \frac{v_2}{R_2} + \cdots + \frac{v_n}{R_n}$$

Since $v_0 = -R_f i_f$, we have

$$v_0 = - \left(\frac{R_f}{R_1} v_1 + \frac{R_f}{R_2} v_2 + \cdots + \frac{R_f}{R_n} v_n \right) \tag{12.13}$$

i.e., we have a *weighted sum* (or a linear combination) of all the input signals. If we take $R_1 = R_2 = \cdots = R_n = R_f$, we simply have

$$v_0 = - \sum_{i=1}^{n} v_i$$

We may also use a potentiometer in place of R_f to have a variable gain summing amplifier.

Example 12.5 Suppose that we have three sources v_1, v_2, v_3, and we need $-2v_1 - 3v_2 - 5v_3$, the corresponding resistances should be

$$R_1 = \frac{1}{2} R_f, \quad R_2 = \frac{1}{3} R_f$$

and

$$R_3 = \frac{1}{5} R_f$$

Exercise 12.4 Design a summing amplifier that accepts two sources whose internal resistances are $10 \, \text{k}\Omega$ and $4 \, \text{k}\Omega$, respectively, and gives out

$$v_0 = -10v_1 - 18v_2$$

Is there any way to reduce the number of components?

To eliminate the negative sign in the sum, we may use another inverting amplifier at the output. We leave it to the reader to study the summing amplifier with a non-inverting amplifier as the building block. Summing amplifiers find applications in audio signal processing such as mixing.

Let us now look at the following circuit in Fig. 12.11.

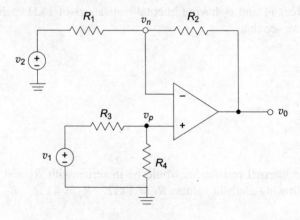

Fig. 12.11

Applying the principle of linearity, we may see that we have cleverly combined the non-inverting and inverting amplifiers. If v_1 is assumed to be the only source acting while v_2 is removed,

$$v_{01} = \frac{R_4}{R_3 + R_4} \left(1 + \frac{R_2}{R_1} \right) v_1$$

Similarly, assuming that v_2 is the only source functioning in the circuit,

$$v_{02} = - \frac{R_2}{R_1} v_2$$

Combining these two partial results, we get

$$v_0 = \frac{R_2}{R_1} \left(\frac{1 + \dfrac{R_1}{R_2}}{1 + \dfrac{R_3}{R_4}} v_1 - v_2 \right) \tag{12.14}$$

Exercise 12.5 Derive Eqn (12.14).

Thus, the circuit shown in Fig. 12.11 may be called the *weighted difference amplifier*. To perform true difference, we choose

$$\frac{R_1}{R_2} = \frac{R_3}{R_4}$$

so that we get

$$v_0 = \frac{R_2}{R_1} (v_1 - v_2) \tag{12.15}$$

Example 12.6 Let us design a circuit such that

$$v_0 = 4v_1 - 2v_2$$

with two sources v_1 and v_2 having internal resistances of $1\,\text{k}\Omega$ each. Looking at Eqn (12.14), we see that

$$\frac{R_2}{R_1} = 2$$

and hence

$$\frac{R_3}{R_4} = \frac{1}{3}$$

Given that the internal resistances would be in series with R_1 and R_3, we may choose the following element values: $R_1 = 1\,\text{k}\Omega$, $R_2 = 4\,\text{k}\Omega$, $R_3 = 1\,\text{k}\Omega$, and $R_4 = 6\,\text{k}\Omega$.

Exercise 12.6 Verify the results of Example 12.6.

It is not difficult to see that each of the sources v_1 and v_2 sees an input impedance of $R_3 + R_4$ and R_1, respectively. In other words, we may have to live with the loading problem while using a weighted difference amplifier.

In many instrumentation and control applications, we need to measure the potential difference between two nodes. In actual practice, the application involves the measurement of pressure or flow which would be transducted into voltages to be measured. The node potentials would be considerably small in magnitudes, and we need to measure the difference sensitively. Owing to the loading problem, we cannot use the difference amplifier directly. We need to overcome this problem somehow. Our experience with the non-inverting amplifier would suggest us to employ one such circuit for each of the node potentials before the difference amplifier. This trick indeed works, and the circuit is called an *instrumentation amplifier*. The circuit is shown in Fig. 12.12.

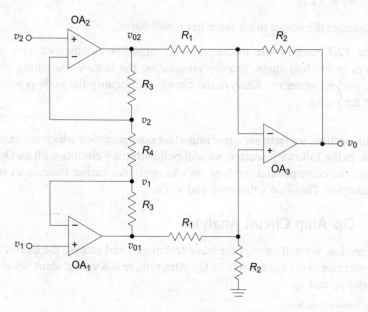

Fig. 12.12

An analysis of this circuit is quite simple. The connections tell us straightaway that, owing to the virtual short between v_p and v_n of the Op-Amps OA_1 and OA_2, the potential difference across the resistance R_4 is, say, $v_2 - v_1$. Hence, a current of $(v_2 - v_1)/R_4$ flows through R_4 from top to bottom. Further, the same current will flow through the two resistances R_3 since neither OA_1 nor OA_2 will draw any current through v_n. Thus, the potential difference $v_{02} - v_{01}$ across the series combination of the three resistances will be

$$v_{02} - v_{01} = (R_4 + 2R_3)\frac{v_2 - v_1}{R_4} = \left(1 + \frac{2R_3}{R_4}\right)(v_2 - v_1)$$

The two Op-Amps and the three resistances make up what is called the *first stage* of the instrumentation amplifier, and this takes care of the loading problem. The second stage is the ordinary difference amplifier, and the final output is

$$v_0 = \frac{R_2}{R_1} \left(1 + \frac{2R_3}{R_4}\right) (v_1 - v_2) \tag{12.16}$$

Example 12.7 Let us design an instrumentation amplifier to obtain

$$v_0 = 10^4 (v_1 - v_2)$$

Owing to the two stages, we can factor the total gain 10^4 in several ways. Arbitrarily, let us choose 10 for the first stage and 10^3 for the second stage. To this end, we may choose

$$R_1 = 1\,\text{k}\Omega, \quad R_2 = 1\,\text{M}\Omega, \quad R_3 = 4.5\,\text{k}\Omega$$

and hence

$$R_4 = 1\,\text{k}\Omega$$

We encourage the reader to try out other possibilities.

Exercise 12.7 Redraw the instrumentation amplifier circuit with two voltage followers in the first stage, thereby eliminating the branch containing resistors R_3, R_4, and R_3 in series. Analyze the circuit and identify the node potentials of each of the nodes.

In this section, we have seen five important configurations which are extremely popular. In the following section we will build arbitrary circuits with an Op-Amp as one of the elements and see how we can apply our earlier techniques such as nodal analysis, Thèvenin's theorem, and so on.

12.3 Op-Amp Circuit Analysis

In this section, we will employ the basic techniques and analyse the circuits. In so doing, we refer to the golden rule of Op-Amps: there is a virtual short between the two nodes v_p and v_n.

Example 12.8 Consider the circuit given in Fig. 12.13.

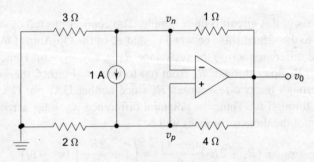

Fig. 12.13

Despite the messy appearance, the circuit has only four nodes, including the ground node. Further, owing to the Op-Amp rule, the two nodes v_p and v_n have the same potential. In other words, we have just two unknowns to be computed. If we have two independent equations, our job is done. Let us proceed with KCR at node v_n, which gives us

$$\frac{v_n}{3\,\Omega} + 1 + \frac{v_n - v_0}{1\,\Omega} = 0$$

Let us apply KCR at node v_p to get

$$\frac{v_p}{2\,\Omega} - 1 + \frac{v_p - v_n}{4\,\Omega} = 0$$

Using the fact that $v_p = v_n$, we have a pair of linear equations of the form

$$\begin{pmatrix} \frac{4}{3} & -1 \\ \frac{3}{4} & -\frac{1}{4} \end{pmatrix} \begin{pmatrix} v_p \\ v_0 \end{pmatrix} = \begin{pmatrix} -1 \\ 1 \end{pmatrix}$$

from which we may solve that $v_p = v_n = 3\,\text{V}$ and $v_0 = 5\,\text{V}$.

An inquisitive reader should have had a question by now. What happens if we apply KCR at node v_0? We leave it to the reader to discover the answer himself.

Exercise 12.8 Verify the results of Example 12.8 and discuss what happens if we choose to apply KCR at node v_0.

Example 12.9 Consider the circuit shown in Fig. 12.14.

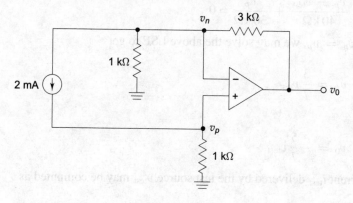

Fig. 12.14

Nodal analysis in this case is even simpler. At node v_p, it is easy to see that

$$\frac{v_p}{1\,\text{k}\Omega} - 2 \times 10^{-3} = 0$$

or $v_p = 2\,\text{V}$ and so is v_n. KCR at node v_n gives us

$$\frac{v_n}{1\,\text{k}\Omega} + 2 \times 10^{-3} + \frac{v_n - v_0}{3\,\text{k}\Omega} = 0$$

from which we may compute $v_0 = 14$ V.

Exercise 12.9 Verify the results of Example 12.9. Repeat the problem with the 1 kΩ between v_n and the ground removed.

Example 12.10 Let us determine the equivalent resistance seen by the input port in the circuit shown in Fig. 12.15.

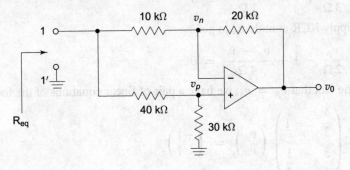

Fig. 12.15

To compute the equivalent resistance, we need to apply a test voltage source v_{test} and compute the current it delivers. Assuming a test source, we get

$$\frac{v_n - v_{\text{test}}}{10\,\text{k}\Omega} + \frac{v_n - v_0}{20\,\text{k}\Omega} = 0$$

$$\frac{v_p - v_{\text{test}}}{40\,\text{k}\Omega} + \frac{v_p}{30\,\text{k}\Omega} = 0$$

Using $v_n = v_p$, we may solve the above LSE to get

$$v_p = \frac{3}{7} v_{\text{test}}$$

and

$$v_0 = -\frac{5}{7} v_{\text{test}}$$

The current i_{test} delivered by the test source v_{test} may be computed as

$$i_{\text{test}} = \frac{v_{\text{test}} - v_n}{10\,\text{k}\Omega} + \frac{v_{\text{test}} - v_p}{40\,\text{k}\Omega} = \frac{1}{14\,\text{k}\Omega} v_s$$

and hence

$$R_{\text{eq}} = 14\,\text{k}\Omega$$

Exercise 12.10 Verify the results of Example 12.10.

Example 12.11 Let us consider the circuit shown in Fig. 12.16.

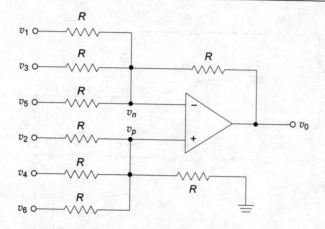

Fig. 12.16

Let us take up the non-inverting node v_p first. Using the principle of linearity, we may easily see that

$$v_p = \frac{R/3}{R + \dfrac{R}{3}} v_2 + \frac{R/3}{R + \dfrac{R}{3}} v_4 + \frac{R/3}{R + \dfrac{R}{3}} v_6$$

$$= \frac{1}{4}(v_2 + v_4 + v_6)$$

Since v_n follows v_p, KCR at node v_n gives us

$$\frac{v_p - v_1}{R} + \frac{v_p - v_3}{R} + \frac{v_p - v_5}{R} + \frac{v_p - v_0}{R} = 0$$

or, equivalently,

$$v_0 = 4\,v_p - (v_1 + v_3 + v_5)$$

$$= (v_2 - v_1) + (v_4 - v_3) + (v_6 - v_5)$$

Exercise 12.11 Verify the results of Example 12.11.

Working with circuits containing an Op-Amp is easy. Isn't it? Let us see the following example that contains two Op-Amps.

Example 12.12 Consider the circuit shown in Fig. 12.17 that contains two Op-Amps.

It looks baffling initially. Nodal analysis does not take us far since we are not sure of the currents at the nodes v_0 and v_{02}, the outputs of the Op-Amps. Let us then begin at the top left corner of the circuit, with the fact that $v_{n1} = v_{p1}$. Since no current enters the Op-Amps, we readily observe that

$$\frac{v_{p1} - v_1}{R_1} + \frac{v_{p1}}{R_2} = 0$$

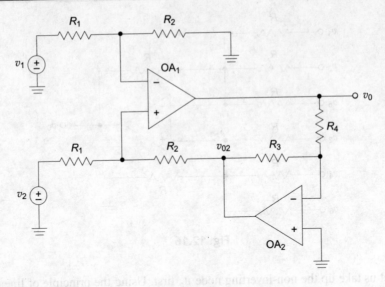

Fig. 12.17

or, equivalently,

$$\left(\frac{1}{R_1} + \frac{1}{R_2}\right) v_{p1} = \frac{1}{R_1} v_1$$

at the node v_{n1}. Similarly, at node v_{p1}, we have

$$\frac{v_{p1} - v_2}{R_1} + \frac{v_{p1} - v_{02}}{R_2} = 0$$

or, equivalently,

$$\frac{1}{R_2} v_{02} = \left(\frac{1}{R_1} + \frac{1}{R_2}\right) v_{p1} - \frac{1}{R_2} v_2$$

$$= \frac{v_1}{R_1} - \frac{v_2}{R_1}$$

$$\Rightarrow \quad v_{02} = \frac{R_2}{R_1} (v_1 - v_2) \tag{12.17}$$

Looking at the connections between v_0 and v_{02} through the Op-Amp OA_2, we may observe that

$$v_{02} = -\frac{R_3}{R_4} v_0$$

since OA_2 is an inverting amplifier. Combining this with the earlier equation, it is easy to see that

$$v_0 = \frac{R_4}{R_3} \frac{R_2}{R_1} (v_2 - v_1)$$

Of course, in this example, we have also made use of the fact that the Op-Amps are basically 'no loading' devices; otherwise, it will not be just an inverting amplifier between v_0 and v_{02} via OA_2.

Exercise 12.12 Obtain the gain v_0/v_i for the circuit shown in Fig. 12.18.

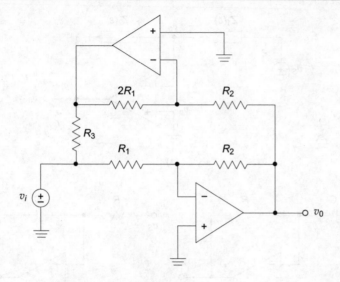

Fig. 12.18

So far, in this section we have confined ourselves to resistive circuits containing Op-Amps. In the following section, we will look at circuits containing energy-storage elements together with the Op-Amps.

12.4 Active Filters using Op-Amps

There are many circuits of practical interest that could be built using Op-Amps in association with one or more energy-storage elements such as capacitors and inductors. Theoretically, though, there is no harm in imagining and working with a circuit that contains an Op-Amp as well as an inductor, integrated circuits quarantine the inductors owing to their size and power requirements. There are some clever circuits that would *emulate* inductors, and we present a few of them in this section. In this section, for the sake of simplicity, let us work with impedances and complex algebra using the Laplace transformation.

Let us first redraw our inverting amplifier with impedances $Z_1(s)$ and $Z_2(s)$ in place of the resistances. This is shown in Fig. 12.19(a).

We quickly observe that

$$\frac{v_0}{v_i} = -\frac{Z_2}{Z_1}$$

with a capacitive reactance $1/sC$ for Z_1 and a resistance R for Z_2 gives us

$$\frac{v_0}{v_i} = -sRC$$

meaning that the output of the Op-Amp is the differentiation of its input, scaled by

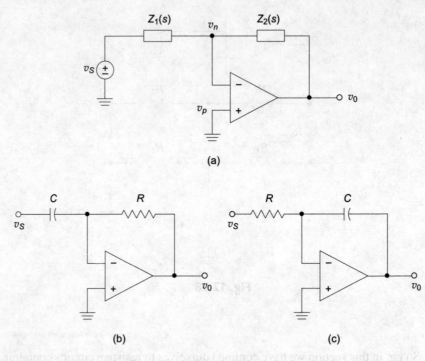

Fig. 12.19

a factor of RC. This circuit is called a *differentiator*, and is shown in Fig. 12.19(b). If the capacitance and the resistance are interchanged, we get

$$\frac{v_0}{v_i} = -\frac{1}{sRC}$$

and the circuit is called an *integrator*, and is shown in Fig. 12.19(c).

Example 12.13 Suppose we have a periodic rectangular waveform (pulse train) as shown in Fig. 12.20(a).

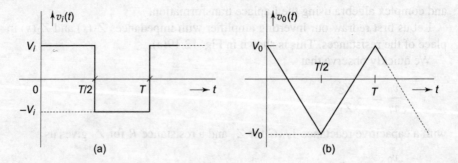

Fig. 12.20

Since the pulse train is a summation of step signals, applying this as an input $v_i(t)$ to an integrator will produce a summation of ramps, i.e., the triangular waveform

$v_0(t)$ shown in Fig. 12.20(b). Notice the $180°$ phase difference—when the input $v_i > 0$, the output is decreasing, and when $v_i < 0$, the output is increasing.

Exercise 12.13 Verify the results of Example 12.13. Explain how the triangular waveform of Fig. 12.20(b), when applied to a differentiator, produces the pulse train of Fig. 12.20(a).

Exercise 12.14 Verify that the circuit in Fig. 12.21 is a non-inverting integrator.

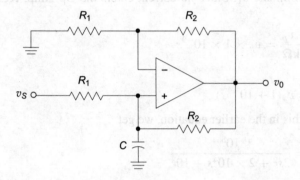

Fig. 12.21

An integrator circuit is an extremely useful one in modern electronics, instrumentation, and control. An analog computer, useful in simulating dynamical systems, may be built using several integrators. Further, one may observe that the transfer function of an integrator has just one pole at the origin. Looking at the frequency response, it is easy to discover that the magnitude is quite high for low-frequency signals, and high-frequency signals are severely attenuated. In other words, an integrator is like a *low-pass filter*. We may modify the passive circuitry around the Op-Amp to improve the filtering characteristics, and we discuss such issues in the following examples.

Example 12.14 Consider the circuit shown in Fig. 12.22.

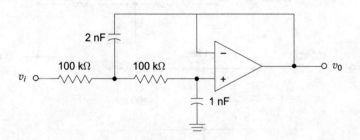

Fig. 12.22

It is evident that

$$v_0 = v_p$$

Now let us determine the relationship between v_i and v_p. Applying KCR at node v_1, we get

$$\frac{v_1 - v_i}{100\,\text{k}\Omega} + (v_1 - v_p)(2 \times 10^{-9}s) + \frac{v_1 - v_p}{100\,\text{k}\Omega} = 0$$

We need to eliminate v_1. Since no current enters the Op-Amp, KCR at node v_p gives us

$$\frac{v_1 - v_p}{100\,\text{k}\Omega} = v_p \times 1 \times 10^{-9}s$$

or

$$v_1 = v_p(1 + 10^{-4}s)$$

Substituting this in the earlier equation, we get

$$\frac{v_0}{v_i} = \frac{10^8}{2s^2 + 2 \times 10^4 s + 10^8}$$

and hence

$$\zeta = \frac{1}{\sqrt{2}} \quad \text{and} \quad \omega_n = \frac{10^4}{\sqrt{2}}$$

It is easy to see that this second-order filter has much better low-pass filtering characteristics.

Exercise 12.15 Obtain the Bode plot of the transfer function of Example 12.14, and verify that the circuit is a unity gain low-pass filter. Suggest a suitable modification so that we may increase the gain of the filter without affecting its frequency response. (***Hint*** In the circuit given in Fig. 12.22, we have employed a voltage follower. What happens if we employ a pair of resistances R_1 and R_2 in the feedback circuit?)

Exercise 12.16 Obtain the transfer function of the circuit shown in Fig. 12.23.

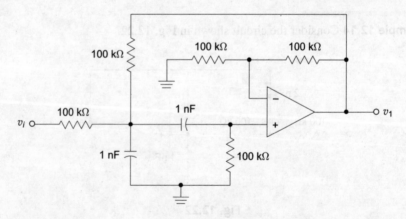

Fig. 12.23

Example 12.15 Consider the circuit shown in Fig. 12.24.

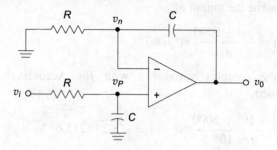

Fig. 12.24

Intuitively, this could be yet another integrator circuit, perhaps a second-order one owing to the presence of two energy-storage elements. Let us see. First, using simple voltage division rule, we find that

$$v_p = \frac{1/sC}{1/sC + R} v_i = \frac{1}{sRC + 1} v_i = v_n$$

Then, KCR at node v_n tells us that

$$\frac{v_p}{R} + (v_p - v_0) sC = 0$$

and hence,

$$v_0 = \frac{sRC + 1}{sRC} v_p = \frac{1}{sRC} v_i$$

Thus, this circuit is a non-inverting integrator, despite the presence of a pair of capacitances. We encourage the reader to compare this circuit with that shown in Fig. 12.21.

Example 12.16 Consider the circuit given in Fig. 12.25.

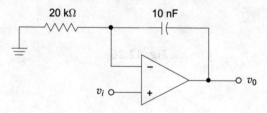

Fig. 12.25

This is a simple circuit with $v_n = v_p = v_i$, and

$$\frac{v_i}{20\,\mathrm{k\Omega}} + (v_i - v_0) \times 10 \times 10^{-9} s = 0$$

Solving for v_0, we find that

$$v_0(s) = \frac{s + 5000}{s} v_i(s)$$

If we are interested in the steady-state response, ignoring the transients for a while, we may write the output as

$$v_0(j\omega) = \frac{j\omega + 5000}{j\omega} v_i(j\omega)$$

by replacing the complex variable s with $j\omega$. Accordingly, if $v_i = 3\cos(10^4 t - 90°)$, then

$$v_0 = \frac{j \times 10^4 + 5000}{j \times 10^4} \times -j3 = \sqrt{3.25} \angle 213.6° \text{ V}$$

Likewise, if we are interested in the step response to see the transients, we simply work in the s-domain itself with $v_i(s) = 1/s$. We leave this as an exercise to the reader.

Exercise 12.17 If $v_i = 3\cos(10^4 t - 90°)$, find the complete response, i.e., the transient response and the steady-state response of the circuit shown in Fig. 12.25, by making use of appropriate Laplace transformation $v_i(s)$.

Exercise 12.18 Show that the circuit in Fig. 12.26 is a non-inverting differetiator. Obtain the Bode plot.

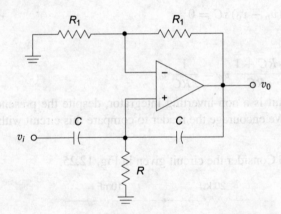

Fig. 12.26

Filter theory is a vast discipline, and there are several textbooks and references exclusively devoted to the subject. The reader may be introduced to the subject in later semesters. In general, any circuit that contains one or more energy-storage elements, in addition to resistances, may be called a filter. Primarily, filters may be built using passive elements alone, such as the ones we have already seen in Chapter 7. The synthesis of circuits in Chapter 11 is also restricted to using passive elements. Alternatively, we may build filters using Op-Amps to achieve several benefits. For instance, the idea of feedback allows us to realize a circuit that exhibits, virtually, any desired response. An even more interesting issue is that we can eliminate inductors altogether, owing to reasons such as size and cost. Let us now see one such clever circuit which emulates an inductor.

Example 12.17 Consider the circuit of Fig. 12.27.

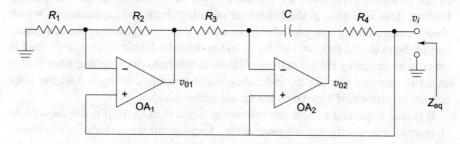

Fig. 12.27

Let us routinely apply a test voltage source v_{test} and compute the current i_{test} it delivers, so that the equivalent impedance as seen by the port 1-1' may be determined. Since no current enters the Op-Amps, we have

$$i_{\text{test}} = \frac{v_{\text{test}} - v_{02}}{R_4}$$

Since v_{p2} of the Op-Amp OA_2 forces v_{n2} to the test voltage v_{test}, KCR at v_{n2} gives us

$$\frac{v_{\text{test}} - v_{02}}{1/sC} + \frac{v_{\text{test}} - v_{01}}{R_3} = 0$$

It is clear that the Op-Amp OA_1 behaves like a non-inverting amplifier, and hence

$$v_{01} = \left(1 + \frac{R_2}{R_1}\right) v_{\text{test}}$$

Eliminating v_{01} and v_{02} from the above three equations, it is easy to see that

$$i_{\text{test}} = \frac{v_{\text{test}}}{s R_1 R_3 R_4 C / R_2}$$

and hence the port 1-1' sees an equivalent impedance of

$$Z_{\text{eq}} = s \left(\frac{R_1 R_3 R_4}{R_2} C \right)$$

implying that it is an inductive impedance with

$$L_{\text{eq}} = \frac{R_1 R_3 R_4}{R_2} C \text{ henrys}$$

Quite obviously, we call this circuit an *inductance simulator*.

Exercise 12.19 Verify the units of L_{eq} in the above equation. Also, for the circuit in Fig. 12.27, determine the resistances and the capacitance required to emulate an inductance of 1 mH.

It is indeed a tricky thing that an Op-Amp displaces an inductor; we may as well build a circuit that displaces a capacitor. We may justify this intuitively as follows. First, let us recall that an amplifier needs external dc supply from which it absorbs energy. This energy is, in turn, dissipated in the surrounding passive circuit elements, particularly the resistances; we have seen this in the examples on loading. This is similar to the behaviour of inductances and capacitances, which absorb energy during one part of a cycle, and release energy during the remaining part of the cycle. In fact, we can force an Op-Amp to deliver more energy than is actually absorbed by the resistances. Owing to this behaviour, an amplifier is also an active element, just like the independent and controlled sources. And, the filter circuits incorporating Op-Amps are called *active filters*.

It is also important to note the following. Since the Op-Amp is the most basic element of an active filter, limitations of Op-Amps dictate the performance of filters. A noteworthy limitation is the restriction of these filter circuits to operate below the megahertz range, which includes the communication systems and instrumentation systems. And, it is precisely at this point that the bulky inductors lose against miniature integrated circuits. Beyond this frequency range, inductors take over again, and their size and cost are justified at very high frequency applications.

In this section, we have seen several circuits that incorporate Op-Amps along with resistances and capacitances. We strongly advise the reader to obtain the Bode plots of the transfer functions of the circuits presented in the examples, exercises, and problems in this chapter. Any further theory of operational amplifiers and their applications is beyond the scope of this textbook.

Summary

In this chapter we have introduced another circuit element called operational amplifier. Though this is a complex circuit containing several transistors, resistors, and capacitors, its size encourages us to think of it as another circuit element. The term operational amplifier, or Op-Amp for short, was coined by J.R. Ragazzini in 1947 to denote a special type of amplifier that could be configured for basic mathematical operations such as addition, subtraction, differentiation, and integration. Analog computers, which basically solve linear differential equations, were the first application of Op-Amps.

It should be remembered that no Op-Amp works without an external dc supply; they require a *constant* source of energy. However, to avoid clutter in drawing circuits, we avoid showing the external power supplies; the student has to pay attention to this aspect while working in the laboratory. A very important concept called *negative feedback* is introduced in this chapter. It is this negative feedback that allows us to build several interesting circuits such as the instrumentation amplifier. We have also seen the rule that governs an Op-Amp: $v_n = v_p$ and no current enters inside; there is a *virtual short* at the input port of the Op-Amp. Subsequently, we have seen the inverting and the non-inverting configurations of an amplifier, which lend themselves to active filters, including inductance simulators. We have also seen that together with the Op-Amp rule, we may utilize our earlier techniques such as nodal analysis, the principle of linearity, and Thèvenin's equivalents for the analysis.

This chapter is intended to provide an introduction to the world of modern electronic circuits. Further details are avoided here, and the reader is encouraged to look into textbooks that are dedicated to Op-Amps. It is hoped that the reader will find this chapter interesting to refer to while doing advanced courses in later semesters.

Problems

12.1 For the circuit shown in Fig. 12.28, compute the output v_0. Also determine the input resistance R_i as seen by the source.

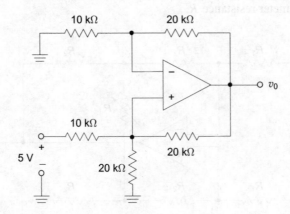

Fig. 12.28

12.2 Compute the gain v_0/v_i of the Op-Amp circuit given in Fig. 12.29. Also, determine the input resistance R_i.

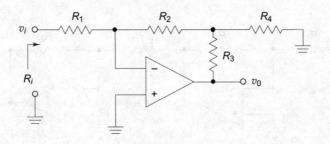

Fig. 12.29

12.3 What is the gain v_0/v_i of the circuit shown in Fig. 12.30?

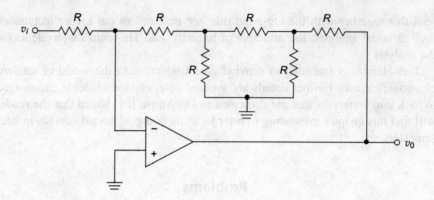

Fig. 12.30

12.4 For the circuit in Fig. 12.31, show that the gain can be varied using the potentiometer resistance R.

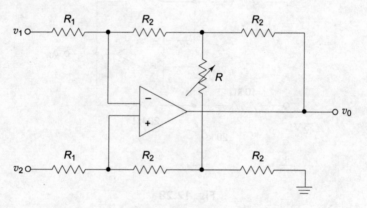

Fig. 12.31

12.5 Determine the conditions under which the circuit in Fig. 12.32 serves as an instrumentation amplifier.

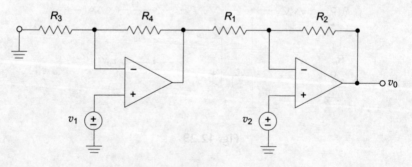

Fig. 12.32

12.6 Show that the gain of the circuit given in Fig. 12.32 can be varied using a potentiometer between the inverting input nodes of both the Op-Amps. Sketch the modified circuit and obtain the conditions under which the circuit serves as an instrumentation amplifier together with an expression for the gain.

12.7 Show that the circuit in Fig. 12.33 is an instrumentation amplifier.

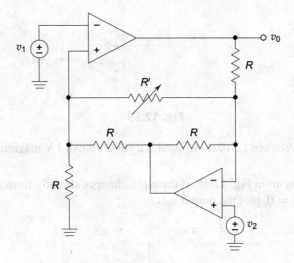

Fig. 12.33

12.8 Consider the first-order circuit given in Fig. 12.34.

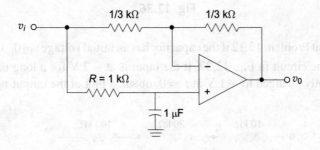

Fig. 12.34

If v_i is a unit step input and the capacitor has an initial voltage of $v_C(0^-) =$ 0.5 V, compute the (i) natural, (ii) forced, (iii) transient, and (iv) steady-state responses.

12.9 Repeat Problem 12.8 for the circuit in Fig. 12.34 with the components R and C interchanged.

12.10 In the circuit shown in Fig. 12.35, if v_i is a single 1 V pulse of width 5 ms, obtain a plot of the output $v_0(t)$.

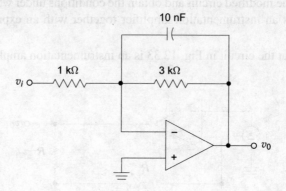

Fig. 12.35

12.11 Repeat Problem 12.10 if the input is a pulse train of 1 V magnitude and 10 ms period.

12.12 In the circuit in Fig. 12.36, if the input changes abruptly from +1 V to −1 V at time $t = 0$, plot the output $v_0(t)$.

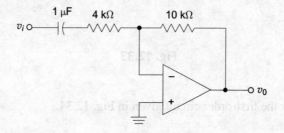

Fig. 12.36

12.13 Repeat Problem 12.12 if the capacitor has an initial voltage $v_C(0^-) = -0.5$ V.

12.14 For the circuit in Fig. 12.37, if the input is at − 2 V for a long time and has abruptly changed to + 1 V at $t = 0$, obtain a plot of the output $v_0(t)$.

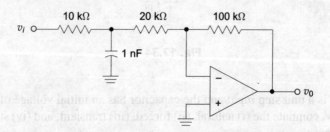

Fig. 12.37

12.15 If the input is a pulse train of magnitude 2 V and period 1 ms, sketch the output $v_0(t)$ for the circuit shown in Fig. 12.38.

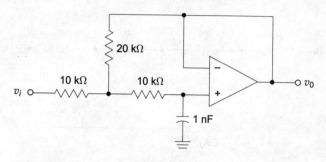

Fig. 12.38

12.16 If the initial capacitor voltage $v_C(0^-) = 1$ mV, plot the output $v_0(t)$ for the circuit shown in Fig. 12.39.

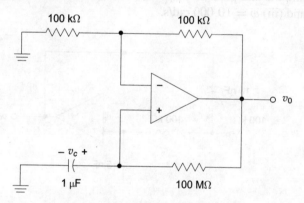

Fig. 12.39

12.17 Obtain the ordinary differential equation relating the input $v_C(t)$ and the output $v_0(t)$ for the circuit of Problem 12.16. What is your observation?

12.18 Show that the circuit in Fig. 12.40 is an integrator of the inverting type.

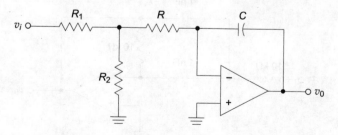

Fig. 12.40

12.19 Show that the circuit in Fig. 12.41 simulates an inductance at its input port.

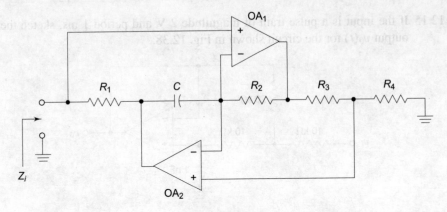

Fig. 12.41

12.20 Consider the circuit shown in Fig. 12.42. If the input is a sinusoid of amplitude 1 V, compute and plot the output $v_0(t)$ if (i) $\omega = 100$ rad/s, (ii) $\omega = 1000$ rad/s, and (iii) $\omega = 10,000$ rad/s.

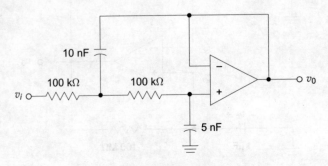

Fig. 12.42

12.21 Repeat Problem 12.20 for the circuit in Fig. 12.43.

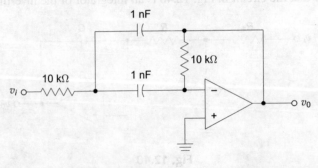

Fig. 12.43

12.22 Obtain the transfer function v_0/v_i of the circuit shown in Fig. 12.44, and hence show that it is a band-pass filter.

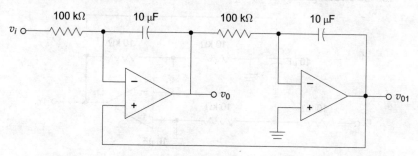

Fig. 12.44

12.23 Find the natural frequency ω_n and the quality factor Q of the circuit shown in Fig. 12.45.

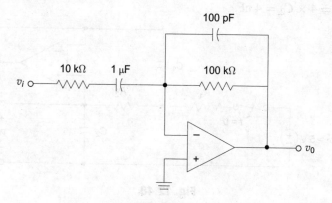

Fig. 12.45

12.24 Repeat Problem 12.23 for the circuit given in Fig. 12.46.

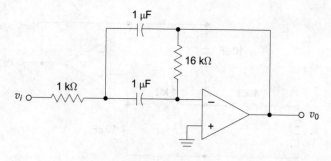

Fig. 12.46

12.25 Obtain the transfer function $H(s)$ and hence the location of zeros and poles for the circuit given in Fig. 12.47.

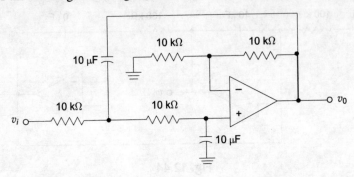

Fig. 12.47

12.26 Assuming that the capacitors are initially discharged, obtain and plot the output $v_0(t)$ for the circuit shown in Fig. 12.48, if $R_1 = 1 R_2 = 100\,\text{k}\Omega$ and $C_1 = 4 \times C_2 = 4\,\text{nF}$.

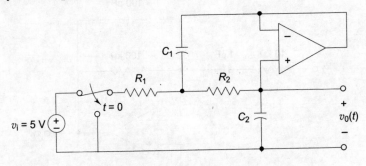

Fig. 12.48

12.27 Repeat Problem 12.26, if (i) $C_2 = 4 \times C_1 = 4\,\text{nF}$.

12.28 Obtain a plot of $v_0(t)$ for the circuit in Fig. 12.49, if the capacitors are initially discharged and

$$v_i(t) = 5[1 + \cos(10,000t)]\ \text{V}$$

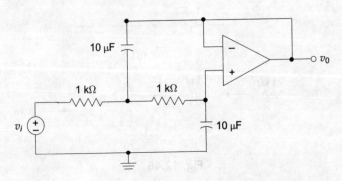

Fig. 12.49

12.29 Repeat Problem 12.28 with the rightmost resistance and capacitance interchanged.

12.30 Sketch the pole-zero plot and the Bode plot of the circuit shown in Fig. 12.50. Repeat the problem with another $100\,k\Omega$ resistance across the capacitance.

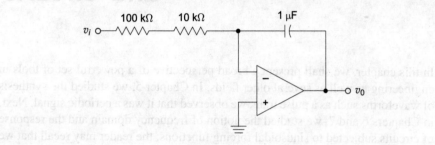

Fig. 12.50

CHAPTER 13

Fourier Analysis of Signals and Circuits

In this chapter, we shall present a broad perspective of a powerful set of tools in engineering as well as several other fields. In Chapter 5, we studied the synthesis of waveforms such as a pulse train; we observed that it was a periodic signal. Next, in Chapters 6 and 7, we studied the notion of frequency domain and the response of circuits subjected to sinusoidal forcing functions; the reader may recall that we have applied the principle of linearity in a restricted sense that all the sources in the circuit were assumed to have the same frequency. Further, we just remarked that it was interesting to study sinusoids since nature has several phenomena, such as the vibrations of a guitar string, that resemble sinusoids; we did not provide any stronger reason to study sinusoidal response. Subsequently, we have generalized the idea of complex exponentials and phasors and studied Laplace transformation in order to fit both transient and steady-state responses in the same framework.

In this chapter, we first show that any arbitrary signal, subject to certain conditions, can be expressed as an aggregation of an infinite number of sinusoids, each of which has a different fundamental frequency. We broadly classify signals as either periodic or non-periodic. We shall then study the steady-state response, ignoring the transients, to an arbitrary input that is ideally assumed to start at $t = 0$ and last forever. In so doing, we shall exploit the linearity property of the circuits and show that the response is also an aggregation of the circuit's responses to the individual sinusoids. The content of this chapter is extremely useful in a number of fields; for instance, in the study of harmonics in power systems and power electronics. We begin with Fourier series, particularly applicable to periodic waveforms. We shall then extend this to non-periodic waveforms and discuss the Fourier integral. Subsequently, we shall look into the applicability of these techniques in circuit analysis.

13.1 The Fourier Series

Historically, in the middle of the eighteenth century much attention was given to the problem of obtaining the mathematical laws governing the motion of a vibrating string with fixed endpoints, such as the one shown in Fig. 13.1.

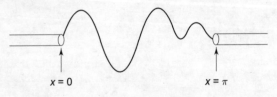

$x = 0$ $x = \pi$

Fig. 13.1

If $y(x, t)$ denotes the ordinate of the string at time t above the point x, then $y(x, t)$ satisfies the so-called *wave equation*:

$$\frac{\partial^2 y}{\partial t^2} = a^2 \frac{\partial^2 y}{\partial x^2}$$

where a is the parameter that depends on the tension of the string.

In 1747 d'Alembert showed that this motion may be represented as

$$y(x, t) = \frac{1}{2} [f(t + x) + g(t - x)]$$

where f and g are 'any' functions of one variable. In some sense, Bernoulli brought in trigonometric functions, particularly the sinusoids, and suggested that

$$y = \sum_{j=1}^{\infty} a_j \sin(jx)\cos(jt)$$

However, there was a controversy in this, since mathematicians of the day believed that d'Alembert's arbitrary function f could not be represented as a sum of 'continuous', 'differentiable', and 'periodic' trigonometric functions. Later, in 1820, this controversy was resolved as a result of two important events. First, it was Fourier[1] (1768–1830) who gave a formal method of expanding an arbitrary function f into a trigonometric series, called the *Fourier series*. He computed some partial sums for some sample f's and verified that they gave very good approximations to f. Then, it was Dirichlet[2] (1805–1859) who proved the first theorem in 1828, giving sufficient and very general conditions for the Fourier series of a function f to converge pointwise to f. He was one of the first to formalize the notions of partial sum and convergence of a series, and his ideas certainly had antecedents in the works of Gauss and Cauchy. In the following sections, we shall delve a little deeper into these ideas with appropriate examples. It is not possible, and it is not the context, to provide more rigorous treatment of Fourier analysis in this textbook. Interested readers may refer to books on *harmonic analysis*.

Fourier analysis is the representation of a signal as an aggregation, i.e., a sum or an integral of its *components* at various frequencies. Fourier series are infinite-series representations[3].

Example 13.1 A bar of length L shown in Fig. 13.2 is heated to an initial temperature profile $f(x)$, $0 \leq x \leq L$, then allowed to cool by conduction only through the ends (with insulation provided around the bar), where the temperature is held at zero for $t > 0$.

---

[1] Jean Baptiste Joseph Fourier was a professor of mathematics at Ecole Polytechnique, Paris.
[2] Peter Gustav Lejeune Dirichlet was a professor of mathematics at Breslau, and later at Göttingen.
[3] This prompts us to think in terms of convergence and divergence.

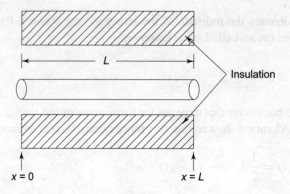

Fig. 13.2

From physics, solving the mentioned-above wave equation with appropriate
boundary conditions, we see that the temperature profile may be expressed as

$$f(x) = \sum_{k=1}^{\infty} b_k \sin\left(\frac{k\pi x}{L}\right)$$

for suitable constants b_k.

In what follows, we will first look at the *series* representation of signals defined
over a finite time interval. Next, we extend this idea to periodic signals. We already
know that a periodic signal is one that satisfies the property

$$f(t \pm qT_p) = f(t)$$

where T_p is called the fundamental period and q is any natural number. Examples
of periodic signals include the sinusoids of the form $A\cos(\omega t + \phi)$ and complex
exponentials. Of course, there are many more periodic signals such as the pulse
train (also called rectangular) and the triangular (also called the saw-tooth owing to
its appearance) signals. These signals are easily produced by function generators
readily available to us in the laboratory. We may immediately observe that there is
no resemblance between sinusoids and other periodic signals. And, isn't it exciting
that we will be able to *express* such periodic signals in terms of sinusoids?

We may define non-periodic signals as those which are *not* periodic. Typically,
speech signals are non-periodic, unless the speaker tends to utter the same
word/sentence/paragraph again and again. The pulse we have built earlier using
two-step signals is a non-periodic signal. After studying periodic signals, we shall
investigate the representation of non-periodic signals.

13.2   Orthogonality Relations

As it is, it might look absurd to talk about the orthogonality (or perpendicularity)
of two (or more) *functions*. We cannot readily visualize such things unlike a pair of
straight lines. Intuitively, what we are referring to here is something like a *cosine*
axis and *sine* axis that are orthogonal to each other (just like the x axis and y axis)

and hence the representation of functions as geometric curves in the cosine-sine space. This is what Fourier could see, and we shall understand this concept carefully in the following sections. Inquisitive readers should refer to good books on linear algebra and functional analysis.

Let us quickly recollect the following results from elementary calculus.

For $m, n = 0, 1, 2, \ldots$,

$$\frac{1}{\pi} \int_{-\pi}^{+\pi} \cos(m\,t)\,\cos(n\,t)\,dt = \begin{cases} 1 & \text{if } m = n \neq 0 \\ 0 & \text{if } m \neq n \end{cases}$$

$$\frac{1}{\pi} \int_{-\pi}^{+\pi} \sin(m\,t)\,\sin(n\,t)\,dt = \begin{cases} 1 & \text{if } m = n \neq 0 \\ 0 & \text{if } m \neq n \end{cases} \qquad (13.1)$$

$$\frac{1}{\pi} \int_{-\pi}^{+\pi} \sin(m\,t)\,\cos(n\,t)\,dt = 0 \ \forall \ m, n$$

Suppose we have a signal $f(t)$ $t \in I = [-\pi, \pi]$. Just as the equation

$$lx + my + n = 0$$

represents the locus of a straight line in the x-y plane, we may represent the signal $f(t)$ in the cosine-sine plane as

$$f(t) = \frac{a_0}{2} + \sum_{k=1}^{\infty} (a_k \cos kt + b_k \sin kt), \quad k \text{ is a natural number} \quad (13.2)$$

where a_0, a_k, and b_k are suitable coefficients called *Fourier coefficients*. It would be useful to view the Fourier representation as a mapping:

$$f(t) \xrightarrow{\text{FS}} \{a_0, \ldots, a_k, \ldots, b_k, \ldots\}$$

i.e., Fourier series map a continuous time function $f(t)$ to a set of real numbers.

13.3 Computing Fourier Coefficients

We shall use the orthogonality relations to derive the coefficients. Multiplying both sides of Eqn (13.2) by cos mt or sin mt and integrating on the interval I, we get[4]

$$a_k = \frac{1}{\pi} \int_{-\pi}^{+\pi} f(t) \cos(k\,t)\,dt, \quad k = 0, 1, 2, \ldots$$

$$b_k = \frac{1}{\pi} \int_{-\pi}^{+\pi} f(t) \sin(k\,t)\,dt, \quad k = 1, 2, \ldots \qquad (13.3)$$

Exercise 13.1 Verify Eqn (13.3) using the orthogonality relations.

---

[4] This is much similar to finding l, m, and n for a given straight line, e.g., $-n/m$ is the intercept on the y axis putting $x = 0$, and $-n/l$ is the intercept on the x axis putting $y = 0$. Notice that putting $x = 0$ or $y = 0$, as the case may be, works since the two axes are mutually orthogonal.

Example 13.2 Let us have a signal

$$f(t) = \begin{cases} 0, & -\pi \le t \le 0 \\ \dfrac{t}{\pi}, & 0 \le t \le \pi \end{cases}$$

shown in Fig. 13.3

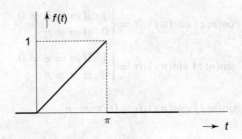

Fig. 13.3

$$\frac{a_0}{2} = \frac{1}{2\pi} \int_0^\pi \frac{t}{\pi} \, dt$$

$$= \frac{1}{4}$$

$$a_k = \frac{1}{\pi} \int_0^{+\pi} \frac{t}{\pi} \cos(k\,t) \, dt$$

$$= \begin{cases} -\dfrac{2}{k^2\pi^2}, & n \text{ odd} \\ 0, & n \text{ even} \end{cases}$$

$$b_k = \frac{1}{\pi} \int_0^{+\pi} \frac{t}{\pi} \sin(k\,t) \, dt$$

$$= \frac{(-1)^{k+1}}{k\pi}$$

Thus,
$$f(t) = \frac{1}{4} + \left(-\frac{2}{\pi^2} \cos t + \frac{1}{\pi} \sin t \right) - \frac{1}{2\pi} \sin 2t$$

$$+ \left(-\frac{2}{9\pi^2} \cos 3t + \frac{1}{3\pi} \sin 3t \right) - \frac{1}{4\pi} \sin 4t + \cdots$$

Exercise 13.2 Verify the results of Example 13.2.

13.3.1 Fourier Series of a Periodic Signal

Definition 13.1
Let the signal $f(\cdot)$ have domain $[-\pi, +\pi]$. The periodic extension of $f(\cdot)$ is the signal defined as follows.

$$f_p(t) \triangleq \begin{cases} f(t), & -\pi \le t < +\pi \\ f(t - 2k\pi), & (2k-1)\pi \le t < (2k+1)\pi \end{cases} \tag{13.4}$$

$$k = \pm 1, \pm 2, \ldots$$

This is shown in Fig. 13.4.

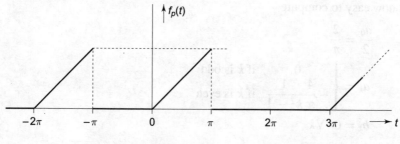

Fig. 13.4

Definition 13.2
The Fourier series for $f_p(\cdot)$ is the same as the Fourier series for $f(\cdot)$, but now with extended domain $-\infty < t < +\infty$.

This definition is justified since each term of the Fourier series is periodic with 2π.

Example 13.3 Let $u(t)$, $t \ge 0$ be a sinusoid of unit amplitude, i.e.,

$$u(t) = \sin(\omega t), \quad t \ge 0$$

The output of a full-wave rectifier, with $u(t)$ as input, will be

$$y(t) = |\sin(\omega t)|, \quad t \ge 0$$

It is useful to obtain a picture of this signal as shown in Fig. 13.5.

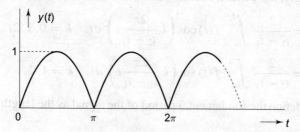

Fig. 13.5

The fundamental period of this signal is π/ω. Hence, we need to *scale* the interval of domain to $[-\pi/\omega, \pi/\omega]$. Accordingly,

$$a_k = \frac{2\omega}{\pi} \int_{-\pi/\omega}^{+\pi/\omega} f(t) \cos(k\,\omega t)\, dt, \quad k = 0, 1, 2, \ldots$$

$$b_k = \frac{2\omega}{\pi} \int_{-\pi/\omega}^{+\pi/\omega} f(t) \sin(k\,\omega t)\, dt, \quad k = 1, 2, \ldots$$

It is now easy to compute

$$\frac{a_0}{2} = \frac{2}{\pi}$$

$$a_k = \begin{cases} 0 & \text{if } k \text{ is odd} \\ -\dfrac{4}{\pi}\dfrac{1}{k^2 - 1} & \text{if } k \text{ is even} \end{cases}$$

$$b_k = 0 \quad \forall k$$

Thus,

$$f(t) = \frac{2}{\pi} - \frac{4}{\pi} \sum_{k=1}^{\infty} \frac{1}{4k^2 - 1} \cos(2k\omega t)$$

We leave it to the reader to reason out why there are no 'sin( )' terms in the series.

Exercise 13.3 Verify the results of Example 13.3.

The above example is a good instance of a periodic signal on an arbitrary finite interval. In general, if $f(t)$ is defined on $[t_1, t_2]$, we may rescale the Fourier coefficients by first rescaling the orthogonality relations. Then, we *define* the Fourier series for $f(t)$ on $[t_1, t_2]$ as

$$f(t) = \frac{a_0}{2} + \sum_{k=1}^{\infty} \left[a_k \cos\left(k\frac{2\pi}{t_2 - t_1} t \right) + b_k \sin\left(k\frac{2\pi}{t_2 - t_1} t \right) \right] \quad (13.5)$$

Accordingly, the general Fourier coefficients are

$$a_k = \frac{2}{t_2 - t_1} \int_{t_1}^{t_2} f(t) \cos\left(k\frac{2\pi}{t_2 - t_1} t \right) dt, \quad k = 0, 1, 2, \ldots$$

$$b_k = \frac{2}{t_2 - t_1} \int_{t_1}^{t_2} f(t) \sin\left(k\frac{2\pi}{t_2 - t_1} t \right) dt, \quad k = 1, 2, \ldots \quad (13.6)$$

We may define the fundamental period of the signal as the length of the interval $[t_1, t_2]$, i.e.,

$$T = t_2 - t_1$$

Exercise 13.4 Verify formally that the expressions just given are correct.

Example 13.4 Consider the periodic rectangular signal shown in Fig. 13.6.

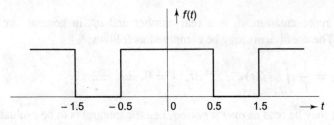

Fig. 13.6

The signal is periodic with fundamental period $T = 2$. Hence, we may obtain the following Fourier series:

$$f(t) = \frac{1}{2} + \sum_{k=1,3,5,\dots}^{\infty} \frac{2}{k\pi} \cos\left[k\pi t + \{(-1)^{(k-1)/2} - 1\}\frac{\pi}{2}\right]$$

Exercise 13.5 Verify the results of Example 13.4. Also, discuss why the coefficients b_k vanish. Under what conditions do the coefficients a_k vanish?

13.4 Complex Exponential Fourier Series

As we have seen in Chapters 6 and 7, for the ease of computation, sinusoidal signals are often expressed in complex exponential form. This is allowed because the exponential system

$$1, \ e^{\pm jt}, \ e^{\pm j2t}, \dots, \quad t \in [-\pi, \pi]$$

is also an *orthogonal* sequence just as the trigonometric system we have been dealing with so far. We use Euler's formula to convert the trigonometric Fourier series into complex exponential Fourier series (CEFS) as follows.

Definition 13.3 Complex exponential Fourier series (CEFS)
Let $f(t)$ be defined on $[t_1, t_2]$. The CEFS of $f(t)$ is the series

$$f(t) = \sum_{k=-\infty}^{+\infty} c_k e^{jk[2\pi/(t_2-t_1)]t}$$

$$= \sum_{k=-\infty}^{+\infty} c_k e^{jk\omega_0 t}, \quad -\infty < t < \infty \qquad (13.7)$$

where ω_0 is the *fundamental frequency* (in rad/s) related to the fundamental period T as

$$\omega_0 = \frac{2\pi}{T} = \frac{2\pi}{t_2 - t_1}$$

In this representation, c_0 is a real number and c_k, in general, are complex numbers. The coefficients may be computed as follows:

$$c_k = \frac{1}{T} \int_{\langle T \rangle} f(t) e^{-jk\omega_0 t} \, dt, \quad k = 0, \pm 1, \pm 2, \ldots \tag{13.8}$$

where $\langle T \rangle$ may be read as *over a period*, i.e., the integral is to be evaluated over a period. Note the appearance of frequency components at the *negative frequencies* $-k\omega_0$. These negative frequencies do not have any physical meaning; rather, they are a result of the mathematical formulation. Sometimes it is useful to call this CEFS as the two-sided Fourier series.

Example 13.5 We revisit Example 13.2. For the CEFS, we have

$$c_k = \begin{cases} \dfrac{1}{4}, & k = 0 \\[2ex] \dfrac{1}{2k^2\pi^2}\left[(-1)^k(1 + jk\pi) - 1\right], & k \neq 0 \end{cases}$$

Notice that the cases $k = 0$ and $k \neq 0$ should be dealt with separately. Comparing with the results in Example 13.2, we find that

$$|c_k| = \sqrt{a_k^2 + b_k^2}, \quad \angle c_k = \arctan(b_k/a_k)$$

Exercise 13.6 Obtain the CEFS of the periodic signal in Example 13.4.

Exercise 13.7 Let $f(t)$ be periodic with T, for $-\infty < t < \infty$. Show that for arbitrary $t_0 \in (-\infty, +\infty)$,

$$\int_{t_0}^{t_0+T} f(t) \, dt = \int_{-T/2}^{+T/2} f(t) \, dt$$

There is no 'rule' to dictate which of FS or CEFS is to be preferred. This is largely a matter of taste and/or convenience; mathematically, the two representations are equivalent.

The Fourier series representation of a periodic signal is a remarkable result. However, we might become restless if we ask ourselves 'Can every (periodic) signal be represented as a Fourier series?' It is no surprise that Fourier himself believed that *any* periodic signal could be expressed as a sum of complex exponentials. Unfortunately, this turned out not to be the case, although virtually all periodic signals of practical interest do have a Fourier series representation. It is hard to judge when Fourier series *converge*. The first correct proof of convergence was done by Dirichlet, almost eight years after Fourier series first came into existence. It is a long story, which we summarize below. The proof is beyond the scope of the present textbook.

13.4.1   Dirichlet's Conditions

A periodic signal $f(t)$ has a Fourier series if it satisfies the following conditions.

1. $f(t)$ is absolutely integrable over any period, i.e.,

$$\int_{t_0}^{t_0+T} |f(t)| \, dt \ < \ \infty$$

for any arbitrary t_0.

2. $f(t)$ has only a finite number of maxima and minima over any period.
3. $f(t)$ has only a finite number of discontinuities over any period.

It is *counter-intuitive* to think of expressing periodic signals, such as the rectangular pulses, with 'corners' as a sum of sinusoids since the sinusoids are infinitely 'smooth' functions. In fact, it was the controversy that followed Bernoulli's work. However, the key here is that the summation is over an infinite number of terms. Thus, for a suitably large value of N, the Fourier series *should be a close* approximation to $f(t)$. It is at this juncture that certain interesting things happen. If $f(t)$ is discontinuous at $t = t_1$, the Fourier summation exhibits an *overshoot* of approximately 9% at t_1^- and t_1^+. In fact, this 9% overshoot is present even in the limit as N approaches ∞. This characteristic was first discovered and mathematically demonstrated by Josiah Willard Gibbs in 1899, and thus the overshoot is referred to as the *Gibbs phenomenon*. More details on this are avoided here for the sake of continuity.

Consider a periodic signal with the complex exponential Fourier series. The frequency components comprising this signal are $0, \omega_0, 2\omega_0, \ldots$ The zero frequency component represents the dc part; ω_0 is called the fundamental, and $k\omega_0$, $k = 2, 3, \ldots$ are called the harmonics. These frequency components may be *displayed* in terms of amplitude $|c_k|$ versus $\omega = k\omega_0$, and phase $\angle c_k$ versus $\omega = k\omega_0$ for all k. These figures are called *magnitude spectrum* and *phase spectrum*, respectively. Strictly speaking, we need to add an adjective—two-sided—since we are interested in the *negative* frequencies also. The word *spectrum* suggests a range of frequency components. In short, given a signal, the spectra readily give us information about the range of frequencies contained together with their relative *strengths*.

It would be helpful to look at these spectra as a mapping:

$$f(t) \ \overset{\text{FS}}{\longrightarrow} \ \{c_0, \pm c_1, \ldots, \pm c_k, \ldots\} \ \longrightarrow \ \text{spectra}$$

Example 13.6 For the CEFS of Example 13.5, the frequency components are displayed in Figs 13.7(a) and 13.7(b).

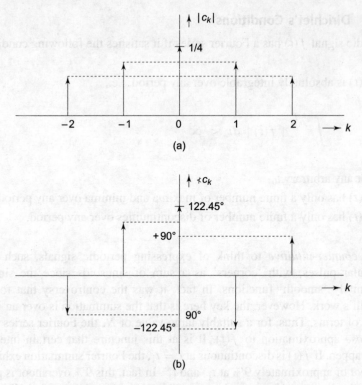

Fig. 13.7

The 'average power' of a periodic signal $f(t)$ is defined by

$$P \triangleq \frac{1}{T} \int_{-T/2}^{+T/2} |f(t)|^2 \, dt$$

Notice that, if $f(t)$ is the voltage across (or the current through) a 1 Ω resistor, the average power delivered to the resistor would be given by the expression above. We say that a signal $f(t)$ is a *power signal* if and only if the average power of the signal is bounded, i.e.,

$$0 < P < \infty$$

In general, periodic signals are power signals.

Parseval's formula The average power of a periodic signal $f(t)$ is given by

$$P = \sum_{k=-\infty}^{+\infty} |c_k|^2 \tag{13.9}$$

In words, Parseval's formula says that the total power of a periodic signal $f(t)$ is the sum of the powers carried by the distinct sinusoidal frequency components of $f(t)$. Parseval's formula is important in applications because it allows one to calculate the power in terms of the Fourier coefficients and accordingly design systems. The proof of this result is omitted here for brevity. The reader may verify the result.

Example 13.7 Let us determine the average power of the signal of Example 13.5.

$$P = \frac{1}{2} \int_{-1/2}^{+1/2} dt = \frac{1}{2} \text{ W}$$

The Fourier coefficients c_k are as follows:

$$c_0 = \frac{1}{2}$$

and

$$c_{(2k+1)} = \pm \frac{1}{(2k+1)\pi} = c_{-(2k+1)}$$

where '+' and '−' appear alternately.
According to Parseval's formula,

$$P = \left(\frac{1}{2}\right)^2 + \frac{2}{\pi^2}\left\{1 + \frac{1}{3^2} + \frac{1}{5^2} + \cdots\right\} \text{ W}$$

We observe that 50% of the average power (= 0.25 W) is concentrated in the dc component itself. An additional component (ω_0) contributes to 90% of the average power; another component ($2\omega_0$) contributes to about 95%, and so on. This suggests that the original signal may be approximated to a very good accuracy by the zero-frequency component plus the fundamental plus two or three harmonics. In other words, the Fourier series of a signal is an indication of how the power associated with a periodic signal distributes among its harmonics. This idea helps us in designing systems in the frequency domain.

In this section we have seen the representation of arbitrary periodic signals as an *infinite* summation of sinusoids of *different* frequencies. Before we move on to the next section to represent non-periodic signals, we shall memorize the following important observation: *the spectrum of a periodic signal consists of an infinite number of discrete frequencies, each of which is an integer multiple of the fundamental frequency.*

Equivalently, a given periodic signal does not contain *all* frequencies, but only a countable infinity of *discrete* frequencies. We say the signal has *discrete* spectra.

Example 13.8 Consider the following sawtooth signal in Fig. 13.8 whose period is 1 s.

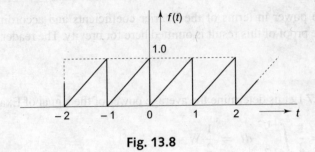

Fig. 13.8

We leave it to the reader to verify that the CEFS of the signal is

$$f(t) = \frac{1}{2} + \sum_{k \neq 0} \frac{j}{2k\pi} e^{j2k\pi t}$$

The discrete spectra are shown in Fig. 13.9. Notice that the phase angle of c_0 is zero, but $\angle c_k, = 90°$ for $k > 0$, and $\angle c_k, = -90°$ for $k < 0$.

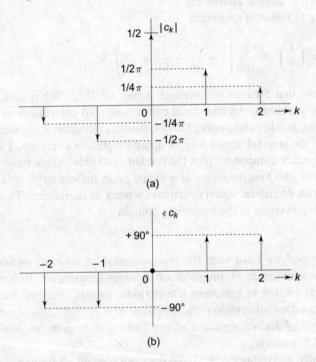

Fig. 13.9

Let us now see what happens if we apply the sawtooth signal of Fig. 13.8 to a simple RC circuit shown in Fig. 13.10. We assume that the capacitor is initially discharged. For the sake of computational simplicity, let us take $\tau = RC = 1/2\pi$.

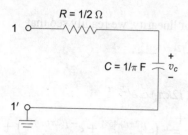

Fig. 13.10

The idea is not to see the input signal as a sawtooth, but as a *superposition* of several sinusoids. We need not, once again, explain that the circuit is a linear circuit and hence the principle of linearity can be applied. We just imagine that across the input port 1-1′ there are infinite number of voltage sources enumerated in terms of the Fourier coefficients c_k. And, we now compute the circuit's response, i.e., the voltage across the capacitor, due to each of the sources.

First, there is a dc source $c_0 = 1/2$ V. Since the capacitor acts like an open circuit in this case, we simply write[5]

$$c_0 = \frac{1}{2}\,V \longrightarrow v_{C_0} = \frac{1}{2}\,V$$

Then, there is a sinusoid with the fundamental frequency $\omega = 2\pi$. From our knowledge of phasors, we may readily write

$$c_1 = \frac{j}{2\pi}e^{j2\pi t} \longrightarrow \mathcal{V}_{C_1} = \frac{\frac{1}{j\omega C}}{R + \frac{1}{j\omega C}}\bigg|_{\omega=2\pi} \quad \frac{1}{2\pi}\angle 90^\circ$$

i.e.,

$$\mathcal{V}_{C_1}(j\omega = j2\pi) = \frac{1}{1 + j1}\frac{j}{2\pi}$$

$$= \frac{1}{2\sqrt{2}\pi}\angle 45^\circ$$

Noting that $c_1 = 1/(2\pi)\angle 90^\circ$, we may rewrite the last expression for $\mathcal{V}_C$ as follows:

$$\mathcal{V}_{C_1}(j2\pi) = \left(\frac{1}{\sqrt{2}}\angle -45^\circ\right)c_1$$

$$= H(j2\pi)\,c_1$$

where $H(j2\pi)$ is the transfer function of the circuit evaluated at $\omega = 2\pi$. Recall that this is exactly what we have done in Chapter 6 using phasors. It is easy to generalize this and write

$$\mathcal{V}_{C_k}(j2k\pi) = H(j2k\pi)c_k \quad \text{for } k = 0, \pm 1, \pm 2, \ldots$$

---

[5] We may read it like an equation in chemistry using the phrase 'produces'!

Applying the principle of linearity, we readily see that

$$v_i(t) = \sum c_k e^{j2k\pi t} \longrightarrow$$

$$v_C(t) = \sum H(j2k\pi) c_k e^{j2k\pi t}$$

$$= \frac{1}{2} + \frac{1}{2\sqrt{2\pi}} \left(e^{j(2\pi t + 45°)} + e^{-j(2\pi t + 45°)} \right) + \cdots$$

$$= \frac{1}{2} + \frac{1}{\sqrt{2\pi}} \cos(2\pi t + 45°) + \cdots$$

i.e., the voltage across the capacitor is also a summation of several sinusoids. It is worthwhile to look at the results graphically, particularly from the frequency and phase points of view. First, the network function has the frequency function as shown in Fig. 13.11.

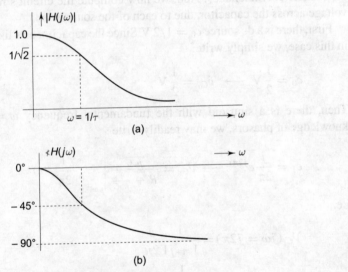

Fig. 13.11

Multiplying the magnitude plots of $H(j\omega)$ and $v_i(j\omega)$ point by point, we get the following magnitude plot of $v_C(j\omega)$ in Fig. 13.12.

Fig. 13.12

We see that the low-frequency components, particularly the dc component, are not affected by the circuit. In other words, the circuit allowed to pass the dc component through it without any change. The fundamental component was allowed to pass through with an attenuation of $1/\sqrt{2} = -3$ dB. Further, the harmonics at integer multiples of the fundamental have been attenuated gradually. The infinite summation is simply a multiplication owing to the fact that the spectrum of the signal $v_i(t)$ is discrete, i.e., the magnitude is zero for non-integral multiples of the fundamental. The phase plot $\angle v_C$ of the output may be obtained by superimposing the individual phase plots.

Exercise 13.8 Verify the results of Example 13.8. Obtain the phase plot.

Exercise 13.9 Repeat Example 13.8 with the voltage across the resistance as the output.

Exercise 13.10 Discuss what happens if we scale up/scale down the time-constant of the RC circuit in Example 13.8 and in Exercise 13.9.

Looking at the magnitude plots alone, it is easy to identify that, in fact, we have performed convolution in the frequency domain. We may also clearly see why the circuit may be called a *low-pass filter*. An ardent reader would question why the output signal is not shown as a function of time. We emphasize that in applications such as filtering the time function of the output is less important than its frequency content; this is what we mean by *Fourier analysis*. We are not just bothered about how the signal looks like with respect to time; perhaps, we would have used Laplace transformation for that. It is this property, i.e., mapping from time-domain to frequency domain, that distinguishes Fourier analysis from Laplace analysis.

13.5   The Fourier Integral

We now look at the Fourier representation of non-periodic signals. We begin the discussion with an intuitive example.

Example 13.9 The CEFS of $f(t) = \cos \omega_0 t$ is given by

$$f(t) = \frac{1}{2} e^{-j\omega_0 t} + \frac{1}{2} e^{j\omega_0 t}$$

and the spectrum appears as shown in Fig. 13.13.

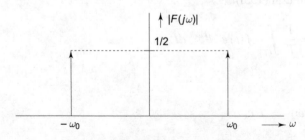

Fig. 13.13

This is a discrete spectrum (why?) displaying the *only* frequency content ω_0 of the given signal. Let us now consider two more sinusoids:

$$f_1(t) = \cos 0.1\, \omega_0\, t \quad \text{and} \quad f_2(t) = \cos 10\, \omega_0\, t$$

The spectrum for each of these signals is the same as shown in Fig. 13.13, except that for $f_1(t)$ the two-sided spectrum is *ten times* closer and for $f_2(t)$ the two-sided spectrum is *ten times* farther. It is easy to see that as $\omega_0 \to 0$, the two-sided spectrum collapses to a single impulse of strength unity at $\omega = 0$. This result is to be expected since as $\omega_0 \to 0$ the signal approaches a constant, i.e., a dc signal with zero frequency.

Let us spend a few minutes on this example asking the following questions.

1. Is a dc signal periodic?
2. Can we represent a dc signal as a Fourier series?

The answer to the first question is rather vague—it can be periodic with any arbitrary finite period, or it can be non-periodic without ever repeating over the entire interval $(-\infty < t < \infty)$. The answer to the second question is a clear 'no' since the signal is *not* absolutely integrable in the sense of Dirichlet. However, practically speaking, we can obtain a CEFS representation (with $c_0 = 1$ and $c_k = 0 \ \forall\, k \neq 0$) by treating the dc signal as a limiting case of a sinusoid.

We apply the same idea, henceforth, and consider non-periodic signals as the limiting cases of periodic signals with the period $T \to \infty$. This extension has practical advantages as well as theoretical interest—because for mathematical modeling purposes it may be much more convenient to consider a 'very large' interval like $[0, T)$ simply as the infinite interval $[0, \infty)$, in order not to have to worry about T at all. In fact, we will take the entire real line $(-\infty, \infty)$ as our interval.

If the fundamental frequency ω_0 is allowed to take arbitrarily small values, i.e., $\omega_0 \to 0$, an immediate question would be—'What happens to the *integer* multiples of ω_0?' The answer is simple—it approaches the continuum, i.e.,

$$\forall\ k = 0,\ \pm 1,\ \pm 2,\ \dots,\quad k\omega_0 \to \omega \in (-\infty, +\infty) \tag{13.10}$$

Let us now look at the extension more formally. We begin with the Fourier coefficients in Eqn (13.8):

$$c_k = \frac{1}{T} \int_{\langle T \rangle} f(t)\, e^{-jk\omega_0 t}\, dt$$

and define

$$F(j\omega) \overset{\Delta}{=} T\, c_k \tag{13.11}$$

Together with Eqn (13.10), we have the following definition:

Definition 13.4 The Fourier Integral

The Fourier integral or the Fourier transform[6] $F(\omega)$ of a signal $f(t)$ is the function

$$F(j\omega) = \mathcal{F}\{f(t)\}$$

$$\stackrel{\Delta}{=} \int_{-\infty}^{+\infty} f(t)\, e^{-j\omega t}\, dt \tag{13.12}$$

defined for all $\omega \in \mathfrak{R}$. We just remark that writing the Fourier transform as $F(\omega)$ instead of $F(j\omega)$ is also acceptable. However, in this textbook we prefer using $j\omega$ as the argument for certain obvious reasons to be discussed later in this section.

Thus, *the frequency components of non-periodic signals are defined for all real frequencies*, and not just for discrete values of ω as in the case of a periodic signal. Equivalently, the spectra for a non-periodic signal are, in general, continuous in ω.

Conversely, we may recompute the original signal $f(t)$ from the transform $F(j\omega)$ by simply plugging in Eqn (13.12) into the CEFS of Eqn (13.7). In so doing, we need to account for the scale of T in view of Eqn (13.11). Accordingly, the summation of sinusoids over the index k becomes an integration with respect to the frequency variable ω over the entire interval $(-\infty, +\infty)$,

$$f(t) = \mathcal{F}^{-1}\{F(j\omega)\}$$

$$\stackrel{\Delta}{=} \frac{1}{2\pi} \int_{-\infty}^{+\infty} F(j\omega)\, e^{j\omega t}\, d\omega \tag{13.13}$$

The relationship in Eqn (13.13) is referred to as the *inverse Fourier transform*. The pair of equations, Eqn (13.12) and Eqn (13.13), are together called the Fourier transform pair:

$$f(t) \leftrightarrow F(j\omega)$$

To obtain the inverse Fourier transform by direct evaluation of the integral in Eqn (13.13) is seldom feasible. Instead, we need to use contour integration in the complex plane. Alternatively, we might find the answer in any table of *integral transforms*. We assure the reader that the time-frequency mapping is one to one so that, in general, essentially distinct functions have distinct Fourier transforms.

In passing, we remark that Fourier transformation has a slightly distinct flavour compared to the Laplace transformation. Laplace transformation is a useful tool to tackle convolution integrals and ordinary differential equations in an easier manner; most of the time we just obtain the solution of a given problem without having to worry much about the physical interpretation. On the other hand, despite the similarity, Fourier transformation helps us to visualize the frequency content of an arbitrary signal. It is a common observation that if 'time' is associated with something, then 'frequency' is also associated. For instance, we go to a bus station to enquire either the time of arrival of a particular bus or the frequency of buses in that particular route. Thus, we look at Fourier transformation as a vehicle that transports us to the frequency domain. It is not very difficult to see that Laplace transformation is, in fact, a *generalization* of Fourier transformation.

---

[6] An inquisitive reader would quickly notice the resemblance of this expression to the Laplace transform. Take a careful note of the limits of integration.

The 'total energy' of a signal $f(t)$ is defined as

$$E \triangleq \int_{-\infty}^{+\infty} |f(t)|^2 \, dt$$

We say that a signal $f(t)$ is an *energy signal* if and only if the total energy of the signal is bounded, i.e., $0 < E < \infty$.

In general, non-periodic signals are energy signals, though one may find exceptions here and there. The energy and power classifications of signals are mutually exclusive. In particular, an energy signal has zero average power, while a power signal has infinite energy.

Exercise 13.11 Verify that an energy signal has zero average power and a power signal has infinite energy.

We will now present, without proof, what is referred to as Parseval's theorem.
Parseval's theorem

$$E = \int_{-\infty}^{+\infty} |f(t)|^2 \, (t) \, dt$$

$$= \frac{1}{2\pi} \int_{-\infty}^{+\infty} |F(j\omega)|^2 \, d\omega \qquad (13.14)$$

$$= \frac{1}{\pi} \int_{0}^{+\infty} |F(j\omega)|^2 \, d\omega$$

i.e., we can compute the total energy by essentially the same recipe (up to the scale factor 2π) in either the time or frequency domain. The proof of this theorem directly follows from the Fourier transform pair. We may interpret

$$\frac{|F(j\omega)|^2}{2\pi}$$

as the 'energy density' of $f(t)$ in the frequency domain.

In passing, we note that the above formulae give us the energy of a signal, in general. However, if we need to compute the energy dissipated by a passive element, we need to take the magnitude of its impedance into account. These relationships have already been discussed in Chapter 8.

13.5.1 Properties of the Fourier Transform

The Fourier transform satisfies a number of properties, similar to the properties of Laplace transform, which help us in computing the transform of arbitrary signals efficiently[7]. We may also use them in several applications as we see in the examples that follow.

---

[7] This is much similar to utilizing the properties of determinants to compute the determinant of a given square matrix efficiently.

1. *Linearity.*

$$\mathcal{F}\{\alpha\, f(t) + \beta\, g(t)\} = \alpha\, \mathcal{F}\{f(t)\} + \beta\, \mathcal{F}\{g(t)\}$$

2. *Shift in time.* For any positive real number c,

$$\mathcal{F}\{f(t \pm c)\} = F(j\omega)\, e^{\pm j\omega c}$$

3. *Time scaling.* For any positive real number a,

$$\mathcal{F}\{f(at)\} = \frac{1}{a}\, F\left(j\frac{\omega}{a}\right)$$

If $a > 1$, you *shrink* the time scale (i.e., speed up the things) or equivalently, you *stretch* the frequency scale. Similarly, if you *stretch* in time scale, you *shrink* the frequency scale.

4. *Time reversal.*

$$\mathcal{F}\{f(-t)\} = F(-j\omega)$$

5. *Multiplication by t^n.*

$$\mathcal{F}\{t^n\, f(t)\} = (j)^n\, \frac{d^n}{d\omega^n}\, F(j\omega)$$

6. *Multiplication by $e^{j\omega_0 t}$.*

$$\mathcal{F}\{e^{\pm j\omega_0 t}\, f(t)\} = F(j(\omega \pm \omega_0))$$

7. *Differentiation in time.*

$$\mathcal{F}\left\{\frac{d^n}{dt^n} f(t)\right\} = (j\omega)^n\, F(j\omega)$$

8. *Convolution in time.*

$$\mathcal{F}\{f(t) * g(t)\} = F(j\omega) \times G(j\omega)$$

9. *Multiplication in time.*

$$\mathcal{F}\{f(t) \times g(t)\} = F(j\omega) * G(j\omega)$$

10. *Duality.* If

$$f(t) \leftrightarrow F(j\omega)$$

is a transform pair, so is

$$F(t) \leftrightarrow 2\pi\, f(-\omega)$$

Exercise 13.12 Verify the above properties of the Fourier transform using the basic definitions in Eqns (13.12) and (13.13).

We can combine the operations of shifting and scaling in a sequence. For instance, if we shift, and then scale, we get

$$\mathcal{F}\{f(at - c)\} = \frac{1}{a}\, e^{-j\omega \frac{c}{a}}\, F\left(j\frac{\omega}{a}\right)$$

Exercise 13.13 A signal $f(t)$ starts at $t = 0$ and lasts exactly 4 min. Tape record it starting at $t = 0$, stop the tape 6 min later, go out for coffee for 8 min, then rewind the tape back at twice the recording speed. What is the Fourier transform $F(\omega)$ of the signal $f(t)$ you will hear during wind-back? Assume that the gadgetry and your ear operate at perfect fidelity at all frequencies and that no disturbing noise is present.

Example 13.10 Let us obtain the Fourier transform of a unit impulse signal.

$$\mathcal{F}\{\delta(t)\} = \int_{-\infty}^{+\infty} \delta(t) e^{-j\omega t} \, dt = 1$$

This is a remarkable result. A unit impulse signal has *all* the frequencies equally strong. The spectrum is shown in Fig. 13.14. This spectrum answers our question—'Why do household gadgets, without any exception, burn when there is a lightning?'.

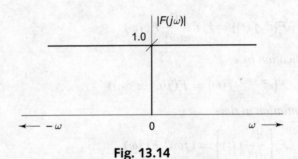

Fig. 13.14

Example 13.11 Using the duality property, we can now obtain the Fourier transform of a dc signal as

$$\mathcal{F}\{f(t) = 1 \ \forall t\} = 2\pi \, \delta(j\omega)$$

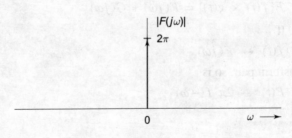

Fig. 13.15

The spectrum shown in Fig. 13.15. This may be expected because a dc signal has only one frequency, i.e., $\omega = 0$. The magnitude 2π suggests that it is strong enough and this answers our question—'Why is direct current sometimes fatal?'.

Example 13.12 Let us now turn our attention to the rectangular pulse shown in Fig. 13.16(a).

$$F(j\omega) = \int_{-a}^{+a} e^{-j\omega t}\, dt = 2a\, \frac{\sin a\omega}{a\omega} = 2a\, \text{sinc}\, \frac{a\omega}{\pi}$$

where we define sinc function as

$$\text{sinc}\,(x) \triangleq \frac{\sin \pi x}{\pi x}$$

The spectrum is shown in Fig. 13.16(b). By duality, we may observe that a pulse signal has a sinc Fourier transform and a sinc signal has a pulse Fourier transform.

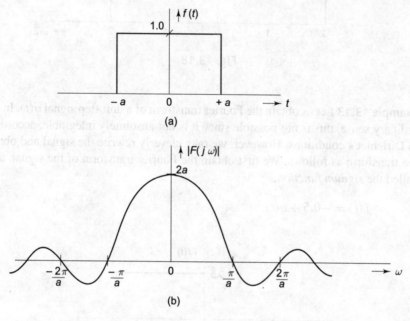

(a)

(b)

Fig. 13.16

Exercise 13.14 Obtain the Fourier transform of the triangular signal shown in Fig. 13.17 and plot the spectra. (***Hint*** Exploit the properties of the Fourier transform.)

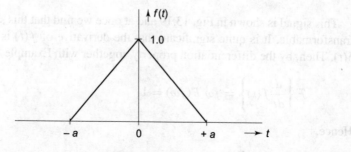

Fig. 13.17

Exercise 13.15 Obtain the inverse Fourier transform of the frequency function shown in Fig. 13.18.

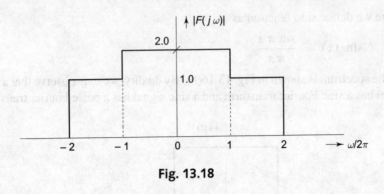

Fig. 13.18

Example 13.13 Let us obtain the Fourier transform of a unit step signal $q(t)$. In the ordinary sense, this is not possible since it is not absolutely integrable, according to Dirichlet's conditions. However, we may cleverly rewrite the signal and obtain the transform as follows. We first obtain the Fourier transform of the signal, also called the *signum function*,

$$f(t) = -0.5 + q(t)$$

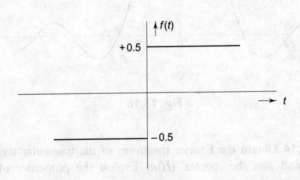

Fig. 13.19

This signal is shown in Fig. 13.19 and at once we find that this signal is Fourier transformable. It is quite significant that the derivative of $f(t)$ is a unit impulse $\delta(t)$. Then, by the differentiation property, together with Example 13.10,

$$\mathcal{F}\left\{\frac{d}{dt}f(t)\right\} = j\omega\, F(j\omega) = 1$$

Hence,

$$F(j\omega) = \frac{1}{j\omega}$$

Let us now apply the linearity property and the result of Example 13.11,

$$- \pi \, \delta(\omega) + \mathcal{F}\{q(t)\} = \frac{1}{j\omega}$$

or

$$\mathcal{F}\{q(t)\} = \pi \, \delta(j\omega) + \frac{1}{j\omega}$$

Exercise 13.16 Using the result of Example 13.13, obtain the Fourier transform of the integral of $f(t)$.

Looking back at the definition, and properties such as linearity, differentiation, and integration, we may readily establish the following relationship between Laplace transformation and Fourier transformation for positive time signals. Of course, we shall not consider the transients

$$F(j\omega) = F(s)|_{s=j\omega} \tag{13.15}$$

i.e., the Fourier transform of $f(t \geq 0)$ is its Laplace transform $F(s)$ evaluated at $s = j\omega$. Accordingly, we may write the generalized Ohm's law as

$$V(j\omega) = Z(j\omega)I(j\omega)$$

and express the general input–output relationship as

$$D(j\omega) = H(j\omega)U(j\omega) \tag{13.16}$$

where $D(j\omega)$ is the Fourier transform of the desired output signal $d(t)$, and $U(j\omega)$ is the Fourier transform of the input signal $u(t)$. $H(j\omega)$ is the network function. What does Eqn (13.16) tell us? In Chapters 6 and 7, we have defined the network function $H(j\omega)$ when the input and hence the output are periodic signals. Clearly, the idea of Fourier transform extends the phasor technique to non-periodic signals. We shall narrate the rest of the story using some examples in the following section.

13.5.2 Circuit Analysis

Example 13.14 Consider the circuit shown in Fig. 13.20(a). Suppose the input $v_S(t)$ is a *signum function* defined by

$$v_S(t) = \begin{cases} -1 & \text{if } t < 0 \\ +1 & \text{if } t \geq 0 \end{cases} = 2 \, \text{sgn}(t)$$

and is shown in Fig. 13.20(b).

The network function may be readily obtained using the generalized voltage division rule as

$$H(j\omega) = \frac{1}{1 + j5\omega} = \frac{1/5}{1/5 + j\omega}$$

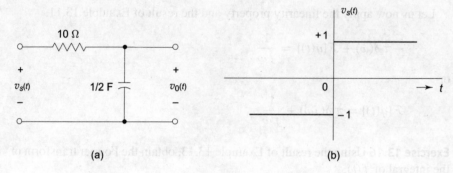

(a)   (b)

Fig. 13.20

The Fourier transform $V_S(j\omega)$ may be found to be

$$V_S(j\omega) = \frac{2}{j\omega}$$

The output is now

$$V_0(j\omega) = H(j\omega)V_S(j\omega) = 2\left(\frac{1}{j\omega} - \frac{1}{\frac{1}{5} + j\omega}\right)$$

where we have split the product into partial fractions. Using the inverse Fourier transforms, together with the linearity property, we get

$$v_0(t) = 2\,\mathrm{sgn}(t) - 2e^{-t/5}q(t)$$

where we have considered the fact that the second term in the partial fraction is Fourier transformable only for $t \geq 0$. In other words,

$$v_0(t) = \begin{cases} -2\text{ V} & \text{if } t < 0 \\ 2 - 2e^{-t/5} & \text{if } t \geq 0 \end{cases}$$

Exercise 13.17 Verify the results of Example 13.14 using time-domain analysis techniques and Laplace transformation. Assume that the capacitor is initially discharged.

Exercise 13.18 Repeat Example 13.14 and Exercise 13.17 with the voltage across the resistance as the output.

Example 13.15 Let us determine the energy dissipated by the 4 Ω resistance in the circuit shown in Fig. 13.21 if the input is

$$v_S(t) = 12\,e^{-2t}$$

It is clear that

$$H(j\omega) = \frac{4}{4 + j\omega}$$

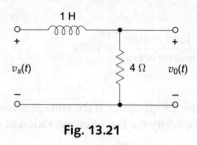

Fig. 13.21

and

$$V_S(j\omega) = \frac{12}{2 + j\omega}$$

and hence,

$$V_0(j\omega) = \frac{48}{(4 + j\omega)(2 + j\omega)} \text{ V}$$

Let us first compute the energy delivered by the voltage source. Using Parseval's theorem,

$$E_S = \frac{1}{\pi} \int_0^\infty |V_S(j\omega)|^2 \, d\omega$$

$$= \frac{144}{\pi} \frac{1}{2} \left[\tan^{-1}\left(\frac{\omega}{2}\right) \right]_0^\infty$$

$$= 36 \text{ J}$$

We encourage the reader to verify this using the time-domain integration. Next, the energy dissipated by the 4 Ω resistance may be computed as

$$E_0 = \frac{\frac{1}{\pi} \int_0^\infty |V_0(j\omega)|^2 \, d\omega}{4 \, \Omega}$$

$$= \frac{48^2}{4\pi} \int_0^\infty \frac{1}{4^2 + \omega^2} \frac{1}{2^2 + \omega^2} \, d\omega$$

$$= \frac{48^2}{4\pi} \left[\int_0^\infty \frac{1/12}{2^2 + \omega^2} \, d\omega - \int_0^\infty \frac{1/12}{4^2 + \omega^2} \, d\omega \right]$$

$$= \frac{48^2}{4\pi \times 12} \left[\frac{1}{2} \tan^{-1}\left(\frac{\omega}{2}\right) - \frac{1}{4} \tan^{-1}\left(\frac{\omega}{4}\right) \right]_0^\infty$$

$$= \frac{48^2}{4\pi \times 12} \frac{\pi}{8}$$

$$= 6 \text{ J}$$

where we have appropriately split $V_0(j\omega)$ into partial fractions and then performed the integration. Thus, only one-sixth, i.e., 16.667% of the source energy is transferred to the load. We may generalize this and define efficiency as follows.

Definition 13.5

$$\text{Efficiency } \eta = \frac{\text{load energy}}{\text{source energy}} \times 100\% \tag{13.17}$$

Exercise 13.19 Determine the efficiency if the voltage across the inductor is taken as the output. And, hence verify the law of conservation of energy.

Exercise 13.20 Obtain the plots of energy density functions $|E_S(j\omega)|^2$ and $|E_0(j\omega)|^2$ and verify that the circuit behaves like a low-pass filter.

We conclude this section with the following observations on Fourier transform vis á vis Laplace transform.

1. Not every signal is Fourier transformable; the signals have to obey Dirichlet's conditions. However, a larger class of signals are Laplace transformable. An immediate example is the unit step signal.

2. By definition, Fourier transformation accommodates all (non-zero) signals over the entire interval $(-\infty, +\infty)$, and hence it is better suited to compute the *steady-state response* quickly. In contrast, Laplace transformation accommodates only the positive time signals, and hence it is better suited to compute the *complete response*, i.e., transient and steady state for $t \geq 0$. It is in this sense we say that Laplace transformation is a generalization of Fourier transformation.

Summary

In this chapter we have attempted to see certain *non-standard* signals such as the triangular and pulse signals. The most significant content of this chapter is that *any* signal can be expressed, subject to certain conditions, in terms of sinusoids. In an earlier chapter we have discussed waveform synthesis pertaining to time-domain. This chapter provided an alternate waveform synthesis in the frequency domain. Broadly, we have classified signals as either periodic or non-periodic. We have learnt that a periodic signal can be expressed as an infinite summation of sinusoids, called the Fourier series, containing the dc, the fundamental, and several harmonics which are essentially integer multiples of the fundamental. We have also seen that this series representation can be either in trigonometric terms or in complex exponential terms. Though there is no specific rule to choose one particular representation, from a generalization point of view the complex exponentials provide an elegant framework. We have suggested that it would be useful to look at the series representation as a map of a continuous time periodic function $f(t)$ to a discrete spectrum. The response of a circuit to an arbitrary periodic signal can be obtained by applying the principle of linearity; loosely speaking what we have done is phasor addition. We have also noted that the frequency content conveys more useful information than the time waveform of the output signal. Periodic signals are power signals, and the average power of a signal can be computed directly from the Fourier coefficients.

Subsequently, we have investigated the frequency content of non-periodic signals. We have found that the Fourier transform is a map of a continuous time non-periodic function to a continuous spectrum. The energy of a signal can be computed using the Fourier transform. We have discovered that the Fourier transform extends the idea of phasors to non-periodic signals. We have seen a few simple examples wherein we make use of Fourier transforms to analyse circuits. More complex circuits are presented in the problems that follow.

The presentation in this chapter is not meant to be exhaustive for, at least, two reasons. First, Fourier analysis is a beautiful subject by itself; no more justice can be done to the subject in a single chapter at the fag end of a book. Secondly, circuit theory, Laplace transformation, Fourier analysis, and linear algebra form a basis for more advanced topics such as digital signal processing, communication systems, and control systems; nowadays they are covered in an intermediate course called signals and systems. We will discuss a few interesting issues in general linear systems in the following chapter.

Problems

13.1 Obtain the trigonometric Fourier series of the periodic signal shown in Fig. 13.22.

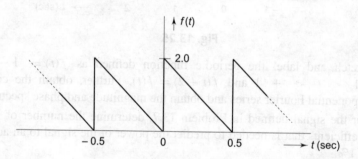

Fig. 13.22

13.2 For the periodic signal of Problem 13.1, obtain the complex exponential Fourier series. Sketch the magnitude and phase spectra.

13.3 Repeat Problem 13.1 for the periodic signal shown in Fig. 13.23.

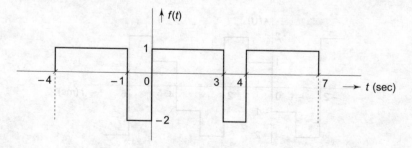

Fig. 13.23

13.4 Repeat Problem 13.2 for the signal of Fig. 13.23.

13.5 Repeat Problem 13.1 for the signal of Fig. 13.24.

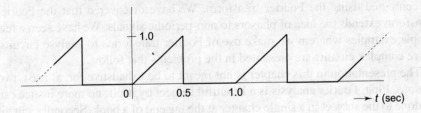

Fig. 13.24

13.6 Repeat Problem 13.1 for the signal of Fig. 13.25.

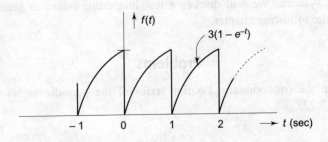

Fig. 13.25

13.7 Sketch and label the periodic function defined as $f(t) = 1 - t^2$ for $-1 < t < +1$ and $f(t+2) = f(t)$. Further, obtain the complex exponential Fourier series and obtain the magnitude and phase spectra.

13.8 For the signal defined in Problem 13.7, determine the number of Fourier coefficients that is needed to predict the power of the signal to an accuracy of 1%.

13.9 Assume that the periodic signal shown in Fig. 13.23 is shifted to its right by 2 ms. Sketch the shifted signal and obtain its Fourier series. Compare your results with those of Problem 13.4.

13.10 Obtain the complex exponential Fourier series of the signal shown in Fig. 13.26.

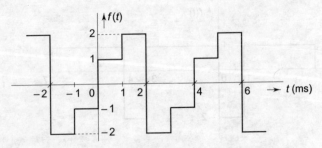

Fig. 13.26

13.11 Consider the periodic current signal shown in Fig. 13.27(a). If this signal is applied to the circuit shown in Fig. 13.27(b), compute the output $v_0(t)$, in terms of its Fourier series. Obtain the discrete spectra of the output.

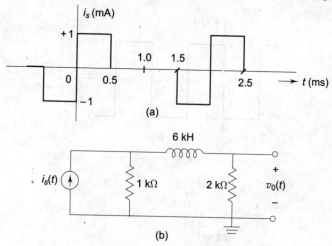

(a)

(b)

Fig. 13.27

13.12 If the voltage $v_S(t)$ of Fig. 13.28(a) is applied to the Op-Amp circuit in Fig. 13.28(b), compute the output $v_0(t)$, and hence obtain the discrete spectra.

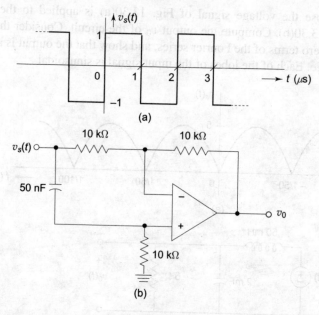

(a)

(b)

Fig. 13.28

13.13 In the chapter on Op-Amps, we have discussed that subjecting an integrator circuit to a rectangular signal yields a triangular signal. Analyse this behaviour using the Fourier series of the input and output signals.

13.14 The rectangular current signal of Fig. 13.29(a) is applied as an input to the circuit shown in Fig. 13.29(b). Determine the output of the circuit, and hence sketch its discrete spectra.

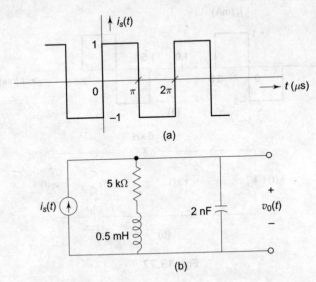

(a)

(b)

Fig. 13.29

13.15 Suppose the voltage signal of Fig. 13.30(a) is applied to the circuit in Fig. 13.30(b). Compute the output v_0 of the circuit. Consider the first four non-zero terms of the Fourier series, and show that the output is almost a dc voltage. Each of the lobes of the input signal is sinusoidal.

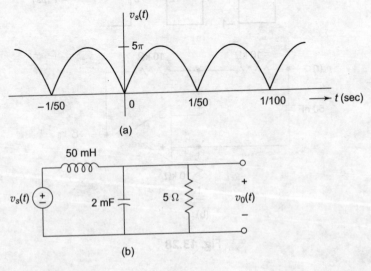

(a)

(b)

Fig. 13.30

13.16 A non-periodic triangle pulse is defined as follows:

$$f(t) = \begin{cases} 1 - \dfrac{|t|}{5} & \text{for } |t| \leq 5 \\ 0 & \text{for } |t| \geq 5 \end{cases}$$

Sketch the signal, find its Fourier transform, and hence obtain the continuous magnitude and phase spectra.

13.17 A non-periodic trapezoidal pulse is shown in Fig. 13.31. Obtain the continuous spectra.

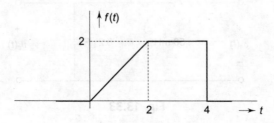

Fig. 13.31

13.18 Given that

$$f(t) = 5e^{-t} \sin t \ \forall \, t \geq 0$$

obtain its Fourier transform and hence its energy contained within (a) the interval $-1 \leq t \leq +1$ s, and (b) the interval $-1 \leq \omega \leq +1$ rad/s.

13.19 Use the linearity property and determine the Fourier transform of the non-periodic signal shown in Fig. 13.32.

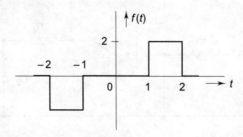

Fig. 13.32

13.20 The signal

$$v_S(t) = 3e^{-at} \ \text{V} \ \forall \, t \geq 0$$

is applied to an *RC* network whose time-constant is $1/2\pi$ s. Determine $a \, (> 0)$ such that the energy of the output voltage $v_C(t)$, across the capacitance, is exactly half that of the input voltage $v_S(t)$.

13.21 The signal

$$v_S(t) = 5e^{-t} \ \text{V} \ \forall \, t \geq 0$$

is applied to an RC network whose time-constant is τ ms. Determine τ such that the energy of the output voltage $v_R(t)$, across the resistance, is exactly half that of the input voltage $v_S(t)$. Also, discuss the case $\eta \to 1$, i.e., the conditions under which the resistance draws the entire energy from source.

13.22 Consider the circuit shown in Fig. 13.33. If $v_S(t) = 2\delta(t)$, compute the percentage of the total energy of the output $v_C(t)$ contained within the interval $0 \le \omega \le 20$ krad/s.

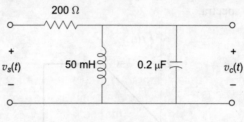

Fig. 13.33

13.23 Repeat Problem 13.22 if the input is a signum function defined in Example 13.14 of this chapter.

13.24 Consider the circuit shown in Fig. 13.34, with the capacitors C_1 and C_2 assumed to be initially discharged. Determine the output $v_0(t)$ if $v_S(t) = 10\cos(100t)$, $R_1 = R_2 = 10\,\text{k}\Omega$, and $C_1 = C_2 = 1\,\mu\text{F}$. Compute the energy dissipated by the capacitance C_2. Verify your results using the techniques discussed in Chapter 12.

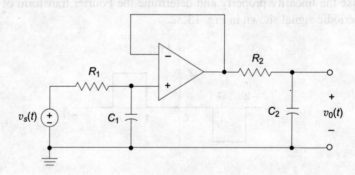

Fig. 13.34

13.25 Repeat Problem 13.24 for the circuit in Fig. 13.34, if $v_S(t) = 10e^{-100t}$, $R_1 = R_2 = 10\,\text{k}\Omega$, and $C_1 = 2C_2 = 1\,\mu\text{F}$.

CHAPTER 14

General Linear Systems

We have devoted thirteen chapters to learn a subject of fundamental importance. In this chapter, we shall attempt to give an extension to this subject. We prefer to do so since circuit theory is not just one of several subjects taught in engineering, it has a beauty and depth of its own.

Broadly speaking, a system may be defined as *anything* that is 'put together'. We shall not provide a formal definition of a *system*, but shall rely on the intuitive concept of a system—such as the central nervous system, solar system, etc.—that the reader has. Several abstract systems[1] may be visualized symbolically as circuits, and the mathematical rigor governing such systems makes sense instantaneously.

14.1 Linear Systems: Examples

All through we have been emphasizing the input–output paradigm and the *linear relationship* between the input and the output in a circuit. A rather immediate application of this theory is in the analysis and design of electronic circuits. A diode may be *modelled* as a resistance R as follows:

- If it is forward biased, then $R \approx R_f \, \Omega$.
- If it is reverse biased, then $R \approx \infty \, \Omega$.

We call this as a 'piecewise linear' model and the *transfer characteristic* looks as shown graphically in Fig. 14.1.

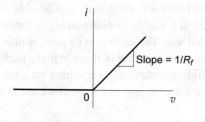

Fig. 14.1

Likewise, we may look at amplifier circuits containing transistors. In general, a transistor is supplied with two distinct input voltages—(i) the dc bias voltages, and (ii) the ac voltages that need to be amplified. The bias voltages force the transistor to behave like a linear element, i.e., with straight line characteristic between the voltage and current. These dc voltages get *added* to the ac signals, i.e., the ac signals

---

[1] Some time back we have even seen a TV commercial calling a detergent powder as a microwashing system!

ride on the dc signals. However, we make use of the linearity property and segregate the *effects* of the two *causes* to focus only on the ac analysis of the transistor amplifier, ignoring the dc voltages. Consequently, we settle for two-port networks with hybrid parameters, which serve to mimic or *model* the *behaviour* of a transistor. Thus, inherently non-linear circuits may be studied as a first approximation using the *linear system theory*. This has been demonstrated in Chapter 12 on operational amplifiers.

Although we have confined ourselves to *electrical* elements, the ideas can be extended very well to other fields. For instance, a charged capacitor may represent a liquid storage tank in a chemical industry: the charge is analogous to the liquid; the voltage developed between the plates is analogous to the pressure difference; and the current flowing through the capacitor is analogous to the water flow. Similarly, what we have called Ohm's law, mechanical engineers refer to as *Hooke's law*; the input–output relationship is given by $y = mx$, a straight line passing through the origin. The mathematical expressions underlying these electrical, chemical, and mechanical systems belong to the same category—linear differential equations[2].

14.2 Systems and Models

Let us look at a bigger picture. In the so-called hard sciences such as physics, biology, etc. the laws and their predictions can be tested by experiments. The purpose of science is to explore nature and discover, say, new elements, genome sequence, etc., and predict the results of new experiments. It is this *prediction* that makes science so useful to us. On the contrary, the goal of engineering may be dubbed as *re-creating* new devices and systems by making maximum use of science. Here we *define* a device as a fundamental building block of a system. For instance, today we have the telephone and its network, communication satellites, computer communication networks with fiber-optic links, electric power systems with nuclear reactors, turbines, etc. the robots and flexible manufacturing systems, and many more. It is even possible to *clone* living organisms[3]. Putting it briefly, engineering is concerned with the invention of new, simple devices, manufacture them cheaply, and interconnect several of them to make useful large-scale systems. This is analogous to building recursively a complex network of sub-networks, each in turn having simpler networks, and so on, until we are left with the sources and the R, L, C elements.

For the purpose of analysis and design, a physical system is replaced by a model, a *mathematical representation* in particular. The model may be quite simple, just meeting the demands of the problem under study. For instance, according to Hooke's law, a spring can be modelled as a linear element in the sense that the elongation is directly proportional to the force applied, provided the force is not too large to tear the spring apart or to crush it. Likewise, Newton's laws are valid

---

[2] If the nth derivative is the highest derivative in the differential equation, we call it an nth-order linear differential equation. Clearly, simple algebraic equations of the form $y = mx + c$ are *zeroth*-order linear differential equations.

[3] That is why the field is called genetic *engineering*.

as long as the bodies travel with velocities comparable to that of light. By a model, we mean a set of equations relevant to the context. The next step is to use this mathematical representation to *analyse* the model, i.e., to determine the properties, strengths, and weaknesses of the system. This is the crux of what is called *system theory*. This analysis helps us to (i) understand the system thoroughly, (ii) specify the features that a system must have in order to meet our demands and requirements, and (iii) use the analysis techniques to design a system.

Many physical laws and relations appear mathematically in the form of differential equations. Therefore, in general, the mathematical representation of any system is a set of differential equations, with time t as an independent variable. There exist several classes of these equations, to which the reader would have been already exposed in an earlier course in engineering mathematics. First, the differential equations could be either an ordinary differential equations (ODE), or a partial differential equations (PDE). Secondly, the coefficients of the differential equations are either constant or varying with time. Thirdly, the differential equations could be linear or non-linear, in terms of the input–output mapping. Most of the times we are interested in a specific class of systems represented by *ordinary differential equations with constant coefficients*. This particular class is called *linear time-invariant lumped* (LTIL) parameter systems. As we have already seen, the RL, RC, and RLC circuits belong to this class. This is the simplest, yet the largest, class of systems. Further, we can solve them easily, either directly or using Laplace transformation. Maxwell's equations, the great laws of electricity that determine the electric field, are linear differential equations; the laws of quantum mechanics, so far as we know, turn out to be linear.

14.2.1 Additional Examples

Although the input–output characteristics of a diode turn out to be non-linear, we can still approximate them to be piecewise linear as in Fig. 14.1; the transistors are biased to operate as linearly as possible; we may even build a complex integrated circuit, the Op-Amp, which exhibits much better linear characteristics than a single transistor. If we have a simple pendulum, the equation for its motion may be derived to be

$$M \frac{d^2}{dt^2} \theta + M \frac{g}{L} \sin \theta = 0 \tag{14.1}$$

where M is the mass of the pendulum, L is its length, and g is the acceleration due to gravity. If we assume that the angular displacement θ (in radians) is small enough, it is easy to see that $\sin \theta \approx \theta$ and the above non-linear equation becomes

$$M\ddot{\theta} + M \frac{g}{L} \theta = 0 \tag{14.2}$$

which is a linear equation. We may also notice quickly that this is like an LC circuit and hence exhibits an oscillatory behaviour. Since this closely matches reality, we accept this *linearized* model.

While the study of a typical LTIL system is modelled mathematically as an ordinary differential equation with constant coefficients, we might use an electrical network as a *physical* model to study the same system in the laboratory. For example, consider a mechanical system, such as an automobile suspension system, with a body of mass M, a spring with constant K, and a damper (or dashpot) with constant D as shown in Fig. 14.2.

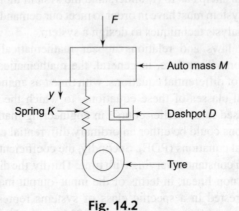

Fig. 14.2

Elementary mechanics tells us to write the equation of motion as

$$M\ddot{y} + D\dot{y} + Ky = F \qquad (14.3)$$

where y is the displacement of the body with reference to some *a priori* fixed coordinates, and F is the force applied. We solve this ODE to learn the relationship among the parameters M, D, K, and F. Typically we wish the shock absorber to give us a cushioning effect after we hit a pothole on the road. We might simulate this mechanical system in an electrical laboratory using an RLC circuit and study the properties. The inductances model the inertia of the wheels, capacitances model the spring constant, and so on. Of course, we need to do some sort of scaling in magnitude as well as frequency. But, the idea is that we can study the dynamical behaviour of a given system in terms of an equivalent system. Do we have to actually change the shock absorber of the automobile and solve the equation again and again? No! it is as simple as changing a capacitance or an inductance or a resistance, and we adjust everything *electrically*. At the end of the experiment, we decide on the actual springs and dampers corresponding to the electrical elements. This is a good idea since we might not be willing to conduct experiments that are expensive as well as risk prone; we prefer a *simulator*, which is easier to build, measure, adjust, and destroy, of course. We call such circuits as *analog computers*.

Let us digress briefly and elaborate on analog computers by taking up the ODE of the mechanical system in Eqn (14.3). With our experience in state variables, we may define two state variables as

$$x_1 \overset{\Delta}{=} y$$

$$x_2 \overset{\Delta}{=} \dot{x}_1 \qquad (14.4)$$

so that the ODE in Eqn (14.3) may be rewritten in terms of x_1 and x_2 as the following pair of first-order differential equations:

$$\dot{x}_1 = x_2 \tag{14.5}$$

$$\dot{x}_2 = -Kx_1 - Dx_2 + F \tag{14.6}$$

We may use a pair of integrators, built using Op-Amps, and use the following scheme (shown in Fig. 14.3) to solve Eqns (14.5) and (14.6). The integrators are shown as rectangular boxes with the integration symbol inside.

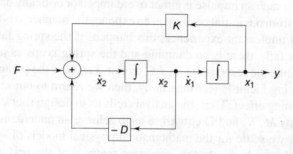

Fig. 14.3

The circle denoting summation is basically a summing amplifier, and the blocks denoting multiplication (with a negative sign) by constant coefficients are inverting amplifiers we have discussed in Chapter 12. We have shown the state variables x_1 and x_2 as the outputs of the integrators. The inputs are, naturally, $\dot{x}_1$ and $\dot{x}_2$. Of course, initial conditions, if any, may also be added. The arrows show the directions, either forward or backward, of signal propagation. Thus, the first integrator receives three inputs, F, $-Dx_2$, and $-Kx_1$, and puts out x_2; the second integrator puts out $x_1 = y$. This schematic representation implemented with the actual hardware such as Op-Amps, resistors, and capacitors is an analog computer. The word 'analog' suggests that the computations are carried out in the continuous-time domain. In contrast, we have general purpose digital computers. We encourage the reader to think of biomedical applications wherein we may need to avoid conducting *real* experiments on humans.

Analysis of a system is solving the equations governing its behaviour. Solutions can be obtained either using techniques such as Laplace transformation, or using analog computers, or simulating the corresponding physical model of the system. The information contained in the solutions has two prominent features—stability and performance. Once we have a model of the system, the first thing we investigate is its stability. Typically, this requires the corresponding characteristic polynomial to be Hurwitz. We have a standard procedure, called Routh–Hurwitz criterion, presented in appendices, to check whether a given polynomial is Hurwitz. This is analogous to checking whether the throttle and brakes of, say, a race car are functioning properly. Next, we examine the performance of the system. Performance is generally given in terms of a set of specifications; for instance the rate of acceleration (pickup), maximum torque, and fuel consumption in the race car example. Performance specifications may be verified by computing certain parameters from the solutions.

Let us illustrate this for the automobile suspension system of Fig. 14.2. We have already learnt, in Chapter 5, that a second-order system exhibits four different solutions under different conditions. They are (i) overdamped, (ii) critically damped, (iii) underdamped, and (iv) undamped responses; this classification is based on the relative size of a parameter called damping ratio ζ. We may study each of these cases and interpret the results as follows. If the automobile hits a pothole, the force F experienced by the shock absorber is a sudden 'jolt' which may be modelled as an impulse $\delta(t)$. If the coefficients M, K, and D are such that the response to such an impulse is either overdamped or critically damped, then we are displaced from our initial position in an exponential manner. It is trivial to notice that this is an unpleasant experience; this happens if the spring fails to function. If the dashpot fails, there is no damping and the spring keeps us jumping up and down; this is like the response of an LC circuit. Lastly, if the parameters are such that the damping factor is less than unity, then we return to our original position with a cushioning effect. Thus, the analysis tells us to design shock absorbers with the coefficients M, K, and D chosen so as to achieve an underdamped response.

In general, we settle for the mathematical/physical models of systems that are simple enough and highlight the important features of the performance of real systems. For instance, we need not study the suspension system anymore, since this is designed to take care of hitting potholes. We will now qualitatively describe the analysis of linear systems when the forcing function is quite complicated. There are two very useful *general* ways in which we can solve the problem. One is this: suppose that, by some tremendous mental effort, it were possible to solve our problem for a special force, the *unit impulse*, i.e., the force is quickly turned on and then off and it is all over. Also, let us suppose that we have an arbitrary force $u(t)$, such as the one shown in Fig. 14.4.

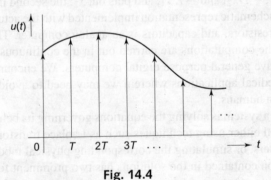

Fig. 14.4

Such a force can be thought of as a succession of blows with a hammer. First there is no force, and all of a sudden there is a steady force—impulse, impulse, impulse, ... In other words, we imagine the continuous (in time) force to be a series of impulses, very close together. Since we assumed that we know the result for an impulse to be some curve, hence the result for a whole series of impulses will be the curve for the first impulse, and then (slightly later) we add to that the curve for the second impulse, the curve for the third impulse, and so on. Thus we can represent, mathematically, the complete solution for arbitrary functions if we know the answer to an impulse; we get the answer to any other force simply by *aggregating*. This method is called *convolution*. We were mentioning about this in earlier chapters. We have also mentioned that Laplace and Fourier transformations tame this messy integral.

Another method is this: suppose that we can solve it for special known forces such as sinusoids of different frequencies. We have seen that it is fairly simple to solve for sinusoids. Now the question is whether we can represent a very complicated force as the sum of two or more simple forces. Fortunately, practically every force can be obtained by adding together *infinite* number of sinusoids of different frequencies. We just have to know how much of each sinusoid to put in to make the given force $u(t)$. And then our answer, $y(t)$, is the corresponding sum of the sine waves, each multiplied by its effective ratio of $y(t)$ to $u(t)$. This method of solution is called the *Fourier analysis*.

It is interesting to note that, historically, the work on 'systems' problems brought together mechanical, electrical, and electronic engineers, and an outcome of this cross-fertilization of ideas was the recognition that neither the frequency response approach used by the communication engineers nor the time response approach favoured by the mechanical engineers was, separately, effective in designing position control systems. This led to the use of Laplace transform techniques. A systems engineer would be immensely benefited by exploiting both the time and the frequency domains for his design. The frequency domain is a kind of hidden companion to our everyday world of time.

The physical principles involved in both these schemes are so simple—involving just the linear equations—that they can be readily understood; but the *mathematical* problems that are involved, the complicated convolution, and so on, appear to be a bit tricky. Nevertheless, they can be tamed with some practice—the idea is very *natural* and simple indeed.

14.3 Concluding Comments

The study of circuit theory provides a basis for understanding many more linear systems. The differential equations, and hence the analog computers, give predictive power to engineering, so that the merits of various design alternatives may be evaluated even before investing labour and material in building them. The linear differential equations is the only large class of differential equations for which a definitive theory exists. This theory is essentially a branch of linear algebra, and that is why we have been emphasizing linear algebra at several places in the previous chapters. We hope the reader is convinced that it is worth spending so much time on linear circuits, for he/she is coached to understand a lot of things. Maxwell's equations for the laws of electricity are linear; the great laws of quantum mechanics turn out, so far as we know, to be linear. Linear systems are so important, at least, because we can solve them.

With a strong background in circuit theory, the reader should enjoy courses such as signal processing, communications systems, and control systems in later semesters, so that the sole purpose of this book is served.

Appendix A
A Tutorial on SPICE

We have introduced the *Simulation Program with Integrated Circuit Emphasis* in Chapter 3. It was in no way an elaborate treatment. We just presented the data structure—an organized way of making the computer understand the circuit. Thereafter, we have not used the program since we emphasized that there is no substitute for understanding the circuit behaviour, which we get only through hand analysis. We have also mentioned that, in an introductory course, we purposefully keep circuit complexity at its lowest level and hence computer power is not mandatory. In this appendix, we demonstrate the capabilities of this simulation program using several examples. We remark that although this program is extremely useful in handling complex circuits, where the time and effort required for hand calculations may be prohibitively large, we must develop an investigative attitude and question the outcome of the program.

The examples we describe in this appendix were successfully run on a PC version of the SPICE program. This program, called Orcad's PSpice 9.2 Lite Edition, can be downloaded free of cost from www.orcad.com or www.PSpice.com. As has been mentioned in Chapter 3, several other versions of the program are available. The reader may inquire the version of SPICE/PSpice available in his institution and quickly learn to run the programs. Further, a typical SPICE program has three parts—circuit description, analysis request, and output request. Circuit description can be either tabular or graphical. In Chapter 3, we have pursued the tabular description. We believe that, in an introductory course, the tabular description is better. There is an associate graphical interface called **Capture** that *draws* circuits using an intuitive menu.

This tutorial is organized in several sections. The reader may use these sections concurrently along with the theory; for instance, the first section on dc analysis may be done along with Chapters 3 and 4. Alternatively, he/she may choose to do the entire tutorial towards the end of the course. We do not provide answers to most of the problems in this tutorial; after all, the reader should run the programs on his/her own computer and he is expected to be equally, or even more, enthusiastic in checking the answers by hand computations.

A.1 Dc Analysis

In this section, we will straightaway get into the analysis of circuits. We suggest the reader to refer to Chapter 3, if required, for writing the PSpice data structures. Let us begin with a simple illustrative example.

Example A.1 Consider the circuit shown in Fig. A.1.

The PSpice data structure for this circuit may be quickly written as follows:

EXAMPLE A.1

R1	1	2	3K Ohm
R2	2	0	6K Ohm
R3	2	3	4K Ohm
R4	3	0	2K Ohm
Vs	1	0	dc	6 V
Is	3	0	dc	5M A
.END			

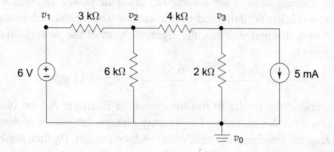

Fig. A.1

Compared to the data structure we have illustrated in Chapter 3, the reader may quickly notice that we have given a name, EXAMPLE A.1, to the program, and we have ended the program formally with a '.END' statement. We have avoided the headings, such as 'Element', 'node v_j', for each of the columns.

It is interesting to note that the data structure above is ready to be read and analysed by the PSpice program. We advise the reader to quickly learn the procedure to run a particular version of the SPICE program accessible to him. The computer produces an output listing shown in the table below.

DC ANALYSIS:

Node	Voltage	Node	Voltage	Node	Voltage
(1)	6.0000	(2)	0.5000	(3)	−6.5000
VOLTAGE	SOURCE	CURRENTS:			
Name			Current		
VS			−1.8333		
TOTAL	POWER	DISSIPATION:	1.10E-02	WATTS	

Exercise A.1 Verify the results of Example A.1 using nodal analysis.

Once we verify the results of the PSpice program using nodal analysis by hand, we notice the following two discrepancies.

1. The node potential v_1 is greater than v_0 (of course, it has to be) and v_2 and we expect the current to flow out of the voltage source. But the output of the PSpice program suggests us otherwise. This is owing to the syntax of PSpice. Recall that when we describe a current source using the PSpice data structure, we describe it as if it were a passive element. In other words, what PSpice computes is the current (including the sign) flowing into the positive terminal of the voltage source and coming out of the negative terminal.

2. The voltage source delivers a power of 11 mW, and the current source delivers a power of 32.5 mW. However, the program computes the total power dissipation as if voltage sources are the only active elements in the circuit; current sources are assumed to dissipate power. In other words, the power, + 32.5 mW, delivered by the current source may be interpreted as −32.5 mW dissipated. Thus, the algebraic sum of the powers dissipated in the circuit is

$$+\frac{5.5^2}{3} + \frac{0.5^2}{6} + \frac{7^2}{4} + \frac{6.5^2}{2} - 6.5 \times 5 = 11 \text{ mW}$$

We recommend the reader to run the circuit of Example A.1 on his/her own computer. Once he/she succeeds, he/she may change the values of some of the elements, rerun the program, and verify the results manually. By then he/she should be confident in running his own PSpice files.

The results of Example A.1 suggest yet another interesting thing—the program computes the current through the voltage sources. We can utilise this idea to compute the current through any branch in a given circuit. The reader would have already got this idea—make a fool of the computer by introducing a *dummy* voltage source of 0 V in the branch of interest. It works! However, the reader should properly identify the branch he/she is interested in, cut it to introduce an additional node, and then put the dummy voltage source, with appropriate polarity. We will redo Example A.1 to compute the current through the 4 kΩ resistance.

Example A.2 The modified input file for the PSpice program is as follows. We may include *non-compilable* statements with an asterisk symbol at the beginning of a line.

EXAMPLE A.2			
R1	1	2	3K Ohm
R2	2	0	6K Ohm
R3	3	4	4K Ohm
R4	4	0	2K Ohm
Vs	1	0	dc	6 V

* DUMMY	VOLTAGE	SOURCE:		
Vd	3	2	dc	0 V

Is	4	0	dc	5M A
.END				

Once this program is run, the output contains the following:

DC ANALYSIS:

Node	Voltage	Node	Voltage	Node	Voltage	Node	Voltage
(1)	6.0000	(2)	0.5000	(3)	0.5000	(4)	−6.5000

VOLT.	SRC	CURR.:					
Name			Current				
Vs			−1.8333				
Vd			−1.7500				
TOTAL	POWER	DISS.:	1.10E-02	WATTS			

(Certain standard words have been abbreviated here.)

Exercise A.2 Discuss why the current through the 4 kΩ branch is computed to be negative by the program.

The PSpice program can be used to conduct a virtual experiment. For instance, if we would like to see how the node potentials of the circuit shown in Fig. A.1 vary as we vary the voltage source and/or the current source, we have the following syntax:

$$.DC \quad SRCNAME \quad SRCSTART \quad SRCSTOP \quad SRCINCR$$

where .DC indicates that it is the dc analysis, SRCNAME is the name of the independent (voltage or current) source, SRCSTART is the starting value in volts/amperes, SRCSTOP is the ending value, and SRCINCR is the incremental value of the source. This is almost similar to the Do or For loops we find in programming languages. Let us do a simple experiment to illustrate the capability of this statement.

Example A.3 Consider the simple circuit shown in Fig. A.2. We wish to vary the source V_S from, say, −1.0 V to +1.0 V and tabulate the voltages across the two resistances, as well as the mesh current.

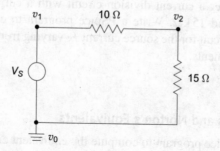

Fig. A.2

The PSpice input file for this circuit may be quickly written as follows:

EXAMPLE A.3			
R1	1	2	10 Ohm
R2	2	0	15 Ohm
Vs	1	0	
.DC	Vs	−1	1	0.5
.PRINT DC	I(Vs)	V(1, 2)	V(2)
.END			

The reader would have noticed another new statement in the input file. He/she would have also guessed its purpose—it prints the results of dc analysis according to the desired format. The general syntax of this statement is as follows:

.PRINT DC   OUTVAR 1   OUTVAR 2 $\cdots$ OUTVAR 8

Here OUTVARN is the desired voltage or current variable. Typically, the PC versions of PSpice allow a maximum of eight variables per statement; if there are more, we may have another statement below this. If we are interested in the voltage across the node pair (j, k) as the output, we write it as $V(j, k)$; if the node k is the reference, it suffices to write $V(j)$. If we are interested in a current variable through an independent voltage source VOLSRC, we write it as $I(VOLSRC)$; as might be expected, this current is assumed to flow through the source from the positive terminal to the negative terminal.

The output of the program is as follows:

Vs	I(Vs)	V(1, 2)	V(2)
−1.0000	−4.0E−02	−0.4000	−0.6000
−0.5000	−2.0E−02	−0.2000	−0.3000
0.0000	0.0E−02	0.0000	0.0000
0.5000	2.0E−02	0.2000	0.3000
1.0000	4.0E−02	0.4000	0.6000

Exercise A.3 Consider a current division circuit with a current source and two resistances $10\,k\Omega$ and $15\,k\Omega$. Write a PSpice program to compute the voltage across the parallel circuit for the source current I_S varying from -2.0 mA to $+12.0$ mA in 0.2 mA increments.

A.1.1 Thèvenin's and Norton's Equivalents

We may use the PSpice program to compute the equivalent circuits. The program has a provision for computing the ratios, which we were generally referring to as transfer functions. We have the following syntax for the *transfer function statement*:

$$.TF \quad \text{OutVar} \quad \text{InSrc}$$

where OutVar is the output variable of interest as defined in connection with the
.PRINT DC statement above. InSrc is any independent source in the circuit.

Example A.4 Let us consider the circuit shown in Fig. A.2 again and write the
following PSpice program.

EXAMPLE A.4				
R1	1	2	10 Ohm	
R2	2	0	15 Ohm	
Vs	1	0	dc	10 V
.TF	V(2)	Vs		
.END				

The first three statements are, by now, trivial. The .TF statement does the
following. It performs the dc analysis and computes the (i) ratio OutVar/InSrc,
(ii) equivalent resistance as seen by InSrc, and (iii) equivalent resistance as seen
by OutVar. Once we run the above program, the output file is as follows:

Node	Voltage	Node	Voltage
(1)	10.0000	(2)	6.0000
V(2)/Vs	= 6.0E-01		
INPUT	RES.	AT Vs	25.0000
OUTPUT	RES.	AT V(2)	6.0000

Exercise A.4 Verify the results of Example A.4.

Example A.5 Consider, once again, the circuit shown in Fig. A.1, redrawn as
Fig. A.3 below. We are now interested in computing Thèvenin's equivalent as
seen by the $4 \text{ k}\Omega$ resistance across port A-B.

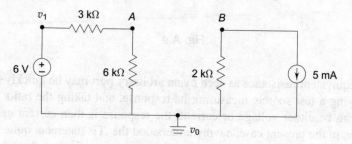

Fig. A.3

The input file may be quickly written as follows:

EXAMPLE A.5

R1	1	2	3 K Ohm	
R2	2	0	6 K Ohm	
R4	3	0	2 K Ohm	
Vs	1	0	dc	6 V
IS	3	0	dc	5M A
.TF	V(2, 3)	Vs		
.END				

The output file contains the following results:

Node	Voltage	Node	Voltage	Node	Voltage
(1)	6.0000	(2)	4.0000	(3)	−10.0000
V(2, 3)/Vs	= 2.3333				
INPUT	RES.	AT Vs	9.0E+03		
OUTPUT	RRES.	AT V(2, 3)	4.0E+03		

Thus, we get

$$v_{Th} = V(2, 3) = 14 \text{ V} \quad \text{and} \quad R_{Th} = \text{Output Res. at } V(2, 3) = 4 \text{k}\Omega$$

Exercise A.5 Verify the results of Example A.5 and compare them with those of Example A.1. Write a PSpice program to compute Norton's equivalent of the circuit shown in Fig. A.3.

Example A.6 Consider the circuit shown in Fig. A.4. Let us determine the equivalent resistance as seen by the node pair *A-B*.

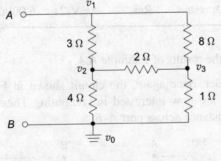

Fig. A.4

The equivalent resistance as seen by an arbitrary port may be quickly obtained by applying a test source, measuring its response, and taking the ratio. The test source can be either voltage or current; the response is then current or voltage. However, in the present case, having understood the .TF statement quite well, we see that it is convenient to apply a test current source of just 1 A and measure the voltage across the port. The transfer function is automatically the equivalent resistance. The PSpice input file for the circuit is as follows.

EXAMPLE A.6
R1	1	2	3 Ohm
R2	2	0	4 Ohm
R3	2	3	2 Ohm
R4	3	0	1 Ohm
R5	1	3	8 Ohm
Is	0	1	dc	1 A
.TF	V(1)	Is	
.END			

Exercise A.6 Run the above PSpice program and verify the results by hand computations.

A.1.2 Controlled Sources

Let us now look at using PSpice to analyse circuits having controlled sources. The syntax has already been introduced in Chapter 3. We will straightaway hit some examples and exercises.

Example A.7 Consider the circuit shown in Fig. A.5 that has a VCCS.

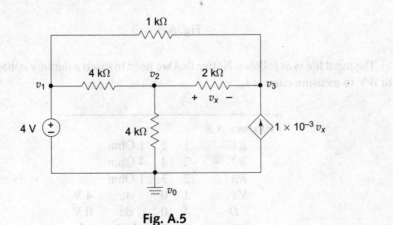

Fig. A.5

The input file is as follows:

EXAMPLE A.7

R1	1	2	4K Ohm
R2	2	0	4K Ohm
R3	2	3	2K Ohm
R4	3	1	1K Ohm
Vs	1	0	dc	4 V
Gs	0	3	2	3	1M A
.END				

The output is as follows:

Node	Voltage	Node	Voltage	Node	Voltage
(1)	4.0000	(2)	2.5714	(3)	3.1428

Exercise A.7 Verify the results of Example A.7.

Example A.8 Consider the circuit shown in Fig. A.6.

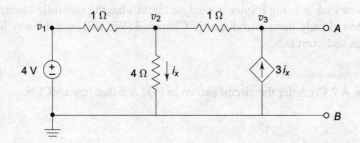

Fig. A.6

The input file is as follows. Notice that we need to insert a dummy voltage source of 0 V to measure current i_x.

EXAMPLE A.8

R1	1	2	1 Ohm
R2	2	4	4 Ohm
R3	2	3	1 Ohm
Vs	1	0	dc	4 V
VD	4	0	dc	0 V
Fs	0	3	VD	2
.END				

The output is as follows:

Node	Voltage	Node	Voltage	Node	Voltage	Node	Voltage
(1)	4.0000	(2)	8.0000	(3)	14.0000	(4)	0.0000

Exercise A.8 Verify the results of Example A.8. Obtain Thèvenin's equivalent as seen by the port A-B using the .TF statement.

Example A.9 Consider the circuit shown in Fig. A.7.

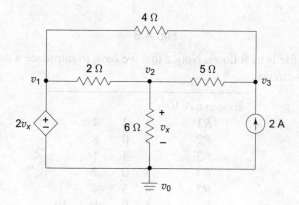

Fig. A.7

The input file is as follows.

```
EXAMPLE A.9
    R1      1   2   2
    R2      2   0   6
    R3      2   3   5
    R4      3   1   4
    Is      0   3   dc  2
    Es      1   0   2   0   2
    .END
```

The output is as follows.

Node	Voltage	Node	Voltage	Node	Voltage
(1)	−4.0000	(2)	−2.0000	(3)	1.3333

Exercise A.9 Verify the results of Example A.9. Modify the above PSpice input file to make use of the .PRINT DC statement to tabulate all the branch voltages and branch currents.

Example A.10 Consider the circuit shown in Fig. A.8.

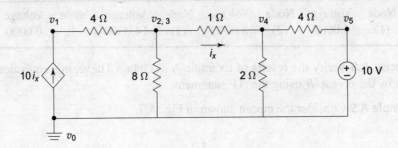

Fig. A.8

The input file is as follows. Notice that we have to introduce a dummy voltage source in this circuit.

EXAMPLE A.10				
R1	1	2	4	
R2	2	0	8	
R3	3	4	1	
R4	4	0	2	
R5	4	5	4	
Vs	5	0	dc	10
VD	2	3	dc	0
Hs	1	0	VD	10
.END				

The output is as follows.

Node	Voltage	Node	Voltage	Node	Voltage	Node	Voltage	Node	Voltage
(1)	20.0000	(2)	8.0000	(3)	8.0000	(4)	6.0000	(5)	10.0000

Exercise A.10 Verify the results of Example A.10. Modify the above PSpice input file to make use of the .PRINT DC statement to tabulate all the branch voltages and branch currents.

A.2 Transient Analysis

We may perform the transient analysis of circuits containing one or more energy storage elements using PSpice. Just as we had a .DC statement, we will now have a .TRAN statement with the following syntax:

.TRAN	TRINCR	TRSTOP	TRSTART	TRMAX	UIC

where TRINCR indicates the plotting increment, in seconds; TRSTOP is the stop time; TTTSTART is the start time, default being 0 s; TRMAX is the maximum step for the analysis, with the default being the smaller of either TRINCR or (TRSTOP-TRSTART)/50. The last one UIC, *use initial conditions*, is optional. Let us look at an example to illustrate the idea.

Example A.11 Consider the first-order circuit shown in Fig. A.9. For the sake of illustration and clarity, we will do this in two steps. In the first step we will compute the initial voltage v_C across the capacitor.

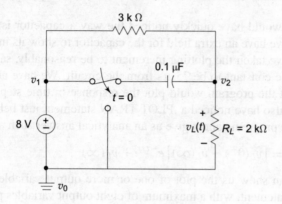

Fig. A.9

The input file is as follows. Notice that we have simply performed the dc analysis.

EXAMPLE A.11				
$R1$	1	2	3K	
$R2$	2	0	1K	
$C1$	1	2	0.1U	
Vs	1	0	dc	8
.DC	Vs	8	8	1
.PRINT DC	V(1, 2)	V(2)		
.END				

The output is as follows.

Vs	V(1, 2)	V(2)
8.0000	6.0000	2.0000

i.e., we have $v_C(0^-) = V(1, 2) = 6$ V and $v_L(0^-) = 2$ V. We will now throw the switch to the ground at $t = 0$ s. Unfortunately the program cannot understand our language. So we tell the computer how the circuit looks like after throwing the switch. However, the program can compute the transient response, given the initial conditions, using the .TRAN statement. We do it as the second step as follows.

EXAMPLE A.11 *contd*

$R1$	1	2	3K	
$R2$	2	0	1K	
$C1$	0	2	0.1U	IC $= 9$
Vs	1	0	dc	8
.TRAN 10U	400U	UIC		
*	TRSTART	TRMAX	are omitted	
.PLOT TRAN	V(2)			
.END				

The reader would have quickly noticed the way a capacitor is represented in the input file; we have an extra field for the capacitor to show its initial condition. Further, we have taken the plotting increment to be reasonably, say, 10 μs, since we get the time constant to be 75 μs from the circuit. We have also omitted the TRMAX so that the program would plot the response in time steps of 8 μs. The reader would also have noticed a .PLOT TRAN statement just below the .TRAN statement. The program cannot give us an analytical answer such as

$$v_L(t) = \left[v_L(0^+) - v_L(\infty)\right] e^{-t/\tau} + v_L(\infty)$$

However, it can show us the plot of one or more output variables, just like the .PRINT DC statement, with a maximum of eight output variables per statement.

Some versions of PSpice, including some PC versions, have an in-built *software oscilloscope*, called the *Probe graphics post-processor*. This may be activated using the .PROBE statement in the input file. Once the program finishes the analysis, the probe processor displays the signals according to our choice via a user-interactive menu. The input file is rewritten as follows.

EXAMPLE A.11 *contd*

$R1$	1	2	3K	
$R2$	2	0	1K	
$C1$	0	2	0.1U	IC $= 9$
Vs	1	0	dc	8
.TRAN 10U	400U	UIC		
.PROBE				
.END				

Exercise A.11 Run the above program on your computer and verify the results by hand computation.

Once we have an oscilloscope on the screen, we cannot resist the temptation to look at the step response of the first-order circuit shown in Fig. A.9. As has been done in the laboratory, we may obtain the step response, indirectly, in terms of the so-called pulse train response. To this end, PSpice allows any independent source to be of pulse type with the following syntax.

VSRCNAME	Node j	Node k	PULSE	(V1	V2	TD	TR	TF	PW	PER)
ISRCNAME	Node j	Node k	PULSE	(I1	I2	TD	TR	TF	PW	PER)

where the fields in the parentheses are illustrated in Fig. A.10 for a typical trapezoidal voltage pulse. Given the syntax above, the reader may easily guess how a current pulse looks like.

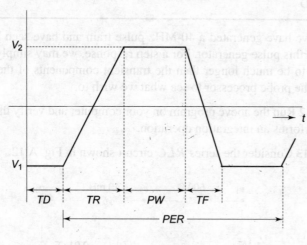

Fig. A.10

Let us now look at an example to use the pulse function.

Example A.12 Consider the circuit shown in Fig. A.11, wherein we are interested in voltage $v_L(t)$ across the load resistance R_L.

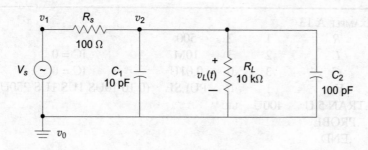

Fig. A.11

The input file is as follows.

EXAMPLE A.12

Rs	1	2	100	
RL	2	0	10K	
C1	2	0	10 P	
C2	2	0	100 P	
Vs	1	0	PULSE	(0.2 3.6 0 3NS 2NS 10NS 25NS)
.TRAN 0.4NS	60NS			
.PROBE				
.END				

Clearly, we have generated a 40-MHz pulse train and have seen how the *RC* circuit loads this pulse generator. For a step response, we may simply specify the pulse width to be much longer than the transient components of the circuit. We may adjust the probe processor to see what we wish to.

Exercise A.12 Run the above program on your computer and verify that the circuit basically performs an integration operation.

Example A.13 Consider the series *RLC* circuit shown in Fig. A.12.

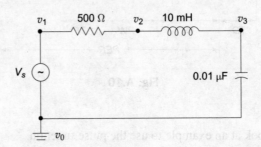

Fig. A.12

We may obtain the transient response across the capacitor using the following PSpice program.

EXAMPLE A.13

R	1	2	500	
L	2	3	10M	IC = 0
C	3	0	0.01U	IC = 0
Vs	1	0	PULSE	(0 10 10US 1US 1US 250US)
.TRAN 5 U	400U	UIC		
.PROBE				
.END				

Exercise A.13 Run the above program on your computer and verify the results of manual computations.

A.3 Ac Analysis

In this section we will see how PSpice can be used to perform the ac analysis. We may expect that the statements used for dc can be slightly modified for the purpose. We will continue doing examples and learn new statements as and when they appear.

Example A.14 Consider the simple circuit shown in Fig. A.13. We will verify the KVR around the mesh.

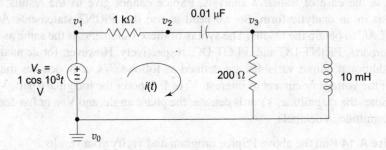

Fig. A.13

We will first write the PSpice program and then explain what each statement means.

```
EXAMPLE A.14
    R1          1          2          1K
    C           2          3          0.01U
    L           3          0          10M
    R2          3          0          200
    Vs          1          0          AC      1      0
    .AC        LIN         1         1000    1000
    .PRINT AC  V(1, 2)   V(2, 3)     V(3)    I(Vs)
    .END
```

The first four lines of the program describe the passive network. An ac source may be described by the following statement in its general form:

SRCNAME node j node k ac MAG PHASE

---

where SRCNAME is the name of the source that begins with V for a voltage source and I for a current source; nodes j and k are the nodes across which the source is connected; ac is a qualifier to indicate that it is an ac type source, and MAG and PHASE indicate the magnitude and phase angle of the sinusoid. If the MAG field is omitted, unity magnitude is taken for default; if the PHASE field is omitted, zero phase angle is taken for default.

Similar to the .DC statement, we have the .AC statement for ac analysis with the following general form:

.AC LIN NL FBEGIN FEND

where the field LIN denotes that the frequency is varied 'lin'early, and NL indicates the number of (discrete) frequency points in the frequency range FBEGIN to FEND. We need to remember that the program understands frequency in Hz only, and not in rad/s.

As in the case of transient analysis, PSpice cannot give us the results of ac analysis in an analytic form. So we need to use the .PRINT statements AC or .PLOT AC to obtain the results. The syntax of these statements is the same as their counterparts .PRINT DC and .PLOT DC, respectively. However, for ac analysis, five additional output variables are defined as follows. VR or IR denotes the real part of the voltage or current of interest; VI or II denotes the imaginary part; VM or IM denotes the magnitude, VP or IP denotes the phase angle, and VDB or IDB denotes the magnitude in decibels.

Exercise A.14 Run the above PSpice program and verify your results.

We will see two more examples to study the above statements and variables for ac analysis.

Example A.15 Consider the circuit shown in Fig. A.14. We will use PSpice to perform the nodal analysis.

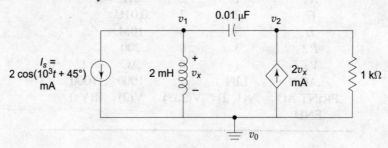

Fig. A.14

Let us straightaway write the PSpice program.

EXAMPLE A.15					
L	1	0	2M		
C	1	2	0.01U		
R	2	0	1K		
Is	1	0	AC	2M	45
Gs	0	2	1	0	2M
.AC	LIN	1	1000	1000	
.PRINT AC	VM(1, 2)	VP(2)	VM(2)		
.END					

Exercise A.15 Run the above program on your computer and verify the results. Also, modify the program to compute all the branch currents.

Example A.16 Consider the circuit shown in Fig. A.15. Let us attempt to plot the mesh current $i(t)$ as we vary the frequency from 1 kHz to 10 kHz, i.e., over a decade.

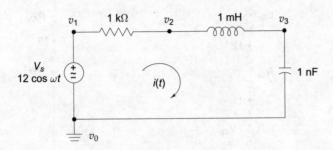

Fig. A.15

We may write the following PSpice program to obtain the frequency response of the mesh current.

EXAMPLE A.16					
R	1	2	1K		
L	2	3	1M		
C	3	0	1N		
Vs	1	0	AC	12	
.AC	DEC	100	1000	10000	
.PLOT AC	IM(Vs)				
.PROBE					
.END					

The reader would have noticed the presence of DEC in the .AC statement. This statement is used when we are interested in performing the ac analysis over a decade of frequencies. We also have another qualifier OCT for varying frequencies over an octave. We will see more about these a while later while dealing with Bode plots.

Exercise A.16 Run the above program on your computer and verify the results of Example A.16.

Let us now see how PSpice can be used to study three-phase systems.

Example A.17 Consider the balanced Y–Y system shown in Fig. A.16.

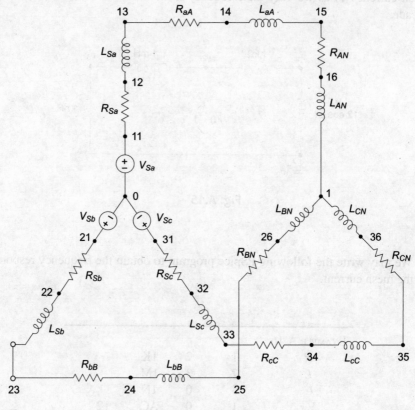

Fig. A.16

Once we identify the nodes, it is not too difficult to write the PSpice program as follows.

EXAMPLE A.17

	Phase 1:			
*	Phase 1:			
Vsa	11	0	AC	10
Rsa	11	12	10	
Lsa	12	13	0.1M	
RaA	13	14	100	
LaA	14	15	0.1M	
RAN	15	16	1K	
LAN	16	1	1M	

(Contd.)

EXAMPLE A.17

*	Phase 2:				
Vsb	21	0	AC	10	-120
Rsb	21	22	10		
Lsb	22	23	0.1M		
RbB	23	24	100		
LbB	24	25	0.1M		
RBN	25	26	1K		
LBN	26	1	1M		
*	Phase 3:				
Vsc	31	0	AC	10	120
Rsc	31	32	10		
Lsc	32	33	0.1M		
RcC	33	34	100		
LcC	34	35	0.1M		
RCN	35	36	1K		
LCN	36	1	1M		
.AC	LIN	1	1000	1000	
.PRINT AC	IM(Vsa)	VM(13)	VM(13,23)	VM(15,25)	VM(15,1)
.PRINT AC	IP(Vsa)	VP(13)	VP(13,23)	VP(15,25)	VP(15,1)
.PRINT AC	IM(Vsb)	VM(23)	VM(23,33)	VM(25,35)	VM(25,1)
.PRINT AC	IP(Vsb)	VP(23)	VP(23,33)	VP(25,35)	VP(25,1)
.PRINT AC	IM(Vsc)	VM(33)	VM(33,13)	VM(35,15)	VM(35,1)
.PRINT AC	IP(Vsc)	VP(33)	VP(33,13)	VP(35,15)	VP(35,1)

The circuit has too many elements, and hence the program appears lengthy. We have enumerated the nodes according to certain convenience. We may also add one more .PRINT statement to see if the potential VM(1) at the neutral of the load is zero. We encourage the reader to run this program and verify the results. PSpice is quite powerful for the analysis of three-phase systems, particularly the unbalanced ones.

Exercise A.17 Run the above program with an unbalanced load and verify your results.

A.4   Generalized Frequency Response

In this section, we will see how PSpice can be used to visualize the natural, complete, and frequency responses of a given circuit.

Example A.18 Consider the circuit shown in Fig. A.17. We will write a program to plot the node potentials v_1 and v_2.

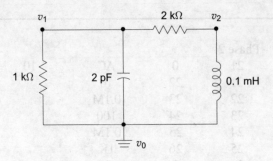

Fig. A.17

The PSpice program is as follows.

EXAMPLE A.18

$R1$	1	0	1K	
C	1	0	2P	IC = 6.7500
$R2$	1	2	2K	
L	2	0	0.1M	IC = 0
.TRAN	0.5U	30U	UIC	
.PROBE				
.END				

Exercise A.18 Run the above program on your computer.

Example A.19 Let us now look at visualizing the sinusoidal (steady-state) response of the circuit shown in Fig. A.18.

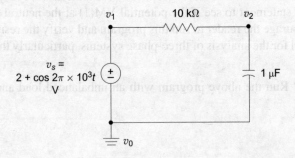

Fig. A.18

The reader would have, by now, tried to guess what could be the statement for an ac source. We describe this as follows. We consider the generalized sinusoid given by

$$u(t) = U_0 + U_m e^{-\alpha(t-t_d)} \sin\left[2\pi(t - t_d) + \theta\right]$$

where all the parameters α, t_d, and θ have already been defined in Chapters 6, 7, and 9. The PSpice statement has the following syntax:

SRCNAME	node j	node k	sin (U_0 U_m Freq t_d α θ)

The field SRCNAME starts with V if it is a voltage source, and starts with I if it is a current source. It should be remembered that the frequency is in Hz.

The PSpice program for the circuit shown in Fig. A.18 is as follows:

EXAMPLE A.19

R	1	2	10K	
C	2	0	1U	IC = 0
Vs	1	0	sin(2 1 1K 0 0 0)	
.TRAN	0.05M	100M	UIC	
.PROBE				
.END				

Here the reader would have noticed that we have used the .TRAN statement to direct the program to compute and display the sinusoidal steady-state response as a function of time. Further, since the time constant of the circuit is 10 ms, we have used a window of 100 ms so that the response will attain its steady state.

Exercise A.19 Run the above program on your computer and verify that the input sinusoid is attenuated by about 20 dB, since the input frequency (1000 Hz) is a decade above the corner frequency (100 Hz) of the circuit.

In the student version PSpice, there is an interesting feature available that allows us to quickly see the Bode plots. We will look at the following example and illustrate the idea.

Example A.20 Suppose we wish to generate the Bode plot of the transfer function

$$H(s) = \frac{(s/2)(s/100 + 1)}{(s/10)(s/10 + 1.2) + 1}$$

where we have explicitly specified the zeros and the poles and hence the corner frequencies of the transfer function. We may use the circuit shown in Fig. A.19 to realize the transfer function above.

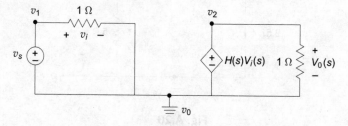

Fig. A.19

The following PSpice program generates the Bode plot.

EXAMPLE A.20

```
R1    1    0          1
R2    2    0          1
E1    2    0          LAPLACE      {V(1,0)} =
+                                  {(s/2) * (s/100 + 1)/ ((s/10) * (s/10 + 1.2) + 1)}
Vs    1    0          AC    1
.AC   DEC   10   0.031   312
.PROBE
.END
```

Notice that we have cleverly used a VCVS to describe the transfer function. Of course, we can also use a VCCS according to the context. The keyword here is LAPLACE. Also, observe that the controlling voltage and the transfer function are enclosed within braces, with an equality sign in between. To continue a statement on to the next line, we use a '+' at the beginning of the line.

Obtaining Bode plots is possible in PSpice used for this appendix, but is not available in SPICE. The reader may check it with other portable versions of SPICE.

Exercise A.20 Run the above program on your computer and verify the results.

A.5 Two-port Parameters

In this section, we will illustrate how PSpice can be used to analyse two-port networks. At the outset, let us recall that we need to perform experiments at either of the ports of the networks. We will write a program and then explain the details.

Example A.21 Consider the two-port network shown in Fig. A.20.

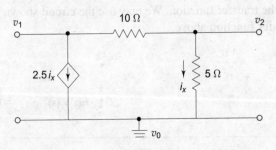

Fig. A.20

To compute the Z parameters, we need to conduct *open-circuit tests*. The corresponding circuit diagrams are shown in Figs A.21(a) and (b).

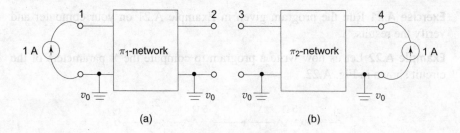

Fig. A.21

Notice that we have applied 1A current sources in each of the two experiments. The idea is quite simple—z_{11} is just the voltage v_1 across the port 1-0, $z_{21} = v_2$, and so on. The reader may wonder whether we need to write separate programs for each of the experiments. Fortunately, we need not. There is a statement, called SUBCKT, which does the job. We will use this directly in the program below, and the reader can quickly understand its syntax, and how it works.

EXAMPLE A.21

.SUBCKT	TWO-PORT	1	2	
$R1$	1	2	10	
$R2$	2	0	5	
$F1$	1	0	Vd	2.5
Vd	3	0	dc	0
.ENDS	TWO-PORT			
$I1$	0	1	dc	1
$I2$	0	2	dc	0
$\Pi1$	1	2	TWO-PORT	
$I3$	0	3	dc	0
$I4$	0	4	dc	1
$\Pi2$	3	4	TWO-PORT	
.END				

Notice that we have also used a dummy source to measure the controlling current of the CCCS.

The output file contains the following:

Node	Vol.	Node	Vol.	Node	Vol.	Node	Vol.
(1)	4.2857	(2)	1.4286	(3)	−5.7143	(4)	1.4286

Since we have used 1A current sources, we may interpret the results as

$$z_{11} = V(1), \quad z_{12} = V(3), \quad z_{21} = V(2), \quad \text{and} \quad z_{22} = V(4)$$

Exercise A.21 Run the program given in Example A.21 on your computer and verify the results.

Example A.22 Let us now write a program to compute the Y parameters of the circuit shown in Fig. A.22.

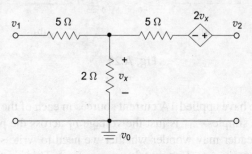

Fig. A.22

To compute the Y parameters, we need to conduct *short-circuit tests*. We will use this directly in the program below. The reader can quickly understand its syntax and how it works.

EXAMPLE A.22					
.SUBCKT	TWO-PORT	1	2		
R1	1	3	5		
R2	3	0	2		
R3	3	4	5		
E1	2	4	3	0	2
.ENDS	TWO-PORT				
V1	1	0	dc	1	
V2	0	2	dc	0	
T1	1	2	TWO-PORT		
V3	0	3	dc	0	
V4	4	0	dc	1	
T2	3	4	TWO-PORT		
.END					

Exercise A.22 Run the above program on your computer and verify that $y_{11} = 11/65$, $y_{12} = -2/65$, $y_{21} = -6/65$, and $y_{22} = 7/65$.

A.6 Laplace and Fourier Analyses

In this section of this appendix we will see how PSpice can be used to perform Laplace and Fourier analyses. Recall that *transfer functions* of the complex variable s help us obtain both the transient and steady-state responses in a unified framework. We have already presented an example program that makes use of the keyword LAPLACE and obtains the frequency response. Here we will give one more example, wherein we obtain the impulse and step responses.

Example A.23 Consider the transfer function

$$H(s) = 12 \frac{(s+2)}{(s+1+j1)(s+1-j1)}$$

We may use the circuit shown in Fig. A.19 (of Example A.20) and write the following program.

```
EXAMPLE A.23
R1   1   0          1
R2   2   0          1
E1   2   0          LAPLACE  {V(1, 0)} =
+                   {12 * (s + 2)/(s * s + 2 * s + 2)}
Vs   1   0          PULSE    (0  1K  0.1 1N  1N  1M  2)
.TRAN  0.1  10
.PROBE
.END
```

The reader might have had some suspicion looking into the argument of PULSE. We have built a unit impulse signal using an arbitrarily narrow (1 ms) pulse with an arbitrarily large magnitude (1 kV) so that the area under the pulse is unity.

Exercise A.23 Run the above program and verify the results.

Fourier analysis may be performed using the following SPICE statement:

.FOUR   FREQ   OUTVAR 1   OUTVAR 2   ···   OUTVAR 8

---

where FREQ is the fundamental frequency in hertz. This statement should be used along with the .TRAN statement. We will illustrate this in the following example.

Example A.24 Suppose a (periodic) square wave signal shown in Fig. A.23(a) is applied to the circuit shown in Fig. A.23(b).

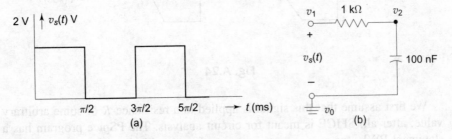

Fig. A.23

We have the following parameters from Fig. A.23(a).

$$\text{FREQ} = \frac{1}{2\pi \text{ ms}} = 159.154 \text{ Hz}$$

The program is as follows:

EXAMPLE A.24

R	1	2	1K
C	2	0	100N

Vs 1   0   PULSE   (15.708 0 1.5708M 1N 1N 3.1416M 6.2832M)
.TRAN   0.05M   12.5664M   0   0.05M
.FOUR   159.154 · V(1)   V(2)
.END

Exercise A.24 Run the above program and verify the results. Your results for the phase angles would differ from the actual phase angles by 90°. Try to reason it out. Further, you may also get acceptable numerical errors owing to the round-off computations, which are beyond your control.

We end the tutorial with the following example.

Example A.25 Let us write a PSpice program to obtain the frequency components of an arbitrary signal shown in Fig. A.24.

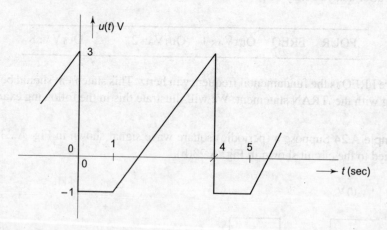

Fig. A.24

We first assume that this signal is applied to a resistance R of some arbitrary value; after all, SPICE is meant for circuit analysis. The PSpice program has a statement PWL, for *piece-wise-linear* waveforms, with the syntax that could be readily understood in the following program.

EXAMPLE A.25

Vs	1	0	PWL(0	-1	1	-1	4	3)
R	1	0			1K			

It may be expected that the PWL statement can have any number of arguments and has the following general form:

SRCNAME	Node j	Node k	PWL(T1	VAL1	T2	VAL2	$\cdots$)

where TK, $K = 1, 2, \ldots$ is the time instant at which the source's magnitude is VALK in volts or amperes depending on whether it is a voltage source or a current source. Between two successive time instants T_K and T_{K+1}, the source value changes *linearly* between Val_K and Val_{K+1}; in other words, the waveform is a straight line with slope

$$\frac{val_{K+1} - val_K}{T_{K+1} - T_K}$$

Exercise A.25 Run the program and verify the results by hand computations.

Concluding Comments

Donald O. Pederson, now a retired professor of University of California at Berkeley, was the brain behind SPICE in 1960s.. He was awarded the prestigious IEEE medal of honour in 1998 for his creation of this simulation program. It is interesting to know that he was with Bell laboratories (after his PhD from Stanford University) when, in 1954, he was invited by the electrical engineering department at the Newark College of Engineering (now the New Jersey Institute of Technology) to teach a course on electrical network theory. After a year's teaching, he found teaching more satisfactory and joined the University of California at Berkeley as an assistant professor, despite a big cut in his salary. We recommend the reader to have a glimpse of the fascinating and motivating history of SPICE, available in IEEE's magazine SPECTRUM dated June 1998.

In this tutorial we have illustrated most of the PSpice statements that are useful for the circuit theory developed in this textbook. PSpice is capable of doing much more, e.g., analysis of electronic circuits, which is beyond the scope of the present text. We have given programs for all the examples in this tutorial. We encourage the reader to first run these programs and verify the results. Thereafter, he may attempt to solve most of the examples/exercises/end-chapter problems given in this textbook.

Appendix B

Laboratory Experiments

In this appendix we give a set of ten experiments that could be performed in the laboratory. We assume that this is one of the first core experiments, in the circuit branches, assigned to the student.

B.1 Expt 1: Know Your Laboratory

Aim To get acquainted with each of the components that make up a practical circuit.

Apparatus Voltage sources, resistors, capacitors, inductors, voltmeters, ammeters, multimeters, connecting wires, and a breadboard.

Procedure A practical circuit in the laboratory is made up of the following components. We suggest the student to make a careful study of each of the components he/she is introduced to.

Voltage sources In the laboratory, we come across only voltage sources. These voltage sources could be either dc type or ac type. They have a set of ratings, such as maximum voltage, maximum current that can be drawn, etc., provided by the manufacturer. We strongly recommend that every student gets adequately familiar with these ratings, and even the make (the manufacturer's name, and the product's name) and the price of the source available in his/her laboratory. Once the ratings have been noted, the student should learn to operate the source, and the method of operation is to be recorded. Attention has to be paid to the 'ground' node of the source.

Resistors In theory, we can have resistance of any value; for instance a 29 Ω resistor. However, when it comes to manufacturing, there are certain limitations and we can avail only standard resistances. The student should first make a note of the size of the component. Then he/she should become familiar with the colour code to identify the value of a given resistor and make a table of the resistances available in his laboratory. Further, each of these resistors come with a *tolerance*. For example, we may pick up a resistor whose colour code indicates that its value is 100 Ω. Along with the colour bands indicating the value, the student would also see bands in either silver or gold colour. These colours indicate the tolerance; usually it is 5% for silver and 10% for gold. What this tolerance tells us is that when we measure the resistance, say with 5% tolerance, we find the value to be 100 ± 5, i.e., the resistance lies in the interval [95, 105]. If the resistance falls below 95, or goes above 105, it may be considered as burnt. In addition to tolerance, a resistor comes with a power rating also. This rating allows us to determine the maximum amount of permissible voltage across, or current through, it without burning it. The student may also make a note of the material used in the manufacture of resistors. Lastly, the student should identify variable resistors called potentiometers, popularly known *pots*, and learn to operate them.

Capacitors The student should also maintain a record of capacitors available in his/her laboratory. They too have tolerances and power ratings. Typically, laboratory capacitors would be available in the range of few microfarads to picofarads; these make use of the same colour code as that of resistors. These capacitors are made of different material, typically electrolytic and polyester, and each has its own advantages and disadvantages.

Inductors Unlike the resistors and capacitors, inductors are quite bulky in practice. In a circuits laboratory, inductors in the range of a few microhenrys are made available since they are lighter in weight, and ordinary connecting wires can be used to connect them to other elements. The student may also see such inductors, typically a copper wire wound on a carbon core, if he/she opens a transistor radio.

Connecting wires Each of the elements has a pair of nodes, and networks can be built by connecting the nodes of elements in an arbitrary manner. To facilitate building circuits, we may use connecting wires which are assumed to be perfect conductors. These connecting wires are available with sleeves so that we can handle them safely. While conducting experiments on circuits, care should be taken to keep the connecting wires straight, without any twists or turns. Before beginning an experiment, the student may verify that these wires offer *almost* zero resistance.

Measuring instruments The physical quantities that are of interest to us in a circuits laboratory are (i) voltage, or potential difference, (ii) current, (iii) resistance, (iv) capacitance, and (v) inductance. The student may find instruments, called meters, that measure each of these quantities. A *voltmeter*, typically available in the range 0–30 V, can be used to measure the potential difference across a branch. Clearly, the voltmeter is to be connected in parallel with the branch of interest. Similarly, an *ammeter*, typically available in the range of microamperes and milliamperes, can be connected in series with a branch to measure the current through it. There is also available a general purpose meter, called the *multimeter*, which measures all the above five quantities, and many more. These multimeters are either a bulky analog type, which in most laboratories are a thing of past, or a pretty digital type. The student is advised to record the ranges of feasible measurements on all these meters.

Breadboard This is a convenient platform on which we build circuits. About a decade ago, the students were trained to *solder* the circuits. Of late breadboards have become popular. On a breadboard the student would see a rectangular *array* of holes divided into four sections; these holes serve as the nodes of the circuit. The passive elements and the connecting wires from the voltage sources and/or meters may be plugged into the holes to build the circuits. Care should be taken to avoid open circuits.

A Simple circuit We encourage the student to build the following bridge circuit in Fig. B.1 on the breadboard.

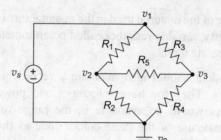

Fig. B.1

Building this circuit on the breadboard is an easy job. First, we need to identify the nodes of the circuit and mark them on the breadboard; there are only four nodes, and one of them is the ground node. Subsequently, we may plug in the elements and connecting wires. To avoid jamming too many elements into a single hole of the breadboard, we may use the topology of the breadboard itself and/or a few more connecting wires to build a circuit that is easy to trace. We cannot emphasize careful planning and *clarity* in circuit building any more.

We encourage the student to literally use his/her five senses to *see, touch, smell, taste,* and *listen to* each of these components in order to get himself thoroughly acquainted with the tools and hence get involved in practical matters. It is really exciting to do so. For instance, smelling is a nice way to check whether the component is working or burnt out; to get some enjoyment out of his own invention, the phonograph, Thomas Alva Edison would bite the speaker horn to feel the bones vibrate in his head; and listening to a *hum* from the electrical equipment makes you feel important in the laboratory!

B.2   Expt 2: Ohm's Law and Voltage Division

Aim To verify that a given resistance obeys Ohm's law, and to study the voltage division rule in a series circuit.

Apparatus (a) A dc voltage source v_S, say 10 V (b) a potentiometer R_p, say 0–10 kΩ (c) resistance R_L, say 10 kΩ (d) a voltmeter, say 0–10 V (e) an ammeter, say 0–3 mA and (f) a breadboard.

Procedure

1. Build the circuit shown in Fig. B.2 on the breadboard. Notice that the ammeter is put next to the potentiometer and the voltmeter is connected across R_L.
2. Switch on the voltage source so that it delivers $v_S = 10$ V to the series network of R_p and R_L.
3. Keep the potentiometer wiper such that R_p is initially 10 kΩ, and then vary it very slowly towards $R_p = 0$ Ω.

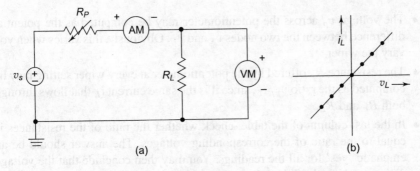

Fig. B.2

4. As you vary the potentiometer wiper, fill up the first three columns of the following table with as many readings of the voltmeter and ammeter as possible. For reference, we call the voltmeter reading as v_L and the ammeter reading as i_L.

S. No.	v_L (V)	i_L (mA)	$v_p =$ $v_S - v_L$	$R_p =$ v_p/i_L	R_L/R_p $\stackrel{?}{=} v_L/v_p$

The student is suggested to record *at least* 20 different $\{v_L, i_L\}$ pairs. If required, you may increase/decrease the voltage v_S supplied by the source and repeat steps 2 and 3 above.

5. Repeat steps 1–3 above with the polarity of the voltage source, and hence the voltmeter and the ammeter, reversed. This will add more readings to the table in step 4. Ideally, you would get the same readings as before with a negative sign.

Results

Once the data are recorded, switch off the voltage source. The experiment is completed and you are ready for an inference. You may begin filling up the remaining columns of the above table as follows.

- On a graph sheet, with voltage on the x axis and current on the y axis, mark the $\{v_L, i_L\}$ pairs, in the first and third quadrants. If there are more readings, the points would be close to each other, otherwise they would be sparse. In any case, join as many points as possible via a single straight line passing through the origin. Verify that the slope of this straight line approximates the reciprocal of the resistance R_L you have chosen for the experiment.

- The straight line passing through the origin suggests that the relationship between the voltage across and the current through a given resistance is

$$v = Ri$$

thereby verifying Ohm's law.

- The voltage v_p across the potentiometer may be computed as the potential difference between the two nodes v_S and v_L. Obviously, this varies when you vary the wiper.

- The resistance R_p offered by the potentiometer at every wiper setting may be computed as the ratio v_p/i_L, since it is the same current i_L that flows through both R_p and R_L.

- In the last column of the table, check whether the ratio of the resistances is equal to the ratio of the corresponding voltages. The answer should be an emphatic 'yes' for all the readings. You may then conclude that the voltage division rule is obeyed by the series circuit since the ratio suggests that

$$\frac{v_L}{R_L} = \frac{v_p}{R_p}$$

$$= \frac{v_L + v_p}{R_L + R_p}$$

$$= \frac{v_S}{R_L + R_p}$$

$$= i_L$$

and hence

$$v_L = \frac{R_L}{R_L + R_p} v_S$$

There could be certain errors beyond your control. The reasons could be tolerances in the components, errors in measurements, and so on. As a general rule of thumb, we allow the experimental results to be close to theoretical results with an error tolerance of not more than 5%.

The student is suggested to record his/her personal experiences while building the circuit, recording the readings, and verifying the results. It is also suggested that you put yourself in the shoes of Ohm and conduct the experiment with enthusiasm, rather than attempting to simply demonstrate somebody's law.

B.3 Expt 3: Kirchhoff's Rules

Aim To verify Kirchhoff's rules in a given network, and understand what a spanning tree is.

Apparatus (a) Dc voltage sources v_{S1} and v_{S2} (b) resistors R_1, R_2, R_3, and R_4 (c) voltmeters, ammeters, and/or multimeters as required and (d) a breadboard.

Procedure

1. Build the circuit shown in Fig. B.3(a) on the breadboard.

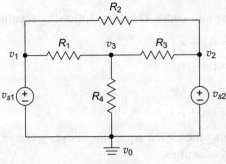

Fig. B.3(a)

2. Identify the (i) nodes, (ii) branches, (iii) meshes, and (iv) loops. There are four nodes, out of which one is the ground node v_0. There are 6 branches; these are the components you have got from the stores. There are seven loops, out of which 3 are meshes.

3. Using a voltmeter of appropriate range, or a multimeter, measure the node potentials v_1, v_2, and v_3 with respect to the ground $v_0 = 0$ V. Obviously, $v_1 = v_{S1}$ and $v_2 = v_{S2}$.

4. Using the node potentials, compute the voltages across each of the four resistances as follows:

$$v_{R_1} = v_1 - v_3, \quad v_{R_2} = v_1 - v_2, \quad v_{R_3} = v_3 - v_2, \quad v_{R_3} = v_3$$

5. Some of these voltages may be negative. At this moment do not worry about that. It just suggests that your meter connections in the circuit have to be reversed.

6. Using an ammeter of appropriate range, or a multimeter, measure the currents i_{01}, i_{13}, i_{12}, i_{30}, i_{32}, and i_{20} through each of the six branches. The following convention is adopted—the current i_{jk} leaves node j, passes through the branch, and enters node k. Again, some of these currents may be negative. Do not worry about that.

Results

- To verify KVR, check the following seven equalities for the seven loops:

$$v_{S1} = v_{R_1} + v_{R_4} \tag{B.1}$$

$$= v_{R_1} + v_{R_3} + v_2 \tag{B.2}$$

$$= v_{R_2} + v_2 \tag{B.3}$$

$$= v_{R_2} + v_{R_3} + v_{R_4} \tag{B.4}$$

$$v_{S2} = v_{R_4} - v_{R_3} \tag{B.5}$$

$$= v_{R_1} - v_{R_2} + v_{R_4} \tag{B.6}$$

$$v_{R2} = v_{R_1} + v_{R_3} \tag{B.7}$$

- At each of the four nodes, check the following equalities to verify KCR:

$$i_{01} = i_{12} + i_{13} \tag{B.8}$$

$$i_{13} = i_{30} + i_{32} \tag{B.9}$$

$$i_{20} = i_{12} + i_{32} \tag{B.10}$$

$$i_{01} = i_{30} + i_{20} \tag{B.11}$$

- While verifying the eleven equations above, preserve the algebraic signs of the measured quantities.
- For each of the four resistors, verify Ohm's law using the measurements.

The idea of a spanning tree The circuit shown in Fig. B.3(a) may be represented using the following graph shown in Fig. B.3(b).

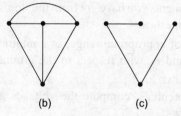

(b) (c)

Fig. B.3

This graph shows the three meshes clearly; each mesh has three branches. Application of KVR around each of the mesh results in an equation of the form

$$v_x + v_y + v_z = 0$$

It is easy to see that if we know two of the voltages, the third may be computed. Now, the question is: what is the minimum number of voltages required to determine the complete set of voltages in the circuit? Since, certain branches are common to adjacent meshes, the voltages across the branches shown in the graph of Fig. B.3(c) are sufficient. Using v_{S1} and v_{R_1}, we may readily compute v_{R_4}. Then, using v_{R_4} and v_{S2}, we may compute v_{R_3}. And, finally, using v_{R_1} and v_{R_3} we may compute v_{R_4}. The graph shown in Fig. B.3(c) is called a *spanning tree*. Algebraically, the set of three independent voltages, which are just sufficient to determine all other unknown voltages, is called a *basis*. Here, v_{S_1} and v_{S_2} are completely independent of each other, and hence they have to be taken into account. The third voltage could be either v_{R_1} or v_{R_3}. The student is encouraged to verify that the same spanning tree gives the minimum set of currents to determine all other branch currents in the circuit.

B.4 Expt 4: Linearity and Superposition

Aim To verify that voltages/currents in a given circuit may be computed using superposition, and understand the principle of linearity.

Apparatus (a) Dc voltage sources v_{S1} and v_{S2} (b) resistors R_1, R_2, R_3, and R_4 (c) voltmeters, ammeters, and/or multimeters as required and (d) a breadboard.

Procedure

1. Build the circuit shown in Fig. B.4(a) on the breadboard.

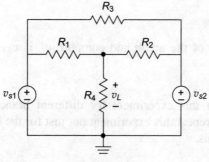

Fig. B.4(a)

2. Fix v_{S2} at, say, -5 V.
3. Vary v_{S1} from, say, -10 V to $+10$ V, and measure v_L for each v_{S1}.
4. Repeat step (3) for different values of v_{S2}, say, -2 V, 0 V, $+3$ V, and $+6$ V.
5. Tabulate your measurements, as follows:

S. No.	$v_{S2} =$		$v_{S2} =$		$v_{S2} =$		$\cdots$
	v_{S1}	v_L	v_{S1}	v_L	v_{S1}	v_L	$\cdots$
							$\cdots$

Results

- For each v_{S2}, in steps (2) and (4), obtain a graph of v_{S1} (on the x axis) versus v_L (on the y axis). You would get a family of parallel straight lines, typically as shown in Fig. B.4(b).

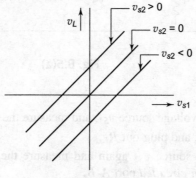

Fig. 4.4(b)

You may infer that the voltages v_{S1}, v_{S2}, and v_L are all *linearly* related, i.e.,

$$v_L = \alpha v_{S1} + \beta v_{S2}$$

and that v_L is a superposition of voltages v_{L1} and v_{L2} due to v_{S1} and v_{S2}, respectively.

Exercise Make use of the graph and compute v_L if $v_{S1} = -12.6$ V and $v_{S2} = +8.2$ V.

You may repeat this experiment for different branch voltages and branch currents. You may repeat this experiment not just for the branch currents but also for the mesh currents.

B.5   Expt 5: Thèvenin's Theorem

Aim To verify Thèvenin's theorem for a given circuit.

Apparatus (a) Dc voltage sources v_{S1} and v_{test}, say, 10 V each (b) resistors $R_1, \ldots$; R_8, and R_L (c) voltmeter, ammeter, and/or multimeter and (d) a breadboard.

Procedure

1. Build the circuit shown in Fig. B.5(a) on the breadboard.

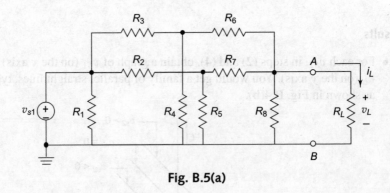

Fig. B.5(a)

2. Switch on the voltage source v_{S1}, and measure the voltage v_L across R_L.
3. Switch off v_{S1}, and plug-out R_L.
4. Switch on the source v_{S1} again and measure the potential difference v_{Th} across the *open-circuited* port A-B.
5. Switch off v_{S1} and compute the equivalent resistance as seen by R_L looking into the port A-B. This can be done in the following two ways.
    (a) Measure the equivalent resistance R_{Th} using a multimeter across the node pair A-B.

(b) Apply a test voltage source v_{test} across the node pair A-B, measure the current i_{test} delivered by the test source, and compute

$$R_{Th} = \frac{v_{test}}{i_{test}}$$

6. Verify that these two methods in step (5) give the same value of R_{Th}.

Results

- The Thèvenin's equivalent circuit of the circuit shown in Fig. B.5(a), as seen by the resistance R_L is shown in Fig. B.5(b).

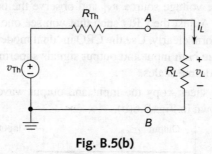

Fig. B.5(b)

In this Thèvenin's equivalent, the voltage source is v_{Th}, which is measured in step (4) of the procedure, and the series resistance is R_{Th}, which is computed in step (5) of the procedure.

- Compute

$$v_L = \frac{R_L}{R_L + R_{Th}} v_{Th}$$

- This 'computed' value of v_L should be the same as that 'measured' in step (2) of the procedure.
- Although theoretically we replace the voltage source by a short circuit, in practice it is necessary to switch it off and keep it in its position. This would ensure that the internal resistance, if any, of the source is taken into account.

B.6   Expt 6: Transients in *RC* Circuits

Aim To study the transient response of a given *RC* circuit.

Apparatus (a) An ac voltage source v_{S1} that can supply a square wave signal (b) a resistor R (c) a capacitor C (d) a CRO, preferably a digital storage oscilloscope and (e) a breadboard

Procedure

1. Fix the time period T of the period square wave, from the source v_{S1}, to be around 10 ms.
2. Choose the components R and C such that the product $\tau = RC = 0.5$ ms.

3. Build the circuit shown in Fig. B.6(a) on the breadboard. The CRO should be connected across the capacitor.

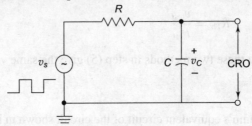

Fig. B.6(a)

4. Switch on the voltage source v_{S1} and observe the output waveform on the CRO screen. Adjust the CRO such that you see one half-cycle of the input square waveform clearly. Use the CRO in 'dual mode', and adjust its settings so that you see both input and output signals superimposed on the same set of voltage and time scales.

5. On a tracing sheet, copy the input and output waveforms. Typically, they appear as shown in Fig. B.6(b).

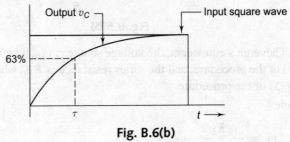

Fig. B.6(b)

Results

- The response of a first-order circuit is governed by the time constant $\tau = RC$. Experimentally, this may be obtained from the output waveform by marking the time at which the output is approximately 63% of the maximum input voltage. Equivalently, you may draw a tangent to the output waveform through the origin and compute its slope; the reciprocal of the slope gives you τ.

The values of the period T and the passive elements R and C are only suggestive.

- You may use a potentiometer in place of R and study the output waveform for the same input waveform for different time constants.

Repeat the experiment for different values of T.

B.7 Expt 7: Transients in *RLC* Circuits

Aim To study the transient response of a given *RLC* circuit.

Apparatus (a) An ac voltage source v_{S1} that would supply a square waveform (b) resistors R_1, R_2, and B_3 (c) a capacitor C (d) an inductor L (e) a CRO, preferably a digital storage oscilloscope and (f) a breadboard.

Procedure

1. Choose the inductor L in the range of microhenrys, and the capacitor in the range of microfarads. Compute the natural frequency

$$\omega_n = \frac{1}{\sqrt{LC}}$$

2. Choose three different resistors such that

$$\zeta = \frac{R_i}{2}\sqrt{\frac{C}{L}} \quad i = 1, 2, 3$$

is (i) < 1, (ii) = 1, and (iii) > 1.

3. Choose the period T of the square waveform, of v_{S1}, in such a way that you can capture clearly one half-cycle on the CRO screen. Adjust the CRO such that you see one half-cycle of the input square waveform clearly. Use the CRO in 'dual mode' and adjust its settings so that you see both input and output signals superimposed on the same set of voltage and time scales.

4. Typically, the passive elements may be chosen such that $\omega_n = 10^6$ krad/s, and $\omega = 10^6$ rad/s. For the overdamped case, the damping factor ζ may be selected to be 1.5, and for the underdamped case $0.6 \leq \zeta \leq 0.8$.

5. Build the series RLC circuit shown in Fig. B.7(a) on the breadboard. The CRO should be connected across the capacitor.

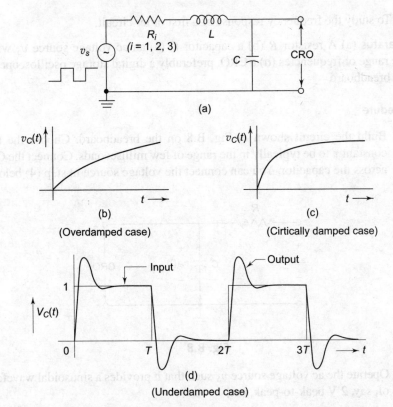

(a)

(b)
(Overdamped case)

(c)
(Critically damped case)

(d)
(Underdamped case)

Fig. B.7

6. Switch on the voltage source v_{S1} and observe the output waveform on the CRO screen for each of the three damping factors.

7. On a tracing sheet, copy the input and output waveforms. Typically, they appear as shown in Figs B.7(b)–(d)

Results

- For the overdamped case, the output is a very slow exponential asymptotically catching up with the input half-cycle.
- For the critically damped case, the output is also an exponential, but slightly faster than the previous case.
- For the underdamped case, you would observe the output to *shoot over* the input and then catch up with the input quickly, exhibiting damped oscillations.

You may use a potentiometer in place of R and study the output waveform for the same input waveform for different values of the damping factor ζ.

Repeat the experiment for different values of T.

B.8 Expt 8: AC Circuit Analysis

Aim To study the frequency response of a first order circuit.

Apparatus (a) A resistor R (b) a capacitor C (c) an ac voltage source v_S with a wide range of frequencies (d) a CRO, preferably a digital storage oscilloscope and (e) a breadboard

Procedure

1. Build the circuit shown in Fig. B.8 on the breadboard. Choose the time constant τ to be typically in the range of few milliseconds. Connect the CRO across the capacitor. You can connect the voltage source in step (4) below.

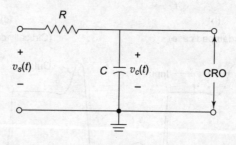

Fig. B.8

2. Operate the ac voltage source v_S such that it provides a sinusoidal waveform of, say, 2 V peak-to-peak, i.e.,

$$v_S(t) = V_S \cos(\omega t) \text{ V}$$

with $V_s = 1$ V.

3. Vary the frequency ω of the voltage source and calibrate the CRO time base and voltage base correctly. Operate the CRO in dual mode so as to observe both input and output waveforms on the same scale.

4. Connect the voltage source to the circuit according to the figure above. Vary the frequency ω of v_S gradually, and for each frequency ω_i observe the sinusoidal voltage

$$v_C(t) = V_C \cos(\omega_i t + \theta) \text{ V}$$

across the capacitor on the CRO. Ensure that $V_S = 1$ V throughout the experiment, without any change.

5. Measure the peak-to-peak amplitude and the phase difference between the output and the input waveforms. Make the measurements for a sufficiently large range of the input frequency ω, say from few rad/s to few hundred krad/s. Tabulate the measurements as follows:

S. No.	ω_i (rad/s)	$|v_C|$ (V)	$20 \log |(V_C/V_S)|$ (dB)	θ (deg)

Observe that the output lags the input, and hence the last column has all negative entries.

6. On a semi-log graph sheet, obtain the Bode plot of the transfer function

$$\frac{v_C(j\omega)}{v_S(j\omega)} = \frac{1}{1 + j\omega RC}$$

Results

Mark the corner frequency. The magnitude at the corner frequency should be $1/\sqrt{2} = -3$ dB, and the phase angle should be $-45°$.

We encourage the student to repeat the experiment with the CRO connected across the resistance and measuring $v_R(t)$.

B.9 Expt 9: Resonance and Maximum Power Transfer

Aim To study resonance in a second-order circuit and verify the maximum power transfer theorem in a given circuit.

Apparatus (a) A resistor R (b) an inductor L (c) a capacitor C (d) a CRO, preferably a digital storage oscilloscope (e) voltage source v_S with a wide range of frequencies and (f) a breadboard.

Procedure

1. Build the circuit shown in Fig. B.9 on the breadboard. Choose L and C such that the natural frequency $(= 1/\sqrt{LC})$ of the circuit is around 10 krad/s.

The CRO is connected across the series combination of the resistor R and the capacitor C.

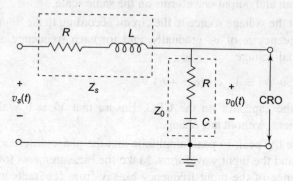

Fig. B.9

2. Calibrate the CRO settings, and operate it in the dual mode.
3. Vary the frequency ω of the voltage source v_S gradually from 1 krad/s to 100 krad/s. Observe the output waveform $v_0(t)$ on the CRO screen along with the input waveform on the same scale.
4. For each frequency ω_i of v_S, measure the amplitude $|v_0|$ and the phase angle θ of the output waveform. Tabulate the measurements as follows:

S. No.	ω_i (rad/s)	$|v_0|$ (V)	$20\log|V_0|$ (dB)	θ (deg)

Notice that, unlike experiment 8, we do not divide the magnitude of the output by that of the input. This is for certain convenience in the following steps.

5. Compute the average power delivered to the load impedance

$$Z_0(j\omega_i) = R + \frac{1}{j\omega_i C}$$

for each of the measurements tabulated above.

6. Obtain the Bode plot of the transfer function:

$$\frac{v_0(j\omega)}{v_s(j\omega)} = \frac{R + 1/j\omega C}{2R + j\omega L + 1/j\omega C}$$

Results

(a) On the Bode plot you observe that the amplitude of the output is maximum at the *resonant frequency* ω_r which equals the natural frequency ω_n.

(b) When the input frequency equals the natural frequency ω_n of the circuit, the following relationship, automatically, holds good:

$$Z_0(j\omega_n) = R + \frac{1}{j\omega_n C} = R + \frac{1}{j(1/\sqrt{LC})C}$$

$$= R - j\sqrt{\frac{L}{C}}$$

$$= R - j\frac{L}{\sqrt{LC}}$$

$$= R - j\,\omega_n L$$

$$= Z_S^*(j\omega_n)$$

which implies that the inductive and capacitive reactances nullify each other at the natural frequency. Clearly, according to the maximum power transfer theorem, the source v_S transfers maximum average power to the load Z_0 at this frequency.

B.10   Expt 10: Two-port Networks

Aim To obtain the Z, Y, T, and h parameters of a given two-port network.

Apparatus (a) Resistors R_1, R_2, R_3, and R_4 (b) a dc voltage source v_S (c) a voltmeter, ammeter, and multimeter and (d) a breadboard.

Procedure
1. Build the circuit of Fig. B.10 on the breadboard.

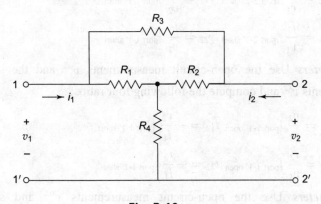

Fig. B.10

2. Measurements
   i. Open-circuit measurements
      (a) Leave port 2-2′ open. Apply v_S across port 1-1′. Measure the current delivered by v_S and the voltage v_2 across port 2-2′. Call this set of three readings v^{o1}:
      $$v^{o1} = \{v_1 (= v_S), i_1, v_2\}$$

(b) Leave port 1-1' open. Apply v_S across port 2-2'. Measure the current delivered by v_S and the voltage v_1 across port 1-1'. Call this set of three readings v^{o2}:
$$v^{o2} = \{v_2 (= v_S), i_2, v_1\}$$

ii. Short-circuit measurements

(a) Leave port 2-2' short circuited. Apply v_S across port 1-1'. Measure the current delivered by v_S and the current i_2 through the short-circuited port 2-2'. Call this set of three readings v^{s1}:
$$v^{s1} = \{v_1 (= v_S), i_1, i_2\}$$

(b) Leave port 1-1' short-circuited. Apply v_S across port 2-2'. Measure the current delivered by v_S and the current i_1 through the short-circuited port 1-1'. Call this set of three readings v^{s2}:
$$v^{s2} = \{v_2 (= v_S), i_2, i_1\}$$

Results

Z parameters Use the open-circuit measurements v^{o1} and v^{o2} and compute the following four ratios:

$$z_{11} = \frac{v_1}{i_1} \Big|_{\text{port 2-2' open}} \quad z_{12} = \frac{v_1}{i_2} \Big|_{\text{port 1-1' open}}$$

$$z_{21} = \frac{v_2}{i_1} \Big|_{\text{port 2-2' open}} \quad z_{22} = \frac{v_2}{i_2} \Big|_{\text{port 1-1' open}}$$

Y parameters Use the short-circuit measurements v^{s1} and v^{s2} and compute the following four ratios.

$$y_{11} = \frac{i_1}{v_1} \Big|_{\text{port 2-2' short}} \quad y_{12} = \frac{i_1}{v_2} \Big|_{\text{port 1-1' short}}$$

$$y_{21} = \frac{i_2}{v_1} \Big|_{\text{port 2-2' short}} \quad y_{22} = \frac{i_2}{v_2} \Big|_{\text{port 1-1' short}}$$

T parameters Use the open-circuit measurements v^{o2} and the short-circuit measurements v^{s2} and compute the following four ratios:

$$t_{11} = \frac{v_2}{v_1} \Big|_{\text{port 1-1' open}} \quad t_{12} = -\frac{v_2}{i_1} \Big|_{\text{port 1-1' short}}$$

$$t_{21} = \frac{i_2}{v_1} \Big|_{\text{port 1-1' open}} \quad t_{22} = -\frac{i_2}{i_1} \Big|_{\text{port 1-1' short}}$$

h parameters Use the open-circuit measurements v^{o2} and short-circuit measurements v^{s1} and compute the following four ratios:

$$h_{11} = \frac{v_1}{i_1} \Big|_{\text{port 2-2' short}} \quad h_{12} = \frac{v_1}{v_2} \Big|_{\text{port 1-1' open}}$$

$$h_{21} = \frac{i_2}{i_1} \Big|_{\text{port 2-2' short}} \quad h_{22} = \frac{i_2}{v_2} \Big|_{\text{port 1-1' open}}$$

- Verify that the Z and the Y parameters are inversely related, i.e.,

$$\begin{bmatrix} z_{11} & z_{12} \\ z_{21} & z_{22} \end{bmatrix} \begin{bmatrix} y_{11} & y_{12} \\ y_{21} & y_{22} \end{bmatrix} = \begin{bmatrix} 1 & 0 \\ 0 & 1 \end{bmatrix}$$

- Verify if the given network is reciprocal.
- Express the T and the h parameters in terms of the Z and Y parameters.
- Repeat the experiment with arbitrary resistors and capacitors and an ac voltage source.

Appendix C

Solution of a Linear System of Equations

Engineering problems, in general, demand the solution of a large set of simultaneous equations. Typically, the number of unknowns, and hence the number of equations, is of the order of few tens. Solving such a large number of equations manually is a time-consuming task; these are better handled using computers. Each of the equations is linear in the variables, and hence these are also known as linear system of equations, LSE for short. In this appendix, we quickly give a summary of the popular techniques used for solving the LSE. To make things simpler, we begin with the following three linear equations:

$$a_{11}x_1 + a_{12}x_2 + a_{13}x_3 = b_1 \qquad (C.1)$$

$$a_{21}x_1 + a_{22}x_2 + a_{23}x_3 = b_2 \qquad (C.2)$$

$$a_{31}x_1 + a_{32}x_2 + a_{33}x_3 = b_3 \qquad (C.3)$$

Let us carefully notice the way the equations are written. On the left-hand side, the unknowns are denoted as $x_j (j = 1, 2, 3)$. The coefficients are written with a 'double subscript' a_{ij} with the second index j coinciding with the subscript of the unknown x_j. On the right hand side are given the known quantities $b_i (i = 1, 2, 3)$. The subscript i coincides with the first index of the coefficients. Each of the above three equations may be written in terms of the product of a row vector and a column vector as:

$$[a_{i1} \; a_{i2} \; a_{i3}] \cdot \begin{bmatrix} x_1 \\ x_2 \\ x_3 \end{bmatrix} = b_i \qquad (C.4)$$

Since the column vector of the unknowns, denoted hereafter as $\bar{x}$, is common to all the three equations, we may nicely *pack* the three equations and obtain the matrix-vector form as

$$[\, a_{ij} \,] \bar{x} = \bar{b} \qquad (C.5)$$

where the vector $\bar{b}$ denotes the column vector obtained by stacking the known quantities b_i. Hereafter, we represent such a LSE as[1]

$$A\bar{x} = \bar{b} \tag{C.6}$$

where we assume that the size of the matrix A is $n \times n$, the size of both the vectors, $\bar{x}$ and $\bar{b}$, is $n \times 1$, i.e., there are n number of unknowns and n number of equations. Typically, the vector of unknowns may be obtained as

$$\bar{x} = A^{-1}\bar{b} \tag{C.7}$$

by simply generalizing our idea of division with numbers. However, there is a very important issue where we must exercise care. We must, naturally of course, check whether matrix A is invertible, i.e., whether the matrix is non-singular. A matrix is invertible if and only if all of its columns (or equivalently, its rows) are linearly independent; in other words, the matrix A has full rank n. If A is invertible, then we say that the LSE is a *consistent* system and hence a *unique* solution exists. Otherwise, the LSE is inconsistent and an infinite number of solutions exist. Oddly enough, first, verifying a $n \times n$ matrix for its rank is not simple. Secondly, inverting a matrix with an eye on the computing resources, particularly the computing time, is indeed a task. We will try to address briefly the second issue in this appendix. We suggest the reader to refer to books on numerical linear algebra for a more elaborate treatment.

C.1 Cramer's Rule

C.1.1 Matrix Inversion

First, we will study the steps involved in computing the inverse of a matrix. By definition,

$$A^{-1} = \frac{1}{\Delta(A)} \operatorname{adj}(A) \tag{C.8}$$

i.e., the ratio of the adjoint to determinant. The determinant is a number, i.e., a scalar, and it involves the following operations.

1. We have a ready-made formula for the second-order determinant:

$$\Delta_2 = \begin{vmatrix} a_{11} & a_{12} \\ a_{21} & a_{22} \end{vmatrix}$$

$$= a_{11}a_{22} - a_{12}a_{21} \tag{C.9}$$

---

[1] Note that the mathematical symbols we use in this appendix have nothing to do with the notation of circuit theory.

2. Using this formula, we may compute the determinant of third order in terms of its *minors* as follows:

$$\Delta_3 = a_{11} \begin{vmatrix} a_{22} & a_{23} \\ a_{32} & a_{33} \end{vmatrix} - a_{12} \begin{vmatrix} a_{21} & a_{23} \\ a_{31} & a_{33} \end{vmatrix} + a_{13} \begin{vmatrix} a_{21} & a_{22} \\ a_{31} & a_{32} \end{vmatrix} \qquad (C.10)$$

$$= a_{11}a_{22}a_{33} + a_{12}a_{23}a_{31} + a_{13}a_{21}a_{32}$$

$$- a_{13}a_{22}a_{31} - a_{23}a_{32}a_{11} - a_{33}a_{21}a_{12}$$

where each of the second-order determinants is the minor of the corresponding coefficient. We may use this formula recursively to compute higher-order determinants. In the expansion of Eqn (C.10), we observe that there is an alternating sign. If we attach this sign with the minor itself, we get the corresponding element's *cofactor*.

3. Replacing each of the elements of matrix A with its cofactor and transposing the result gives us the adjoint adj(A).

4. Thus, a determinant of order n is equal to the sum of the product of the elements of any row (or column) multiplied by their corresponding $(n-1)$-order cofactors. The minor determinants can be expanded by the same rule recursively, and it is easy to see that the determinant Δ_n is obtained as a sum of $n \times n!$ product factors, i.e., $(n-1) \times n!$ multiplications. This number explodes with increasing n. For instance, if $n = 2$, we had just 2 multiplications; if $n = 3$, we had 12 multiplications. If $n = 4$, we will have 72 multiplications, and so on.

To compute the unknown vector $\overline{x}$ of n elements, we need another n^2 multiplications with $\overline{b}$. This makes things even worse. Thus, it is not *efficient*[2] to compute the solution of a LSE using adjoints and determinants. However, for LSEs up to third order, we can derive a formula, popularly known as *Cramer's rule*. We illustrate this algorithm with the following example.

Example C.1

Consider the LSE:

$$\begin{bmatrix} 2 & -1 & -1 \\ 4 & -3 & 2 \\ -3 & 1 & 1 \end{bmatrix} \overline{x} = \begin{bmatrix} -5 \\ -8 \\ 6 \end{bmatrix}$$

The determinant may be easily computed, using Eqn (C.10), to be

$$\Delta(A) = 5$$

---

[2] Although we use this word 'efficient' loosely, it has a very precise definition in the theory of algorithms.

The adjoint matrix may be computed to be

$$
\begin{bmatrix}
-5 & 0 & -5 \\
-10 & -1 & -8 \\
-5 & 1 & -2
\end{bmatrix}
$$

We make the following observations:

1. We observe that each of the unknowns x_j, $j = 1, 2, 3$ is a result of multiplying the jth row of A^{-1} with vector $\bar{b}$.

2. The elements of the jth row of A^{-1}, i.e., the elements of the jth row of adj(A) divided by $\Delta(A)$, are the cofactors of the elements of the jth column of A, also divided by $\Delta(A)$. For instance, the elements of the third row

$$
[-5 \quad 1 \quad -2]
$$

of adj(A) are the cofactors of the third column

$$
\begin{bmatrix}
-1 \\
2 \\
1
\end{bmatrix}
$$

of matrix A.

3. Then, how do we compute these cofactors? Clearly, for each of the elements a_{ij}, we discard the jth column and the ith row that intersects the position of the element $a_{ij} \in A$. In other words, we simply do not bother about the elements of the jth column in this computation. We will now compute $\Delta(A) \cdot x_3$ as follows:

$$
\Delta(A) \cdot x_3 = [-5 \quad 1 \quad -2] \cdot \begin{bmatrix} -5 \\ -8 \\ 6 \end{bmatrix}
$$

$$
= \text{cofactor}(-1) \cdot -5 + \text{cofactor}(2) \cdot -8 + \text{cofactor}(1) \cdot 6
$$

$$
= (-5) \times (-1)^{1+3} \times \begin{vmatrix} 4 & -3 \\ -3 & 1 \end{vmatrix}
$$

$$
+ (-8) \times (-1)^{2+3} \times \begin{vmatrix} 2 & -1 \\ -3 & 1 \end{vmatrix}
$$

$$
+ (6) \times (-1)^{3+3} \times \begin{vmatrix} 2 & -1 \\ 4 & -3 \end{vmatrix}
$$

4. It is easy to see that we get the same result if we substitute the jth column with the vector $\bar{b}$ itself and expand the determinant along this column.

5. Let us now quickly recollect that a determinant is computed as a sum of products of all the elements along a column (or row) with their corresponding cofactors. In this sense, we define

$$
\Delta_j(A) \triangleq \text{the determinant} \Delta(A)
$$

$$
\text{with the } j\text{th column replaced by the vector } \bar{b} \qquad \text{(C.11)}
$$

so that

$$x_j = \frac{\Delta_j(A)}{\Delta(A)}, \quad j = 1, 2, \ldots, n \tag{C.12}$$

This is called Cramer's rule. We now present the complete solution of the LSE in this example as follows:

$$x_1 = \frac{1}{5} \begin{vmatrix} -5 & -1 & -1 \\ -8 & -3 & 2 \\ 6 & 1 & 1 \end{vmatrix} = -1$$

$$x_2 = \frac{1}{5} \begin{vmatrix} 2 & -5 & -1 \\ 4 & -8 & 2 \\ -3 & 6 & 1 \end{vmatrix} = 2$$

and

$$x_3 = \frac{1}{5} \begin{vmatrix} 2 & -1 & -5 \\ 4 & -3 & -8 \\ -3 & 1 & 6 \end{vmatrix} = 1$$

As has been mentioned earlier, for LSEs of fourth or higher orders Cramer's rule is inefficient owing to its demand for an exploding number of multiplications. In the next section we will explore a more sensible algorithm.

C.2 Gaussian Elimination

The basic idea behind Gaussian elimination is none other than the crude way of solving a set of simultaneous equations we have learnt during high school. However, here we polish that method and present it in a better form. We will take up the same example for illustration.

Example C.2

Let us form a composite matrix with vector $\bar{b}$ put as the fourth column of A as follows:

$$A' = \begin{bmatrix} 2 & -1 & -1 & -5 \\ 4 & -3 & 2 & -8 \\ -3 & 1 & 1 & 6 \end{bmatrix} \tag{C.13}$$

Let us now perform a series of elementary row operations on each of the rows.

1. We will make the elements a'_{21} and a'_{31} equal to zero. For instance, if we multiply the elements a'_{1*} of the first row by -2 and add it to the second row, we get

$$A' = \begin{bmatrix} 2 & -1 & -1 & -5 \\ 0 & -1 & 4 & 2 \\ -3 & 1 & 1 & 6 \end{bmatrix} \tag{C.14}$$

Similarly, if we multiply the elements a'_{1*} of the first row by 1.5 and add it to the third row, we get

$$A' = \begin{bmatrix} 2 & -1 & -1 & -5 \\ 0 & -1 & 4 & 2 \\ 0 & -1/2 & -1/2 & -3/2 \end{bmatrix} \qquad (C.15)$$

We overwrite the original A' matrix with this new matrix.

2. We next make the element a'_{32} equal to zero. This can be simply done by multiplying the elements of the current second row by $-1/2$ and adding it to the current third row. The result is

$$A' = \begin{bmatrix} 2 & -1 & -1 & -5 \\ 0 & -1 & 4 & 2 \\ 0 & 0 & -5/2 & -5/2 \end{bmatrix} \qquad (C.16)$$

3. What did we achieve in the last step? The unknown x_3 is immediately solved. We may now substitute this into the second equation and solve for x_2, and finally using these two partial solutions we may solve for x_1 in the first equation.

This algorithm is called Gaussian elimination. The basic idea is the *systematic* elimination of the variables until matrix A looks like a triangular matrix. In each iteration, we eliminate the leftmost element of the matrix starting with the second row. We may overwrite the original matrix with the modified one. We encourage the reader to write a computer program for this algorithm. It may be verified that this algorithm requires a maximum of $n^3/3$ multiplications, and hence it is much more efficient than Cramer's rule. Further, numerical errors that might arise while working with digital computers can also be controlled in Gaussian elimination. For more details, we recommend the reader to refer to any good book on numerical analysis.

Appendix D
Routh–Hurwitz Criterion

The stability problem was originally posed by Maxwell[1]. He showed that, by examining the coefficients of the differential equations, the stability of the system could be determined. He was also able to give necessary and sufficient conditions for equations up to fourth-order; for fifth-order equations he gave two necessary conditions. However, Edward John Routh[2] and Adolf Hurwitz[3] have solved the problem completely. The result involves computing a sequence of determinants from the coefficients of the characteristic polynomial. The 'Routh–Hurwitz' criterion may be formally developed as follows.

First, we consider the characteristic polynomial with the coefficients a_j:

$$D(s) = a_0 s^n + a_1 s^{n-1} + a_2 s^{n-2} + \cdots + a_{n-1} s + a_n$$

We define the following **Routh–Hurwitz (R–H) array**:

s^n row:	a_0	a_2	a_4	a_6	$\cdots$
s^{n-1} row:	a_1	a_3	a_5	a_7	$\cdots$
s^{n-2} row:	b_2	b_4	b_6	b_8	$\cdots$
s^{n-3} row:	c_3	c_5	c_7	c_9	$\cdots$
$\vdots$	$\vdots$	$\vdots$			
s^1 row:	d_n				
s^0 row:	e_{n+1}				

Observe that the elements in s^n and s^{n-1} rows of the array are appropriately ordered coefficients of $D(s)$. The elements of the third s^{n-2} row are formed using the preceding two rows as follows:

$$b_2 = -\frac{1}{a_1} \begin{vmatrix} a_0 & a_2 \\ a_1 & a_3 \end{vmatrix} = a_2 - \frac{a_0}{a_1} a_3$$

and the general formula for b_k, $k = 2, 4, \ldots$, is

$$b_k = -\frac{1}{a_1} \begin{vmatrix} a_0 & a_k \\ a_1 & a_{k+1} \end{vmatrix} = a_k - \frac{a_0}{a_1} a_{k+1}$$

Notice that the first column remains same in all the determinants.

[1] J.C. Maxwell, 'On Governors', *Proc. Royal Society*, vol. 16, 1867/1868, pp. 270–83.
[2] E.J. Routh, *A Treatise on the Stability of a Given State of Motion*, Macmillan, London, 1877.
[3] A. Hurwitz, Über die Bedingungen unter welchen eine Gleichung nur Wurzeln mit negativen reelen Teilen besitzt," *Mathematische Annalen*, vol. 46, 1895, pp. 273–80.

The fourth s^{n-3} row, similar to the third row, is computed as follows:

$$c_k = -\frac{1}{b_2}\begin{vmatrix} a_1 & a_k \\ b_2 & b_{k+1} \end{vmatrix} = a_k - \frac{a_1}{b_2}b_{k+1} \quad k = 3, 5, 7, \ldots$$

The reader would have, by now, observed a nice pattern in the computations. Each subsequent row of the $(n+1)$-row Routh–Hurwitz array is formed from its preceding two rows in an analogous fashion. Zeros can be inserted for undefined elements, as required, and any row can be scaled by a positive scalar to simplify the computation of a subsequent row. The following example serves as an illustration.

Example D.1

Let $D(s) = s^6 + 5s^5 + 15s^4 + 55s^3 + 154s^2 + 210s + 100$. We form the first two rows immediately:

s^6 row:	1	15	154	100
s^5 row:	5	55	210	0

We may divide the s^5 row with 5 and replace it with the new expression. Accordingly, we get the s^4 row as follows:

s^4 row:	4	112	100	0

We may divide this s^4 row by 4. We proceed analogously to compute the subsequent rows and the completed array is computed as follows:

s^6 row:	1	15	154	100
s^5 row:	5	55	210	0
s^4 row:	1	28	25	0
s^3 row:	−1	1	0	
s^2 row:	29	25		
s^1 row:	1.862			
s^0 row:	25			

Exercise D.1 Write a computer program that would generate the Routh–Hurwitz array given the coefficients $a_0, \ldots, a_n$ of a polynomial $D(s)$.

Routh–Hurwitz criterion The number of roots of $D(s)$ with positive real parts is equal to the number of sign changes in the first column of the Routh–Hurwitz array.

We may readily infer that the system is stable if and only if *all* the elements in the first column of the Routh–Hurwitz array have a same sign.

Example D.2

In the previous example, there are exactly two sign changes in the first column—the first due to the sign change from $+1$ to -1 and the second due to the sign change from -1 to $+29$. It follows that the system has two poles with positive real parts and hence it is unstable.

Exercise D.2 Obtain the roots of the polynomial in Example D.1 and verify the results of Example D.2.

There are two special cases of this criterion.

Case 1 If a row has a zero in the first column and has at least one non-zero element, we replace the zero in the first column with ϵ, an infinitesimally small *positive* number, and continue building the array.

We will illustrate this in the following example.

Example D.3

Let $D(s) = -2s^4 - 4s^3 - 4s^2 - 8s - 1$. Note that the necessary condition is satisfied since all of the coefficients are present and have the same (negative) sign. The first two rows of the R–H array are

s^4 row:	-2	-4	-1
s^3 row:	-4	-8	0

We may compute the s^2 row as

s^2 row:	0	-1

We now replace this row with an ϵ in place of 0.

s^2 row:	ϵ	-1

The complete Routh–Hurwitz array is as follows:

s^4 row:	-2	-4	-1
s^3 row:	-4	-8	0
s^2 row:	ϵ	-1	
s^1 row:	$\approx \dfrac{-4}{\epsilon}$		
s^0 row:	-1		

Since there are exactly two sign changes in the first column (independent of whether ϵ is positive or negative), it follows that there are exactly two roots with positive real parts and hence the system is unstable.

Case 2 An entire row of zeros may be encountered sometimes. It is interesting to note that this happens only in the s^{2k+1} row corresponding to the odd powers of s. This implies the presence of mirror-image roots relative to the imaginary axis or one or more pairs of imaginary-conjugate roots of the form $\pm j\omega$.

In such a situation, we use the preceding row s^{2k} to form an *auxiliary* polynomial in even powers as follows:

$$A(s) = A_{2k}s^{2k} + A_{2k-2}s^{2k-2} + \cdots + A_0$$

where A_i are the elements of the s^h row.

The entries of the s^{h+1} row are just the coefficients of the derivative of $A(s)$.

There is a bonus in this last case: the roots of $A(s)$ are also the roots of the original polynomial $D(s)$.

We will illustrate this in the following example.

Example D.4

Let $D(s) = s^6 + 2s^5 - 9s^4 - 12s^3 + 43s^2 + 50s - 75$.

(Notice that in this case even the necessary condition fails.)

The s^4 row is as follows:

s^4 row:	-1	6	-25

The s^3 row is then

s^3 row:	0	0	0

At this point, we define the auxiliary polynomial

$$A(s) = -s^4 + 6s^2 - 25$$

We can now replace the entire zero s^3 row with the coefficients of

$$\frac{d}{ds}A(s) = -4s^3 + 12s$$

Thus, the s^3 row is

s^3 row:	-4	12	0

We may continue to build the R–H array.

What is more interesting in this case is that the auxiliary polynomial is a factor of the given polynomial. In fact,

$$D(s) = A(s)(-s^2 - 2s + 3)$$

and the four roots of $A(s)$ are $s = \pm 2 \pm j1$, which are located symmetrically about the origin.

Exercise D.3 Verify the results of Examples D.3 and D.4.

While the Routh–Hurwitz criterion appears to be quite simple and attractive, the proof of this criterion is a bit involved and involves using the theory of complex variables. However, it is stimulating to note that there is considerable interest in the research community discussing simpler proofs that could be accessible to undergraduate students. We encourage the reader to ask for a proof of this R–H criterion when he/she encounters it formally in a later semester.

We end this appendix with another interesting note. While this criterion is considered most useful, there are certain drawbacks that have been uncovered in the recent past. For instance, if the polynomial $D(s)$ has multiple poles on the imaginary axis of the s plane, the R–H criterion will not reveal this. We leave the following exercise to the reader to play with.

Exercise D.4 Verify that the R–H criterion applied to the following polynomial does not reveal the presence of multiple roots on the imaginary axis.

$$D(s) = s^5 + s^4 + 2s^3 + 2s^2 + s + 1$$

Bibliography

Bode, H.W., *Network Analysis and Feedback Amplifier Design*, D. Van Nostrand, New Jersey, 1945.

Cauer, W., *Synthesis of Linear Communication Networks*, McGraw Hill, New York, 1958.

de Bruin, F.F.G., and H. Vos, 'A basic course in network analysis: Parts I & II', *IEEE Tr. Edn.*, vol. 38, no. 1, 1995, pp. 1–6.

DeCarlo, R.A. and Pen-Min Lin, *Linear Circuit Analysis*, 2nd edn, OUP Indian Edition, 2003.

Franco, S., *Electric Circuit Fundamentals*, Saunders College Publishing, 1995 (available in India with OUP).

Gottling, J.G., 'Node and mesh analysis by inspection', *IEEE Tr. Edn.*, vol. 38, no. 4, 1995, pp. 312–316.

Guillemin, E.A., *The Mathematics of Circuit Analysis*, MIT Press/John Wiley, New York, 1949.

Guillemin, E.A., *Introductory Circuit Theory*, John Wiley, New York, 1953.

Guillemin, E.A., *Synthesis of Passive Networks*, John Wiley, New York, 1957.

Harary, F., *Graph Theory*, Addison Wesley, 1969 (available in India with Narosa Publishing House).

Hayt, W.H., Jr., J.E., Kemmerley, and S.M. Durbin, *Engineering Circuit Analysis*, 6/e, McGraw Hill, New York, 2002 (available in India with Tata McGraw Hill).

Huelsman, L.P., *Circuits, Matrices, and Linear Vector Spaces*, McGraw Hill, New York 1963.

Kreyszig, E., *Advanced Engineering Mathematics*, 7th edn, John Wiley, 2001.

Rashid, M.H., *Spice for Circuits and Electronics Using PSPICE*, 2nd edn, Pearson Education, New Jersey, 1995 (available in India with Prentice Hall of India).

Roberts, G.W., and A.S. Sedra, *SPICE*, 2nd edn, Oxford University Press, New York, 1997.

Truxal, J.G., *Automatic Feedback Control System Synthesis*, McGraw Hill, New York, 1955.

Tuttle, D.F., Jr, *Network Synthesis*, vol. 1, John Wiley, New York, 1958.

Van Valkenberg, M.E., *Network Analysis*, 3rd edn, Prentice Hall, Englewood Cliffs, NJ, 1974 (also available as an Indian Edition with Prentice Hall of India).

Van Valkenburg, M.E., *Introduction to Modern Network Synthesis*, John Wiley, New York, 1960. An Indian Edition was earlier available with Wiley Eastern; now it might be out of print.

Williams, J.H., Jr., *Fundamentals of Applied Dynamics*, John Wiley, New York, 1996.

Index